Computational Group Theory
and the Theory of Groups

CONTEMPORARY MATHEMATICS

470

Computational Group Theory and the Theory of Groups

AMS Special Session
on Computational Group Theory
March 3–4, 2007
Davidson College, Davidson, North Carolina

Luise-Charlotte Kappe
Arturo Magidin
Robert Fitzgerald Morse
Editors

American Mathematical Society
Providence, Rhode Island

2000 *Mathematics Subject Classification.* Primary 20–06, 20B40, 20C40, 20D08, 20D15, 20D35, 20F12, 20F18, 20J05.

Library of Congress Cataloging-in-Publication Data

AMS Special Session on Computational Group Theory (2007 : Davidson College)
    Computational group theory and the theory of groups : AMS Special Session on Computational Group Theory, March 3–4, 2007, Davidson, North Carolina / Luise-Charlotte Kappe, Arturo Magidin, and Robert Fitzgerald Morse, editors.
        p. cm. — (Contemporary mathematics ; v. 470)
    Includes bibliographical references.
    ISBN 978-0-8218-4365-9 (alk. paper)
    1. Group theory—Congresses. 2. Computational complexity—Congresses. I. Kappe, Luise-Charlotte. II. Magidin, Arturo. III. Morse, Robert Fitzgerald. IV. Title.
QA174.A673   2007
512′.2–dc22   2008019987

Contents

Preface

The power and availability of general purpose computational algebra systems such as GAP, MAGMA, and SAGE have increased rapidly in recent years. These systems encapsulate very sophisticated algorithms in Computational Group Theory (CGT) and make them available to the mathematics community. Given such computational tools, how can they be applied to assist in pure research in Group Theory? The purpose of this volume is to provide illustrative answers to this question.

This volume consists of contributions by researchers who were invited to the AMS Special Session on Computational Group Theory held at Davidson College, March 3-4, 2007. The main focus of the special session was on applications of CGT to a wide range of theoretical aspects of Group Theory. The session showcased instances of using CGT to solve important and interesting problems such as the construction of simple groups, the classification of p-groups via coclass, problems from representation theory, and structure results for finite and infinite nilpotent groups and permutation groups. Many of these applications are featured in the research articles in this volume. This volume also includes an introductory article by R.F. Morse highlighting applications of CGT capabilities in the Theory of Groups and two survey articles, one by G. Ellis that explores the use of CGT for homological computations, and another by D. Joyner and D. Kohel that introduces the CGT component of SAGE. These three articles serve both as an invitation to those doing pure research in Group Theory to consider what CGT can do for them, and to suggest possible directions for future work and expansion to those who are already using CGT in their research.

It is our hope that this volume will encourage researchers and graduate students to think about ways in which they can incorporate CGT in their own research; that by seeing many different applications of CGT to traditional problems in Group Theory, readers will recognize CGT as yet another tool at their disposal.

The second editor was supported in part by an grant from the Louisiana Board of Regents. The third editor was supported by the Institute for Global Enterprise in Indiana. The first and third editors thank the second editor for his work managing and editing the submissions. The three of us are grateful to all the participants in the Special Session and to all our authors. We are also very grateful to the American Mathematical Society for their help in the publication of this volume, particularly Christine M. Thivierge for her patience and help.

Luise-Charlotte Kappe
Arturo Magidin
Robert Fitzgerald Morse

Contemporary Mathematics
Volume **470**, 2008

On the application of computational group theory to the theory of groups

Robert Fitzgerald Morse

Dedicated to Jerold J. McCloy

ABSTRACT. The availability of computational group theory systems such as GAP (Groups, Algorithms, and Programming) and MAGMA is assisting in the development of theoretical results in group theory. This paper documents some capabilities of these computational group theory systems and how these capabilities have been applied to find new results in the theory of groups.

1. Introduction

The ubiquity of personal computers and the general availability of computational systems for computational group theory have combined to afford nearly anyone with the ability to make very sophisticated computations with both finite and infinite groups. Group theorists can search through databases of all groups up to order 2000 checking for examples of and counterexamples to specific conditions. Various classes of finite and infinite groups can also be searched, such as the infinite metacyclic groups and p-groups up to order p^7. Infinite and finite nilpotent quotients can be computed for a given finitely presented group. Computations like these can be done using a standard personal computer (or laptop) running readily available computational group theory software, for example, GAP (Group, Algorithms, and Programming) [8] and MAGMA [3]. More and more mathematicians are using these computational tools to aid them in obtaining new general results in the theory of groups

Until the early 1990's those working in computational group theory did so without much interest from group theorists in general. The average group theorist had limited access to the necessary hardware, and the software available for making computations with groups was often written for specialized use. This began to

---

2000 *Mathematics Subject Classification.* 20-02, 20F05, 20J06, 20D99.
This paper is based in part on my talk at the AMS Special Session on Computational Group Theory, Davidson College, Davidson, North Carolina, March 3, 2007.
This research was supported in part by a grant from the Institute for Global Enterprise of Indiana.

change with the introduction of systems such as Cayley (now MAGMA) and GAP whose intended goal was to provide a system for general use.

The following excerpt is from Joachim Neubüser's 1988 announcement of the first public release of GAP:

> Up to this decade the interest of pure mathematicians in computational group theory was stirred by, but also mostly confined to, the information that group theoretical software produced on their special research problems – and hampered by the uneasy feeling that one was using black boxes of uncontrollable reliability. However the last years have seen a rapid spread of interest in the understanding, design and even implementation of group theoretical algorithms, which are gradually getting accepted both as tools for a working group theoretician like standard methods of proof and as worthwhile objects of study like connections between notions expressed in theorems.

Neubüser's elegant talk "An Invitation to Computational Group Theory" at Groups St. Andrews in Galway in 1993 [21] and the subsequent workshop on GAP the second week of that conference started a mainstream interest in the use of systems like GAP. During the hands-on portion of the workshop, a number of exercises were given out that demonstrated the power that computational methods have in answering real questions that arise in doing research. Examples of exercises given at the workshop are presented below. Solutions to these exercises are given in Section 5.

(1) Let $D(2,3,n)$ be the Von Dyck group

$$D(2,3,n) = \langle a, b \mid a^2, b^3, (ab)^n \rangle.$$

Show that $D(2,3,5)$ is finite of order 60, and that $D(2,3,6)$ and $D(2,3,7)$ are infinite.

(2) Let g be the finitely presented group $\langle a, b \mid a^2, b^3 \rangle$. Show that the derived subgroup of g is a free group of rank 2.

Hints Use `PresentationNormalClosure`.

(3) Let a_4 be the alternating group on 4 points. Let

$$H(a_4) = \langle x \to x^u \cdot v \in \mathrm{Map}(a_4, a_4) \mid u \in \mathrm{Aut}(a_4), v \in a_4 \rangle$$

be the holomorph of a_4.
  1. Determine a permutation representation of degree 12 for $H(a_4)$.
  2. Determine the size of $H(a_4)$.
  3. Construct two minimal normal subgroups of order 4 of $H(a_4)$.

Hints Use `PermList`.

Today most new personal computers have two or more processors and gigabytes of memory and disk space. Algorithmic advances in computing with finite and infinite groups are continually being encapsulated into systems like GAP and MAGMA. Hence the power of both current hardware and software allows for a greater number of mathematicians to make high end group theoretical computations. As Joachim Neubüser envisioned in his announcement of GAP in 1988, computational methods are becoming mainstream in the work of group theorists.

The purpose of the Special Session on Computational Group Theory at the Sectional Meeting of the American Mathematical Society hosted by Davidson College, March 3–6, 2007 was to acknowledge this influence and to share some applications of computational group theory to current "traditional" research in group theory.

In this paper we outline some of the capabilities available to the working group theorist via GAP, one of several computational group theory systems, and give some examples of how these capabilities have been applied to develop general results in group theory. In Section 2 we discuss how one can obtain results from libraries of groups and in Section 3 we discuss how quotient algorithms can be used to create groups from classes of groups and how they can be applied to compute the non-abelian tensor squares of infinite nilpotent groups. In Section 4 we discuss coset enumeration and rewriting techniques used to show that 4-Engel groups are nilpotent. This outline is necessarily incomplete and follows the author's own research interests.

Ultimately the desired outcome of the Special Session and this paper is to encourage the creative potential of the reader to apply the computational power now readily available to solve problems in group theory.

2. Libraries and classes of groups

In this section we look at how libraries of groups encapsulated in GAP or MAGMA can help in obtaining general group theoretic results and constructions. To illustrate this consider the problem of determining whether the derived subgroup of a group contains a noncommutator. Let G be a group. We denote the set of commutators of G by $K(G) = \{[x, y] \mid x, y \in G\}$. The derived subgroup G' is generated by $K(G)$. It is an open problem in general to give necessary and sufficient conditions for when $G' \neq K(G)$. One necessary condition for when $G' \neq K(G)$ is that the group G must have sufficiently large order. This naturally leads to the question: What is the group H of smallest order such that $H' \neq K(H)$?

Fite [6] showed there exists a group H of order 256 such that $H' \neq K(H)$. This puts an upper bound on our search. Using the library of small groups in GAP, which consists of all groups up to order 2000 except for the groups of order 1024, we search all the nonabelian groups up to order 255 looking for the first group in which $H' \neq K(H)$.

```
gap> First(AllGroups(Size,[2..255],IsAbelian,false), g->
>           Size(Set(List(Cartesian(g,g),LeftNormedComm)))<>
>           Size(DerivedSubgroup(g))
           );
<pc group of size 96 with 6 generators>
```

We then look at all groups of order 96 and find two nonisomorphic groups of this order whose derived subgroups contain a noncommutator.

```
gap> gps := Filtered(AllGroups(Size,96,IsAbelian,false), g->
> Size(Set(List(Cartesian(g,g),LeftNormedComm)))<>
> Size(DerivedSubgroup(g)));
[ <pc group of size 96 with 6 generators>,
  <pc group of size 96 with 6 generators> ]
```

Interpreting the given output is a common problem encountered when using computational tools. In the example above, we know there are two groups of order 96 in which the derived subgroup contains a noncommutator and no smaller examples exist that satisfy this condition. The answer is given but very little information as to why it is an answer is provided. A proof for the order bound requires a systematic analysis of each group of order up to 95 to show that the derived subgroup of each of these groups does not contain a noncommutator. This systematic analysis is aided by several results listed below that give sufficient conditions for when $G' = K(G)$.

Nearly all groups of order less than 96 can be eliminated by the following result of Guralnick.

THEOREM 1 ([**10**]). *If* $|G'|$ *is divisible by at most three primes (counting multiplicity), then* $G' = K(G)$.

Using the small groups library we find the following eleven groups of order at most 95 whose derived subgroups are divisible by more than three primes (counting multiplicity):

```
gap> grps := AllGroups(Size,[2..95],
>   x->Length(Factors(Size(DerivedSubgroup(x)))))>3,true);
[ <pc group of size 48 with 5 generators>,
  <pc group of size 48 with 5 generators>,
  <pc group of size 48 with 5 generators>,
  <pc group of size 48 with 5 generators>,
  Group([ (1,2,3,4,5), (1,2,3) ]),
  <pc group of size 64 with 6 generators>,
  <pc group of size 64 with 6 generators>,
  <pc group of size 64 with 6 generators>,
  <pc group of size 72 with 5 generators>,
  <pc group of size 72 with 5 generators>,
  <pc group of size 80 with 5 generators> ]
```

We can eliminate six of these groups by the following two theorems dealing with metabelian groups.

THEOREM 2 ([**24**]). *If* G *is nilpotent and* G' *is cyclic, then* $G' = K(G)$.

THEOREM 3 ([**27**]). *Suppose the group* G *contains a normal abelian subgroup* A *with cyclic factor group* G/A. *Then* $K(G) = G'$.

We first check directly if the groups meet the conditions of Theorem 2 or the conditions of Theorem 3 with $A = G'$.

```
gap> Filtered(grps, x->
> not(IsNilpotent(x) and IsCyclic(DerivedSubgroup(x))) and
> not(IsAbelian(DerivedSubgroup(x)) and
> IsCyclic(x/DerivedSubgroup(x))));
[ <pc group of size 48 with 5 generators>,
  <pc group of size 48 with 5 generators>,
  Group([ (1,2,3,4,5), (1,2,3) ]),
  <pc group of size 72 with 5 generators>,
  <pc group of size 72 with 5 generators> ]
```

None of the remaining five groups are metabelian so we do not need to consider any other abelian subgroups A in Theorem 3. Of these remaining groups, A_5, the third group, can be eliminated by the following theorem, since A_5 is perfect and is generated by its Sylow p-subgroups for $p = 2, 3, 5$, each of which can be generated by 2 generators or fewer.

THEOREM 4 ([10]). *Let P be a Sylow p-subgroup of G with $P^* = P \cap G'$ abelian and P^* is generated with minimal cardinality generating set of cardinality at most 2. Then $P^* \subseteq K(G)$.*

Hence we are left with four groups that need to be examined to show their derived subgroups do not contain a noncommutator. Our filtering of all the groups of order less than 96 in the GAP library using known theory directs a computer-free proof that all such groups have the property $G' = K(G)$.

One can compute the structure of the two groups of order 96 stored in the GAP variable gps defined above using the following command:

```
gap> List(gps,StructureDescription);
[ "((C4 x C2) : C4) : C3", "(C2 x C2 x Q8) : C3" ]
```

We can examine the structure of these groups to see if either of them have some property that can be generalized to a class of groups such that the derived subgroup of each group in the class contains a noncommutator. Guralnick [9] forms such a class of groups, whose smallest member is the group $(C_2 \times C_2 \times Q_8) \rtimes C_3$ of order 96.

PROPOSITION 5 ([9]). *Let $G_1 = \mathrm{SL}(2,3) = \langle H_1, x \rangle$, where $H_1 = G_1'$ and $x^3 = 1$. Let G_2 be any group with a normal abelian subgroup H_2 such that G_2/H_2 has order 3. Hence $G_2 = \langle H_2, y \rangle$, where $y^3 \in H_2$. Let G be the subgroup of $G_1 \times G_2$ generated by the set $H_1 \times H_2 \cup \{(x, y)\}$. Then $G' \neq K(G)$.*

A particular group in this class of groups can be constructed by the following GAP function:

```
grpclass :=
    function(G2,H2,y)
        local G1,D,H1,x;

        ## Construct G1, H1 and x
        ##
        G1 := SL(2,3);
        H1 := DerivedSubgroup(G1);
        x  := [[0*Z(3),Z(3)^0],[Z(3),Z(3)]];

        ## Construct the direct product G1 x G2
        ##
        D  := DirectProduct(G1,G2);

        ## Return the subgroup generated by the image
        ## of H_1, H_2 in D and (x,1)*(1,y)
        ##
        return Subgroup(D, Concatenation(
            GeneratorsOfGroup(Image(Embedding(D,1),H1)),
```

```
      GeneratorsOfGroup(Image(Embedding(D,2),H2)),
      [Image(Embedding(D,1),x)*Image(Embedding(D,2),y)]));
  end;
```

The example of order 96 is constructed with $G_2 = A_4$. However, we can choose any group for G_2 that meets the requirements of Proposition 5. For example, choose G_2 to be the extraspecial group of order 27 generated by y and b with $H_2 = \langle b, [y, b] \rangle$. In GAP we have

```
gap> F  := FreeGroup("a","b");; a:= F.1;; b:=F.2;;
gap> R  := [a^3,b^3,(a*b)^3, (a*b^-1)^3];;
gap> G2 := F/R;; y := G2.1;; b:=G2.2;;
gap> H2 := Subgroup(G2,[b,Comm(y,b)]);;
```

Using our GAP function `grpclass` we obtain a group G of order 216 such that $G' \neq K(G)$.

```
gap> G := grpclass(G2,H2,y);
<group with 5 generators>
gap> Size(G);
216
gap> Size(DerivedSubgroup(G))=
>         Size(Set(List(Cartesian(G,G),LeftNormedComm)));
false
```

The analysis showing that all finite groups of order less than 96 have the property $G' = K(G)$ was completed without computer assistance by Robert Guralnick in his Ph.D. dissertation. These results were subsequently published in [9] and [10].

Macdonald [19] gives an example of a group G of order p^{10}, for arbitrary prime p, such that $G' \neq K(G)$. This leads to the following question, answered by Kappe and Morse [15]: What is the smallest n such that a group G of order p^n has $G' \neq K(G)$?

For $p = 2$, a lower bound for n is 7, since we have shown that all groups of order 64 have $G' = K(G)$ and for $p \geq 3$, n must be at least 6 by Theorem 1. Searching the group libraries in GAP we find 52 groups of order 2^7, 12 groups of order 3^6, and 76 groups of order 5^6 such that $G' \neq K(G)$. Careful examination reveals structural similarities among these groups. From this analysis three classes of groups for $p = 2$, $p = 3$, and $p > 3$ were defined such that all groups in these classes have the property that $G' \neq K(G)$. These classes are listed in the following proposition.

PROPOSITION 6 ([15]).
  (i) Let $H = \langle a, b, c \rangle$ be a nilpotent group of class precisely 3. If the elements a^4, b^2, c^2, $[a, c]$, $[b, c]$ and $(ab)^2$ all lie in $Z(H)$, then $K(H) \neq H'$.
  (ii) Let $H = \langle a, b \rangle$ be a nilpotent group of class exactly 4 with a^3, b^9, and $[b, a, b] \in Z(H)$. Then $K(H) \neq H'$.
  (iii) Let $H = \langle a, b \rangle$ be a nilpotent group of class exactly 4 with $[b, a, b] \in Z(H)$ and $\exp(H') = p$, where $p \geq 5$. Then $K(H) \neq H'$.

Although it is not proven in [15], all examples with $G' \neq K(G)$ of orders 2^7, 3^6, and 5^6 fall into one of the classes of groups in Proposition 6. It is conjectured that all p-groups, $p > 3$, with $G' \neq K(G)$ fall into the class of groups defined in Proposition 6 (iii).

The ability to examine all groups in a particular class of group such as all group of order less than 256 or all groups of order 5^6 for some particular property can give great insight in how such a property expresses itself in groups. Our example property in this section has been when the derived subgroup contains a noncommutator. From the information derived from searching through all groups within the class we were able to create classes of groups such that the derived subgroup contains a noncommutator. These classes contain an infinite number of groups. In the next section we show how quotient methods can help with constructing groups within such classes.

3. Quotient Algorithms

In this section we consider how one can apply quotient algorithms to obtain a specific group from a class of groups, and how to change the representation of a group so that it is more amenable to computational methods. The latter application plays a part in computing the nonabelian tensor square of a nilpotent group.

Consider the class of groups from Proposition 6 (i). We can create the freest group in the class using a nilpotent quotient algorithm such as the one implemented in the nq package [22] for GAP. This package, along with its supporting package polycyclic [4], takes a finitely presented group H and finds the largest nilpotent image Q with a specified upper bound on the nilpotency class of the image. The image Q can be infinite. Assuming the packages nq and polycyclic are available, we create a finitely presented group H whose relations match those of Proposition 6 (i) and compute the group G that is the largest nilpotent quotient of H with nilpotency class 3 with the following GAP commands:

```
gap> F := FreeGroup("a","b","c");;
gap> AssignGeneratorVariables(F);;
gap> R := [Comm(a^4,b), Comm(a^4,c), Comm(b^2,a), Comm(b^2,c),
>          Comm(c^2,a), Comm(c^2,b),
>          Comm(Comm(a,c),a), Comm(Comm(a,c),b), Comm(Comm(a,c),c),
>          Comm(Comm(b,c),a), Comm(Comm(b,c),b), Comm(Comm(b,c),c),
>          Comm((a*b)^2,a), Comm((a*b)^2,b), Comm((a*b)^2,c)];;
gap> H := F/R;;
gap> G := NilpotentQuotient(H,3);
Pcp-group with orders [ 0, 0, 0, 2, 2, 2, 2 ]
```

The group G constructed with these commands is an infinite group with a consistent polycyclic presentation. Many computations can be made with infinite polycyclic groups with consistent polycyclic presentations. For example, we can compute their derived subgroups, centers, and quotient groups:

```
gap> DerivedSubgroup(G);
Pcp-group with orders [ 2, 2, 2, 2 ]
gap> Centre(G);
Pcp-group with orders [ 0, 0, 0, 2, 2, 2 ]
gap> Size(DerivedSubgroup(G));
16
gap> Size(G/Centre(G));
16
```

Our computations show that G has a finite derived subgroup G' and that the quotient $G/Z(G)$ is finite. In this case we can compute $K(G)$ by computing a right transversal T of $Z(G)$ in G. Then $K(G)$ is the set $\{[a, b] \mid a$ and b in $T\}$. This set is computed with GAP as follows:

```
gap> nat := NaturalHomomorphism(G,Centre(G));;
gap> T := List(Image(nat),x->Representative(PreImages(nat,x)));;
gap> Length(T);
16
gap> K := Set(List(Cartesian(T,T),LeftNormedComm));;
gap> Size(K);
15
```

We see that $K(G)$ has cardinality 15 and $|G'| = 16$. Hence $K(G) \neq G'$, as Proposition 6 asserts.

To find the largest 2-quotient of the finitely presented group H of class 3 we use the ANUPQ [7] package to find a p-quotient for a given prime p. The only p-groups in this class are 2-groups. Using ANUPQ we have:

```
gap> P := Pq(H:Prime:=2);;
gap> NilpotencyClassOfGroup(P);
3
gap> Collected(Factors(Size(P)));
[ [ 2, 193 ] ]
gap> Collected(Factors(Exponent(P)));
[ [ 2, 63 ] ]
```

Hence P is a 2-group of nilpotency class 3, order 2^{193}, and exponent 2^{63}. Using the same technique we used for the infinite group G above, we find $K(P) \neq P'$.

By adding order relations involving the generators a, b, and c to the list of relations implied in Proposition 6 (i), the class 3 nilpotent quotient of the resulting finitely presented group will be finite. In this case we can use either the GAP library function EpimorphismNilpotentQuotient or the GAP package nq to compute the nilpotent quotient. For example, adding the order relations a^{12}, b^6, and c^6 to the relations stored in the GAP variable R above, we compute the nilpotent quotient of the resulting finitely presented group using both methods as follows:

```
gap> # Concatenate the needed relations to the base relations R
gap> # and create the finitely presented group H
gap> H := F/Concatenation(R,[a^12,b^6,c^6]);;
gap> # Find the nilpotent quotient of H, which must be finite
gap> G := NilpotentQuotient(H,3);
Pcp-group with orders [ 12, 6, 6, 2, 2, 2, 2 ]
gap> Size(G);
6912
gap> Factors(Size(G));
[ 2, 2, 2, 2, 2, 2, 2, 2, 3, 3, 3 ]
gap> # Now use the GAP library function to find the
gap> # nilpotent quotient of H
gap> J := Image(EpimorphismNilpotentQuotient(H,3));
<pc group of size 6912 with 11 generators>
```

We see that finite groups in the class of groups defined in Proposition 6 (i) need not be 2-groups. However, we may be interested in the 2-quotient for the finitely presented group H stored in the GAP variable H above. Again we can use the GAP library function EpimorphismPGroup or the GAP package ANUPQ. Below we use the library function to find the largest 2-quotient of H.

```
gap> G := Image(EpimorphismPGroup(H,2));
<pc group of size 256 with 8 generators>
gap> Size(DerivedSubgroup(G))=
>         Size(Set(List(Cartesian(G,G),LeftNormedComm)));
false
```

Two polycyclic quotient algorithms for infinite quotients are described in [18] and [5]. These algorithms are limited but can be useful in certain cases, particularly if you know that the group defined by some finite presentation is a polycyclic group. In such a case these algorithms can find a consistent polycyclic presentation for the group, allowing for effective computations with the group. We give a short example using the experimental GAP package ipcq, which implements the algorithm developed in [5]. Consider the group

$$Q = \langle a, b, c \mid a^2 = b, \ c^a = c^{-1}, \ b^c = b \rangle,$$

which is a torsion-free extension of a free abelian group of rank 2 by the cyclic group of order 2. The group Q is polycyclic since it is an extension of two polycyclic groups. Using the GAP package ipcq we find the largest polycyclic quotient of G as follows:

```
gap> F := FreeGroup("a","b","c");;
gap> AssignGeneratorVariables(F);
gap> rels := [a^2/b, c^a/c^-1, Comm(b,c)];
[ a^2*b^-1, a^-1*b*a*b^-1, a^-1*c*a*c, b^-1*c^-1*b*c ]
gap> H := F/rels;
<fp group on the generators [ a, b, c ]>
gap> Q := PolycyclicQuotient(H);
Pcp-group with orders [ 2, 0, 2, 0 ]
```

The GAP object Q represents a group that is a torsion-free extension of $C_0 \times C_0$, a direct product of infinite cyclic groups, by the cyclic group C_2:

```
gap> IsTorsionFree(Q);
true
gap> # The Fitting subgroup is free abelian of rank 2
gap> L:=FittingSubgroup(Q);
Pcp-group with orders [ 0, 2, 0 ]
gap> IsAbelian(L);
true
gap> AbelianInvariants(L);
[ 0, 0 ]
gap> StructureDescription(Q/L);
"C2"
```

Having obtained a consistent polycyclic presentation for Q, we can use this presentation for further computations.

GAP also has a solvable quotient algorithm that takes as input a finite solvable group with a finite presentation and returns a polycyclic group. As an example of its application, consider the largest finite generalized triangle group R, called the Rosenberger Monster [20]. This group has the following presentation:

$$(1) \qquad R = \langle a, b \mid a^2, \ b^3, \ (ababab ab^2 ab^2 abab^2 ab^2)^2 \rangle.$$

Most of what we know about the group R has been determined by computational methods. Using GAP, R was shown [16] to be a finite nonsolvable group of order $2^{20} \cdot 3^4 \cdot 5$ and to be a central product of a solvable group A of order $2^{18} \cdot 3^3$ and $\mathrm{SL}(2,5)$ with amalgamated central subgroup of order 2. In [20] it was shown the subgroup A of R is generated by

$$\langle a, \ b[a, b^2], \ b^2[a, b^2][a, b]^2, \ b^2(ab)^{12} \rangle.$$

We can find a polycyclic presentation for the subgroup A of R using the solvable quotient algorithm found in the GAP library. We first create R and the subgroup A.

```
gap> F    := FreeGroup("a","b");; a:=F.1;; b:=F.2;;
gap> rels := [a^2, b^3,
>             (a*b*a*b*a*b*a*b^2*a*b^2*a*b*a*b^2*a*b^2)^2];;
gap> # Construct the finitely presented group R
gap> R := F/rels;
<fp group on the generators [ a, b ]>
gap> a:= R.1;; b:=R.2;; # Remember generators for later use
gap>
gap> A := Subgroup(R,[a,b*Comm(a,b^2), b^2*Comm(a,b^2)*Comm(a,b)^2,
>                     b^2*(a*b)^12]);;
```

The subgroup in the GAP variable A is defined only in terms of its generators. We now find a group Afp isomorphic to A that is finitely presented, from which we can compute its solvable quotient.

```
gap> # Obtain a finite presentation for  A
gap> Afp := Image(IsomorphismFpGroup(A));;
gap> # Obtain a polycyclic presentation for A
gap> Apc := Image(EpimorphismSolvableQuotient(Afp,2^18*3^3));;
gap> Collected(Factors(Size(Apc)));
[ [ 2, 18 ], [ 3, 3 ] ]
gap> DerivedLength(Apc);
4
```

Our previous example shows that quotient algorithms can help determine the structure of a subgroup of a finitely presented group. We can apply quotient algorithms in the opposite direction by obtaining a representation of a larger group amenable to computer calculations, in which the group of interest is a subgroup. As an example, consider the following group construction. Let G be a group. Then the nonabelian tensor square of G, denoted by $G \otimes G$, is the group generated by the symbols $g \otimes h$ for all $g, h \in G$ and subject to the relations

$$gh \otimes h' = (^g h \otimes {}^g h')(g \otimes h') \text{ and } g \otimes hh' = (g \otimes h)(^h g \otimes {}^h h'),$$

where $^x y = xyx^{-1}$. There exists a group $\nu(G)$ such that $G \otimes G$ is isomorphic to a subgroup of $\nu(G)'$. If G is a polycyclic then $\nu(G)$ is polycyclic [2] and if G is nilpotent then $\nu(G)$ is nilpotent [23]. So in particular if G is a free nilpotent group of finite rank then $\nu(G)$ is a finitely generated nilpotent group. Given a poly-cyclic presentation of the nilpotent group G we can compute a finite presentation for $\nu(G)$ [2]. This presentation is not necessarily a consistent polycyclic presentation for $\nu(G)$. The nilpotent quotient program nq can be used to find a consistent polycyclic presentation for $\nu(G)$. From this consistent polycyclic presentation we can compute the subgroup of $\nu(G)$ isomorphic to $G \otimes G$ using standard algorithms for polycyclic groups. The GAP function found in [2] constructs a finite presentation for $\nu(G)$, finds a consistent polycyclic presentation for $\nu(G)$ by applying the nilpotent quotient algorithm and then computes the subgroup of $\nu(G)$ isomorphic to $G \otimes G$.

Determining a specific group using computational methods, such as finding $G \otimes G$ in $\nu(G)$ for a given G as above, can be relatively straightforward. However, characterizing the structure of the group computed may not be so clear. This is an example of the more general problem of how one maps the output from GAP to the mathematics one is doing. What is returned by a computational program may not have a natural interpretation to the symbolic manipulation we are interested in doing, and finding the proper interpretation can be challenging.

In [1] we investigated the structure of the derived subgroups of the free nilpotent groups of finite rank. This investigation was assisted by the use of quotient methods. The main difficulty was mapping the computational output back to our symbolic work so we might be able to obtain general results. We conclude this section with a description of how this mapping was achieved.

Using nq a specific free nilpotent group of fixed rank and class can easily be constructed. In the following we construct G to be the free nilpotent group of class 4 and rank 4.

```
gap> G := NilpotentQuotient(FreeGroup(4),4);;
gap> IsTorsionFree(G);
true
gap> NilpotencyClassOfGroup(G);
4
gap> HirschLength(G);
90
gap> HirschLength(DerivedSubgroup(G));
86
```

The GAP variable G contains a presentation for the desired free nilpotent group. We see that its Hirsch length is 90, which fits the theory as

$$\sum_{i=1}^{4} M(4, i) = 90,$$

where $M(n, w)$ counts the number of basic commutators of weight w in n generators. (The definition of a basic commutator is given below.) The value of $M(n, w)$ is given by the Witt formula

$$M(n, w) = \frac{1}{w} \sum_{d|w} \mu(d) n^{m/d},$$

where μ is the Möbius function. The polycyclic generating sequence obtained by the nq program is a nilpotent generating sequence consisting of simple left normed commutators. All of the generators in the polycyclic generating sequence have the form

$[[\cdots[g_1, g_2], g_3], \cdots, g_n]$. Notationally, we assume that all of the brackets are from the left unless otherwise specified and hence this commutator is written $[g_1, \ldots, g_n]$.

After some initial investigation, our conjecture was that certain types of basic commutators generate two direct factors of the derived subgroup of a free nilpotent group. The problem was to create a basic sequence of commutators that we could use to compute our examples of free nilpotent groups to support our conjecture.

A basic commutator is a commutator that arises in the collection process (see [11]). Only certain commutators arise in this way. Let F be a free group of rank n generated by $x_1, \ldots, x_n$. Fixing the order of the generators of F, then $c_i = x_i$, for $i = 1, \ldots, n$ are basic commutators of weight one. Having defined the basic commutators of weight less than m, the basic commutators of weight m are $c_k = [c_i, c_j]$, where c_i and c_j are basic commutators of weights w_i and w_j respectively, $w_i + w_j = m$, $c_i > c_j$, and if $c_i = [c_s, c_t]$ then $c_j \geq c_t$. We let w_k be the weight of c_k. The basic commutators of weight m follow those of weight less than m and are ordered arbitrarily with respect to each other.

The commutators $c_1, \ldots, c_n, \ldots$ form an infinite sequence and, as Sims [26] points out, "basic" is an attribute of the sequence, not an attribute of particular commutators. If we fix the ordering of the generators of F there is one basic sequence of commutators such that the ordering is lexicographical. That is, if $c_i = [c_r, c_s]$ and $c_j = [c_t, c_u]$ are both of weight m then $c_i < c_j$ if $c_r < c_t$ or if $c_r = c_t$ and $c_s < c_u$. We now show how to construct a lexicographical sequence from the order given for the generating symbols in GAP.

A fully bracketed commutator is defined as follows. If c has weight 1 then $[c]$ is fully bracketed. Suppose that x and y are fully bracketed commutators. Then $[x, y]$ is fully bracketed. We can represent a fully bracketed commutator in GAP as a list of lists in the symbol set. Suppose the symbol set is $\{a, b, c\}$ and $[[a, b], [c]]$ is a fully bracketed commutator. This translates directly into GAP:

```
gap> fbc := [[a,b],[c]]; # Assumes the symbols a,b,c are defined
```

Using this representation we can define a weight function.

```
## Compute the weight of a fully bracketed commutator
## represented as a list of lists.
##
weight :=
    function(x)
        if Length(x)=1 then return 1; fi;
        return weight(x[1])+weight(x[2]);
    end;
```

The following function tests whether a fully bracketed commutator is a commutator in the lexicographical basic sequence.

```
## Test whether a fully bracketed commutator represented as a list
## of lists represents a commutator in the lexicographical basic
## sequence.
##
IsBasicComm :=
    function(x)
        if weight(x)=1 then return true; fi;
        if x[1]<=x[2] then return false; fi;
```

```
      if weight(x[1])>1 then return x[1][2]<=x[2]; fi;
      return true;
   end;
```

Now we have the functions to create the lexicographic basic sequence of fully bracketed commutators up to a specified weight. We use the indices of the symbols list as our underlying symbol set so that we can use list comparison functions in GAP.

```
## Create the lexicographical basic sequence of commutators based on
## the ordering of the symbol list up to a given weight.
##
## symb -- List of symbols   wgt  -- Weight bound
##
BasicSequence :=
    function(symb,wgt)
        local bc,    ## List of basic commutators
              r,     ## Index variable
              sub;   ## Local helper function

        ## Local function to substitute the indices in the
        ## commutators for the elements of symb
        ##
        sub := function(l)
            if Length(l)=1 then return symb{l}; fi;
            return Concatenation([sub(l[1])], [sub(l[2])]);
        end;

        ## We use the indices [1..Length(symb)] as a symbol set
        ## to build the basic sequences so that we can use the
        ## native lexicographical comparison operators in GAP
        ## on lists of lists of integers.

        ## Basic sequence of commutators of weight 1
        ##
        bc := List([1..Length(symb)],x->[x]);

        ## Use the basic sequence of weight 1..r-1 to create
        ## the next commutators in the sequence of weight r.
        ##
        for r in [2..wgt] do
            Append(bc,Filtered(Cartesian(bc,bc),
                        x-> weight(x)=r and IsBasicComm(x)));
        od;

        ## Substitute back the values of symb into the commutators
        ##
        return List(bc,sub);
    end;
```

Consider the free nilpotent group of rank 4 and class 4 stored in the GAP variable G. It has a minimal generating set that we can use as our symbol set. We create a list of basic commutators of weight up to 4.

```
gap> gens := MinimalGeneratingSet(G);
[ g1, g2, g3, g4 ]
gap> bc    := BasicSequence(gens,4);
[ [ g1 ], [ g2 ], [ g3 ], [ g4 ],
  [ [ g2 ], [ g1 ] ],  [ [ g3 ], [ g1 ] ], [ [ g3 ], [ g2 ] ],
  [ [ g4 ], [ g1 ] ], [ [ g4 ], [ g2 ] ], [ [ g4 ], [ g3 ] ], ....
gap> Collected(List(bc, x->weight(x)));
[ [ 1, 4 ], [ 2, 6 ], [ 3, 20 ], [ 4, 60 ] ]
```

We have now obtained a symbolic representation of these commutators and we next put them to work. If the symbol set consists of group elements as above then we can evaluate the lists above to elements in the group G.

```
## Create the group element of a fully bracketed commutator 'fb'.
## This function requires that the objects used to build the
## commutator are group elements.
##
EvaluateComm :=
    function(fb)
        if Length(fb)=1 then return LeftNormedComm(fb); fi;
        return Comm(EvaluateComm(fb[1]),EvaluateComm(fb[2]));
    end;
```

Let H be a free nilpotent group of finite rank. We know by the Hall Basis Theorem that H' is generated by all of the basic commutators of weight greater than 1. We conjectured that certain simple basic commutators generate an abelian subgroup that is a direct factor of H'. A commutator c_k in a basic sequence is simple if $w_k = 1$ or if $w_k > 1$ and $c_k = [c_i, c_j]$ then $w_j = 1$. We exploit this property in the following function:

```
## Test to determine whether a commutator in a basic sequence
## is simple.
##
IsSimpleComm :=
    function(x)
        if weight(x)=1 then return true; fi;
        return weight(x[2])=1;
    end;
```

We are now able to test our conjecture on the free nilpotent group of rank 4 and class 4 we have stored in the GAP variable G.

```
gap> gens := MinimalGeneratingSet(G);
[ g1, g2, g3, g4 ]
gap> # Compute all of the basic commutators of weight > 1
gap> bs := Filtered(BasicSequence(gens,4), x->weight(x)>1);;
gap> # Compute the simple commutators of weight > 2
gap> bcsim := Filtered(bs, x->IsSimpleComm(x) and weight(x)>2);;
```

```
gap> bcrest:= Filtered(bs, x->not x in bcsim);
gap> A := Subgroup(G,List(bcsim,x->EvaluateComm(x)));;
gap> N := Subgroup(G,List(bcrest,x->EvaluateComm(x)));;
gap> Intersection(A,N);
Pcp-group with orders [  ]
gap> Subgroup(G,Concatenation(GeneratorsOfGroup(A),
> GeneratorsOfGroup(N)))=DerivedSubgroup(G);
true
gap> IsAbelian(A);
true
gap> Collected(AbelianInvariants(A));
[ [ 0, 65 ] ]
gap> Length(bcsim);
65
gap> NilpotencyClassOfGroup(N);
2
```

The calculations above show that the derived subgroup of the free nilpotent group of class 4 and rank 4 is a direct product of a nilpotent of class 2 group and a free abelian group. Moreover, we were able to see that the generators of these direct factors can be characterized by the structure of their basic commutators. From similar analysis on several other free nilpotent groups we developed the following theorem.

THEOREM 7 ([1]). *Let $\mathcal{N}_{n,c}$ be the free nilpotent group of class $c \geq 1$ and rank $n \geq 1$. If $n = 1$ or $c = 1$ then $\mathcal{N}_{n,c}$ is abelian and $\mathcal{N}'_{n,c}$ is trivial. If $n > 1$ and $c = 2$ then $\mathcal{N}'_{n,c}$ is free abelian of rank $M(n,2) = \binom{n}{2}$. If $n > 1$ and $c > 2$ then*

$$\mathcal{N}'_{n,c} \cong N \times F_f^{\mathrm{ab}},$$

where N is a nilpotent group of class $\lfloor \frac{c}{2} \rfloor$ and where the rank f of the free abelian group F_f^{ab} is the number of simple basic commutators in $\mathcal{N}_{n,c}$ of weight greater than $c - 2$.

We have seen quotient methods used in three distinct ways. First, given a class of groups defined by generators and relations we can obtain examples of groups in this class, even infinite examples, using nilpotent, solvable and p quotient algorithms. Second, quotient methods provide ways to examine substructure within a given group. To illustrate this, we showed how to find solvable subgroups within the largest finite generalized triangle group. Lastly, we used the nilpotent algorithm to construct free nilpotent groups with a consistent polycyclic presentation. With this polycyclic presentation we were able to compute subgroups and intersections of subgroups to determine the structure of their derived subgroups. Moreover, we have shown how to map our mathematics to the computations to aid in obtaining a general result.

4. Coset Enumeration and String Rewriting

In this section we give a brief account of the resolution of the conjecture that 4-Engel groups are locally nilpotent [12]. The proof of this result uses computational methods extensively, namely coset enumeration and the Knuth-Bendix procedure.

The variety of n-Engel groups consists of all groups satisfying the two variable commutator law $[x, y, \ldots, y]$, where the variable y is repeated n times and the commutator is left normed. It is an open question whether all n-Engel groups are locally nilpotent. Levi [17] proved that the 2-Engel groups are nilpotent of class at most 3 and Heineken [14] showed that 3-Engel groups are locally nilpotent. Investigations into 4-Engel groups were started in earnest by Traustason [28]. He shows in particular that if 4-Engel groups of exponent p are locally finite, then 4-Engel p-groups are locally finite. Vaughan-Lee [30] proves that 4-Engel groups of exponent 5 are locally finite and hence all 4-Engel 5-groups are locally finite. Vaughan-Lee's proof reduces the problem to a 3-generator 4-Engel group G and he gives a finite presentation for three subgroups that have finite index in G. The demonstration that these subgroups have finite index uses coset enumeration. Havas and Vaughan-Lee ([12], [13]) generalize this computation to primes up to 31, giving evidence that 4-Engel p-groups are locally finite.

To complete the proof that 4-Engel groups are locally nilpotent, Havas and Vaughan-Lee needed to prove that a particular finitely presented 4-Engel group T is nilpotent. To do this they use a method described by Sims [25]; essentially they construct a polycyclic presentation for a nilpotent quotient N of T and use rewriting techniques to show that the kernel of the mapping from T to N is trivial. Subsequent to the computer-assisted proof of Havas and Vaughan-Lee [12] and motivated by their findings, Traustason [29] gives a computer-free proof that the group T above is nilpotent.

Concluding remarks

While the topics and details in this paper reflect the author's interests, it can be seen that the ability to make very sophisticated computations can greatly aid purely theoretical results in group theory. Computations can give hints so that either generalizations can be proven or traditional proofs of the result might be found. For example, without computational assistance the question of the local nilpotence of 4-Engel groups would probably be still open.

The application of computational group theory to the theory of groups is aiding our ability to prove far reaching results in group theory. The examples presented in this paper as well as the articles in this volume illustrate this interaction.

5. Solutions to the exercises

In this section we provide solutions to the problems presented in Section 1. There is no claim to uniqueness in how these problems are solved. There may be other more efficient or clever ways to solve these problems.

For the first problem we create the Von Dyke groups $D(2, 3, 5)$, $D(2, 3, 6)$, and $D(2, 3, 7)$.

```
gap> F := FreeGroup("a","b");;
gap> AssignGeneratorVariables(F);;
gap> R := [a^2,b^3];;
gap> D235 := F/[a^2,b^3,(a*b)^5];;
gap> D236 := F/[a^2,b^3,(a*b)^6];;
gap> D237 := F/[a^2,b^3,(a*b)^7];;
```

Direct calculation gives:

```
gap> Size(D235);
60
```

For $D(2,3,6)$ we see that its abelianization is finite but the abelianization of its derived subgroup has an infinite cyclic factor.

```
gap> AbelianInvariants(D236);
[ 2, 3 ]
gap> AbelianInvariants(DerivedSubgroup(D236));
[ 0, 0 ]
```

Hence $D(2,3,6)$ is infinite. This approach does not work for $D(2,3,7)$ since it is perfect.

```
gap> AbelianInvariants(D237);
[ ]
```

Therefore, we need look for a subgroup of $D(2,3,7)$ whose abelianization has an infinite cyclic group. We create subgroups of $D(2,3,7)$ using the function that computes subgroups of low index.

```
gap> # Find the first subgroup with index 8
gap> S := Filtered(LowIndexSubgroupsFpGroup(D237,8),
>                      x->Index(D237,x)=8)[1];;
gap> # Make S an FP group
gap> GeneratorsOfGroup(S);;
gap> # Find all subgroups of index 7 of S
gap> i7 := Filtered(LowIndexSubgroupsFpGroup(S,7), x->Index(S,x)=7);
[Group(<fp, no generators known>),Group(<fp, no generators known>),
 Group(<fp, no generators known>), Group(<fp, no generators known>),
 Group(<fp, no generators known>), Group(<fp, no generators known>)]
gap> List(i7,AbelianInvariants);
[ [ 3, 3, 3, 3 ], [ 0, 0, 3 ], [ 0, 0, 3 ], [ 3, 3, 3, 3 ],
  [ 0, 0, 3 ], [ 0, 0, 3 ] ]
```

We see that subgroups of S of index 7 have abelian invariants, indicating an infinite homomorphic image.

The second problem is solved by obtaining a presentation for the derived subgroup and observing that the presentation given is that of a free group.

```
gap> F := FreeGroup("a","b");
<free group on the generators [ a, b ]>
gap> AssignGeneratorVariables(F);;
gap> R := [a^2,b^3];;
gap> G := F/R;
<fp group on the generators [ a, b ]>
gap> PresentationNormalClosure(G,DerivedSubgroup(G));
<presentation with 2 gens and 0 rels of total length 0>
```

To solve the last problem we compute the action of each pair (α, p), where $\alpha \in \mathrm{Aut}(A_4)$ and $p \in A_4$, on a list containing the elements of A_4. This action is represented as a permutation of the 12 positions representing the elements of A_4.

```
gap> a4 := AlternatingGroup(4);;
gap> act := List(Cartesian(AutomorphismGroup(a4),a4),
>             x->List(a4, y->Image(x[1],y)*x[2]));;
gap> lst := List(act, x->List(x, y->Position(Elements(a4),y)));;
gap> gens := List(lst,
>             x->PermList(List(x, y->Position(Elements(a4),y))));;
gap> g := Group(gens);;
gap> Size(g);
288
gap> Number(NormalSubgroups(g), x->Size(x)=2);
0
gap> Number(NormalSubgroups(g), x->Size(x)=4);
2
```

Acknowledgements

The author is grateful for the help he received from Russell Blyth and Arturo Magidin. They carefully read drafts of this paper and greatly helped with its exposition. The author would also like to thank Arturo for his attention to detail in managing and editing this entire volume. His work is greatly appreciated.

References

[1] Russell D. Blyth, Primož Moravec, and Robert Fitzgerald Morse. On the derived subgroup of the free nilpotent group and its applications. In *Aspects of Infinite Groups*, volume 1 of *Algebra and Discrete Mathematics*. World Scientific Publishing Co. Pte. Ltd., Hackensack, NJ, 2008.

[2] Russell D. Blyth and Robert Fitzgerald Morse. On computing the nonabelian tensor square of polycyclic groups. Submitted.

[3] Wieb Bosma, John Cannon, and Catherine Playoust. The Magma algebra system. I. The user language. *J. Symbolic Comput.*, 24(3-4):235–265, 1997. Computational algebra and number theory (London, 1993).

[4] B. Eick and W. Nickel. *Polycyclic – Computing with polycyclic groups*, 2002. A GAP Package, see [8].

[5] Bettina Eick, Alice C. Niemeyer, and Oreste Panaia. A polycyclic quotient algorithm. Preprint.

[6] William Benjamin Fite. On metabelian groups. *Trans. Amer. Math. Soc.*, 3(3):331–353, 1902.

[7] Greg Gamble, Werner Nickel, and Eamonn O'Brien. *ANUPQ – p-quotient*, 2006. A GAP Package, see [8].

[8] The GAP Group. GAP – *Groups, Algorithms, and Programming*, Version 4.4, 2005. (http://www.gap-system.org).

[9] Robert M. Guralnick. Expressing group elements as commutators. *Rocky Mountain J. Math.*, 10(3):651–654, 1980.

[10] Robert M. Guralnick. Commutators and commutator subgroups. *Adv. in Math.*, 45(3):319–330, 1982.

[11] Marshall Hall, Jr. *The theory of groups*. The Macmillan Co., New York, N.Y., 1959.

[12] George Havas and M. R. Vaughan-Lee. 4-Engel groups are locally nilpotent. *Internat. J. Algebra Comput.*, 15(4):649–682, 2005.

[13] George Havas and M. R. Vaughan-Lee. Computing with 4-Engel groups. In *Groups St. Andrews 2005. Vol. 2*, volume 340 of *London Math. Soc. Lecture Note Ser.*, pages 475–485. Cambridge Univ. Press, Cambridge, 2007.

[14] Hermann Heineken. Engelsche Elemente der Länge drei. *Illinois J. Math.*, 5:681–707, 1961.

[15] Luise-Charlotte Kappe and Robert Fitzgerald Morse. On commutators in p-groups. *J. Group Theory*, 8(4):415–429, 2005.

[16] L. Lévai, G. Rosenberger, and B. Souvignier. All finite generalized triangle groups. *Trans. Amer. Math. Soc.*, 347(9):3625–3627, 1995.

[17] F. W. Levi. Groups in which the commutator operation satisfies certain algebraic conditions. *J. Indian Math. Soc. (N.S.)*, 6:87–97, 1942.

[18] Eddie H. Lo. A polycyclic quotient algorithm. *J. Symbolic Comput.*, 25(1):61–97, 1998.

[19] I. D. Macdonald. Commutators and their products. *Amer. Math. Monthly*, 93(6):440–444, 1986.

[20] Robert Fitzgerald Morse. On the Rosenberger monster. In *Combinatorial group theory, discrete groups, and number theory*, volume 421 of *Contemp. Math.*, pages 251–260. Amer. Math. Soc., Providence, RI, 2006.

[21] J. Neubüser. An invitation to computational group theory. In *Groups '93 Galway/St. Andrews, Vol. 2*, volume 212 of *London Math. Soc. Lecture Note Ser.*, pages 457–475. Cambridge Univ. Press, Cambridge, 1995.

[22] W. Nickel. *nq – Nilpotent Quotients of Finitely Presented Groups*, 2003. A GAP Package, see [8].

[23] N. R. Rocco. On a construction related to the nonabelian tensor square of a group. *Bol. Soc. Brasil. Mat. (N.S.)*, 22(1):63–79, 1991.

[24] D. M. Rodney. On cyclic derived subgroups. *J. London Math. Soc. (2)*, 8:642–646, 1974.

[25] Charles C. Sims. Verifying nilpotence. *J. Symbolic Comput.*, 3(3):231–247, 1987.

[26] Charles C. Sims. *Computation with finitely presented groups*, volume 48 of *Encyclopedia of Mathematics and its Applications*. Cambridge University Press, Cambridge, 1994.

[27] Eugene Spiegel. Calculating commutators in groups. *Math. Mag.*, 49(4):192–194, 1976.

[28] Gunnar Traustason. On 4-Engel groups. *J. Algebra*, 178(2):414–429, 1995.

[29] Gunnar Traustason. A note on the local nilpotence of 4-Engel groups. *Internat. J. Algebra Comput.*, 15(4):757–764, 2005.

[30] Michael Vaughan-Lee. Engel-4 groups of exponent 5. *Proc. London Math. Soc. (3)*, 74(2):306–334, 1997.

DEPARTMENT OF ELECTRICAL ENGINEERING AND COMPUTER SCIENCE, UNIVERSITY OF EVANSVILLE, EVANSVILLE IN 47722 USA

E-mail address: `rfmorse@evansville.edu`

URL: `faculty.evansville.edu/rm43`

Contemporary Mathematics
Volume **470**, 2008

A Classification of Certain Maximal Subgroups of Alternating Groups

Bret Benesh

ABSTRACT. This paper addresses an extension of Problem 12.82 of the Kourovka notebook, which asks for all ordered pairs (n, m) such that the symmetric group S_n embeds in S_m as a maximal subgroup. Problem 12.82 was answered in a previous paper by the author and Benjamin Newton. In this paper, we will consider the extension problem where we allow either or both of the groups from the ordered pair to be an alternating group.

1. Introduction

While graduate students enrolled in a computational group theory course, the author and Benjamin Newton encountered problem 12.82 of the Kourovka Notebook [**5**]. This problem, submitted by V. I. Suschanskiĭ, poses the question of describing the set $\mathcal{M}$ of all pairs of positive integers (n, m) such that the symmetric group S_m contains a maximal subgroup isomorphic to S_n. One obvious family of such pairs is

$$\{(n, n + 1) \mid n \geq 1\}.$$

The goal of the course was to provide an answer to this question with the help of the computational group theory system MAGMA [**1**]. A review of the literature indicated that a second family [**2**, **3**] was known:

$$\left\{(n, m) \;\middle|\; m = \binom{n}{k},\; 2 \leq k \leq n/2 - 1,\; \binom{n-2}{k-1} \text{ is odd}\right\}.$$

MAGMA was used to check the maximal subgroups of symmetric groups of small degree, and it was determined that these two families did not constitute a complete solution to Suschanskiĭ's question. The data generated by MAGMA led to a discovery of a third family [**6**]:

$$\left\{\left(kr, \frac{(kr)!}{(r!)^k k!}\right) \;\middle|\; k, r > 1,\; k + r \geq 6,\; (k, r) \in \mathcal{C}\right\},$$

where $\mathcal{C}$ is defined to be the set of all ordered pairs of the form $(2, 2^d + 1), (3, 2^e + 1)$, or $(2l, 2)$ for $d \geq 0, e \geq 1$, and $l \geq 2$. It was proved in [**6**] that these are the only three possible families, and all such ordered pairs lie in one of the three families. In

---

2000 *Mathematics Subject Classification.* 20B35, 20E28.
Key words and phrases. symmetric group, permutation groups, maximal subgroups.

this paper, we examine an extension of the question answered in the computational group theory course: the case when one or both of the groups in Suschanskiĭ's question is allowed to be an alternating group.

2. Preliminaries

We begin by stating the following three questions:

Q1: For what ordered pairs (n, m) does S_m have a maximal subgroup that is isomorphic to A_n?

Q2: For what ordered pairs (n, m) does A_m have a maximal subgroup that is isomorphic to S_n?

Q3: For what ordered pairs (n, m) does A_m have a maximal subgroup that is isomorphic to A_n?

We can answer the first question immediately with the following easy proposition.

PROPOSITION 2.1. *Suppose that a symmetric group S_n has a subgroup H that can be generated by a subset that only contains elements of odd order. Then H is a subgroup of the alternating group A_n.* $\square$

Since any group isomorphic to an alternating group can be generated by the images of 3-cycles, this proves that the only time S_m has a maximal subgroup isomorphic to A_n is if $n = m$.

To answer the remaining two questions, we simply need to look at maximal subgroups of the alternating group A_m. We will answer these questions by finding families of ordered pairs, and then showing that there can be no other ordered pairs outside of these families. There will be seven families that compose the answer to Q2, and these will be denoted $\mathcal{F}(S)_i$; the families that answer Q3 will be denoted $\mathcal{F}(A)_i$.

We reviewie a few basic facts about the maximal subgroups of symmetric and alternating groups. The following is well-known, and is not difficult to show.

PROPOSITION 2.2. *Let $m > 2$, X_m be either S_m or A_m, and M be a maximal subgroup of X_m. Then one of the following holds:*

(a) *M acts intransitively on $\{1, \ldots, m\}$ and $M \cong (S_k \times S_{m-k}) \cap X_m$, where $k \neq \frac{m}{2}$.*

(b) *M acts transitively but imprimitively on $\{1, \ldots, m\}$, $M \cong (S_r \wr S_k) \cap X_m$, where $kr = m$ and $k, r > 1$.*

(c) *M acts primitively on $\{1, \ldots, m\}$.* $\square$

The cases where the maximal subgroup does not act primitively are relatively easy and can be dealt with immediately. Suppose that A_m has a maximal subgroup M that is isomorphic to S_n, and that M acts intransitively on $\{1, \ldots, m\}$. Then M has the structure from Proposition 2.2(a), and it is an easy exercise to see that M must lie in the following family:

$$\mathcal{F}(S)_1 := \{(n, n + 2) \mid n \geq 3\}.$$

The only case where A_m has a maximal subgroup that is isomorphic to a symmetric group S_n that acts transitively but imprimitively on $\{1, \ldots, m\}$ is when $(n, m) = (4, 6)$. This ordered pair is already in $\mathcal{F}(S)_1$, although that instance represented an intransitive maximal subgroup isomorphic to S_4.

Now suppose that A_m has a maximal subgroup M that is isomorphic to A_n, and that M does not act transitively on $\{1, \ldots, m\}$. Then M has the structure from part (a) of Proposition 2.2, and it is again an easy exercise to see that M must lie in the following family:

$$\mathcal{F}(A)_1 := \{(n, n+1) \mid n \geq 3\}.$$

Finally, note that an alternating group A_n for $n \geq 5$ can never have the form of the wreath product from part (b) of Proposition 2.2, since such a wreath product is not simple. For $n < 5$, we may check the cases individually to see that there are no maximal subgroups of the form described in part (b) that answer Q3.

3. The primitive case

For the remainder of the paper, we will be considering a subgroup X_n of A_m such that X_n is isomorphic to S_n or A_n, and that acts primitively on $\{1, \ldots, m\}$. Then X_n is in family (f) from [4] (all of the families (a) through (f) are listed after the following paragraph), and is therefore maximal in A_m unless one of following holds:

(1) $n = 6$ and $X_n < M \leq Aut(S_n)$, where M also embeds into A_m.
(2) $X_n \cong A_n$, and X_n is contained in the image of S_n in A_m.
(3) The pair (n, m) is explicitly listed as an exception in [4].

It remains to determine exactly when S_n and A_n act primitively on a set of cardinality $m \neq n$. To do this, we assume X_n acts primitively and we look at a point stabilizer H in X_n. Because the action of X_n on $\{1, \ldots, m\}$ is primitive, H is maximal in X_n, and $m = |X_n : H|$. The possibilities for H were enumerated in [4]:

(a) $H \cong (S_k \times S_{n-k}) \cap X_n$ where $k \neq \frac{n}{2}$ (the intransitive case).
(b) $H \cong (S_r \wr S_k) \cap X_n$ where $n = kr$ and $k, r > 1$ (the imprimitive case).
(c) $H \cong AGL(k, p) \cap X_n$ where $n = p^k$ and p prime (the affine case).
(d) $H \cong (T^k.(Out(T) \times S_k) \cap X_n$ where T is a nonabelian simple group, $k \geq 2$, and $n = |T|^{k-1}$ (the diagonal case).
(e) $H \cong (S_r \wr S_k) \cap X_n$ where $n = r^k$, $r \geq 5$, and $k > 1$ (the wreath case).
(f) $T \triangleleft H \leq Aut(T)$ with T a nonabelian simple group, $T \neq A_n$, and H acts primitively on $\{1, \ldots, n\}$ (the almost simple case).

Moreover, [4] states that any subgroup of X_n of one of these forms is maximal, save for a list of explicit exceptions. The action of X_n on the cosets of a maximal subgroup yields a primitive action, and so we may simply consider the action of X_n on subgroups of the six forms listed above. We now only need to determine the values of n where S_n (respectively A_n) has a maximal subgroup of each type, taking into account the exceptions listed in [4]. Once again, MAGMA was useful in working with these exceptions.

Note that for cases (c)–(f), S_n embeds in A_m rather than the more general S_m by results from [6]. Therefore, if H is of one of these four types with $H \not\leq A_n$, then the image of A_n will always be contained in the image of S_n in A_m. In this case, the image of A_n will never be maximal. Similarly, we may conclude in cases (a) and (b) that A_m has no maximal subgroup isomorphic to A_n if the image of S_n embeds into A_n.

We will examine the six cases for H individually.

3.1. The intransitive case. Suppose now that H is intransitive. If $X_n \cong S_n$, then $H \cong S_k \times S_{n-k}$ ($k \neq n/2$), and $m = |S_n : H| = \binom{n}{k}$. The results from [**2, 3**] and the exceptions from [**4**] tell us that we get the following maximal embeddings of S_n into A_m:

$$\mathcal{F}(S)_2 := \left\{ \left(n, \binom{n}{k}\right) \; \middle| \; \begin{array}{l} 2 \leq k \leq \dfrac{n}{2} - 1, \; \binom{n-2}{k-1} \text{ is even,} \\[2mm] \left(n, \binom{n}{k}\right) \neq (6, 15), (10, 120), (12, 495) \end{array} \right\}.$$

If $X_n \cong A_n$, then $H \cong (S_k \times S_{n-k}) \cap A_n$ and $m = |A_n : H| = \binom{n}{k}$. This will not be maximal if the image of S_n is contained in A_m, which leaves:

$$\mathcal{F}(A)_2 := \left\{ \left(n, \binom{n}{k}\right) \; \middle| \; 2 \leq k \leq \dfrac{n}{2} - 1, \; \binom{n-2}{k-1} \text{ is odd} \right\}.$$

3.2. The imprimitive case. Suppose now that H is transitive but imprimitive. If $X_n \cong S_n$, then $H \cong S_r \wr S_k$ ($kr = n$, and $k, r > 1$), and $m = |S_n : H| = \frac{(kr)!}{(r!)^k k!}$. The results from [**6**] and the exceptions from [**4**] tell us that we get the following maximal embeddings of S_n into A_m:

$$\mathcal{F}(S)_3 := \left\{ \left(kr, \dfrac{(kr)!}{(r!)^k k!}\right) \; \middle| \; k, r > 1, \; k + r \geq 6, \; (k, r) \notin \mathcal{C} \right\},$$

where $\mathcal{C}$ was defined in the introduction.

If $X_n \cong A_n$, then $H \cong (S_r \wr S_k) \cap A_n$ ($kr = n$, and $k, r > 1$), and

$$m = |A_n : H| = \dfrac{(kr)!/2}{(r!)^k k!/2} = \dfrac{(kr)!}{(r!)^k k!}.$$

This will not be maximal if the image of S_n is contained in A_m, which leaves:

$$\mathcal{F}(A)_3 := \left\{ \left(kr, \dfrac{(kr)!}{(r!)^k k!}\right) \; \middle| \; k, r > 1, \; k + r \geq 6, \; (k, r) \in \mathcal{C} \right\}.$$

3.3. The affine case. Suppose now that $H \cong AGL(k, p)$ is affine; then n is equal to p^k. We will use the fact that the affine general linear group $AGL(k, p)$ is contained A_{p^k} iff $p = 2$.

If $X_n \cong S_{p^k}$, then S_{p^k} only has a maximal affine subgroup if p is odd. There is an infinite family of exceptions listed in [**4**] that occur when $k = 1$. This gives us an infinite family:

$$\mathcal{F}(S)_4 := \left\{ \left(p^k, |S_{p^k} : AGL(k, p)|\right) \; \middle| \; p \text{ is an odd prime, } k > 1 \right\}.$$

If $X_n \cong A_{p^k}$, then X_n will always be contained in the image of S_{p^k} unless $p = 2$. Then $X_n \cong A_{2^k}$ embeds maximally in A_m in the following conditions:

$$\mathcal{F}(A)_4 := \left\{ \left(2^k, |A_{2^k} : AGL(k, 2)|\right) \; \middle| \; k \geq 3 \right\}.$$

3.4. The diagonal case. Suppose now that H is diagonal, let T be a nonabelian simple group, let $k \geq 2$, and let $D = T^k.(Out(T) \times S_k)$. Then $n = |T|^{k-1}$ and $H \cong D$. We would like to be able to determine exactly when this H is contained in A_n, but we only have an incomplete answer. It is known that H can lie outside of A_n iff one of the following occurs:

(1) $k = 2$ and $|\{t \in T \mid t^2 = 1\}| \equiv 2 \pmod 4$.

(2) $k > 2$, and $Out(T)$ contains an automorphism of T that acts of T^{k-1} via an odd permutation.

If $X_n \cong S_{|T|^{k-1}}$ and D contains an odd permutation of T^{k-1}, then $m = |S_n : D|$. This yields:

$$\mathcal{F}(S)_5 := \left\{ \left(|T|^{k-1}, |S_{|T|^{k-1}} : D| \right) \;\middle|\; k \geq 2,\, D \text{ contains an odd permutation} \right\}.$$

If $X_n \cong A_{|T|^{k-1}}$ and D contains only even permutations of T^{k-1}, then m is equal to $|A_n : D|$. This yields:

$$\mathcal{F}(A)_5 := \left\{ \left(|T|^{k-1}, |A_{|T|^{k-1}} : D| \right) \;\middle|\; k \geq 2,\, D \text{ contains only even permutations} \right\}.$$

There were no exceptions in [4] for this case.

3.5. The wreath case. Suppose now that $H \cong S_r \wr S_k$ for $n = r^k$, $r \geq 5$, $k > 1$. Then H always contains an odd permutation, and is never a subgroup of A_n; then the image of A_n will always be contained in the image of S_n and will never be maximal. We conclude that $m = |S_n : H| = \frac{(r^k)!}{(r!)^k k!}$, and we get the following two families of ordered pairs:

$$\mathcal{F}(S)_6 \;\; := \;\; \left\{ \left(r^k, \frac{(r^k)!}{(r!)^k k!} \right) \;\middle|\; r \geq 5,\, k > 1 \right\}$$

$$\mathcal{F}(A)_6 \;\; := \;\; \emptyset.$$

There were no exceptions in [4] for this case.

3.6. The almost simple case. Suppose now that H is almost simple. We would like to determine when $H < A_n$, but this is currently an intractable problem. However, we can provide an implicit solution. Note that all simple groups are generated by their elements of odd order. Then by Proposition 2.1, all simple groups must be contained in alternating groups. Other almost simple groups, however, can lie outside of the alternating group.

We now consider the exceptions from [4] for the almost simple case. Define the four sets of exceptions as follows:

$$\mathcal{E}(S) \;\; := \;\; \left\{ (8, 120), (10, 2520), (22, |A_{24} : M_{24}|) \right\}$$

$$X(A) \;\; := \;\; \left\{ \begin{array}{l} (7, 15), (9, 120), (11, 2520), (23, |A_{24} : M_{24}|), \\ (175, |A_{176} : HS|), (275, |A_{276} : \mathrm{Co}_3|) \end{array} \right\}$$

$$\mathcal{I}(A)_1 \;\; := \;\; \left\{ \left(c - 1, |A_c : \mathrm{Sp}(2d, 2)| \right) \;\middle|\; c = 2^{2d-1} \pm 2^{d+1},\, d \geq 3 \right\}$$

$$\mathcal{I}(A)_2 \;\; := \;\; \left\{ \left(2^d - 1, |A_{2^d} : \mathrm{AGL}(d, 2)| \right) \;\middle|\; d \geq 3 \right\}.$$

For convenience, define $\mathcal{E}(A) = X(A) \cup \mathcal{I}(A)_1 \cup \mathcal{I}(A)_2$. Then we get two more families of ordered pairs:

$$\mathcal{F}(S)_7 \;\; := \;\; \left\{ \left(n, |S_n : H| \right) \;\middle|\; \begin{array}{l} H \text{ almost simple, primitive,} \\ H \not\leq A_n,\, (n, |S_n : H|) \notin \mathcal{E}(S) \end{array} \right\},$$

$$\mathcal{F}(A)_7 \;\; := \;\; \left\{ \left(n, |A_n : H| \right) \;\middle|\; \begin{array}{l} H \text{ almost simple, primitive,} \\ H \leq A_n,\, (n, |A_n : H|) \notin \mathcal{E}(A) \end{array} \right\}.$$

4. Conclusion

Save for the implicit definitions the diagonal and almost simple cases, we may now explicitly answer the original three questions posed in this paper. These are complete lists of solutions to Q2 and Q3 because we have exhausted every possible type of maximal subgroup of symmetric groups and alternating groups by [**4**].

THEOREM 4.1. *The set of all ordered pairs (n, m) such that S_m has a maximal subgroup that is isomorphic to A_n is exactly*

$$\{(n, n) \mid n \geq 2\}.$$

THEOREM 4.2. *The set of all ordered pairs (n, m) such that A_m has a maximal subgroup that is isomorphic to S_n is exactly*

$$\bigcup_{i=1}^{7} \mathcal{F}(S)_i.$$

THEOREM 4.3. *The set of all ordered pairs (n, m) such that A_m has a maximal subgroup that is isomorphic to A_n is exactly*

$$\bigcup_{i=1}^{7} \mathcal{F}(A)_i.$$

5. Acknowledgements

This work was inspired by a computational group theory course taught by Professor Nigel Boston at the University of Wisconsin-Madison, and the author would like to thank him for his guidance and encouragement. Additionally, the author would like to thank Jack Schmidt for generalizing the author's original version of Proposition 2.1; this generalization was helpful in the almost simple case. Finally, the author would like to thank Benjamin Newton, who was extremely helpful in getting these results.

References

[1] W. Bosma, J. Cannon, and C. Playoust. *The* MAGMA *algebra system. I. The user language.* J. Symbolic Comput., **24**(3-4):235–265, 1997.

[2] E. Halberstadt, *On Certain Maximal Subgroups of Symmetric or Alternating Groups*, Math Z. **151** (1976), 117–125.

[3] L. A. Kalužnin and M. H. Klin, *Certain Maximal Subgroups of Symmetric and Alternating Groups*, Math. Sb. **87** (1972), 91–121.

[4] M. W. Liebeck, C. E. Praeger, J. Saxl, *A Classification of the Maximal Subgroups of the Finite Alternating and Symmetric Groups*, J. Algebra **111** (1987), 365–383.

[5] V. D. Mazurov and E. I. Khukhro, *Unsolved Problems in Group Theory: The Kourovka Notebook*, 16th Ed., Novosibirsk, 2006.

[6] B. Newton and B. Benesh, *A Classification of Certain Maximal Subgroups of Symmetric Groups*, J. Algebra **304** (2006), 1108–1113.

DEPARTMENT OF MATHEMATICS, HARVARD UNIVERSITY, ONE OXFORD STREET, CAMBRIDGE, MASSACHUSETTS 02138
E-mail address: benesh@math.harvard.edu

Contemporary Mathematics
Volume **470**, 2008

On the nonabelian tensor squares of free nilpotent groups of finite rank

Russell D. Blyth, Primož Moravec, and Robert Fitzgerald Morse

ABSTRACT. We determine the nonabelian tensor squares and related homological functors of the free nilpotent groups of finite rank.

1. Introduction

Let G be any group. Then the group $G \otimes G$ generated by the symbols $g \otimes h$, where $g, h \in G$, subject to the relations

$$gh \otimes k = ({}^g h \otimes {}^g k)(g \otimes k) \quad \text{and} \quad g \otimes hk = (g \otimes h)({}^h g \otimes {}^h k)$$

for all g, h, and k in G, where ${}^x y = xyx^{-1}$ for $x, y \in G$, is called the *nonabelian tensor square of* G. Let $\nabla(G)$ be the subgroup of $G \otimes G$ generated by the set $\{g \otimes g \mid g \in G\}$. The group $\nabla(G)$ is a central subgroup of $G \otimes G$ [**8**]. The factor group $G \otimes G / \nabla(G)$ is called the *nonabelian exterior square* of G, denoted by $G \wedge G$. For elements g and h in G, the coset $(g \otimes h)\nabla(G)$ is denoted $g \wedge h$.

In his paper [**13**] C. Miller gives a group theoretic interpretation of the Schur multiplier of a group G or, equivalently, $H_2(G)$, the second integral homology group of G. Miller shows that $H_2(G)$ is the group that contains all relations satisfied by the commutators in G modulo those commutator relations which are trivially, or universally, satisfied by G. He interprets $H_2(G)$ to be a measure of the extent to which relations among commutators in G fail to be consequences of universal commutator relations. A relation is universally satisfied if it holds in the free group. We list some of these in (2.1).

For the free group F_n of rank n, the group $H_2(F_n)$ is trivial since F_n does not satisfy any relations other than the universal commutator relations. Let $\mathcal{N}_{n,c} \cong F_n/\gamma_{c+1}(F_n)$ be the free nilpotent group of class c and rank n. By Theorem 1 of [**17**], $H_2(G)$ is isomorphic to the free abelian group of rank $M(n, c + 1)$, where $M(n, c)$ is the number of basic commutators in n symbols of weight c. This matches

2000 *Mathematics Subject Classification.* 20F05, 20F12, 20F18, 20J06.

Key words and phrases. Free nilpotent groups, Nonabelian tensor squares, Schur multiplier.

The second and third authors thank the Institute for Global Enterprise in Indiana for its financial support of this research. The second author thanks the Ministry of Science of Slovenia for supporting his postdoctorial leave to visit the University of Evansville and the Institute for Global Enterprise in Indiana for its generous hospitality while visiting there.

Miller's interpretation of $H_2(G)$, as it captures the commutator relations of $\mathcal{N}_{n,c}$ that are not consequences of universal commutator relations.

Implicit in Miller's work is that $H_2(G)$ is the kernel of the commutator mapping

$$(1.1) \qquad 1 \longrightarrow H_2(G) \longrightarrow G \wedge G \xrightarrow{\ \kappa'\ } G' \longrightarrow 1,$$

where $\kappa'(g \wedge h) = [g, h]$ for all g, h in G. R. K. Dennis in his preprint "In Search of New 'Homology' Functors Having a Close Relationship to K-theory" [9] makes note of (1.1) and extends the nonabelian exterior square to the nonabelian tensor square. Dennis considers the commutator map

$$1 \longrightarrow J_2(G) \longrightarrow G \otimes G \xrightarrow{\ \kappa\ } G' \longrightarrow 1,$$

where $\kappa(g \otimes h) = [g, h]$ for all g, h, in G, and investigates its kernel $J_2(G)$. Brown and Loday in [8] show that $J_2(G)$ is isomorphic to $\pi_3 SK(G, 1)$, the third homotopy group of the suspension of an Eilenberg MacLane space $K(G, 1)$.

In the same paper, Brown and Loday introduce the nonabelian tensor product $G \otimes H$ of two groups G and H. This product is defined if the two groups act on each other in a compatible way. The nonabelian tensor square $G \otimes G$ can be considered a specialization of the nonabelian tensor product, where the actions are taken to be conjugation. The nonabelian tensor square of a group is always defined.

The study of $G \otimes G$ from a group theoretic point of view was started by Brown, Johnson, and Robertson in their seminal paper "Some Computations of Non-Abelian Tensor Products of Groups" [7]. One focus of their paper is to "compute" the nonabelian tensor square for various groups. By computing the non-abelian tensor square of a group G, we mean finding a simple or standard form for expressing $G \otimes G$. The definition of the nonabelian tensor square gives no insight as to the group it describes or its structure. Starting in [7], methods in computational group theory have been invoked to investigate this problem.

In [7] the approach to computing the nonabelian tensor square for a finite group G is to form the finite presentation given in the definition and to use a computer program to perform Tietze transformations to simplify the presentation. This simplified presentation is then examined to determine the isomorphism type of $G \otimes G$. This technique was applied to all of the nonabelian groups of order up to 30. This method becomes impractical for large finite groups since one starts with $|G|^2$ generators and $2|G|^3$ relations.

To compute some examples of the nonabelian tensor product for finite groups, Ellis and Leonard [11] construct a group in which the nonabelian tensor product naturally embeds. In the specialized case of the nonabelian tensor square, we denote this group by $\nu(G)$, following Rocco [15], who independently investigated its properties.

In the following, we fix G to be an arbitrary group with presentation $\langle \mathcal{G} | \mathcal{R} \rangle$.

DEFINITION 1.1. Let G be a group with presentation $\langle \mathcal{G} | \mathcal{R} \rangle$ and let G^φ be an isomorphic copy of G via the mapping $\varphi : g \to g^\varphi$ for all $g \in G$. We define the group $\nu(G)$ to be

$$\nu(G) = \langle \mathcal{G}, \mathcal{G}^\varphi | \mathcal{R}, \mathcal{R}^\varphi, {}^x[g, h^\varphi] = [{}^x g, ({}^x h)^\varphi] = {}^{x^\varphi}[g, h^\varphi], \forall x, g, h \in G \rangle.$$

The motivation for considering $\nu(G)$ relative to the nonabelian tensor square is the following theorem given in [11] and [15].

THEOREM 1.2. *Let G be a group. The map $\phi : G \otimes G \to [G, G^\varphi] \lhd \nu(G)$ defined by $\phi(g \otimes h) = [g, h^\varphi]$ for all g and h in G is an isomorphism.*

It is clear from the definition of $\nu(G)$ that it is generated by $2|\mathcal{G}|$ elements, which is significantly smaller than the number of generators given in the definition of the nonabelian tensor square. Ellis and Leonard [11] show that the relations for $\nu(G)$ can be significantly pruned depending on the size and structure of the center of G. Hence the computational strategy is to construct a relatively small finite presentation of $\nu(G)$, compute a concrete presentation for $\nu(G)$, and then apply standard computational group theory methods to find the subgroup $[G, G^\varphi]$. Ellis and Leonard were able to compute the nonabelian tensor squares for some large finite p-groups, such as the Burnside group of exponent 4 and rank 2, which has order 2^{12}, by applying a p-quotient algorithm to find a power-conjugate presentation of $\nu(G)$, from which the subgroup $[G, G^\varphi]$ can easily be determined. This computation is essentially impossible using the Tietze transformations method.

For an infinite group the definition of the nonabelian tensor square leads to an infinite presentation. The standard technique for computing the nonabelian tensor square for infinite groups is to find a mapping $\Phi : G \times G \to L$ for some group L. If Φ satisfies certain conditions then we call Φ a crossed pairing. If Φ is a crossed pairing then it lifts to a unique homomorphism $\Phi^* : G \otimes G \to L$. Hence to compute the nonabelian tensor square one proposes a group L that one intends to show is isomorphic to $G \otimes G$, devises a crossed pairing Φ, and shows that the lift Φ^* is actually an isomorphism. This method has been used to compute the nonabelian tensor squares for the free nilpotent groups of class 2 of finite rank [1] and the infinite metacyclic groups [3]. In each of these cases the nonabelian tensor square is abelian. The crossed pairing method was also used to compute the nonabelian tensor squares of the free 2-Engel groups of finite rank (see [2] and [6]). An appropriate group L for the free 2-Engel group of rank n was suggested by using the computational techniques of Ellis and Leonard [11] to compute the nonabelian tensor square of a finite image of the free 2-Engel group of rank n, namely the Burnside group of exponent 3 and rank n. In the free 2-Engel case, where the nonabelian tensor squares are not abelian, the computations were overwhelming, and the viability of this method for general use seems limited.

To overcome the limitations of the crossed pairing method when the nonabelian tensor square is not abelian, Blyth and Morse [5] extend the method used by Ellis and Leonard [11] to infinite groups and, in particular, to polycyclic groups. If G is a polycyclic group then $G \otimes G$ is polycyclic and so is $\nu(G)$ [5]. Hence for finite and infinite polycyclic groups both $G \otimes G$ and $\nu(G)$ are finitely presented. A finite presentation of $\nu(G)$ can be described in terms of a polycyclic generating sequence of G. Using a polycyclic quotient algorithm, one is able to compute a polycyclic representation for $\nu(G)$ and use standard algorithms for polycyclic groups to compute the subgroup $[G, G^\varphi]$. Such standard algorithms are implemented in the GAP [12] package Polycyclic [10]. For nilpotent groups this method works well since there exist fast and effective nilpotent quotient algorithms, for example, nq [14], for computing a polycyclic presentation for $\nu(G)$. A simple GAP program that computes the nonabelian tensor square for nilpotent groups is given in [5]. This program creates a finite presentation for $\nu(G)$ using the polycyclic presentation of G and then organizes a series of function calls to compute a polycyclic presentation for $\nu(G)$ and to compute the subgroup $[G, G^\varphi]$ of $\nu(G)$.

Rocco [15] initiated the development of a commutator calculus associated with the subgroup $[G, G^\varphi]$ of $\nu(G)$ that allows for general computations in this subgroup. This commutator calculus is extended in [5]. In this paper we extend it further by providing some new identities (see Lemma 2.1, identities (ii) and (iii)). The commutator calculus is used in [5] to compute the nonabelian tensor square of the free nilpotent group of class 3. We use this calculus in proving Theorem 1.6, the main theorem of this paper.

In [5] a group $\tau(G)$ is defined that is analogous to the group $\nu(G)$ in that the subgroup $[G, G^\varphi]$ of $\tau(G)$ is isomorphic to the nonabelian exterior square, $G \wedge G$, of G.

DEFINITION 1.3. Let G be any group. Then we define $\tau(G)$ to be the quotient group $\nu(G)/\phi(\nabla(G))$, where $\phi : G \otimes G \to [G, G^\varphi]$ is as defined in Theorem 1.2.

Since ϕ isomorphically embeds $\nabla(G)$ into $[G, G^\varphi]$, it follows that

$$[G, G^\varphi]/\phi(\nabla(G)) \cong G \wedge G.$$

We henceforth denote $[G, G^\varphi]/\phi(\nabla(G))$ by $[G, G^\varphi]_{\tau(G)}$. The following proposition is now evident.

PROPOSITION 1.4. Let G be any group. The map

$$\hat{\phi} : G \wedge G \to [G, G^\varphi]_{\tau(G)} \lhd \tau(G)$$

defined by $\hat{\phi}(g \wedge h) = [g, h^\varphi]_{\tau(G)}$ is an isomorphism.

Assembling the maps together we obtain the following sequence of mappings:

$$(1.2) \qquad G \otimes G \xrightarrow{\phi} [G, G^\varphi] \xrightarrow{\sigma} [G, G^\varphi]_{\tau(G)} \xrightarrow{\hat{\phi}^{-1}} G \wedge G$$

where ϕ and $\hat{\phi}^{-1}$ are isomorphisms and σ is an epimorphism that is the restriction of the canonical epimorphism $\nu(G) \to \tau(G)$ to the subgroup $[G, G^\varphi]$. Using (1.2) an arbitrary generator $g \otimes h$ of $G \otimes G$ is mapped to the element $g \wedge h$ in $G \wedge G$ by

$$(1.3) \qquad \hat{\phi}^{-1}(\sigma(\phi(g \otimes h))) = g \wedge h.$$

We use this composition of homomorphisms in the proof of Theorem 1.6 below.

To date only the nonabelian tensor squares of the free nilpotent groups of class 2 and 3 with finite rank have been computed.

THEOREM 1.5 ([1],[5]). Denote the free nilpotent group of class c and rank $n > 1$ by $\mathcal{N}_{n,c}$ and the free abelian group of rank n by F_n^{ab}.

(i) For $c = 2$, $\mathcal{N}_{n,2} \otimes \mathcal{N}_{n,2} \cong F_{f(n)}^{\mathrm{ab}}$, where

$$f(n) = \frac{n(n^2 + 2n - 1)}{3}.$$

(ii) For $c = 3$, $\mathcal{N}_{n,3} \otimes \mathcal{N}_{n,3}$ is the direct product of W_n and $F_{h(n)}^{\mathrm{ab}}$, where W_n is nilpotent of class 2, minimally generated by $n(n-1)$ elements and

$$h(n) = \frac{n(3n^3 + 14n^2 - 3n + 10)}{24}.$$

The following commutative diagram is found in [7]:

(1.4)

$$
\begin{array}{ccccccc}
& & 0 & & 0 & & \\
& & \downarrow & & \downarrow & & \\
H_3(G) & \longrightarrow & \Gamma(G^{\mathrm{ab}}) & \overset{\psi}{\longrightarrow} & J_2(G) & \longrightarrow & H_2(G) & \longrightarrow & 0 \\
& & \downarrow & & \downarrow & & \downarrow \\
0 & \longrightarrow & \nabla(G) & \longrightarrow & G \otimes G & \longrightarrow & G \wedge G & \longrightarrow & 1 \\
& & & & \kappa \downarrow & & \kappa' \downarrow \\
& & & & G' & = & G' \\
& & & & \downarrow & & \downarrow \\
& & & & 1 & & 1
\end{array}
$$

All sequences in this diagram are exact and the short exact sequences are central. The group $\Gamma(G^{\mathrm{ab}})$ is the Whitehead quadratic functor found in [18].

The purpose of this paper is to compute the nonabelian tensor squares of the free nilpotent groups of class c and rank n as well as most of the other homological functors in Diagram (1.4). Our main theorem is the following:

THEOREM 1.6. *Let* $G = \mathcal{N}_{n,c}$ *be the free nilpotent group of class* c *and rank* $n > 1$. *Then*

$$
G \otimes G \cong \Gamma(G^{ab}) \times G \wedge G.
$$

A covering group $\hat{G}$ of a group G is a central extension

$$
1 \longrightarrow H_2(G) \overset{\iota}{\longrightarrow} \hat{G} \longrightarrow G \longrightarrow 1,
$$

where the image of ι is a subset of $\hat{G}'$. If G is a finitely generated group then $\hat{G}'$ is isomorphic to $G \wedge G$ by [7, Corollary 2].

Suppose $G = \mathcal{N}_{n,c}$. Then $\hat{G} \cong \mathcal{N}_{n,c+1}$ is a covering group for G. Since G is a finitely presented group, $G \wedge G$ is isomorphic to $\mathcal{N}'_{n,c+1}$. In Section 3 we prove that $\nabla(G)$ is isomorphic to $\Gamma(G^{\mathrm{ab}})$ (Corollary 3.2). Since G^{ab} is isomorphic to F_n^{ab}, by a result of Whitehead [18, Section 5], $\Gamma(G^{\mathrm{ab}}) \cong F_{\binom{n+1}{2}}^{\mathrm{ab}}$.

From these observations we obtain the following corollary.

COROLLARY 1.7. *Let* $G = \mathcal{N}_{n,c}$ *be the free nilpotent group of class* c *and rank* $n > 1$. *Then*

$$
G \otimes G \cong \mathcal{N}'_{n,c+1} \times F_{\binom{n+1}{2}}^{\mathrm{ab}}.
$$

In the case when $c = 3$, the subgroup W_n of Theorem 1.5 is not isomorphic to $\mathcal{N}'_{n,4}$, as the free abelian factor has rank larger than $\binom{n+1}{2}$. The group W_n is in fact a direct product of a free nilpotent group of class 2 and rank $M(n,2)$, and a free abelian group of rank $\binom{n}{2}$. This case suggests an investigation into the structure of $\mathcal{N}'_{n,4}$. We will give a detailed structural description of $\mathcal{N}'_{n,c+1}$, the derived subgroup of the free nilpotent group of class $c + 1 > 3$ and rank n, in a companion paper [4]. However, to illustrate the application of Corollary 1.7, we give a complete structure description for the $c = 3$ case in Section 3.

Theorem 1.6 is motivated by exploring examples computed using GAP [**12**]. An illustrative description of how these examples were computed is given in Section 5. Our proof of Theorem 1.6 given in Section 4 relies on knowledge of $\nabla(G)$. The structure of the group $\nabla(G)$ and most of the groups in Diagram 1.4, except the nonabelian tensor square, is given in Section 3.

2. A Commutator Calculus

In this section we introduce and extend a commutator calculus for the subgroup $[G, G^\varphi]$ of $\nu(G)$. An account of this calculus can be found in [**5**], which is based in part on [**15**]. The identities found in [**15**] use right actions and are restated using left actions both in [**5**] and this paper. The identities listed for the tensor square in Proposition 3 of [**7**], which use left actions, are now naturally reflected in the identities found in this calculus. Since all conjugation and commutation in this paper is done using left actions, we include a few basic commutator identities for the convenience of the reader. Let G be any group and x, y and z be elements of G. Then

$$^x y = [x, y] \cdot y;$$
$$[xy, z] = {}^x[y, z] \cdot [x, z];$$
(2.1)
$$[x, yz] = [x, y] \cdot {}^y[x, z];$$
$$[x^{-1}, y] = {}^{x^{-1}}[x, y]^{-1} = [x^{-1}, [x, y]^{-1}] \cdot [x, y]^{-1};$$
$$[x, y^{-1}] = {}^{y^{-1}}[x, y]^{-1} = [y^{-1}, [x, y]^{-1}] \cdot [x, y]^{-1};$$

and
$$[x^{-1}, y^{-1}] = [x^{-1}, [y^{-1}, [x, y]]] \cdot [y^{-1}, [x, y]] \cdot [x^{-1}, [x, y]] \cdot [x, y].$$

The following lemma records some basic identities used in this paper.

LEMMA 2.1. *Let G be a group. The following relations hold in $\nu(G)$:*

(i) $^{[g_3, g_4^\varphi]}[g_1, g_2^\varphi] = {}^{[g_3, g_4]}[g_1, g_2^\varphi]$ and $^{[g_3^\varphi, g_4]}[g_1, g_2^\varphi] = {}^{[g_3, g_4]}[g_1, g_2^\varphi]$ *for all g_1, g_2, g_3, g_4 in G;*

(ii) $[g_1^\varphi, g_2, g_3] = [g_1, g_2, g_3^\varphi] = [g_1^\varphi, g_2, g_3^\varphi] = [g_1, g_2^\varphi, g_3] = [g_1^\varphi, g_2^\varphi, g_3] = [g_1, g_2^\varphi, g_3^\varphi]$ *all g_1, g_2, g_3 in G;*

(iii) $[g_1, [g_2, g_3]^\varphi] = [g_2, g_3, g_1^\varphi]^{-1};$

(iv) $[g, g^\varphi]$ *is central in $\nu(G)$ for all g in G;*

(v) $[g_1, g_2^\varphi][g_2, g_1^\varphi]$ *is central in $\nu(G)$ for all g_1, g_2 in G;*

(vi) $[g, g^\varphi] = 1$ *for all g in G'.*

PROOF. All of the identities can be found in [**5**] except (ii) and (iii). In [**15**] it was shown that

(2.2)
$$[g_1^\varphi, g_2, g_3] = [g_1, g_2, g_3^\varphi] = [g_1^\varphi, g_2, g_3^\varphi]$$

and

(2.3)
$$[g_1, g_2^\varphi, g_3] = [g_1^\varphi, g_2^\varphi, g_3] = [g_1, g_2^\varphi, g_3^\varphi].$$

Using these identities we have

$$
\begin{aligned}
[g_1, g_2^\varphi, g_3] &= [[g_2^\varphi, g_1]^{-1}, g_3] \\
&= [g_2^\varphi, g_1]^{-1}[g_2^\varphi, g_1, g_3]^{-1} \\
&= [g_2, g_1]^{-1}[g_2^\varphi, g_1, g_3]^{-1} && \text{by Lemma 2.1(i)} \\
&= [g_1, g_2][g_2, g_1, g_3^\varphi]^{-1} && \text{by (2.2)} \\
&= [g_1, g_2][g_1, g_2]^{-1}[g_1, g_2, g_3^\varphi] \\
&= [g_1, g_2, g_3^\varphi].
\end{aligned}
$$

Hence it follows that all six commutators in (2.2) and (2.3) are equal.

Identity (iii) is a simple consequence of (ii):

$$
[g_1, [g_2, g_3]^\varphi] = [g_2^\varphi, g_3^\varphi, g_1]^{-1} = [g_2, g_3, g_1^\varphi]^{-1}. \qquad \square
$$

We represent a generator $g \otimes h$ of $G \otimes G$ as $[g, h^\varphi]$ using the isomorphism ϕ of Theorem 1.2. The commutator identities of Lemma 2.1 allow us to make nonabelian tensor computations with familiar commutator calculations.

3. Structure of Homological Functors

In this section we determine the groups $\nabla(\mathcal{N}_{n,c})$, $\Gamma(\mathcal{N}_{n,c}^{ab})$, and $J_2(\mathcal{N}_{n,c})$. Let A be an abelian group. The Whitehead quadratic functor, $\Gamma(A)$, is an abelian group with generators $\gamma(a)$, where $a \in A$, with the following relations:

$$
\gamma(a^{-1}) = \gamma(a),
$$
$$
\gamma(abc)\gamma(a)\gamma(b)\gamma(c) = \gamma(ab)\gamma(bc)\gamma(ca),
$$

for $a, b, c \in A$. There is a well defined homomorphism

$$
\psi : \Gamma(G^{ab}) \to G \otimes G
$$

such that $\psi(\gamma(g)G') = g \otimes g$ [**7**, page 181]. The image of ψ is $\nabla(G)$.

The projection map $\pi : G \otimes G \to G^{ab} \otimes G^{ab}$, where $G^{ab} \otimes G^{ab}$ is an ordinary tensor product [**7**, Remark 2] abelianizes $G \otimes G$. Suppose G^{ab} has a basis $\{a_1, \ldots, a_n\}$. Then $G^{ab} \otimes G^{ab}$ is an abelian group with a basis

$$
(3.1) \qquad \{a_i \otimes a_i, a_i \otimes a_j, (a_i \otimes a_j)(a_j \otimes a_i) \mid 1 \le i, j \le n, i < j\}.
$$

PROPOSITION 3.1. *Let G be a group whose abelianization is free abelian of finite rank. Then $\nabla(G) \cong \Gamma(G^{ab})$.*

PROOF. The abelianization G^{ab} of G is isomorphic to F_n^{ab} for some n. Let $\{a_i \mid 1 \le i \le n\}$ be a basis for G^{ab}. By a result of Whitehead [**18**, page 62] there is a basis B for $\Gamma(G^{ab})$ consisting of $\gamma(a_i)$ and $(a_i, a_j) = \gamma(a_i a_j)\gamma(a_i)^{-1}\gamma(a_j)^{-1}$ for $1 \le i, j \le n$ and $i < j$. Define $\Phi = \pi\psi$, where ψ and π are defined above. Then we have that $\Phi(\gamma(a_i)) = a_i \otimes a_i$ and

$$
\begin{aligned}
\Phi((a_i, a_j)) &= \Phi(\gamma(a_i a_j))\Phi(\gamma(a_i)^{-1})\Phi(\gamma(a_j)^{-1}) \\
&= \Phi(\gamma(a_i a_j))\Phi(\gamma(a_i))^{-1}\Phi(\gamma(a_j))^{-1} \\
&= (a_i a_j \otimes a_i a_j)(a_i \otimes a_i)^{-1}(a_j \otimes a_j)^{-1} \\
&= (a_i \otimes a_j)(a_j \otimes a_i)(a_j \otimes a_j)(a_i \otimes a_i)(a_i \otimes a_i)^{-1}(a_j \otimes a_j)^{-1} \\
&= (a_i \otimes a_j)(a_j \otimes a_i).
\end{aligned}
$$

The last two equalities hold as we are computing in the usual abelian tensor product, $G^{\mathrm{ab}} \otimes G^{\mathrm{ab}}$. The images of the elements of the basis B for $\Gamma(G^{\mathrm{ab}})$ under Φ are part of the basis (3.1) for $G^{\mathrm{ab}} \otimes G^{\mathrm{ab}}$. We conclude that ψ is injective. Since ψ is also surjective, it is bijective and $\Gamma(G^{\mathrm{ab}}) \cong \nabla(G)$. $\square$

COROLLARY 3.2. *Let $G = \mathcal{N}_{n,c}$ be the free nilpotent group of class c and rank $n > 1$. Then $\nabla(G) \cong \Gamma(G^{\mathrm{ab}})$.*

Using Corollary 1.7 we can give a complete structure description of $\mathcal{N}_{n,3} \otimes \mathcal{N}_{n,3}$.

PROPOSITION 3.3. *Let $G = \mathcal{N}_{n,3}$ be the free nilpotent group of class 3 and rank $n > 1$. Then $G \otimes G \cong \mathcal{N}_{M(n,2),2} \times F^{\mathrm{ab}}_{f(n)}$, where*

$$f(n) = \frac{n(n+1)(3n^2 + 11n - 2)}{24}.$$

PROOF. By Corollary 1.7, $G \otimes G \cong \mathcal{N}'_{n,4} \times F^{\mathrm{ab}}_{\binom{n+1}{2}}$. In [5, Lemma 32] it was shown that

$$\mathcal{N}'_{n,4} \cong \mathcal{N}_{M(n,2),2} \times F^{\mathrm{ab}}_{g(n)},$$

where $g(n) = M(n,3) + M(n,4) - M(M(n,2),2)$. Using the Witt-Hall identity, $M(n,2) = n(n-1)/2$, $M(n,3) = n(n^2-1)/3$, and $M(n,4) = n^2(n^2-1)/4$. Therefore,

$$g(n) = \frac{n(n^2-1)}{3} + \frac{n^2(n^2-1)}{4} - \frac{n^4 - 2n^3 - n^2 + 2n}{8}$$
$$= 3n^4 + 14n^3 - 3n^2 - 14n24.$$

It follows that

$$f(n) = \binom{n+1}{2} + g(n)$$
$$= \binom{n+1}{2} + \frac{3n^4 + 14n^3 - 3n^2 - 14n}{24}$$
$$= \frac{n(n+1)(3n^2 + 11n - 2)}{24}.$$

$\square$

The description of $\mathcal{N}_{n,3} \otimes \mathcal{N}_{n,3}$ in Theorem 1.5 (ii) fails to include all of $\nabla(\mathcal{N}_{n,3})$ in the free abelian factor of the direct product. The group $\nabla(\mathcal{N}_{n,c})$ has n independent generators of the form $g_i \otimes g_i$ and $\binom{n}{2}$ independent generators of the form $(g_i \otimes g_j)(g_j \otimes g_i)$. This matches the result of Corollary 3.2: $\nabla(\mathcal{N}_{n,c}) \cong \Gamma(\mathcal{N}^{\mathrm{ab}}_{n,c})$, which is free abelian of rank $\binom{n+1}{2} = n + \binom{n}{2}$. The direct factor W_n of the direct product in Theorem 1.5 (ii) includes the $\binom{n}{2}$ generators of $\nabla(\mathcal{N}_{n,3})$ of the form $(g_i \otimes g_j)(g_j \otimes g_i)$. Hence the rank of the free abelian direct factor of Theorem 1.5 is

$$n + g(n) = n + M(n,3) + M(n,4) - M(M(n,2),2)$$
$$(3.2) \qquad = \frac{n(3n^3 + 14n^2 - 3n + 10)}{24}.$$

Adding in the generators of $\nabla(\mathcal{N}_{n,3})$ included in W_n to (3.2), we obtain $f(n)$ from Proposition 3.3.

The discussion above depends on the proof of Theorem 1.6, of course. We begin with some preliminary results that are used in the proof of Theorem 1.6.

We start by formally introducing the notion of a basic sequence of commutators. Our exposition follows Sims [16].

A basic sequence of commutators in the free group F_n of rank n is an infinite sequence $c_1, c_2, \ldots$ of elements of F_n, where each c_i has associated with it a positive integer w_i called its weight. The sequence is defined as follows. The c_i are ordered by weight i.e. if $j > i$ then $w_j \geq w_i$. The $c_1, \ldots, c_n$ are the free generators of F_n arranged in some order. If $w_k > 1$ then c_k is described explicitly by $[c_j, c_i]$, where $j > i$ and $w_j + w_i = w_k$. If $w_j > 1$, so that c_j is described by $[c_q, c_p]$ with $q > p$, then $p \leq i$. Lastly, for each $j > i$ such that either $w_j = 1$ or $w_j > 1$ and c_j is described as $[c_q, c_p]$ with $p \leq i$, there is a unique index k such that c_k is described as $[c_j, c_i]$. We fix one basic sequence of commutators in the free group F_n and denote it by $\mathcal{C}_n$. The elements of $\mathcal{C}_n$ are referred to as basic commutators. We denote the subsequence of commutators of $\mathcal{C}_n$ whose weight is at most w by $\mathcal{C}_{n,w}$. The elements of $\mathcal{C}_{n,c}$ map to $\mathcal{N}_{n,c}$, the free nilpotent group of class c and rank n, via the natural homomorphism $F_n \rightarrow F_n/\gamma_{c+1}(F_n) \cong \mathcal{N}_{n,c}$. We will consider elements of $\mathcal{C}_{n,c}$ as the same as their images in $\mathcal{N}_{n,c}$.

The following proposition, found in [16], will be used in the next section.

PROPOSITION 3.4. *The subsequence $\mathcal{C}_{n,c}$ of $\mathcal{C}_n$ forms a polycyclic generating sequence for $\mathcal{N}_{n,c}$.*

We conclude this section by showing that $J_2(\mathcal{N}_{n,c})$ splits. This follows from the following more general statement.

PROPOSITION 3.5. *Let G be a polycyclic group whose abelianization and second homology are both free abelian groups. Then $J_2(G) \cong \Gamma(G^{\mathrm{ab}}) \times H_2(G)$.*

PROOF. Let G be a polycyclic group, generated by $\{g_i \mid 1 \leq i \leq n\}$. The central subgroup $J_2(G)$ of $G \otimes G$ is the kernel of the commutator mapping $\kappa : G \otimes G \rightarrow G'$, defined by $\kappa(g \otimes h) = [g, h]$. The tensor square $G \otimes G$ is a polycyclic group [5]. Hence $J_2(G)$ is a finitely generated abelian group, and therefore is a direct product $F \times T$, where F is a free abelian group of finite rank and T is a finite abelian group. The group $\Gamma(G^{\mathrm{ab}})$ is free abelian of rank $\binom{n+1}{2}$, while the Schur multiplier $H_2(G)$ is free abelian of finite rank, say d. In the exact sequence

$$(3.3) \qquad \Gamma(G^{\mathrm{ab}}) \xrightarrow{\ \psi\ } J_2(G) \xrightarrow{\ \beta\ } H_2(G) \longrightarrow 0$$

from Diagram (1.4), the kernel of β is isomorphic to $\nabla(G)$, which in turn is isomorphic to $\Gamma(G^{\mathrm{ab}})$ by Proposition 3.1. Hence the kernel of β is free abelian of rank $\binom{n+1}{2}$. But $\beta(T) = 1$, since $H_2(G)$ is torsion free, so $T \subset \ker(\beta)$. We conclude that $T = 1$, and that $J_2(G)$ is free abelian of rank $\binom{n+1}{2} + d$ and hence is isomorphic to $\Gamma(G^{\mathrm{ab}}) \times H_2(G)$. $\qquad\square$

COROLLARY 3.6. *Let $G = \mathcal{N}_{n,c}$ be the free nilpotent group of class c and rank $n > 1$. Then $J_2(G) \cong \Gamma(G^{\mathrm{ab}}) \times H_2(G)$ is free abelian of rank $\binom{n+1}{2} + M(n, c+1)$.*

4. Structure of the Tensor Square

In this section we prove Theorem 1.6. Our proof relies extensively on the commutator calculus given in Section 2.

If G is nilpotent of class c then $\nu(G)$ is nilpotent of class at most $c + 1$ [15]. We make use of a general observation about nilpotent groups that we apply to $\nu(G)$. If X is a set of elements of a nilpotent group of class c, then any commutator of any weight at least 2 with entries from $X \cup X^{-1}$ can be written as a product of commutators all of whose entries lie in X. This fact is proved by induction on the weight k of a commutator, with base case $k = c$, using the commutator identities (2.1). Consequently, the following result for $\nu(G)$ holds.

LEMMA 4.1. *Let G be a nilpotent group of class c. Let u and v be commutators of weight $i \geq 1$ and $j \geq 1$ respectively. Then in $\nu(G)$ the commutators $[u^{-1}, v^{\varphi}]$, $[u, v^{-\varphi}]$ and $[u^{-1}, v^{-\varphi}]$ can all be expressed as products of commutators whose entries are positive words in u, u^{φ}, v and v^{φ}.*

By Proposition 3.4 in Section 3, the sequence $C_{n,c} = \{c_1, \ldots, c_t\}$ is a polycyclic generating sequence for $\mathcal{N}_{n,c}$. We denote the elements $c_1, \ldots, c_n$ in $C_{n,c}$ of weight 1 by $g_1, \ldots, g_n$. By [5, Proposition 25] and Lemma 4.1 the subgroup $[\mathcal{N}_{n,c}, \mathcal{N}_{n,c}^{\varphi}]$ of $\nu(\mathcal{N}_{n,c})$ is generated by the elements

$$(4.1) \qquad \{[c_i, c_j^{\varphi}] \mid c_i, c_j \in C_{n,c}\}.$$

Our goal is to prune this set of generators for $[\mathcal{N}_{n,c}, \mathcal{N}_{n,c}^{\varphi}]$ so that a one-to-one correspondence between a generating set for $[\mathcal{N}_{n,c}, \mathcal{N}_{n,c}^{\varphi}]$ and the set of generators of the factors in the direct product of Theorem 1.6 can be realized.

LEMMA 4.2. *Let $G = \mathcal{N}_{n,c}$ be the free nilpotent group of class c and rank $n > 1$, with polycyclic generating sequence $C_{n,c} = \{c_1, \ldots, c_t\}$. Then $[G, G^{\varphi}]$ is generated by*

(i) *$[g_i, g_i^{\varphi}]$ for $i = 1, \ldots, n$;*
(ii) *$[g_i, g_j^{\varphi}]$ for $1 \leq i < j \leq t$;*
(iii) *$[c_j, c_i^{\varphi}]$ for $1 \leq i < j \leq t$, where $w_j + w_i \leq c + 1$.*

PROOF. All generators of $[G, G^{\varphi}]$ of the form $[c_i, c_i^{\varphi}]$ for $i > n$ are trivial by Lemma 2.1 (vi). This leaves only the generators of the form $[g_i, g_i^{\varphi}]$ for $i = 1, \ldots, n$ as possibly nontrivial generators of the form $[c_i, c_i^{\varphi}]$. These generators are listed in (i).

Suppose $i < j$ and $w_i + w_j \geq 3$ with $w_i \geq 2$. Then $c_i = [c_q, c_p]$ for some q, p such that $q > p$. By Lemma 2.1 (iii) we have

$$[c_i, c_j^{\varphi}] = [[c_q, c_p], c_j^{\varphi}] = [c_j, [c_q, c_p]^{\varphi}]^{-1} = [c_j, c_i^{\varphi}]^{-1}.$$

Similarly, the equality holds if $w_j \geq 2$. Hence all generators of the form $[c_i, c_j^{\varphi}]$ with $w_i + w_j \geq 3$ and $i < j$ can be expressed in terms of elements of (iii). However, this argument does not eliminate those generators with $w_i = w_j = 1$ and $i < j$; these generators are listed in (ii).

Since $\nu(G)$ is nilpotent of class at most $c + 1$ all generators of the form $[c_j, c_i^{\varphi}]$, where $j > i$ and $w_j + w_i > c + 1$ are trivial. Hence the upper weight restriction of (iii) holds. $\qquad \square$

Our analysis of $\mathcal{N}_{n,c} \otimes \mathcal{N}_{n,c}$ now focuses on the subgroup generated by the elements listed in Lemma 4.2 (iii).

PROPOSITION 4.3. *Let $G = \mathcal{N}_{n,c}$ be the free nilpotent group of class c and rank $n > 1$, with the basic sequence of commutators $C_{n,c} = \{c_1, \ldots, c_t\}$ as its polycyclic generating sequence. The subgroup*

$$N = \langle [c_j, c_i^{\varphi}] \mid j > i, w_j + w_i \leq c + 1 \rangle$$

is a normal subgroup of $[G, G^{\varphi}]$ isomorphic to $G \wedge G$.

PROOF. The elements $[g_i, g_i^\varphi]$ for $i = 1, \ldots, n$ are in the center of $\nu(G)$. Hence we need only show that

$$[g_i, g_j^\varphi][c_k, c_m^\varphi]$$

is an element of N when $i < j$ and $k > m$. Now

$$
\begin{aligned}
{}^{[g_i, g_j^\varphi]}[c_k, c_m^\varphi] &= {}^{[g_i^\varphi, g_j]}[c_k, c_m^\varphi] && \text{by Lemma 2.1(i)} \\
&= {}^{[g_j, g_i^\varphi]^{-1}}[c_k, c_m^\varphi] \\
&= [g_j, g_i^\varphi]^{-1} \cdot [c_k, c_m^\varphi] \cdot [g_j, g_i^\varphi],
\end{aligned}
$$

which is an element of N.

To show that N is isomorphic to $G \wedge G$, we recall that $G \wedge G$ is isomorphic to the derived subgroup of $\mathcal{N}_{n,c+1}$, which is generated by the commutators

$$\mathcal{C} = \{c_i \in \mathcal{C}_{n,c+1} \mid w_i > 1\}.$$

Every element c_k in $\mathcal{C}$ is uniquely expressed by $c_k = [c_j, c_i]$ for some $c_j, c_i \in \mathcal{C}_{n,c}$, where $j > i$. The isomorphism from $G \wedge G$ to $\mathcal{N}'_{n,c+1}$ is realized by the map $c_j \wedge c_i \mapsto [c_j, c_i] = c_k$. The isomorphism from $[G, G^\varphi]_{\tau(G)}$ to $\mathcal{N}'_{n,c+1}$ is defined by $\hat{\phi}^{-1}([c_j, c_i^\varphi]_{\tau(G)}) = [c_j, c_i]$. Similarly we can set up a mapping of generators from N to $\mathcal{N}'_{n,c+1}$ by $[c_j, c_i^\varphi] \mapsto [c_j, c_i]$.

We have now the following version of a short exact sequence from Diagram (1.4):

$$
(4.2) \qquad 0 \longrightarrow \nabla(G) \overset{\iota}{\longrightarrow} [G, G^\varphi] \overset{\sigma}{\longrightarrow} \mathcal{N}'_{n,c+1} \longrightarrow 0
$$

Suppose $x \in \iota(\nabla(G)) \cap N$. Then using the Hall collection process, x may be written as a product of powers of the generators of N in order of increasing commutator weight. Let d be the least weight of a factor that appears nontrivially in this expression for x. Then the induced map $\sigma^* : [G, G^\varphi] \mapsto \gamma_d(\mathcal{N}_{n,c+1})/\gamma_{d+1}(\mathcal{N}_{n,c+1})$ maps x to a product of powers of basic commutators of weight d. Since the quotient $\gamma_d(\mathcal{N}_{n,c+1})/\gamma_{d+1}(\mathcal{N}_{n,c+1})$ is a free abelian group with basis the set of basic commutators of weight d, and $\sigma(x) = 1$ (as $x \in \iota(\nabla(G)) = \ker(\sigma)$) we obtain a contradiction unless $x = 1$. Hence $\iota(\nabla(G)) \cap N = 1$.

The mapping from N to $\mathcal{N}'_{n,c+1}$ is an epimorphism, and since $\ker(\sigma) \cap N = 1$, we conclude that N is isomorphic to $\mathcal{N}'_{n,c+1}$, and hence the result follows. $\qquad \square$

PROOF OF THEOREM 1.6. Since $\nabla(G)$ is a central subgroup of $[G, G^\varphi]$ it is normal in $[G, G^\varphi]$. The subgroup $N \cong G \wedge G$ is normal in $[G, G^\varphi]$ by Proposition 4.3. As was shown in the proof of Proposition 4.3, $\nabla(G) \cap N = 1$. Now $[g_i, g_j^\varphi]$ for $i < j$ are the only generators of $[G, G^\varphi]$ not obviously in either N or $\nabla(G)$. However

$$[g_i, g_j^\varphi] = ([g_i, g_j^\varphi][g_j, g_i^\varphi]) \cdot [g_j, g_i^\varphi]^{-1}$$

is a product of elements of $\nabla(G)$ and N. Hence $[G, G^\varphi] = \nabla(G)N$. Therefore we conclude that

$$G \otimes G \cong [G, G^\varphi] = \nabla(G) \times N \cong \nabla(G) \times G \wedge G \cong \Gamma(G^{ab}) \times G \wedge G.$$

The last isomorphism holds by Corollary 3.2. $\qquad \square$

5. Computational Interplay

In this section we provide an account of how Theorem 1.6 was motivated by our use of computational methods, some of which are outlined in Section 1. Specifically, the computed examples provided evidence that the nonabelian tensor square of the free nilpotent group $\mathcal{N}_{n,c}$ is a direct product of its nonabelian exterior square and a free abelian group whose rank depends on the rank n of the group. This observation is not immediately obvious. This fact was missed in three earlier publications: Bacon, Kappe and Morse [2]; Blyth, Morse and Redden [6]; and Blyth and Morse [5]. All three of these papers also used computer examples to help formulate their final general results.

One problem with making computer calculations is interpreting the output given by the computer. Moreover, relating this output to the symbolic manipulations required can be a challenge. Our strategy is to provide a GAP representation of the symbolic or abstract objects we are working with and to then map the GAP symbolic objects to the computer generated output. In our particular case we were interested in mapping a basic sequence of commutators to the generators of $G \otimes G$. So we first represent basic commutators in GAP and then relate them to the polycyclic groups we construct. The purpose of this section is to explicitly demonstrate how we accomplished this.

We start with the following GAP functions that create objects that symbolically represent a basic sequence of commutators. These functions are straightforward to implement:

```
BasicSeq(<symset>,<maxweight>);
ComEval(<comm>);
Weight(<comm>);
```

The function **BasicSeq** generates a basic sequence of commutators in the symbol set <symset> of weights 1 to <maxweight>. The output of **BasicSeq** is a list of lists. Each list in the list represents a fully bracketed commutator. If s is in the symbol set then $[s]$ is a fully bracketed commutator of length 1. Suppose c and d are fully bracketed commutators of weights w_c and w_d respectively. Then $[c, d]$ is a fully bracketed commutator of weight $w_c + w_d$. If the elements from the symbol set are group elements then we can form the element of the group represented by a fully bracketed commutator. The function **ComEval** forms this group element from a fully bracketed commutator represented as a list. The **Weight** function computes the weight of the commutator one of these lists represents. Below is a example whose symbol set consists of the generators of a free group of rank 3.

```
gap> F := FreeGroup(3);;
gap> b := BasicSeq(GeneratorsOfGroup(F),3);;
gap> PrintArray(b);
[ [                   f1 ],
  [                   f2 ],
  [                   f3 ],
  [               [ f2 ],          [ f1 ] ],
  [               [ f3 ],          [ f1 ] ],
  [               [ f3 ],          [ f2 ] ],
  [  [ [ f2 ], [ f1 ] ],          [ f1 ] ],
  [  [ [ f2 ], [ f1 ] ],          [ f2 ] ],
```

```
[  [ [ f2 ], [ f1 ] ],                [ f3 ] ],
[  [ [ f3 ], [ f1 ] ],                [ f1 ] ],
[  [ [ f3 ], [ f1 ] ],                [ f2 ] ],
[  [ [ f3 ], [ f1 ] ],                [ f3 ] ],
[  [ [ f3 ], [ f2 ] ],                [ f2 ] ],
[  [ [ f3 ], [ f2 ] ],                [ f3 ] ] ]
```

```
gap> ## Compute the weights of each commutator represented.
gap> ## The number of commutators of each weight
gap> ## corresponds to the Witt-Hall formula.
gap> ##
gap> List(b,Weight);
[ 1, 1, 1, 2, 2, 2, 3, 3, 3, 3, 3, 3, 3, 3 ]

gap> ## Evaluate each commutator as an element in the free
gap> ## group F
gap> ##
gap> List(b,ComEval);
[ f1, f2, f3, f2^-1*f1^-1*f2*f1, f3^-1*f1^-1*f3*f1,
  ...
  ...
  ...
  f2^-1*f3^-1*f2*f3^-1*f2^-1*f3*f2*f3 ]
```

The goal is to map the elements of a basic sequence of commutators to elements of $G \otimes G$. To do this we need the subgroups G and G^φ, which are the left and right isomorphic embeddings of G in $\nu(G)$. The GAP program listed in [5] to compute the nonabelian tensor square computes these values and returns $[G, G^\varphi] \cong G \otimes G$. This GAP program can be modified to return a record with the fields lbg (left base group), and rbg (right base group). These GAP variables correspond to the mathematical objects G and G^φ respectively. The nonabelian tensor square is computed by the command CommutatorSubgroup(lbg,rbg). Nothing from the program listed in [5] is modified except we are returning different computed values in the form of a record. We rename this function BaseGroups from the name TensorSquare given in [5] to reflect the different returned values. We fix our example group to be $\mathcal{N}_{3,3} = G$. The GAP object G below corresponds to G. We compute lbg, and rbg for G and compute the tensor square, ts, from them.

```
gap> ## Create the free nilpotent of group of class 3
gap> ## and rank 3 and compute the base groups.
gap> ##
gap> G   := NilpotentQuotient(FreeGroup(3),3);;
gap> r   := BaseGroups(G);;

gap> ## Save the base groups for later use
gap> lbg := r.lbg;
Pcp-group with orders [ 0, 0, 0, 0, 0, 0, 0, 0, 0, 0, 0, 0, 0, 0 ]
gap> rbg := r.rbg;
Pcp-group with orders [ 0, 0, 0, 0, 0, 0, 0, 0, 0, 0, 0, 0, 0, 0 ]
```

```
gap> ## Compute the tensor square from the returned values
gap> ##
gap> ts := CommutatorSubgroup(lbg,rbg);
Pcp-group with orders [ 0, 0, 0, 0, 0, 0, 0, 0, 0, 0, 0,
                        0, 0, 0, 0, 0, 0, 0, 0, 0, 0, 0,
                        0, 0, 0, 0, 0, 0, 0, 0, 0, 0, 0,
                        0, 0 ]

gap> ## List the induced generating sequence of ts
gap> ##
gap> Igs(ts);
[ g10, g11, g12, g13, g14, g15, g17, g18, g19, g30, g31,
  g34, g35, g36, g40, g41, g42, g46, g47, g48, g49, g50,
  g51, g52, g53, g54, g55, g56, g57, g58, g59, g60, g61,
  g62, g63 ]
```

How can we show that **ts** above is a direct product of two subgroups? How can we identify the generators of the two subgroups as specific elements in **ts** that can be used to guide us in our proof? The following is one method for doing this using the basic commutator routines introduced above.

```
gap> ## Record the minimal generators of F, lbg and rbg
gap> ##
gap> Fm := GeneratorsOfGroup(F);;
gap> Lm := MinimalGeneratingSet(lbg);;
gap> Rm := MinimalGeneratingSet(rbg);;

gap> ## Create the basic sequence of commutators of weight
gap> ## at most four and then prune them to those of
gap> ## weight greater than 1.
gap> ##
gap> b2 := Filtered(BasicSeq(GeneratorsOfGroup(F),4),
                    x->Weight(x)>1);;
gap> PrintArray(b2);
[ [                               [ f2 ],                  [ f1 ] ],
  [                               [ f3 ],                  [ f1 ] ],
  [                               [ f3 ],                  [ f2 ] ],
  [                 [ [ f2 ], [ f1 ] ],                    [ f1 ] ],
  [                 [ [ f2 ], [ f1 ] ],                    [ f2 ] ],
  [                 [ [ f2 ], [ f1 ] ],                    [ f3 ] ],
  [                 [ [ f3 ], [ f1 ] ],                    [ f1 ] ],
  [                 [ [ f3 ], [ f1 ] ],                    [ f2 ] ],
  [                 [ [ f3 ], [ f1 ] ],                    [ f3 ] ],
  [                 [ [ f3 ], [ f2 ] ],                    [ f2 ] ],
  [                 [ [ f3 ], [ f2 ] ],                    [ f3 ] ],
  [                 [ [ f3 ], [ f1 ] ], [ [ f2 ], [ f1 ] ] ],
  [                 [ [ f3 ], [ f2 ] ], [ [ f2 ], [ f1 ] ] ],
  [                 [ [ f3 ], [ f2 ] ], [ [ f3 ], [ f1 ] ] ],
  [   [ [ [ f2 ], [ f1 ] ], [ f1 ] ],                    [ f1 ] ],
  [   [ [ [ f2 ], [ f1 ] ], [ f1 ] ],                    [ f2 ] ],
```

```
[ [ [ [ f2 ], [ f1 ] ], [ f1 ] ],                    [ f3 ] ],
[ [ [ [ f2 ], [ f1 ] ], [ f2 ] ],                    [ f2 ] ],
[ [ [ [ f2 ], [ f1 ] ], [ f2 ] ],                    [ f3 ] ],
[ [ [ [ f2 ], [ f1 ] ], [ f3 ] ],                    [ f3 ] ],
[ [ [ [ f3 ], [ f1 ] ], [ f1 ] ],                    [ f1 ] ],
[ [ [ [ f3 ], [ f1 ] ], [ f1 ] ],                    [ f2 ] ],
[ [ [ [ f3 ], [ f1 ] ], [ f1 ] ],                    [ f3 ] ],
[ [ [ [ f3 ], [ f1 ] ], [ f2 ] ],                    [ f2 ] ],
[ [ [ [ f3 ], [ f1 ] ], [ f2 ] ],                    [ f3 ] ],
[ [ [ [ f3 ], [ f1 ] ], [ f3 ] ],                    [ f3 ] ],
[ [ [ [ f3 ], [ f2 ] ], [ f2 ] ],                    [ f2 ] ],
[ [ [ [ f3 ], [ f2 ] ], [ f2 ] ],                    [ f3 ] ],
[ [ [ [ f3 ], [ f2 ] ], [ f3 ] ],                    [ f3 ] ] ]
```

The elements in the GAP object b2 represent all of the commutators in the basic sequence of the form $[c_i, c_j]$, where c_i and c_j are elements in the sequence and $i > j$. These commutators generate $\mathcal{N}'_{3,4} \cong \gamma_2(F_3/\gamma_5(F_3))$, which we know is isomorphic to $G \wedge G$.

Our objective now is to map the elements of this basic sequence of commutators to elements in ts and see if the resulting subgroup is normal in ts and, if so, whether or not it is a direct factor.

We evaluate each c_i and c_j of $[c_i, c_j]$ to obtain pairs of words in the free generators Fm.

```
gap> ## The elements of b2 are lists that represent
gap> ## a fully bracketed commutator of weight at least
gap> ## two. Hence it is a list of the form [left, right],
gap> ## where left and right are lists representing
gap> ## commutators in our basic sequence.
gap> ## We create a word in the free group F for the left
gap> ## and right commutator using our ComEval function.
gap>
gap> e2 := List(b2,x->[ComEval(x[1]),ComEval(x[2])]);
[ [ f2, f1 ], [ f3, f1 ], [ f3, f2 ],
  [ f2^-1*f1^-1*f2*f1, f1 ], [ f2^-1*f1^-1*f2*f1, f2 ],
  [ f2^-1*f1^-1*f2*f1, f3 ], [ f3^-1*f1^-1*f3*f1, f1 ],
  ...
  ...
  ...
  [ f2^-1*f3^-1*f2*f3*f2^-1*f3^-1*f2^-1*f3*f2^2, f3 ],
  [ f2^-1*f3^-1*f2*f3^-1*f2^-1*f3*f2*f3, f3 ] ]
```

For each element of e2, we substitute the free generator symbols of the left element with the generators of the left base group whose GAP object is Lm and we substitute the free generators of the right element with the generators of the right base group Rm.

```
gap> t2 := List(e2,x->
          [MappedWord(x[1],Fm,Lm), MappedWord(x[2],Fm,Rm)]);
[ [ g2, g4 ],  [ g3, g4 ],  [ g3, g5 ],  [ g7, g4 ],
  [ g7, g5 ],  [ g7, g6 ],  [ g8, g4 ],  [ g8, g5 ],
  [ g8, g6 ],  [ g9, g5 ],  [ g9, g6 ],  [ g8, g16 ],
```

```
[ g9, g16 ], [ g9, g20 ], [ g22, g4 ], [ g22, g5 ],
[ g22, g6 ], [ g23, g5 ], [ g23, g6 ], [ g25*g27^-1, g6 ],
[ g24, g4 ], [ g24, g5 ], [ g24, g6 ], [ g25, g5 ],
[ g25, g6 ], [ g26, g6 ], [ g28, g5 ], [ g28, g6 ],
[ g29, g6 ] ]
```

We now create the subgroup `ts` generated by the commutators represented by the elements of `t2` and check to see that it is normal in `ts`.

```
gap> N := Subgroup(ts,List(t2,LeftNormedComm));;
gap> IsNormal(ts,N);
true
```

By [5, Lemma 21] $\nabla(G)$ is generated by the elements $[g_i, g_i^\varphi]$ for $i = 1, \ldots, n$ and $[g_i, g_j^\varphi][g_j, g_i^\varphi]$, where $1 \leq i, j \leq n$ and $i \neq j$. In our example, $n = 3$. Hence $\nabla(G)$ is generated by only six generators. We enumerate these generators and create the subgroup generated by them.

```
gap> A := Subgroup(ts,[Comm(Lm[1],Rm[1]),Comm(Lm[2],Rm[2]),
                        Comm(Lm[3],Rm[3]),
                        Comm(Lm[1],Rm[2])*Comm(Lm[2],Rm[1]),
                        Comm(Lm[1],Rm[3])*Comm(Lm[3],Rm[1]),
                        Comm(Lm[2],Rm[3])*Comm(Lm[3],Rm[2])]
                    );;
gap> IsSubgroup(Centre(ts),A);
true
```

The subgroups `N` and `A` are normal in `ts`. They also have trivial intersection:

```
gap> IsTrivial(Intersection(N,A));
true
```

The nonabelian tensor square `ts` is generated by `N` and `A`:

```
gap> ts = Subgroup(ts,Concatenation(GeneratorsOfGroup(N),
                                     GeneratorsOfGroup(A)));
true
```

We conclude that `ts` is the direct product of `N` and `A`.

Finally we show that `N` is isomorphic to $\mathcal{N}'_{3,4} \cong G \wedge G$.

```
gap> ## Form the derived subgroup of the free nilpotent group
gap> ## of class 4 and rank 3.
gap> ##
gap> H := DerivedSubgroup(NilpotentQuotient(FreeGroup(3),4));;

gap> ## Form a mapping from N to H and check to see if it is
gap> ## a homomorphism
gap> ##
gap> hom := GroupGeneralMappingByImages(N,H,
              MinimalGeneratingSet(N),MinimalGeneratingSet(H));;
gap> IsPcpGroupHomomorphism(hom);
true
gap> ## Check to see if hom is an isomorphism
gap> ##
gap> IsTrivial(Kernel(hom));
true
```

While computing examples like this does not constitute a proof, such examples gave us direction in formulating and proving Theorem 1.6.

References

1. Michael R. Bacon. On the nonabelian tensor square of a nilpotent group of class two. *Glasgow Math. J.*, **36**(3):291–296, 1994.
2. Michael R. Bacon, Luise-Charlotte Kappe, and Robert Fitzgerald Morse. On the nonabelian tensor square of a 2-Engel group. *Arch. Math. (Basel)*, **69**(5):353–364, 1997.
3. James R. Beuerle and Luise-Charlotte Kappe. Infinite metacyclic groups and their non-abelian tensor squares. *Proc. Edinburgh Math. Soc. (2)*, **43**(3):651–662, 2000.
4. Russell D. Blyth, Primož Moravec, and Robert Fitzgerald Morse. On the derived subgroup of the free nilpotent group and its applications. To appear.
5. Russell D. Blyth and Robert Fitzgerald Morse. Computing the nonabelian tensor square of polycyclic groups. Submitted.
6. Russell D. Blyth, Robert Fitzgerald Morse, and Joanne L. Redden. On computing the non-abelian tensor squares of the free 2-Engel groups. *Proc. Edinb. Math. Soc. (2)*, **47**(2):305–323, 2004.
7. R. Brown, D. L. Johnson, and E. F. Robertson. Some computations of nonabelian tensor products of groups. *J. Algebra*, **111**(1):177–202, 1987.
8. Ronald Brown and Jean-Louis Loday. Van Kampen theorems for diagrams of spaces. *Topology*, **26**(3):311–335, 1987. With an appendix by M. Zisman.
9. R. Keith Dennis. In Search of New "Homology" Functors having a Close Relationship to *K*-theory. Unpublished preprint.
10. B. Eick and W. Nickel. *Polycyclic – Computing with polycyclic groups*, 2002. A GAP Package, see [**12**].
11. Graham Ellis and Frank Leonard. Computing Schur multipliers and tensor products of finite groups. *Proc. Roy. Irish Acad. Sect. A*, **95**(2):137–147, 1995.
12. The GAP Group. *GAP – Groups, Algorithms, and Programming, Version 4.4*, 2005. (http://www.gap-system.org).
13. Clair Miller. The second homology group of a group; relations among commutators. *Proc. Amer. Math. Soc.*, **3**:588–595, 1952.
14. W. Nickel. *nq – Nilpotent Quotients of Finitely Presented Groups*, 2003. A GAP Package, see [**12**].
15. N. R. Rocco. On a construction related to the nonabelian tensor square of a group. *Bol. Soc. Brasil. Mat. (N.S.)*, **22**(1):63–79, 1991.
16. Charles C. Sims. *Computation with finitely presented groups*, volume 48 of *Encyclopedia of Mathematics and its Applications*. Cambridge University Press, Cambridge, 1994.
17. Ursula Martin Webb. The Schur multiplier of a nilpotent group. *Trans. Amer. Math. Soc.*, **291**(2):755–763, 1985.
18. J. H. C. Whitehead. A certain exact sequence. *Ann. of Math. (2)*, **52**:51–110, 1950.

DEPARTMENT OF MATHEMATICS AND COMPUTER SCIENCE, SAINT LOUIS UNIVERSITY, ST. LOUIS, MO 63103, USA
    E-mail address: blythrd@slu.edu

FAKULTETA ZA MATEMATIKO IN FIZIKO, UNIVERZA V LJUBLJANI, JADRANSKA 19, 1000 LJUBLJANA, SLOVENIA
    E-mail address: primoz.moravec@fmf.uni-lj.si

DEPARTMENT OF ELECTRICAL ENGINEERING AND COMPUTER SCIENCE, UNIVERSITY OF EVANSVILLE, EVANSVILLE IN 47722 USA
    E-mail address: rfmorse@evansville.edu
    URL: faculty.evansville.edu/rm43

Contemporary Mathematics
Volume **470**, 2008

Investigating p-groups by coclass with GAP

Heiko Dietrich, Bettina Eick, and Dörte Feichtenschlager

ABSTRACT. We show how the computer algebra system GAP can be applied in the investigation of p-groups by coclass. We outline various experimental results; these underpin many of the more recent results in coclass theory and they lead to new conjectures.

1. Introduction

The coclass of a finite p-group of order p^n and nilpotency class c is defined as $\mathrm{cc}(G) = n - c$. Leedham-Green and Newman [**24**] suggested the use of coclass as the primary invariant to investigate and classify p-groups and they proposed five conjectures on the structure of the p-groups of a fixed coclass. This suggestion and the related conjectures initiated a deep and interesting research project.

A first milestone in coclass theory was the complete proof of all five coclass conjectures of [**24**]. Various results have been obtained along the way, until finally two independent proofs of these conjectures emerged, see [**30**] and [**19**]. A full account of these proofs including further details and references is given in [**25**].

The proofs of the coclass conjectures yield a significant first insight into the structure of the p-groups of coclass r. The coclass project now continues to investigate the situation in more detail. A conjecture underpinning this research project is following:

Conjecture I: *Let p be a prime and $r \in \mathbb{N}$. The p-groups of coclass r can be divided into finitely many* coclass families *such that the structure of the groups in a family can be described in a uniform way. In particular:*

(I) *All groups in a family can be defined by a single parameterized presentation.*

(II) *Many structural invariants of the groups in a family can be exhibited in a uniform way. For example, the following invariants of the groups in a family can be described by a single parameterized presentation:*

    (a) *their Schur multiplicators,*

    (b) *their automorphism groups, and*

    (c) *their cohomology rings over a ring R.*

2000 *Mathematics Subject Classification*. Primary 20D15, Secondary 20D45.

The first author was supported by the Studienstiftung des deutschen Volkes and the Braunschweigischer Hochschulbund e.V. The third author was supported by the DAAD.

This conjecture is rather vague, as no precise definition of 'coclass family' is given. A concrete definition of this term would already be a major step forward in this research area and perhaps also a significant step ahead in proving the conjecture. Nonetheless, one can observe that if this conjecture is true, then a classification of p-groups by coclass is possible and would be a very powerful tool in the understanding of p-groups.

The prime $p = 2$ plays a special role in this theory. For this case, Eick and Leedham-Green [12] introduced a definition of 'coclass family' and proved Conjecture I.I. Hence in the case $p = 2$, the focus of coclass theory is on investigating and proving instances of Conjecture I.II. See [10], [11] and [4] for first steps towards this goal. For all other primes, Conjectures I.I and I.II are both wide open at present.

Computer experiments have been used extensively in coclass theory. In particular, many computer experiments were made before Conjecture I could be introduced and they have also been extremely useful for many of the more recent results in this research area. It is the aim of this paper to describe some of the underlying algorithms and show how their implementation in GAP can be used to support Conjecture I.

For this purpose we first recall in Section 2 the definition and some of the fundamental features of the coclass graph $\mathcal{G}(p, r)$ and we exhibit its close connection to Conjecture I. Then in Section 3 we describe our algorithms which can be used to draw significant parts of $\mathcal{G}(p, r)$ and thus to investigate and underpin Conjecture I.

The remaining part of this paper is then devoted to exhibiting various sample applications of our algorithm with a view towards Conjecture I. First, we investigate certain cohomology rings of the 2-groups of coclass at most 2 and obtain a conjecture on their structure which supports Conjecture I.II(c). Then we consider the 5-groups of coclass 1. We provide evidence that Conjecture I.I holds for these groups and indicate how the corresponding coclass families could be obtained. Also, we investigate their Schur multiplicators and automorphism groups obtaining evidence supporting Conjecture I.II for these groups.

2. The coclass graph

A fundamental tool in coclass theory is the graph $\mathcal{G}(p, r)$: its vertices correspond to the isomorphism types of p-groups of coclass r, and two groups G and H are joined by a (directed) edge $G \to H$ if there exists a normal subgroup N in H with $|N| = p$ and $H/N \cong G$. It is not difficult to show that N must be the last non-trivial term of the lower central series of H and hence G is the unique ancestor for H. Thus $\mathcal{G}(p, r)$ is a forest. We investigate this graph in more detail in the following sections.

2.1. Coclass trees in $\mathcal{G}(p, r)$. A group H is a *descendant* of G in $\mathcal{G}(p, r)$ if there exists a path from G to H in $\mathcal{G}(p, r)$. We define $\mathcal{T}_G$ as the subgraph of $\mathcal{G}(p, r)$ generated by all descendants of G and so $\mathcal{T}_G$ is a tree with root G. We say that $\mathcal{T}_G$ is a *coclass tree* if it contains exactly one infinite path with root G and is maximal with this property; that is, there is no tree $\mathcal{T}_H$ containing G such that $\mathcal{T}_H$ has just one infinite path with root H. This notation allows one to state the following deep and fundamental result on $\mathcal{G}(p, r)$ which was obtained in the course of proving the coclass conjectures. We refer to [25] for references and background.

THEOREM 2.1. *The graph $\mathcal{G}(p, r)$ consists of finitely many coclass trees and finitely many groups not contained in a coclass tree.*

One option is now to investigate every individual coclass tree by itself. Note that it would be sufficient to prove Conjecture I for every coclass tree to obtain the same result for all of $\mathcal{G}(p, r)$. As a first step towards investigating coclass trees, we introduce some further notation.

Let $\mathcal{T} = \mathcal{T}_G$ be a coclass tree in $\mathcal{G}(p, r)$ and let $G = G_0, G_1, \ldots$ denote the groups on its unique maximal infinite path which is also called the *main line* of $\mathcal{T}$. The subgraph $\mathcal{B}_i$ of $\mathcal{T}$ generated by all descendants of G_i which are not also descendants of G_{i+1} is the *ith branch* of $\mathcal{T}$. Hence every branch $\mathcal{B}_i$ is a finite tree with root G_i. The tree $\mathcal{T}$ consists of its branches $\mathcal{B}_0, \mathcal{B}_1, \ldots$ and these are all connected by the main line of $\mathcal{T}$. To investigate the structure of $\mathcal{T}$, one needs to understand its branches $\mathcal{B}_0, \mathcal{B}_1, \ldots$.

The *depth* of a branch $\mathcal{B}_i$ is the length of a longest path in $\mathcal{B}_i$ starting at G_i. The *width* of $\mathcal{B}_i$ is the maximum number of groups of the same order in $\mathcal{B}_i$. We say that the coclass tree $\mathcal{T}$ has *bounded depth* if the depths of its branches $\mathcal{B}_0, \mathcal{B}_1, \ldots$ are bounded and, similarly, we say that $\mathcal{T}$ has *bounded width* if the widths of its branches are bounded. Furthermore, a coclass tree with bounded width and bounded depth is called *bounded*.

The depths and widths of branches describe the complexity of a coclass tree $\mathcal{T}$ to some extend. It follows from [12] that a coclass tree $\mathcal{T}$ of bounded depth also has bounded width. The following remark gives some examples on depth and width in $\mathcal{G}(p, r)$ for some small cases on p and r.

REMARK 2.2. *The following can be deduced from [23], see also [25].*
(a) The coclass trees in $\mathcal{G}(2, r)$, $r \in \mathbb{N}$, and in $\mathcal{G}(3, 1)$ have bounded depth.
(b) The coclass tree in $\mathcal{G}(5, 1)$ has bounded width, but unbounded depth.
(c) The coclass tree in $\mathcal{G}(p, 1)$, $p \geq 7$ prime, has unbounded depth and width.

2.2. Pro-p-groups of finite coclass. Let $\mathcal{T} = \mathcal{T}_G$ be a coclass tree in $\mathcal{G}(p, r)$. Then the inverse limit of the groups on the main line of $\mathcal{T}$ is a pro-p-group S. The structure of the resulting pro-p-groups has been investigated in detail in the course of the proof of the coclass conjectures. We refer to [25] for various details, background and references.

Here we note that it is known that S has the structure of a *pre-space group*; that is, there is a normal subgroup $T \trianglelefteq S$ such that T is isomorphic to the d-dimensional p-adic module $\mathbb{Z}_p^d$ and S/T is a finite p-group. The subgroup T is not unique for S, but its rank d is. This rank d is also called the *dimension* of S and of the tree $\mathcal{T}$.

Further, we note that all but finitely many of the lower central series quotients $S/\gamma_i(S)$ have coclass r and thus are main line groups in $\mathcal{T}$. This also shows that S is a group of coclass r, where the coclass of S is defined by $\mathrm{cc}(S) = \lim_{i \to \infty} \mathrm{cc}(S/\gamma_i(S))$.

2.3. Construction rules and coclass families. Computational experiments suggest that every coclass tree $\mathcal{T}_G$ in $\mathcal{G}(p, r)$ can be described by starting from a finite subtree with root G and then applying certain constructions. In a very loose form this is captured by the following conjecture.

Conjecture II: *Let $\mathcal{T}$ be a coclass tree of dimension d in $\mathcal{G}(p, r)$. Then there exists an $f \in \mathbb{N}$ so that for every $i \geq f$ the branch $\mathcal{B}_{i+d}$ can be constructed from $\mathcal{B}_i$ using certain rules so that these rules define a surjective map $\varphi_i : \mathcal{B}_{i+d} \to \mathcal{B}_i$.*

Similar to Conjecture I, Conjecture II is rather vague, as it does not specify the 'construction rules'. A central idea is that the 'construction rules' of Conjecture II are closely related to the 'coclass families' of Conjecture I. We choose a suitable $m \geq f$ and, for every group $G \in \mathcal{B}_i$ with $m \leq i < m + d$, we define an associated coclass family $\mathcal{F}_G$ which contains G and all iterated preimages of G under the maps φ_{i+j} for $j \in \mathbb{N}$. It remains open what values of m are suitable with respect to Conjecture I.

This setup allows to divide every coclass tree into finitely many (infinite) coclass families and finitely many groups lying outside these families. Thus also the full graph $\mathcal{G}(p, r)$ can then be divided into finitely many coclass families and finitely many groups lying outside these families. The latter ones are considered as *sporadic groups* and are usually ignored in the following.

2.4. Coclass trees of bounded depth, in particular $\mathcal{G}(2, r)$. In the special case of a bounded coclass tree, a precise definition for the construction maps in Conjecture II is known. We recall the corresponding theorem as follows. Note that this was conjectured by Newman and O'Brien [**26**] for $p = 2$ and proved by Du Sautoy [**6**] and, independently, by Eick and Leedham-Green [**12**]. The latter result also contains the connection to the coclass families and Conjecture I.

THEOREM 2.3. *Let $\mathcal{T}$ be a bounded coclass tree of dimension d. Then there exists an $f \in \mathbb{N}$ so that for every $i \geq f$ there is an isomorphism $\mathcal{B}_{i+d} \to \mathcal{B}_i$.*

As it is known that every coclass tree in $\mathcal{G}(2, r)$ has bounded depth, it now follows that $\mathcal{G}(2, r)$ is quite well understood and coclass families are available in this case. We include a brief explicit outline for $r = 1$ and $r = 2$ in the following.

The graph $\mathcal{G}(2, 1)$ is the easiest to understand of all coclass trees. It was first determined by Blackburn [**3**] and we recall it in Figure 1.

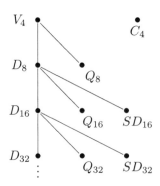

FIGURE 1. The graph $\mathcal{G}(2, 1)$.

The graph $\mathcal{G}(2, 1)$ consists of a single coclass tree $\mathcal{T} = \mathcal{T}_{V_4}$ and an isolated point C_4. The tree $\mathcal{T}$ has dimension 1 and it satisfies $\mathcal{B}_i \cong \mathcal{B}_{i+1}$ for all $i \geq 1$. The branch $\mathcal{B}_1$ contains 3 groups: D_8, Q_{16} and SD_{16}. Correspondingly, $\mathcal{G}(2, 1)$ contains

3 coclass families and 3 sporadic groups (C_4, V_4 and Q_8). Each of the 3 coclass families can be defined by a parameterized presentation on 1 parameter as follows:

$$\begin{aligned}
D_{2^n} &= \langle x, y \,|\, x^{2^{n-1}} = 1,\ y^2 = 1,\ x^y = x^{-1} \rangle && (n \geq 3), \\
SD_{2^n} &= \langle x, y \,|\, x^{2^{n-1}} = 1,\ y^2 = 1,\ x^y = x^{2^{n-2}-1} \rangle && (n \geq 4), \\
Q_{2^n} &= \langle x, y \,|\, x^{2^{n-1}} = 1,\ y^2 = x^{2^{n-2}},\ x^y = x^{-1} \rangle && (n \geq 4).
\end{aligned}$$

The graph $\mathcal{G}(2,2)$ is more complex, see [17, 18], and [26]. It contains five coclass trees which we denote by $\mathcal{T}_1(2,2), \ldots, \mathcal{T}_5(2,2)$ with altogether 51 coclass families and 27 sporadic groups. We briefly summarize the five trees by exhibiting their dimensions and their roots G_0 using their identifications in the SMALLGROUPS library [2]. The smallest f with $B_i \cong B_{i+d}$ for all $i \geq f$ has been determined experimentally for each of the five trees and we also add the so-called periodic root G_f to the summary below. Further, the table contains the numbers of coclass families arising from each of the trees.

	$\mathcal{T}_1(2,2)$	$\mathcal{T}_2(2,2)$	$\mathcal{T}_3(2,2)$	$\mathcal{T}_4(2,2)$	$\mathcal{T}_5(2,2)$
dim	2	2	1	1	1
G_0	(64,34)	(64,32)	(16,4)	(32,9)	(8,5)
# families	19	16	4	6	6
G_f	(64,34)	(64,32)	(16,4)	(32,9)	(16,11)

TABLE 1. Coclass trees of $\mathcal{G}(2,2)$.

2.5. Coclass trees of bounded width, in particular $\mathcal{G}(5,1)$. Every graph $\mathcal{G}(p,1)$ of p-groups of coclass 1 consists of a single isolated point C_{p^2} and a single coclass tree $\mathcal{T}_{C_p \times C_p}$ of dimension $p-1$. The structure of the coclass trees has been investigated in detail by Leedham-Green and McKay [20, 21, 22, 23]; see also [25].

As noted in Remark 2.2, the tree in $\mathcal{G}(5,1)$ is an interesting example from a theoretical and computational point of view: it has unbounded depth so that Conjectures I and II are open for it, but it has bounded width; the latter implies that it is still accessible to computer experiments. The first experiments for this tree were undertaken by Newman [27]. Below we outline various new investigations of the coclass tree in $\mathcal{G}(5,1)$ and we show how our experiments support Conjectures I and II.

3. Computing coclass trees with GAP

The coclass graph $\mathcal{G}(p,r)$ plays a fundamental role in coclass theory and the computation of significant finite sections of $\mathcal{G}(p,r)$ is also one of our most important tools. We describe some of the algorithms used for this purpose in this section.

As a first step in visualizing $\mathcal{G}(p,r)$, we usually determine some or all of the pro-p-groups of coclass r up to isomorphism. These correspond 1-1 to the coclass trees of $\mathcal{G}(p,r)$. As a second step we then investigate the coclass tree corresponding to a given pro-p-group of coclass r in more detail. Algorithms for these two steps are described in the following two sections.

3.1. The pro-p-groups of coclass r. For $r = 1$ there is just one pro-p-group of coclass 1 for every prime p. This has the form $C_p \ltimes \mathbb{Z}_p^{p-1}$, where $\mathbb{Z}_p$ denotes the p-adic numbers and C_p is the cyclic group of order p. For $r > 1$ there is no complete classification of the pro-p-groups of coclass r available, but computational methods can be used to determine all (or some) of the pro-p-groups of coclass r for small p and r.

A group G is a *uniserial p-adic space group of dimension d* if G is an extension of $T = \mathbb{Z}_p^d$ by a finite p-group P so that P acts *uniserially* on T; that is, the series defined by $T_0 = T$ and $T_{i+1} = [T_i, P]$ satisfies $[T_i : T_{i+1}] = p$ for all $i \in \mathbb{N}_0$. Every uniserial p-adic space group is a pro-p-group of finite coclass and these groups form a significant part of all pro-p-groups of a certain coclass. In [8] an algorithm for constructing all uniserial p-adic space groups of a given coclass r for odd primes p is described. This together with the results for $p = 2$ in [26] yields the following numbers of uniserial p-adic space groups of coclass r.

r	2	3	4
$p = 2$	2	22	
$p = 3$	10	1271	137299952383
$p = 5$	95	1110136753555665	
$p = 7$	4575		

Once the space groups are given, one can construct all pro-p-groups of coclass r as central extensions of space groups (see also [26] for $p = 2$). This approach yields the following numbers of pro-p-groups of coclass r.

- 5 pro-2-groups of coclass 2,
- 54 pro-2-groups of coclass 3,
- 16 pro-3-groups of coclass 2.

The table shows that the number of pro-p-groups of coclass r grows heavily with p and r. Asymptotic bounds for these numbers can be found in [9]. We conclude that the full tree $\mathcal{G}(p, r)$ can only be investigated for very small p and r; in other cases, we can investigate just parts of the tree.

3.2. Computing the tree corresponding to a pro-p-group. In this section we assume that we are given a pro-p-group S of coclass r and we want to investigate its corresponding coclass tree $\mathcal{T}$; that is, we want to determine finite parts of $\mathcal{T}$ with a view towards Conjecture II and Theorem 2.3.

Our main algorithm is a method to determine all *immediate descendants* of a p-group G; that is, all descendants H of G having distance 1 to G. Using an iterated application of this method, we can thus construct the descendant tree $\mathcal{T}_G$ for a given group G in $\mathcal{T}$ down to a given depth. Similarly, we can compute the branch $\mathcal{B}_i$ of $\mathcal{T}$ for a given i by constructing all descendants of its root G_i which are not descendants of G_{i+1}. Note that if i is large enough, then there exists a $t \in \mathbb{N}$ such that $G_i = S/\gamma_t(S)$. We visualize the resulting trees using the XGAP Package [5] of GAP.

The ANUPQ Package [29] of GAP contains a method to construct the immediate p-descendants of a given group G; that is, all groups H such that $H/\lambda_c(H) \cong G$ and $|\lambda_c(H)| = p$, where $\lambda_j(H)$ is the jth subgroup of the lower exponent-p central series and c is the exponent-p class of G. Further, it is shown in [26, Theorem 6.1] that the immediate p-descendants of a p-group G of coclass r coincide with its immediate descendants, if G has order at least $p^{8p^r + r}$. However, this bound on the order of G is frequently too large to be practical for our later applications. Thus we introduce below a new algorithm for computing the immediate descendants of an arbitrary p-group. This algorithm is a variation of a method introduced in [13].

Recall that the immediate descendants of a p-group G coincide with the groups H of class $c + 1$ so that $H/\gamma_c(H) \cong G$ and $|\gamma_c(H)| = p$, where c is the class of G. Thus every immediate descendant H of G is a central extension of C_p by G and corresponds to one (or several) elements of $H^2(G, \mathbb{F}_p)$. Our aim is to determine the immediate descendants of G by constructing those elements of $H^2(G, \mathbb{F}_p)$ which correspond to immediate descendants and then solving the isomorphism problem for them.

For $\alpha \in Z^2(G, \mathbb{F}_p)$ we define $\alpha^*(g, h) = \alpha(g, h) + \alpha(g^{-1}, h^{-1}) + \alpha(g^{-1}h^{-1}, gh) - \alpha(g, g^{-1}) - \alpha(h, h^{-1})$. Further, let $U = \{\alpha \in Z^2(G, \mathbb{F}_p) \mid \alpha^*(g, h) = 0$ for all $g \in G$ and $h \in \gamma_c(G)\}$ and let $\overline{U}$ be its image in $H^2(G, \mathbb{F}_p)$. Then $\overline{U}$ is a subspace of the $\mathbb{F}_p$-vector space $H^2(G, \mathbb{F}_p)$. Let $\mathcal{W}$ denote the set of all 1-dimensional subspaces of $H^2(G, \mathbb{F}_p)$ which are not contained in $\overline{U}$.

The group $\mathrm{Aut}(G)$ acts naturally on the vector space $Z^2(G, \mathbb{F}_p)$ via $\alpha^\beta(g, h) = \alpha(g^\beta, h^\beta)$. This action induces an action of $\mathrm{Aut}(G)$ on $H^2(G, \mathbb{F}_p)$ and thus on the set $\mathcal{W}$. We use this action to solve our considered problem as in the following theorem.

THEOREM 3.1. *The $\mathrm{Aut}(G)$-orbits on $\mathcal{W}$ correspond 1-1 to a set of isomorphism representatives of immediate descendants of G.*

PROOF. Let E_α denote the extension of C_p by G with respect to $\alpha \in Z^2(G, \mathbb{F}_p)$. We write the elements of E_α as tuples (g, v) with $g \in G$ and $v \in \mathbb{F}_p$. The multiplication in E_α then translates to $(g, v)(h, w) = (gh, v + w + \alpha(v, w))$ and $(g, v)^{-1} = (g^{-1}, -v - \alpha(g, g^{-1}))$. Our proof proceeds in two steps.

(1) We show that E_α is an immediate descendant of G if and only if $\alpha \notin U$. Suppose that E_α is an immediate descendant of G. Then it has class $c + 1$ and $\gamma_{c+1}(E_\alpha) = [\gamma_c(E_\alpha), E_\alpha] = C_p$, the module of the extension. Hence there exist $(g, v) \in E_\alpha$ and $(h, w) \in \gamma_c(E_\alpha)$ with $[(g, v), (h, w)] \neq (1, 0)$. A straightforward computation shows that this is equivalent to $\alpha(g, h) + \alpha(g^{-1}, h^{-1}) + \alpha(g^{-1}h^{-1}, gh) \neq \alpha(g, g^{-1}) + \alpha(h, h^{-1})$ and thus we obtain that $\alpha \notin U$. The converse follows by a similar argument.

(2) Applying Theorem 3.6 of [1], it follows that there is a 1-1 correspondence between a set of isomorphism representatives of immediate descendants of G and the $\mathrm{Aut}(G) \times \mathrm{Aut}(C_p)$-orbits of elements of $H^2(G, \mathbb{F}_p) \setminus \overline{U}$. The action of $\mathrm{Aut}(C_p)$ on the elements of $H^2(G, \mathbb{F}_p)$ has as orbits exactly the 1-dimensional subspaces. Thus the theorem follows.   □

Theorem 3.1 can be implemented readily in GAP, as GAP contains already a useful machinery for computing with second cohomology groups and handling extensions as well as a method for solving orbit-stabilizer problems. Hence we obtain an efficient method for constructing immediate descendants.

4. Application: Cohomology of 2-groups

In this section we investigate the cohomology of the 2-groups of coclass at most 2. The following theorem has been proved by Carlson [4]:

THEOREM 4.1. *Let k be a field of characteristic 2. For each fixed r there exist only finitely many isomorphism types of graded commutative k-algebras R with $R \cong H^*(G, k)$ where G is a 2-group of coclass r.*

Carlson's proof contains a counting argument and does not specify which groups have isomorphic cohomology rings. Eick and Leedham-Green [7] conjectured that the coclass families might provide the environment for finding infinite families of groups with isomorphic cohomology rings. This resulted in the following specialization of Conjecture I.II(c).

Conjecture III: *Let $\mathcal{F}$ be a coclass family of p-groups and k a field of characteristic p. Then $H^*(G, k) \cong H^*(H, k)$ for all $G, H \in \mathcal{F}$.*

For the coclass families in $\mathcal{G}(2, 1)$ it is known that Conjecture III holds. More precisely, the following theorem is available for this case, see [4], [28] and [15]:

THEOREM 4.2. *Let $i \geq 4$ ($i \geq 3$ for (a)).*
(a) $H^*(D_{2^i}, \mathbb{F}_2) \cong \mathbb{F}_2[z, y, x]/(zy)$ *with z and y of degree 1 and x of degree 2.*
(b) $H^*(Q_{2^i}, \mathbb{F}_2) \cong \mathbb{F}_2[z, y, x]/(z^2 + zy, y^3)$ *with z and y of degree 1, x of degree 4.*
(c) $H^*(SD_{2^i}, \mathbb{F}_2) \cong \mathbb{F}_2[z, y, x, w]/(zy, y^3, yx, z^2w + x^2)$ *with z and y of degree 1, x of degree 3, and w of degree 4.*

A similar theorem is available for the integral cohomology of the 2-groups of coclass 1. It was proved by Thomas, Hayami and Sanada, Evens and Priddy (see [32], [16] and [15]). Again, it supports Conjecture I.II(c).

THEOREM 4.3. *Let $i \geq 4$ ($i \geq 3$ for (a)).*
(a) $H^*(D_{2^i}, \mathbb{Z}) \cong \mathbb{Z}[z, y, x, w]/(2z, 2y, 2w, 2^{i-1}x, w^2 - zx, y^2 - zy)$ *with z and y of degree 2, x of degree 4, and w of degree 3.*
(b) $H^*(Q_{2^i}, \mathbb{Z}) \cong \mathbb{Z}[z, y, x]/(2z, 2y, 2^ix, z^2, y^2 - 2^{i-1}x, zy - 2^{i-1}x)$ *with z and y of degree 2 and x of degree 4.*
(c) $H^*(SD_{2^i}, \mathbb{Z}) \cong \mathbb{Z}[z, y, x, w]/(2z, 2y, 2w, 2^{i-1}x, y^2, zy, yw, w^2 - z^3x)$ *with z and y of degree 2, x of degree 4, and w of degree 5.*

However, we know much less about the cohomology of the 2-groups of coclass 2. We used the algorithm of Section 3 to investigate the cohomology of the groups in $\mathcal{G}(2, 2)$. For this purpose we considered each of the five coclass trees $\mathcal{T}_1(2, 2), \ldots, \mathcal{T}_5(2, 2)$ introduced in Section 2.4 in turn and computed $H^i(G, \mathbb{F}_2)$ for $1 \leq i \leq 4$ using the HAP Package [14] of GAP for all 2-groups of order up to 2^{11} and some groups of order 2^{13} in them. We summarize our results briefly as follows.

- For the trees $\mathcal{T}_1(2, 2), \mathcal{T}_3(2, 2), \mathcal{T}_4(2, 2)$, and $\mathcal{T}_5(2, 2)$ we obtained the cohomology information for every group in $\mathcal{B}_f, \ldots, \mathcal{B}_{f+2d-1}$ and observed that $\mathcal{B}_{f+i} \cong \mathcal{B}_{f+i+d}$. Furthermore, the investigated   mod 2-cohomology groups of groups in the same coclass family are isomorphic. This supports Conjecture III starting from the periodic roots G_f of the trees.
- For the tree $\mathcal{T}_2(2, 2)$ we have that $f = 0$ and $d = 2$. We determined cohomology groups for every group in $\mathcal{B}_0, \ldots, \mathcal{B}_4$ and found that these are isomorphic for groups which correspond under the isomorphisms $\mathcal{B}_{m+i} \cong \mathcal{B}_{m+i+d}$ with $1 = m > f = 0$ and $i \in \{0, 1\}$. This supports Conjecture III, but we have to take the coclass families in this tree starting from $\mathcal{B}_1$ and not $\mathcal{B}_0$.

We introduce some more notation to visualize a coclass tree and its cohomology information in a compact form. For each of the trees we draw a finite part starting at its periodic root G_f as listed in Section 2.4. For the trees $\mathcal{T}_1(2, 2), \mathcal{T}_3(2, 2), \mathcal{T}_4(2, 2)$, and $\mathcal{T}_5(2, 2)$ we draw $\mathcal{B}_f, \ldots \mathcal{B}_{f+d-1}$ and the main line. For the tree $\mathcal{T}_2(2, 2)$ we draw $\mathcal{B}_f, \ldots, \mathcal{B}_{m+d-1}$. If Conjecture III holds, then this yields a complete picture about this tree and its cohomology information by Theorem 2.3.

Every group in a graph is labeled by its cohomology information as computed by HAP and GAP. This information is a 4-tuple $[e_1, \ldots, e_4]$ with $e_i = \dim_{\mathbb{F}_2} H^i(G, \mathbb{F}_2)$ for $1 \leq i \leq 4$. Further, we draw the graphs in compacted form by collecting all immediate descendants which are leaves to a single vertex. This vertex then has a multiple label so that every label corresponds to one group. If a label occurs n times, then we display it as $n[e_1, \ldots, e_4]$.

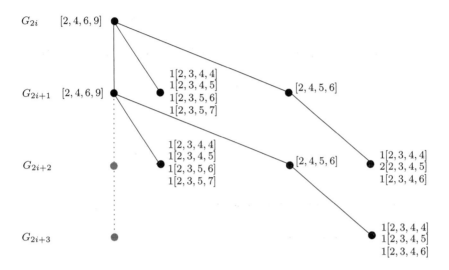

FIGURE 2.   Conjectured cohomology for $\mathcal{B}_{2i}$ and $\mathcal{B}_{2i+1}$ for $i \in \mathbb{N}_0$ in $\mathcal{T}_1(2,2)$.

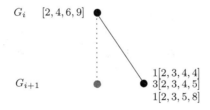

FIGURE 3.   Conjectured cohomology for $\mathcal{B}_i$ for $i \in \mathbb{N}_0$ in $\mathcal{T}_4(2,2)$.

5. Application: 5-groups of coclass 1

Let $\mathcal{T}$ be the coclass tree in $\mathcal{G}(5,1)$ and recall that $\mathcal{T}$ has bounded width, but unbounded depth. Experimental evidence suggests that Conjecture II holds for $\mathcal{T}$; we describe the necessary 'construction rules' below.

For $i \geq 0$ let $\mathcal{B}_i$ denote the ith branch of $\mathcal{T}$ so that $\mathcal{B}_i$ has a root G_i of order p^{i+2}. We say that a group in $\mathcal{B}_i$ is *capable* if it has an immediate descendant and it is *terminal* otherwise. For $k \in \mathbb{N}$ we denote by $\mathcal{B}_i(k)$ the *shaved subtree* of $\mathcal{B}_i$: this is the subtree generated by all capable groups of depth at most k in $\mathcal{B}_i$ together with all their terminal immediate descendants. Similarly, for $l \geq k$ we denote by

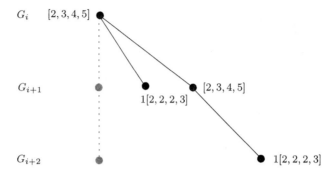

FIGURE 4.   Conjectured cohomology for $\mathcal{B}_i$ for $i \in \mathbb{N}_0$ in $\mathcal{T}_3(2,2)$.

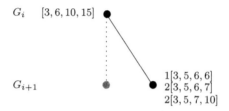

FIGURE 5.   Conjectured cohomology for $\mathcal{B}_i$ for $i \in \mathbb{N}$ in $\mathcal{T}_5(2,2)$.

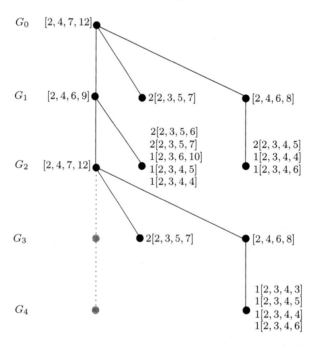

FIGURE 6.   Cohomology for $\mathcal{B}_0, \mathcal{B}_1, \mathcal{B}_2$ in $\mathcal{T}_2(2,2)$.

$\mathcal{B}_i(l, k)$ the *collar* of $\mathcal{B}_i$: this is the subgraph consisting of all capable groups of depth j with $l \leq j \leq k$ together with all their terminal immediate descendents.

Our computational investigations of $\mathcal{B}_i$ with $8 \leq i \leq 30$ yield the following conjecture.

Conjecture IV: *Let $i \geq 8$ and write $i = 8 + 4x + y$ with $0 \leq y \leq 3$ and $x \geq 0$.*

- *The branch $\mathcal{B}_i$ of the coclass tree of $\mathcal{G}(5, 1)$ has depth $i + 2$. It consists of*
- *a head* $\quad\quad H(i) \quad = \quad \mathcal{B}_i(5 + y),$
- *x collars* $\quad C(i, j) \quad = \quad \mathcal{B}_i(6 + y + 4j, 9 + y + 4j) \quad$ *with $0 \leq j \leq x - 1$, and*
- *a tail* $\quad\quad T(i) \quad = \quad \mathcal{B}_i(i - 2, i + 1).$
- *The heads and tails satisfy $H(i) \cong H(i + 4)$ and $T(i) \cong T(i + 4)$, respectively.*
- *The collars satisfy $C(i, j) \cong C(i + 4, j)$, and $C(i, j) \cong C(i, j - 1)$ for all $j \geq 1$.*

Conjecture IV is visualized in Figures 7 – 10 using the following notation. Each figure shows a branch $\mathcal{B}_i$ by exhibiting its head, one of its x collars, and its tail. All terminal immediate descendants of a group are represented by only one vertex which is labeled with the number of groups that it represents.

Conjecture IV supports Conjecture II and we obtain three types of 'construction rules' for the branches in $\mathcal{T}$. Thus Conjecture IV also supports Conjecture I.I and we would obtain three types of coclass families from the construction rules: those families arising from a head are families based on 1 parameter, those arising from a collar are families based on 2 parameters and those arising from a tail are families based on 1 parameter. In particular, the groups in $H(i)$, $T(i)$, and $C(i, 0)$ with $12 \leq i \leq 15$ would define disjoint coclass families. Their union would contain all groups in $\mathcal{G}(5, 1)$ which are descendants of G_{12}. The resulting numbers of conjectured coclass families are listed in Table 2.

# families	$y = 0$	$y = 1$	$y = 2$	$y = 3$	Σ
in the heads	366	578	741	953	2638
in the collars	748	756	748	756	3008
in the tails	730	735	730	737	2932
Σ	1844	2069	2219	2446	8578

TABLE 2. Conjectured coclass families in $\mathcal{G}(5, 1)$.

Thus our conjectures lead us the assumption that $\mathcal{G}(5, 1)$ is partitioned into 8578 coclass families, one sporadic (isolated) group C_{25}, and 8399 further sporadic groups lying in $\mathcal{B}_0, \dots, \mathcal{B}_{11}$.

5.1. Schur multiplicators. In this section we consider the coclass tree in $\mathcal{G}(5, 1)$ together with information on the Schur multiplicators of the underlying groups. First, we recall that by [11] we know that for every $s \in \mathbb{N}$ there are at most finitely many groups G in $\mathcal{G}(5, 1)$ with $|M(G)| \leq 5^s$. Hence we have to expect a growing order of the Schur multiplicators when going downwards in the tree. Further, we recall the following result from [11] to obtain some background on Schur multiplicators of descendants.

LEMMA 5.1. *Let H be an immediate descendant of G in $\mathcal{G}(p, 1)$. Then there exists a short exact sequence*

$$1 \to K(H, G) \to M(H) \to M(G) \to C_p \to 1$$

and $K(H, G)$ is elementary abelian of order $1, p$ or p^2.

The following conjecture is obtained by the computation of the Schur multiplicators of all 53216 groups in the branches $\mathcal{B}_i$ with $8 \leq i \leq 26$ of $\mathcal{G}(5,1)$. We denote the abelian invariants of a Schur multiplicator $M(G)$ by $I(M(G))$.

Conjecture V: *For $n \in \mathbb{N}$ write $n = 4s_n + r_n$ with $1 \leq r_n \leq 4$.*

(a) *Let G be a capable group of order p^n in a branch $\mathcal{B}_i$ with $i \geq 8$ in $\mathcal{G}(5,1)$. Then*

$$I(M(G)) = \begin{cases} (5, \quad 5^{s_n}, \quad 5^{s_n}) & \text{if } r_n = 1, 2, \\ (5, \quad 5^{s_n}, \quad 5^{s_n+1}) & \text{if } r_n = 3, 4, \end{cases}$$

(b) *Let H be a terminal immediate descendants of G with G as in (a). Then*

$$I(M(H)) = \begin{cases} (5^{s_n}, \quad 5^{s_n}) & \text{if } r_n = 1, 2, \\ (5^{s_n}, \quad 5^{s_n+1}) & \text{if } r_n = 3, 4. \end{cases}$$

This conjecture translates to the coclass families in $\mathcal{G}(5,1)$ as follows. For some $i \in \{12, \ldots, 15\}$ let G be a group in a branch $\mathcal{B}_i$, so that G is the stem group of a coclass family. Note that $\mathcal{B}_i$ consists of a head $H(i)$, a single collar $C(i,0)$ and a tail $T(i)$ in this case. Let K be a group in the coclass family defined by G. Then K is contained in a branch $\mathcal{B}_j$ with $j - i = 4l$ and K is capable if and only if G is capable. Writing $I(M(G)) = (5^a, 5^b, 5^c)$ with $a \in \{0, 1\}$, we obtain the following conjectured behaviour of the Schur multiplicators.

- If $G \in H(i)$, then $I(M(K)) = (5^a, 5^{b+l}, 5^{c+l})$.
- If $G \in T(i)$, then $I(M(K)) = (5^a, 5^{b+2l}, 5^{c+2l})$.
- If $G \in C(i,0)$, then $K \in C(j,k)$ for some k and $I(M(K)) = (5^a, 5^{b+l+k}, 5^{c+l+k})$.

If our conjectures are true, then they imply that there is a generic formula for the abelian invariants of the Schur multiplicators of the groups in a coclass family of $\mathcal{G}(5,1)$. This formula is a 1-parameter formula if G is a group in a head or tail, and it is a 2-parameter formula if G is a group in a collar.

5.2. Automorphism groups. In this section we consider the coclass tree in $\mathcal{G}(5,1)$ together with information on the (outer) automorphism group of the underlying groups. Our computational investigation of the groups in the branches $\mathcal{B}_i$ with $12 \leq i \leq 26$ in the graph $\mathcal{G}(5,1)$ can be summarized by the following conjecture.

Conjecture VI: *Let $\mathcal{F}$ be a coclass family in $\mathcal{G}(5,1)$ and let $\mathcal{B}_i$ be a branch of $\mathcal{G}(5,1)$ with $i \geq 12$. Let G be a group in $\mathcal{F} \cap \mathcal{B}_i$. Then there exist u and v depending on $\mathcal{F}$, but not on G or i, so that*

$$|Out(G)| = u5^{i+v}.$$

Moreover,

(a) *if $\mathcal{F}$ arises from a head, then $u \in \{1, 2, 4, 16\}$ and $v \in \{-1, 0, 1, 2, 3\}$.*
(b) *if $\mathcal{F}$ arises from a tail or collar, then $u \in \{1, 2, 4\}$ and $v \in \{2, 3\}$.*

If this conjecture is true, then it implies a generic formula for the order of the outer automorphism groups of the groups in a coclass family. This supports Conjecture I.II(b). Note that the formula is a 1-parameter formula only, even if the family is a 2-parameter family.

Further, our calculations underpin a conjecture of Newman [27] saying that more than 90% of all groups in $\mathcal{G}(5,1)$ have a 5-group as automorphism group.

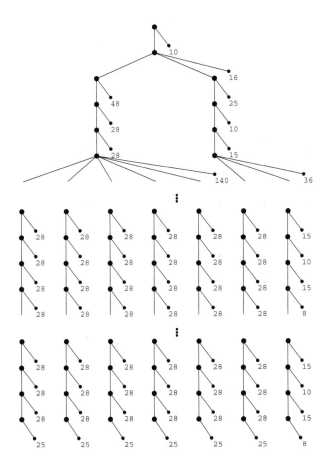

FIGURE 7. The conjectured branches $\mathcal{B}_i$ with $i = 8 + 4j$ and $j \geq 0$.

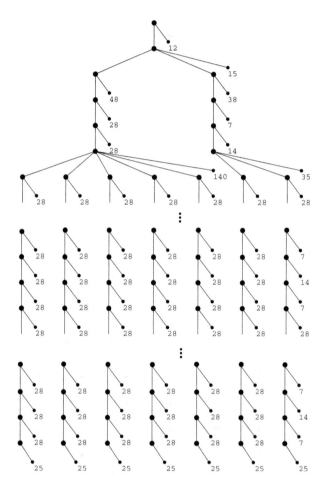

FIGURE 8. The conjectured branches $\mathcal{B}_i$ with $i = 8 + 4j + 1$ and $j \geq 0$.

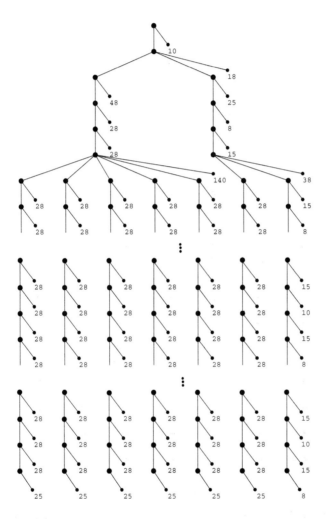

FIGURE 9. The conjectured branches $\mathcal{B}_i$ with $i = 8 + 4j + 2$ and $j \geq 0$.

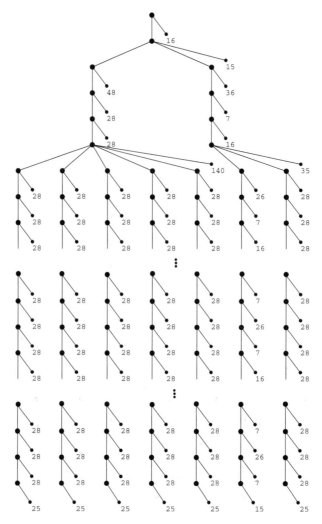

FIGURE 10. The conjectured branches $\mathcal{B}_i$ with $i = 8 + 4j + 3$ and $j \geq 0$.

References

1. H. U. Besche and B. Eick, *Construction of finite groups*, J. Symb. Comput. **27** (1999), 387 – 404.
2. H. U. Besche, B. Eick, and E. A. O'Brien, *Smallgroups - a library of groups of small order*, 2005, A refereed GAP 4 package, see [**31**].
3. N. Blackburn, *On a special class of p-groups*, Acta. Math. **100** (1958), 45 – 92.
4. J. F. Carlson, *Coclass and cohomology*, J. of Pure and Applied Algebra **200** (2005), no. 3, 251 – 266.
5. F. Celler and M. Neunhöffer, *Xgap - a graphical interface to gap*, 2000, A refereed GAP 4 package, see [**31**].
6. M. du Sautoy, *Counting p-groups and nilpotent groups*, Inst. Hautes Etudes Sci. Publ. Math. **92** (2001), 63 – 112.
7. B. Eick and C. R. Leedham-Green, Private communication.
8. B. Eick, *Determination of the uniserial space groups with a given coclass*, J. London Math. Soc. **71** (2005), 622 – 642.

9. _____, *On the number of infinite branches in the graph of p-groups of coclass r*, J. Group Theory **8** (2005), 687 – 700.

10. _____, *Automorphism groups of 2-groups*, J. Algebra **300** (2006), 91 – 101.

11. _____, *Schur multiplicators of finite p-groups with fixed coclass*, Accepted by Israel J. Math. (2007).

12. B. Eick and C. R. Leedham-Green, *Periodic patterns in coclass graphs and the classification of 2-groups by coclass*, Submitted. (www.tu-bs.de/~beick/pl.html) (2006).

13. B. Eick and E. A. O'Brien, *Enumerating p-groups*, J. Austral. Math. Soc. (Series A) **67** (1999), 191 – 205.

14. G. Ellis, *HAP - homological algebra programming*, 2006, A refereed GAP 4 package, see [**31**].

15. L. Evens and S. Priddy, *The cohomology of the semidihedral group*, Contemp. Math. **37** (1985), 61 – 72.

16. T. Hayami and K. Sanada, *Cohomology ring of the generalized quaternion group with coefficients in an order*, Comm. Algebra **30** (2002), no. 8, 3611 – 3628.

17. R. James, *2-groups of almost maximal class*, J. Austral. Math. Soc. **19** (1975), 343 – 357.

18. _____, *Corrigendum: "2-groups of almost maximal class"*, J. Austral. Math. Soc. Ser. A **35** (1983), no. 3, 307.

19. C. R. Leedham-Green, *The structure of finite p-groups*, J. London Math. Soc. (2) **50** (1994), 49 – 67.

20. C. R. Leedham-Green and S. McKay, *On p-groups of maximal class I*, Quart. J. Math. Oxfort (2) **27** (1976), 297–311.

21. _____, *On p-groups of maximal class II.*, Quart. J. Math. Oxfort (2) **29** (1978), 175–186.

22. _____, *On p-groups of maximal class III.*, Quart. J. Math. Oxfort (2) **29** (1978), 281–299.

23. C. R. Leedham-Green and S. McKay, *On the classification of p-groups of maximal class*, Quart. J. of Math. Oxford **35** (1984), 293 – 304.

24. C. R. Leedham-Green and M. F. Newman, *Space groups and groups of prime-power order I*, Archiv der Mathematik **35** (1980), 193 – 203.

25. C. R. Leedham-Green and S. McKay, *The structure of groups of prime power order*, London Mathematical Society Monographs, Oxford Science Publications, 2002.

26. M. F. Newman and E. A. O'Brien, *Classifying 2-groups by coclass*, Trans. Amer. Math. Soc. **351** (1999), 131 – 169.

27. M. F. Newman, *Groups of prime-power order*, Groups-Canberra 1989, Lecture notes in mathematics, vol. 1456, Springer, 1990, pp. 49 – 62.

28. S. A. Mitchell and S. B. Priddy, *Symmetric product spectra and splittings of classifying spaces*, Amer. J. Math. **106** (1984), no. 1, 219–232.

29. E. A. O'Brien, ANUPQ- *the anu p-quotient algorithm*, 1990, Also available in MAGMA and as GAP package.

30. A. Shalev, *The structure of finite p-groups: Effective proofs of the coclass conjectures*, Invent. Math. **115** (1994), 315 – 345.

31. The GAP Group, GAP – *Groups, Algorithms and Programming, Version 4.4*, Available from http://www.gap-system.org, 2005.

32. C. B. Thomas, *Chern classes and metacyclic groups*, Mathematika **18** (1971), 196 – 200.

TU BRAUNSCHWEIG, INSTITUT COMPUTATIONAL MATHEMATICS, POCKELSSTR. 14, 38106 BRAUNSCHWEIG, GERMANY
E-mail address: h.dietrich@tu-bs.de

TU BRAUNSCHWEIG, INSTITUT COMPUTATIONAL MATHEMATICS, POCKELSSTR. 14, 38106 BRAUNSCHWEIG, GERMANY
E-mail address: beick@tu-bs.de

TU BRAUNSCHWEIG, INSTITUT COMPUTATIONAL MATHEMATICS, POCKELSSTR. 14, 38106 BRAUNSCHWEIG, GERMANY
E-mail address: d.feichtenschlager@tu-bs.de

Contemporary Mathematics
Volume **470**, 2008

Homological Algebra Programming

Graham Ellis

1. Introduction

This article is an overview of some attempts at using GAP [**11**] and Polymake [**12**] software for basic calculations in group cohomology. Section 2 contains examples, obtained using the HAP package [**13**] for the GAP system, which illustrate the potential of computational techniques in this area. Subsequent sections briefly outline the methods underlying these examples. Section 3 explains how an element of choice, present in many homological algebra constructions, can be made algorithmic using the notion of contracting homotopy. Section 4 describes a linear algebraic approach to computing the cohomology of small groups. Section 5 discusses five direct geometric methods for obtaining the integral cohomology of a range of finite and infinte groups. The final section considers two methods for computing the cohomology of a group G from that of certain subgroups of G.

We begin with some basic definitions. Let k be an integral domain and let $A = k[x_1, \ldots, x_n]/I$ be the quotient of a free associative ring by a two-sided ideal I. Let M be an A-module. A sequence of A-module homomorphisms

$$\cdots \xrightarrow{d_4} R_3 \xrightarrow{d_3} R_2 \xrightarrow{d_2} R_1 \xrightarrow{d_1} R_0$$

is said to be a *free A-resolution* of M if

- *(Exactness)* $\ker d_n = \operatorname{image} d_{n+1}$ for all $n \geq 1$,
- *(Freeness)* R_n is a free A-module for all $n \geq 0$,
- *(Augmentation)* the cokernel of d_1 is isomorphic to the module M.

Given such a resolution R_* and an A-module N one defines

$$\operatorname{Ext}_A^n(M, N) = \frac{\ker(\operatorname{Hom}_A(R_n, N) \to \operatorname{Hom}_A(R_{n+1}, N))}{\operatorname{image}(\operatorname{Hom}_A(R_{n-1}, N) \to \operatorname{Hom}_A(R_n, N))}$$

and

$$\operatorname{Tor}_n^A(M, N) = \frac{\ker(R_n \otimes_A N \to R_{n-1} \otimes_A N)}{\operatorname{image}(R_{n+1} \otimes_A N \to R_n \otimes_A N)}.$$

It can be shown that, up to isomorphism, the functors $\operatorname{Ext}_A^n(M, N)$ and $\operatorname{Tor}_n^A(M, N)$ do not depend on the choice of resolution R_*.

2000 *Mathematics Subject Classification*. Primary 20J05.
This work was partially supported by Marie Curie grant MTKD-CT-2006-042685.

Commutative algebra software such as CoCoA [9], Macaulay [18] and Singular [22] contains a range of Gröbner basis methods for computing these functors in the case where k is a field and the ring A is commutative. The Plural [22] extension to Singular handles certain non-commutative rings A.

To define the cohomology of a group G one takes the ring of integers $k = \mathbb{Z}$, the module $M = \mathbb{Z}$ with trivial G-action, the group ring $A = \mathbb{Z}G$, and sets

$$H^n(G, N) = \operatorname{Ext}^n_{\mathbb{Z}G}(\mathbb{Z}, N), \quad H_n(G, N) = \operatorname{Tor}^{\mathbb{Z}G}_n(\mathbb{Z}, N).$$

The GAP and MAGMA computational algebra systems contain methods for computing group cohomology in dimensions $n = 1$ and $n = 2$ for a range of groups G and modules N. MAGMA also has methods for the higher-dimensional cohomology of small p-groups with coefficients in the field of p elements, i.e., $N = \operatorname{GF}(p)$.

The computation of group cohomology involves two computationally expensive but independent tasks: the computation of a free resolution, and the computation of the homology of a chain complex. For the latter one can use GAP's internal Smith Normal Form function or more specialized Smith Normal Form algorithms avaiable in the GAP packages EDIM [17] and SIMPHOM [10]. Sections 3–6 below focus on the former task.

2. Example computations

The following automated proofs can be reproduced using the computer algebra system GAP with the HAP package loaded.

THEOREM 2.1 ([19]). *The Mathieu group M_{23} has trivial integral homology $H_n(M_{23}, \mathbb{Z}) = 0$ in dimensions $n = 1, 2, 3$.*

PROOF.
```
gap> GroupHomology(MathieuGroup(23),1);
[ ]
gap> GroupHomology(MathieuGroup(23),2);
[ ]
gap> GroupHomology(MathieuGroup(23),3);
[ ]
```

Explanation of proof. The HAP package includes a function GroupHomology(G,n) which inputs a group G and integer $n \geq 1$. The function checks to see if G is finite, infinite, abelian, small order, nilpotent, crystallographic, Artin etc. On the basis of this crude data the function tries to decide on an appropriate method for constructing $(n + 1)$-terms of a free $\mathbb{Z}G$-resolution. It applies the method, then tensors the constructed resolution with the trivial module $\mathbb{Z}$, and finally chooses a Smith Normal Form algorithm to compute the n-th homology of the resulting chain complex. This homology is, by definition, the integral homology of G. The homology is returned as a list of its abelian invariants. In the above proof the lists are empty because the homology is trivial.   □

Most groups G can be viewed in many diferent ways and the choice of the most appropriate methods for computing homology is a difficult one and often needs experimentation. The command GroupHomology(G,n) is a composite of several more basic HAP functions and attempts, in a fairly crude way, to make reasonable choices for parameters in the calculation of group homology. For any particular group G better results can usually be obtained by using the more basic functions directly with the user's choice of parameters.

Theorem 2.1 was originally proved in [19] as a counter-example to a long standing conjecture (attributed in [19] to J.-L. Loday) that if a finite group has trivial integral homology in the first three dimensions then the group must be trivial. The paper [19] also shows that $H_4(M_{23}, \mathbb{Z}) = 0$.

THEOREM 2.2. [4, 20] (i) The group $K_3 = \ker(\mathsf{SL}_2(\mathbb{Z}_{3^3}) \to \mathsf{SL}_2(\mathbb{Z}_3))$ has third integral homology group of exponent 27. (ii) In dimensions $n \neq 3$, $1 \leq n \leq 6$ the group K_3 has integral homology of exponent at most 9.

PROOF.
```
gap> K3:=MaximalSubgroups(
>               SylowSubgroup(SL(2,Integers mod 3^3),3))[2];;
gap> K3:=Image(IsomorphismPcGroup(K3));;
gap> Display(List([1..4],n->GroupHomology(K3,n)));
[ [   3,    3,    3 ],
  [   3,    3,    3 ],
  [   3,    3,    3,    3,    3,    3,   27 ],
  [   3,    3,    3,    3,    3,    3,    3,    3 ],
  [   3,    3,    3,    3,    3,    3,    3,    3,    3,    3,    9,
        9,    9 ],
  [   3,    3,    3,    3,    3,    3,    3,    3,    3,    3,    9,
        9,    9,    9,    9 ] ]
```
□

Theorem 2.2(i) was originally proved in [4] with the prime 3 replaced by any odd prime. The statement of Theorem 2.2(ii) was proved for all dimensions $n \neq 3$ in [20] using a technique of Ian Leary. The papers [4, 20] provided a counter-example to A. Adem's long standing conjecture that if the integral homology of a finite p-group has exponent e in some dimension then it has exponent e in infinitely many dimensions.

THEOREM 2.3. [25] The mod 2 cohomology $H^n(M_{11}, \mathbb{Z}_2)$ of the Mathieu group M_{11} is a vector space of dimension equal to the coefficients of x^n in the Poincaré series $(x^4 - x^3 + x^2 - x + 1)/(x^6 - x^5 + x^4 - 2*x^3 + x^2 - x + 1)$ for all $n \leq 20$.

PROOF.
```
gap> PoincareSeriesPrimePart(MathieuGroup(11),2,20);
(x^4-x^3+x^2-x+1)/(x^6-x^5+x^4-2*x^3+x^2-x+1)
```

Explanation of proof. The HAP function PoincareSeriesPrimePart(G,p,N) inputs a finite group G, a prime p and a positive integer N. It returns a series $p(x)$ in which the coefficient of x^n equals the dimension of $H^n(G, \mathbb{Z}_p)$ for all $n \leq N$. This equality is not guaranteed for $n > N$. (However, since the cohomology ring $H^*(G, \mathbb{Z}_p)$ is finitely generated, there definitely exists some unkown value N_0 such that if the coefficients of the polynomial $p(x)$ are correct for all $n < N_0$ then the coefficients of $p(x)$ are correct for all n.) □

Theorem 2.3 was originally proved in [25] for all $n \geq 1$. We should emphasize that, currently, HAP can only prove Theorem 2.3 for values of n in some finite range such as $0 \leq n \leq 20$.

THEOREM 2.4. [**23**] *The symmetric group* $G = S_3$ *admits a periodic free* $\mathbb{Z}G$-*resolution of* $\mathbb{Z}$ *of period 4.*

PROOF.
```
gap> F:=FreeGroup(2);; x:=F.1;; y:=F.2;;
gap> S3:=F/[ x^2, x*y*x^-1*y^-2 ];;
gap> R:=ResolutionSmallFpGroup(S3,5);;
gap> List([1..5],i->R!.dimension(i));
[ 2, 2, 1, 1, 2 ]
gap> R!.boundary(5,1)=R!.boundary(1,1);
true
gap> R!.boundary(5,2)=R!.boundary(1,2);
true
```

Explanation of proof. The HAP function `ResolutionSmallFpGroup(G,n)` inputs a finitely presented group G and a positive integer n. It returns a record containing the description of n terms of a free $\mathbb{Z}G$-resolution of $\mathbb{Z}$. The resolution corresponds to the cellular chain complex of the universal cover of a classifying CW-space $X = K(G, 1)$; the 2-skeleton of X is the usual 2-complex associated to the given presentation of G. The last two commands in the proof show that the fifth boundary map of the resolution equals the first. The resolution can thus be extended to a periodic one. $\qquad\qquad\qquad\qquad\qquad\qquad\qquad\qquad\qquad\qquad\qquad\qquad\quad\square$

Theorem 2.4 was first proved in [**23**]. Groups which act freely on spheres admit periodic resolutions. The interest in the group S_3 is that it does not act freely on a sphere.

THEOREM 2.5. [**6**] *The quaternion group* Q *of order 8 admits a classifying CW-space with one cell in dimension* $4n$, *three cells in dimension* $4n + 1$, *four cells in dimension* $4n + 2$, *and two cells in dimension* $4n + 3$. *The 2-skeleton of* X *corresponds to the presentation* $Q = \langle i, j, k : ij = k, jk = i, ki = j, ikj = 1 \rangle$.

PROOF.
```
gap> A:=[[ 0,-1,0,0,],[1,0,0,0,],[0,0,0,1],[0,0,-1,0]];;
gap> B:=[[ 0,0,-1,0],[0,0,0,-1],[1,0,0,0],[0,1,0,0]];;
gap> Q:=Group(A,B);; P:=PolytopalComplex(Q,[1,0,0,0]);;
gap> ranks:=List([0..3],n->Dimension(P)(n));
[ 1, 3, 4, 2 ]
gap> List([1..3],n->List([1..ranks[n+1]],
>              k->Order(P!.stabilizer(n,k))));
[ [ 1, 1, 1 ], [ 1, 1, 1, 1 ], [ 1, 1 ] ]
gap> PresentationOfResolution(P);
rec( freeGroup := <free group on the generators [ f1, f2, f3 ]>,
  relators := [ f2*f3^-1*f1^-1, f3*f2*f1^-1,
                    f1*f2*f3, f1*f3^-1*f2 ] )
```

Explanation of proof. The group Q is entered as a group generated by two 4×4 matrices. The HAP function `PolytopalComplex(G,v)` inputs a finite group G of $n \times n$ rational matrices and a rational vector v of length n. The function returns a record describing the cellular chain complex of the convex hull of the orbit of v under the action of G. The group G acts on this convex hull, and the record contains the stabilizer groups of the various faces in this polytope. If the stabilizers are trivial then the cellular chain complex forms part of a periodic free $\mathbb{Z}G$-resolution of $\mathbb{Z}$. The HAP function `PresentationOfResolution(R)` returns the group presentation corresponding to the first two dimensions of a geometric resolution R. □

Theorem 2.5 yields a slightly different periodic free $\mathbb{Z}Q$-resolution of $\mathbb{Z}$ to that given by Cartan and Eilenberg [6]. In particular, its 2-skeleton corresponds to the 'natural' presentation of the quaternions.

THEOREM 2.6. *The symmetric group S_4 admits a classifying CW-space whose 2-skeleton corresponds to the Coxeter presentation*

$$S_4 = \langle x, y, z : x^2 = y^2 = z^2 = (xz)^2 = (xy)^3 = (yz)^3 = 1 \rangle$$

and which has precisely 97 cells in dimension 20.

PROOF.
```
gap> R:=ResolutionFiniteGroup(SymmetricGroup(4), 20);;
gap> P:=PresentationOfResolution(R);
[ <free group on the generators [ f1, f2, f3 ]>,
  [ f1^2, f2^2, f3^2, f3*f1*f3*f1,
    f1*f2*f1*f2*f1*f2, f2*f3*f2*f3*f2*f3 ] ]
gap> Dimension(R)(20);
97
```

Explanation of proof. The HAP function `ResolutionFiniteGroup(G,n)` constructs the first n terms of a free $\mathbb{Z}G$-resolution of a finite group G; the resolution is isomorphic to the cellular chain complex of the universal covering space of some classifying CW-space for G. □

Theorem 2.6 answers negatively a question in [21] which asks whether any classifying space for an n generator Coxeter group G, whose 2-skeleton corresponds to the standard Coxeter presentation of G, must have at least $\frac{(n+k-1)!}{(n-1)!k!}$ k-dimensional cells.

THEOREM 2.7. *(i) [15] The free nilpotent group G of class two on 4 generators has integral cohomology groups*

$$H^1(G, \mathbb{Z}) \cong \mathbb{Z}^4, \quad H^2(G, \mathbb{Z}) \cong \mathbb{Z}^{20}, \qquad H^3(G, \mathbb{Z}) \cong \mathbb{Z}^{56},$$

$$H^4(G, \mathbb{Z}) \cong \mathbb{Z}^{84}, \quad H^5(G, \mathbb{Z}) \cong \mathbb{Z}_3^4 \oplus \mathbb{Z}^{90}, \quad H^6(G, \mathbb{Z}) \cong \mathbb{Z}_3^4 \oplus \mathbb{Z}^{84},$$

$$H^7(G, \mathbb{Z}) \cong \mathbb{Z}^{56}, \quad H^8(G, \mathbb{Z}) \cong \mathbb{Z}^{20}, \qquad H^9(G, \mathbb{Z}) \cong \mathbb{Z}^4,$$

$$H^{10}(G, \mathbb{Z}) \cong \mathbb{Z}, \quad H^n(G, \mathbb{Z}) = 0 \ (n \geq 11).$$

(ii) The ring $H^*(G, \mathbb{Z})$ is generated by 4 classes in degree 1, 20 classes in degree 2, 36 classes in degree 3, and 20 classes in degree 4.

PROOF. To save space we show how to prove the completely analogous result for the free nilpotent group of class two on $n = 3$ generators. The only modification needed to get the above theorem is to set $n = 4$ and $m = 11$.

```
gap> n:=3;;m:=7;;
gap> F:=FreeGroup(3);;G:=NilpotentQuotient(F,2);;
gap> R:=ResolutionNilpotentGroup(G,10);;
gap> for n in [1..m-1] do
```

```
> Print(``Cohomology in dimension '',n,`` = '',
> Cohomology(HomToIntegers(R),n),``\n''); od;
Cohomology in dimension 1 = [ 0, 0, 0 ]
Cohomology in dimension 2 = [ 0, 0, 0, 0, 0, 0, 0, 0 ]
Cohomology in dimension 3 = [ 0, 0, 0, 0, 0, 0, 0, 0, 0, 0, 0, 0 ]
Cohomology in dimension 4 = [ 0, 0, 0, 0, 0, 0, 0, 0 ]
Cohomology in dimension 5 = [ 0, 0, 0 ]
Cohomology in dimension 6 = [ 0 ]
gap> Dimension(R)(7);
0
gap> List([1..m-1],n->Length(IntegralRingGenerators(R,n)));
[ 3, 8, 6, 0, 0, 0]
```

$\square$

Theorem 2.7(i) was originally proved in [15]. Theorem 2.7(ii) appears to be new.

THEOREM 2.8. [16] *The finite-type Artin group A corresponding to the exceptional finite reflection group F_4 has integral cohomology*

$$H^1(A,\mathbb{Z}) \cong \mathbb{Z}^2, \qquad H^2(A,\mathbb{Z}) \cong \mathbb{Z}^2, \qquad H^3(A,\mathbb{Z}) \cong \mathbb{Z}^2,$$

$$H^4(A,\mathbb{Z}) \cong \mathbb{Z}, \qquad H^n(A,\mathbb{Z}) \cong 0 \ (n \geq 5).$$

PROOF.
```
gap> D:=[[1,[2,3]],[2,[3,4]],[3,[4,3]]];;
gap> Display(List([1..5],i->GroupCohomology(D,i)));
[ [ 0, 0 ], [ 0, 0 ], [ 0, 0 ], [ 0 ], [ ] ]
```

Explanation of proof. An Artin group is represented in HAP by a list D which describes the group's Coxeter diagram. The Coxeter diagram for F_4 has four vertices and three labelled edges. The first and second vertices are connected by an edge labelled by 3; the second and third vertices are connected by an edge labelled by 4, the third and final vertices are connected by an edge labelled 3. The HAP function GroupCohomology() can be applied directly to D. $\square$

Theorem 2.8 was originally proved in [16] where the cohomology of each of the seven exceptional finite-type Artin groups was calculated.

THEOREM 2.9. [1, 8] *The 3-generator affine braid group $A = \langle x, y, z : xyx = yxy, yzy = zyz, xzx = zxz \rangle$ admits a 2-dimensional classifying space, namely the CW-space associated to its presentation.*

PROOF. We must show that the 2-dimensional CW-space B associated to the presentation is aspherical. Asphericity is obviously a homotopy invariant. So we can alternatively test the asphericity of the homotopy equivalent space B' arising from the presentation of A obtained by adding generators a, b, c, adding relations $a = xy$, $b = yz$, $c = zx$ and replacing all occurrences of xy, yz, and zx.

```
gap> F:=FreeGroup(6);;
gap> x:=F.1;;y:=F.2;;z:=F.3;;a:=F.4;;b:=F.5;;c:=F.6;;
gap> rels:=[a^-1*x*y, b^-1*y*z, c^-1*z*x, a*x*(y*a)^-1,
> b*y*(z*b)^-1, c*z*(x*c)^-1];;
```

```
gap> IsAspherical(F,rels);
Presentation is aspherical.
true
```

☐

Theorem 2.9 is a particular example of a result of Appel and Schupp [1] who proved that all Artin groups of large type have 2-dimensional classifying spaces. Theorem 2.9 is also a special case of a result of Charney and Peiffer [8] who showed that the $(n + 1)$-generator affine braid group admits an n-dimensional classifying space.

PROPOSITION 2.10. *The mod 2 cohomology ring* $H^*(D_{64}, \mathbb{Z}_2)$ *of the dihedral group of order 64 is generated by two elements in degree 1, one element in degree 2, and possibly (though not very likely) some generators of degree greater than 20.*

PROOF.
```
gap> A:=ModPCohomologyRing(DihedralGroup(64),20);
gap> List(ModPRingGenerators(A),a->A!.degree(a));
[ 0, 1, 1, 2 ]
```

☐

PROPOSITION 2.11. *The amalgamated free product* $S_5 *_{S_3} S_4$ *of the symmetric groups* S_5 *and* S_4 *over* S_3 *has seventh integral homology* $H_7(S_5 *_{S_3} S_4, \mathbb{Z}) = (\mathbb{Z}_2)^3 \oplus \mathbb{Z}_4 \oplus \mathbb{Z}_{60}$.

PROOF.
```
gap> S5:=SymmetricGroup(5);; S4:=SymmetricGroup(4);;
gap> S3:=SymmetricGroup(3);;
gap> S3S5:=GroupHomomorphismByFunction(S3,S5,x->x);;
gap> S3S4:=GroupHomomorphismByFunction(S3,S4,x->x);;
gap> D:=[S5,S4,[S3S5,S3S4]];;
gap> R:=ResolutionGraphOfGroups(D,8);;
gap> Homology(TensorWithIntegers(R),7);
[ 2, 2, 2, 4, 60 ]
```

Explanation of proof. An amalgamated free product is a special case of a graph of groups. The function ResolutionGraphOfGroups(D,n) inputs a list D describing a graph of groups, together with a positive integer n. It returns n terms of a free $\mathbb{Z}G$-resolution for the group G described by the graph of groups.   ☐

PROPOSITION 2.12. *Let* $G = \mathrm{GL}_4(3)$ *be the group of invertible* 4×4 *matrices over the field of three elements; this is a non-perfect, non-solvable group of order* 24261120. *Let* $SK(G, 1)$ *be the suspension of a classifying space for* G. *Then this suspension space has third homotopy group* $\pi_3(SK(G, 1)) = \mathbb{Z}_2$.

PROOF. The following commands are based on the isomorphism

$$\pi_3(SK(G, 1)) = \ker(G \otimes G \to G),$$

involving the nonabelian tensor square [5].

```
gap> G:=Image(IsomorphismPermGroup(GL(4,3)));;
gap> ThirdHomotopyGroupOfSuspensionB_alt(G);
[ [ ], [ 2 ] ]
```

$\square$

3. The role of contracting homotopies

A free kG-resolution R_* is represented in HAP as a record R with various components such as R!.boundary(n,i) which gives the image of the ith free generator e_i^n of R_n under the boundary map $d_n: R_n \to R_{n-1}$, and R!.group which gives the group G. One of the less obvious components R!.homotopy(n,[i,g]) returns, for $g \in G$, the image of $g \cdot e_i^n$ under a contracting homotopy $h_n: R_n \to R_{n+1}$.

Recall that a contracting homotopy on R_* is a family of abelian group homomorphisms $h_n: R_n \to R_{n+1}$ $(n \geq 0)$ satisfying $d_{n+1}h_n(x) + h_{n-1}d_n(x) = x$ for all $x \in R_n$ (where $h_{-1} = 0$). Since the h_n are not G-equivariant one needs to specify $h_n(x)$ on a set of abelian group generators for R_n. Having specified a contracting homotopy, one can use it to make constructive the following element of choice which frequently occurs in homological algebra:

> For each $x \in \ker(d_n: R_n \to R_{n-1})$ choose an element $\tilde{x} \in R_{n+1}$ such that $d_{n+1}(\tilde{x}) = x$.

The choice can be made by setting $\tilde{x} = h_n(x)$.

Thus contracting homotopies are used in HAP for: (i) constructing homology homomorphisms $H_*(G, N) \to H_*(G', N')$ induced by group homomorphisms $\phi: G \to G'$ and ϕ-equivariant module homomorphisms $N \to N'$; (ii) constructing cup products $H^p(G, k) \otimes H^q(G, k) \to H^{p+q}(G, k)$; (iii) applying perturbation techniques which input free kH-resolutions for various subgroups H in G and output a free kG-resolution.

4. Linear algebra techniques

For a small group G and integral domain k (typically $k = \mathbb{Z}$ or k a finite field) one can naively represent an element in the group ring kG as a vector of length $|G|$ over k. An element in a free kG-module $(kG)^t$ can be represented as a vector of length $t \times |G|$ over k. To compute a free kG-resolution

$$R_*: \quad \cdots \to R_n \to R_{n-1} \to \cdots \to R_0,$$

one can set $R_0 = kG$, define $d_0: kG \to k$, $\Sigma\lambda_g g \mapsto \Sigma\lambda_g$, and then recursively

(1) determine $\ker d_n$ (using gaussian elimination if k is a field, and Smith Normal Form if $k = \mathbb{Z}$),
(2) determine a small subset $\{v_1, \ldots, v_t\} \subset \ker d_n$ whose kG-span equals $\ker d_n$,
(3) set $R_{n+1} = (kG)^t$,
(4) define $d_{n+1}: (kG)^t \to R_n$ by sending the ith free generator to v_i.

The construction of a contracting homotopy $h_n: R_n \to R_{n+1}, x \mapsto \tilde{x}$ essentially boils down to solving a matrix equation $d_{n+1}(\tilde{x}) = x$ where $\tilde{x}$ is unknown and x is a known vector in the image of d_{n+1}. The most costly part of the recursive procedure is Step 2. If k is the field of p elements and G a p-group then the radical of $\ker d_n$ is the vector space spanned by vectors $v - g \cdot v$ where v ranges over a k-basis for

$\ker d_n$ and g ranges over generators for G. Any basis for the complement of the radical yields a minimal set $\{v_1, \cdots v_t\}$ with kG-span equal to $\ker d_n$. Even when G is not a p-group this method can be used to find a set whose kP-span equals $\ker d_n$, where P is a Sylow p-subgroup of G. Given a non-minimal set with kG-span equal to $\ker d_n$ one can use naive methods to find a minimal subset, even in the case where $k = \mathbb{Z}$.

These linear algebraic techniques are used in the proofs of Theorems 2.3, 2.4 and Proposition 2.10.

5. Geometric techniques

Naive linear algebra can obviously only be applied to extremely small groups. For larger groups, or for integral coefficients, one can make use of the following fact from elementary algebraic topology.

> If a group G acts fixed-point freely and cellularly on a contractible CW-space X then the cellular chain complex
> $$C_*(X) : \cdots \to C_2(X) \to C_1(X) \to C_0(X)$$
> is a free $\mathbb{Z}G$-resolution of $\mathbb{Z}$.

The quotient X/G of such a space X obtained by killing the action of G is a *classifying space* for the group G. Recall that $C_n(X)$ is the free abelian group whose free generators correspond to the n-cells of X. The boundary homomorphism $d_n \colon C_n(X) \to C_{n-1}(X)$ can be obtained directly from an explicit description of X.

For a finite group G one can recursively construct the n-skeleta
$$X^0 \subset X^1 \subset X^2 \subset \cdots \subset X$$

of a suitable CW-space X by:

(1) setting $X^0 = G$,
(2) constructing X^{n+1} by attaching just enough $(n+1)$-cells to X^n so that
    (a) $\pi_n(X^{n+1}) = 0$,
    (b) G freely permutes the cells of X^{n+1}.

This technique is used in the proofs of Theorems 2.1, 2.2, 2.6 and Proposition 2.11.

One can also use Polymake computational geometry software to obtain a contractible space X with free G-action. For example, suppose we have a faithful linear representation $\alpha \colon G \to \mathrm{GL}_n(\mathbb{R})$ of a finite group G. We can choose a vector $v \in \mathbb{R}^n$ with trivial stabilizer group in G and define the *orbit polytope* $P(G)$ as
$$P(G) = \text{Convex hull}\{\alpha g(v) : g \in G\}.$$

The polytope $P(Q)$ is a contractible m-dimensional CW-space ($m \le n$) and G acts on it by permuting cells. Consequently its cellular chain complex $C_* = C_*(P(G))$ is an exact sequence of $\mathbb{Z}G$-modules with $H_0(C_*) = \mathbb{Z}$ and $C_m = \mathbb{Z}$. Copies of C_* can be spliced together to form a periodic $\mathbb{Z}G$-resolution
$$\cdots \to C_1 \to C_0 \to C_{m-1} \to \cdots \to C_1 \to C_0 \to C_{m-1} \to \cdots \to C_1 \to C_0$$

of period m. If G acts freely on the $(m-1)$-skeleton of $P(G)$ then this is a free periodic resolution. This polytopal resolution is used in the proof of Theorem 2.5 where the relevant orbit polytope is computed using Polymake software [12].

If G does not act freely on the $(m-1)$-skeleton of $P(G)$ then the perturbation techniques described in the next section can be used to convert $C_*(P(G))$ into a free resolution.

Polymake software can also be used to construct resolutions for infinite groups such as Bieberbach groups G. Here one uses convex hull computations to construct a Dirichlet-Voronoi fundamental domain for the action of G on Euclidean space $\mathbb{R}^n$. This yields a tessellation (and hence CW-structure) on the contractible space $\mathbb{R}^n$. The resulting cellular chain complex $C_*(\mathbb{R}^n)$ is a $\mathbb{Z}G$-resolution of $\mathbb{Z}$.

A third use of Polymake is made in the proof of Theorem 2.9. The idea here is to show that a certain space B can not contain a 2-sphere. An argument based on Euler characteristics is encoded as a linear programming problem and then solved using Polymake.

A theoretically constructed polytopal classifying space can be used to compute the cohomology of certain Artin groups. To explain this we recall that a *Coxeter matrix* is a symmetric $n \times n$ matrix each of whose entries $m(i, j)$ is a positive integer or ∞, with $m(i, j) = 1$ if and only if $i = j$. Such a matrix can be represented by the *Coxeter graph* D with n vertices, and with a labelled edge joining vertices i and j if $m(i, j) \geq 3$. An *Artin group* A_D and *Coxeter group* W_D is assigned to each Coxeter graph as follows. The Artin group A_D is generated by a set of elements $S = \{x_1, \cdots x_n\}$ subject to the relations $(x_i x_j)_{m(i,j)} = (x_j x_i)_{m(i,j)}$ for all $i \neq j$, where $(xy)_m$ denotes the word $xyxyx \cdots$ of length m. The Coxeter group W_D is the group satisfying the additional relations $x^2 = 1$ for $x \in S$. Denote by $\bar{x}$ the image in W_D of the generator x.

A finite Coxeter group W_D can be realized as a group of orthogonal transformations of $\mathbb{R}^n$ with generators $\bar{x}$ equal to reflections [14]. For any point v in $\mathbb{R}^n$ lying in the complement of the generating mirrors we denote by P_D the convex hull of the orbit of v under the action of W_D. It is readily shown that the face lattice of the n-dimensional convex polytope P_D depends only on the graph D, and that the polytope is simple (i.e. each vertex touches precisely n edges). Simplicity implies that the face lattice of the polytope is in fact determined by its 1-skeleton [2]. Label each edge in P_D by the unique generator $\bar{x}$ stabilizing it. Define the *length* of an element g in W_D to be the shortest length of a word in the generators representing it. It is possible to orient each edge in P_D so that its initial vertex gv and final vertex $g'v$ are such that the length of g is less than the length of g'. With this orientation the boundaries of 2-faces in P_D spell words corresponding to the relators of the Artin group A_D.

For an arbitrary Coxeter graph D we define a CW-space B_D as follows. For each full subgraph D_i in D, that is maximal with respect to the property that W_{D_i} is a finite subgroup, we construct the oriented and labelled convex polytope W_{D_i}. The space B_D is obtained from the union of the polytopes W_{D_i} by identifying in a canonical fashion all faces with the same labelled and directed edge graph. It is conjectured that B_D is a classifying space for the Artin group A_D [7]. In cases where this conjecture is known, the cellular chain complex $C_*(\tilde{B}_D)$ of the universal cover of B_D is a free $\mathbb{Z}A_D$-resolution. This free resolution is used, for example, in our proof of Theorem 2.8.

6. Using subgroups

We mention two ways in which subgroups of G can be brought into the calculation of the (co)homology of G.

The first technique concerns a finite group G with Sylow p-subgroup P. There is a surjection $H_n(P, \mathbb{Z}) \to H_n(G, \mathbb{Z})_{(p)}$ from the homology of P onto the p-part of the homology of G. The kernel of this surjection is described in terms of double coset representatives

$$G = \bigcup_x P x P$$

and induced homomorphisms $H_n(P, \mathbb{Z}) \to H_n(x P x^{-1}, \mathbb{Z})$. by the classical Cartan-Eilenberg double coset formula. Thus, the homology of a large finite group can be deduced from that of its Sylow subgroups. This technique is used, for example, in our proof of Theorems 2.1 and 2.3.

The second technique concerns a group G which may be infinite. Suppose we have a $\mathbb{Z}G$-resolution of $\mathbb{Z}$

$$C_* : \cdots \to C_n \to C_{n-1} \to \cdots \to C_0 \to \mathbb{Z}.$$

but that C_* is not free. Suppose, however, that for each m we have a free $\mathbb{Z}G$-resolution of the module C_m

$$D_{m*} : \to D_{m,n} \to D_{m,n-1} \to \cdots \to D_{m,0} \to C_m.$$

THEOREM 6.1. [**24**] *There is a free $\mathbb{Z}G$-resolution $R_* \to \mathbb{Z}$ with*

$$R_n = \bigoplus_{p+q=n} D_{p,q}$$

The proof of this Theorem given in [**24**] can be made constructive by using contracting homotopies on the resolutions D_{m*}. Furthermore, a contracting homotopy on R_* can be constructed by a formula involving contracting homotopies on the D_{m*} and on C_*.

The following three scenarios are covered by Theorem 6.1:

(1) C_* is the cellular chain complex of an orbit polytope;
(2) C_* is the cellular chain complex of a graph of groups;
(3) C_* is a free $\mathbb{Z}(G/N)$-resolution.

Theorem 6.1 is used for example in our proofs of Theorem 2.2 and Proposition 2.11.

References

[1] K.J. Appel and P.E. Schupp, "Artin groups and infinite Coxeter groups", *Invent. Math.*, 72 (1983), 201-220.
[2] R. Blind and P. Mani-Levitska, "Puzzles and polytope isomorphisms", *Aequationes Math.*, 34 (1987), 287-297.
[3] W. Bosma, J. Cannon, and C. Playoust. The MAGMA algebra system. I. The user language. *J. Symbolic Comput.*, 24(3-4) (1997), 235-265.
[4] W. Browder and J. Pakianathan, "Cohomology of uniformly powerful p-groups", *Trans. Amer. Math. Soc.* 352 (2000), no. 6, 2659–2688.
[5] R. Brown and J.-L. Loday, "Van Kampen theorems for diagrams of spaces", with an appendix by M. Zisman, *Topology* 26 (1987), no. 3, 311–335.
[6] H. Cartan and S. Eilenberg, *Homological algebra*, Princeton University Press, Princeton, N. J., 1956. xv+390 pp.

[7] R. Charney and M.W. Davis, "Finite $K(\pi,1)$s for Artin groups", *Prospects in topology (Princeton, NJ, 1994)*, 110–124, Ann. of Math. Stud., 138, Princeton Univ. Press, Princeton, NJ, 1995.

[8] R. Charney and D. Peifer, "The $K(\pi,1)$-conjecture for the affine braid groups", *Comment. Math. Helv.*, 78 no. 3 (2003), 584–600.

[9] CoCoATeam, CoCoA: a system for doing Computations in Commutative Algebra (http://cocoa.dima.unige.it)

[10] J.-G- Dumas, F. Heckenbach, D. Saunders and V. Welkmar, Simplicial Homology, a package for the GAP computational algebra system. (http://www.gap-system.org/Packages/undep.html)

[11] The GAP Group, GAP – Groups, Algorithms, and Programming, Version 4.4.9; 2006. (http://www.gap-system.org)

[12] E. Gawrilow and M. Joswig, "Polymake: a framework for analyzing convex polytopes", 43-74, ed. Gil Kalai and Günter M. Ziegler, *Polytopes — Combinatorics and Computation*, Birkhäuser, 2000

[13] Homological Algebra programming, Version 1.8 (2007), a package for the GAP computational algebra system. (http://www.gap-system.org/Packages/hap.html)

[14] J.E. Humphreys, *Reflection groups and Coxeter groups*, Cambridge studies in advanced mathematics 29 (CUP, 1990).

[15] L.A. Lambe, "Cohomology of principal G-bundles over a torus when $H^*(BG;R)$ is polynomial", *Bull. Soc. Math. Belg.* ser. A 38 (1986), 247–264 (1987).

[16] C. Landi, "Cohomology rings of Artin groups", *Atti Accad. Naz. Lincei Cl. Sci. Fis. Mat. Natur. Rend. Lincei (9) Mat. Appl.*, 11 no. 1 (2000), 41-65.

[17] F. Lübeck, Elementary Divisors of Integer Matrices, Version 1.2.3 (2006), a package for the GAP computational algebra system. (http://www.gap-system.org/Packages/edim.html)

[18] Macaulay: a system for computation in algebraic geometry and commutative algebra. (http://www.math.columbia.edu/ bayer/Macaulay)

[19] R.J. Milgram, "The cohomology of the Mathieu group M_{23}", *J. Group Theory* 3 (2000), no. 1, 7–26.

[20] J. Pakianathan, "Exponents and the cohomology of finite groups", *Proc. Amer. Math. Soc.* 128 (2000), no. 7, 1893–1897.

[21] M. Salvetti, "Cohomology of Coxeter groups", Arrangements in Boston: a Conference on Hyperplane Arrangements (1999), Topology Appl. 118 (2002), no. 1-2, 199–208.

[22] Singular Team, Singular: a computer algebra system for polynomial based calculatons. (http://www.singular.uni-kl.de)

[23] R.G. Swan, 'Periodic resolutions for finite groups', *Ann. of Math. (2)* 72 (1960), 267-291.

[24] C.T.C. Wall, "Resolutions of extensions of groups", *Proc. Cambridge Philos. Soc.* 57 (1961), 251-255.

[25] P.J. Webb, "A local method in group cohomology" Comment. Math. Helv. 62 (1987), no. 1, 135–167.

NATIONAL UNIVERSITY OF IRELAND, GALWAY.

Contemporary Mathematics
Volume **470**, 2008

Groups with a Finite Covering by Isomorphic Abelian Subgroups

Tuval S. Foguel and Matthew F. Ragland

Dedicated to Dr. James C. Beidleman on the occasion of his seventieth birthday.

ABSTRACT. In this paper, we look at groups with a finite covering by proper isomorphic abelian subgroups (CIA-groups). Our main focus will be on finite groups with such a covering. In particular, we will see that there are no simple CIA-groups, but on the other hand, that every finite group is a direct factor of a CIA-group. A complete characterization of finite abelian CIA-groups is given and, modulo their Sylow structure, a characterization of finite nilpotent CIA-groups is given. We also show that a CIA-group G must contain an element whose order is the exponent of G. It is trivial that groups where the exponent of the group and the exponent of the center coincide satisfy this property. Hence we consider the class of CIA-groups G whose exponent is greater than the exponent of $Z(G)$. We say such groups have a "small" center. A question we leave open is whether or not there exist centerless CIA-groups. Such a group must have a "small" center providing us with further motivation to study CIA-groups with small center. Lastly, the role of GAP in this work is discussed.

1. Introduction

A group is said to have a covering by subgroups if it is the set-theoretic union of proper subgroups, and, if the set of subgroups is finite, we say the covering is finite. To develop our theme, let us briefly look at the background and history of group coverings. Results on finite coverings by subgroups first appeared in a book by Scorza [19] with an emphasis on coverings by a small number of subgroups. Bernhard Neumann in [15] and [16] investigated coverings by cosets. The following theorem, often called Neumann's Lemma, is a key to many group theoretic results. In particular, a characterization of groups having finite coverings is stated as a corollary of the following theorem.

THEOREM 1.1. *Let $G = \bigcup_{i=1}^{k} g_i H_i$, where $H_1, \ldots, H_k$ are (not necessarily distinct) subgroups of G. Then, if we omit from the union any cosets $g_i H_i$ for which $[G : H_i]$ is infinite, the union of the remaining cosets is still all of G.*

2000 *Mathematics Subject Classification.* 20K01, 20D15, 20E34.
Key words and phrases. Group Theory, Group Coverings.

COROLLARY 1.2. *A group has a finite covering by subgroups if and only if it has a finite non-cyclic homomorphic image.*

The following unpublished result by Reinhold Baer (see Theorem 4.6 in [**17**]) leads to the investigation of finite coverings by special subgroups as can be found in [**5**] and [**13**].

THEOREM 1.3. *A group is central-by-finite if and only if it is the union of finitely many abelian subgroups.*

Coverings have been widely studied in groups, and recently, analogous coverings for rings, semigroups, and loops have been discussed in [**1**], [**14**], and [**7**], respectively. In the first author's study of loops covered by subgroups (see [**7**], [**8**], and [**9**]), a family of loops that are covered by isomorphic abelian subgroups was encountered.

DEFINITION 1.4. Given $(\mathbb{F}, +, \cdot)$ a field, and a finite idempotent quasigroup $(Q, \odot)$, let $\mathcal{L}^{(Q)}(\mathbb{F}) = \{a_q(x) : x \in \mathbb{F}^* \text{ and } q \in Q\} \cup \{\mathbf{1}\}$ (i.e. each element of the form $a_q(x)$ in this set is double indexed by q and x) be the loop whose binary operation is defined as follows:

    i. For any $l \in \mathcal{L}^{(Q)}(\mathbb{F})$, $\mathbf{1}l = l\mathbf{1} = l$.
    ii. For $x, y \in \mathbb{F}^*$,

$$a_q(x)a_q(y) = \begin{cases} a_q(x+y) & \text{if } x + y \neq 0 \\ \mathbf{1} & \text{otherwise} \end{cases}$$

    iii. For $x, y \in \mathbb{F}^*$, $a_{q_1}(x)a_{q_2}(y) = a_{q_1 \odot q_2}(xy)$ for $q_1 \neq q_2$.

The loops in this family are covered by $|Q|$ copies of $(\mathbb{F}, +)$. These loops can even be simple. In view of this family of loops and Theorem 1.3, it is only natural to ask what one can say about groups which are covered by isomorphic abelian subgroups. This is the primary aim of this work.

In this paper, our main focus will be on finite groups with a finite covering by proper isomorphic abelian subgroups. We call such groups CIA-groups. In Section 2, we give some basic motivating examples and show that direct products of CIA-groups are CIA-groups. Section 3 is devoted to a complete characterization of finite abelian CIA-groups and a characterization of finite nilpotent CIA-groups modulo their Sylow structure. In particular, we will see that a group G is a finite abelian CIA-group if and only if G has a direct factor which is a direct product of two isomorphic cyclic subgroups. The finite nilpotent CIA-groups will be seen to be those finite nilpotent groups whose Sylow subgroups are abelian or CIA-groups with at least one Sylow subgroup being a CIA-group. We will see in Section 4 that the groups of square-free order, the dihedral groups (excluding the Klein four group), the Frobenius groups, the symmetric groups, the alternating groups, and all simple groups are examples of non-CIA-groups. However, in Theorem 5.1, it is shown that every finite group is a direct factor of a CIA-group. This result gives rise to several examples of CIA-groups.

In Section 5, we give examples of finite CIA-groups which have "small" centers, that is, CIA-groups where the exponent of the group is greater than the exponent of its center. This section asks the question, "When is the center of a CIA-group 'small'?" Let us give some motivation for this question. As one will see, every example we give of a CIA-group will have a nontrivial center. We have been unable to prove that this is always the case but if a centerless CIA-group exists, then it certainly has a small center. This was our initial motivation for searching for group with a small center. In Theorem 4.1, we show that a finite CIA-group G must contain an element whose order is the exponent of G. One way to guarantee that a finite group has such an element is to force the exponent of the group and its center to be the same. CIA-groups with a small center will certainly satisfy Theorem 4.1, but, in some sense, in a nontrivial fashion. Lastly, we will see that groups constructed using Theorem 5.1 will not have a small center. So CIA-groups with a small center will need to be constructed in some other manner. How one can construct such CIA-groups with a small center takes up the remainder of Section 5.

It should be pointed out that this work would not have been possible without the use of GAP [10]. In Section 6, we give the details of our GAP documentation, but let us briefly elaborate on the elementary, yet crucial, role GAP played in this research. Without knowing much about CIA-groups, we looked to examples to help us understand how CIA-groups arise and to understand their structure. GAP was used to determine which groups in GAP's "SmallGroups" [2] library were CIA-groups. A function called IsUnionOfIsomorphicSubgroups was defined for GAP which we used to check which groups in the "SmallGroups" library were CIA-groups. The GAP command StructureDescription was then used to give us a decomposition of the CIA-groups found into various direct and semidirect products of common groups. Nearly all of the results in the above paragraphs were first conjectured by looking at pages of GAP output consisting of different CIA-groups and their structure descriptions as given by GAP.

2. Preliminaries

In this section, we will look at some basic motivating examples. We will also prove a basic lemma concerning direct products of groups covered by isomorphic abelian subgroups.

Throughout, we will denote the cyclic group of order n by C_n. Also, $\exp(G)$ will be used to denote the exponent of the group G.

DEFINITION 2.1. A group G has a finite covering by proper isomorphic abelian subgroups if $G = \bigcup_{i=1}^{n} A_i$ where the A_i's are proper isomorphic abelian subgroups of G. We will call such groups CIA-groups.

EXAMPLE 2.2. Let G be a finite group. If $\exp(G) = p$ where p is a prime and G is not cyclic, then G is a CIA-group since G is covered by all subgroups of order p.

EXAMPLE 2.3. Let $G = A_4 \times C_3$ where A_4 denotes the alternating group on 4 elements. One can check that G is not a CIA-group. Note that G contains the Klein four group K_4 as a normal subgroup. However K_4 and $G/K_4 \simeq C_3 \times C_3$ are both CIA-groups. So the class of CIA-groups is not closed with respect to forming extensions.

EXAMPLE 2.4. The abelian group $C_4 \times C_2$ is not a CIA-group.

EXAMPLE 2.5. The quaternion group Q_8 is a CIA-group, since Q_8 is covered by all subgroups of order 4.

EXAMPLE 2.6. Let G be a finite group. If $\exp(G) = p^2$ where p is a prime, then $G \times C_{p^2}$ is a CIA-group, since $G \times C_{p^2}$ is covered by all subgroups isomorphic to $C_{p^2} \times C_p$.

If G is a nonabelian group of $\exp(G) = p^2$ and $|G| = p^3$ where p is an odd prime, then G is not a CIA-group since there are elements of order p that do not commute with any element of order p^2. So one can easily see that CIA-groups are not closed with respect to taking subgroups or quotients.

However, it is the case that CIA-groups are closed with respect to taking direct products. Also worth noting is the fact that the direct product of a CIA-group with an abelian group is again a CIA-group. This tells us that every abelian group is a direct factor of a CIA-group. We will see in Section 5 that this is true of any finite group.

LEMMA 2.7. *Let H be a CIA-group. Suppose K is either a CIA-group or an abelian group. Then $H \times K$ is a CIA-group.*

PROOF. Suppose first that K is a CIA-group. Write $H = \bigcup_{i=1}^{h} H_i$ and $K = \bigcup_{j=1}^{k} K_j$ each as unions of isomorphic abelian groups. Then

$$H \times K = \bigcup_{\substack{1 \leq i \leq h \\ 1 \leq j \leq k}} H_i \times K_j$$

is a covering of $H \times K$ by isomorphic abelian groups.

Simply replace each K_j above with K and one can easily see that $H \times K$ is a CIA-group when K is abelian. □

3. Abelian and Nilpotent CIA-groups

In this section, we will examine the structure of finite abelian CIA-groups and finite nilpotent CIA-groups. Throughout, superscripts on groups are used for indexation only. We begin with a definition that will be important in characterizing the finite abelian CIA-groups which are of prime power order.

DEFINITION 3.1. Given a finite group G, $g \in G$ is a *maximum root of G* provided whenever $h^n = g$ we have $|g| = |h|$. Let $\pi_{me}(G)$ be used to denote the following:

$$\pi_{me}(G) = \{|g| : g \text{ is a maximum root in } G\}.$$

First let us examine the homocyclic abelian p-groups. These groups are CIA-groups provided they have more than one component.

LEMMA 3.2. *Let $P = P_1 \times P_2 \times \cdots \times P_t$ be homocyclic where each P_i is a cyclic group of order p^α. Then P is the union of subgroups isomorphic to C_{p^α}. Moreover, if $t > 1$, then P is a CIA-group.*

PROOF. Let $x \in P$. We will show $x \in H \simeq C_{p^\alpha}$ for some $H \leq P$. Suppose each P_i is generated by the element z_i. Then $x = z_1^{\gamma_1} \cdots z_t^{\gamma_t}$ where $0 \leq \gamma_i < p^\alpha$ for all i. We can suppose p divides γ_i for all i or else $|x| = p^\alpha$ in which case we let $H = \langle x \rangle$. Let p^δ be the largest power of p for which p^δ divides γ_i for all i. Then $z_1^{\gamma_1/p^\delta} \cdots z_t^{\gamma_t/p^\delta}$ has order p^α. Now $x \in H = \langle z_1^{\gamma_1/p^\delta} \cdots z_t^{\gamma_t/p^\delta} \rangle \simeq C_{p^\alpha}$ and the desired result follows. □

We are now in a position to characterize finite abelian CIA-groups of prime power order.

THEOREM 3.3. *Let P be an abelian p-group with order p^β. Using the Frobenius-Stickelberger Theorem, write $P = P_1 \times \cdots \times P_t$ with $P_i = C^1_{p^{\beta_i}} \times \cdots \times C^{n_i}_{p^{\beta_i}}$ where $C^k_{p^{\beta_i}} \simeq C_{p^{\beta_i}}$ for all k and i, $\beta_i \neq \beta_j$ for $i \neq j$, and $\sum_{i=1}^t n_i \beta_i = \beta$. Then the following hold:*

(i) *P equals the union of subgroups isomorphic to $C_{p^{\beta_1}} \times \cdots \times C_{p^{\beta_t}}$;*

(ii) *P is a CIA-group if and only if $n_i > 1$ for some i.*

PROOF. (i) If $n_i = 1$ for all i, then $P \simeq C_{p^{\beta_1}} \times \cdots \times C_{p^{\beta_t}}$ and the result is trivial. If $t = 1$, then P is homocyclic and Lemma 3.2 gives the desired result. So we can suppose $n_i \neq 1$ for some i and $t \neq 1$. Without loss of generality, let us suppose $n_1 \neq 1$. Let $\overline{P}_1 = C^2_{p^{\beta_1}} \times \cdots \times C^{n_1}_{p^{\beta_1}}$ and let $\overline{P} = \overline{P}_1 \times P_2 \times \cdots \times P_t$. Note that $P = C^1_{p^{\beta_1}} \times \overline{P}$.

Let $x \in P$. We will show x is an element of some subgroup of P isomorphic to $C = C_{p^{\beta_1}} \times \cdots \times C_{p^{\beta_t}}$. By induction on $|P|$, we have that $\overline{P}$ can be written as the union of subgroups of $\overline{P}$ isomorphic to C. So if $x \in \overline{P}$, then x is in some subgroup of $\overline{P}$, and hence a subgroup of P, isomorphic to C. Clearly if $x \in C^1_{p^{\beta_1}}$, then x is in some subgroup of P isomorphic to C. So we can suppose that $x = zy$ where z and y are nontrivial elements of $C^1_{p^{\beta_1}}$ and $\overline{P}$, respectively. Write $y = y_1 y_2$ with $y_1 \in \overline{P}_1$ and $y_2 \in P_2 \times \cdots \times P_t$. By induction, y_2 is in some subgroup of $P_2 \times \cdots \times P_t$ isomorphic to $C_{p^{\beta_2}} \times \cdots \times C_{p^{\beta_t}}$. By Lemma 3.2, zy_1 is in some subgroup of P_1 isomorphic to $C_{p^{\beta_1}}$. Hence $x = zy_1 y_2$ is an element of some subgroup of P isomorphic to $C_{p^{\beta_1}} \times \cdots \times C_{p^{\beta_t}}$.

(ii) If $n_i > 1$, then (i) says P is a CIA-group.

Suppose $n_i = 1$ for all i. Then, we may write $P = C_{p^{\beta_1}} \times \cdots \times C_{p^{\beta_t}}$. Supposing P is a CIA-group, write $P = \bigcup_{i=1}^m A_i$ where each $A_i \simeq C_{p^{\alpha_1}} \times \cdots \times C_{p^{\alpha_t}}$ with $1 \leq p^{\alpha_i} \leq p^{\beta_i}$ for all i and $p^{\alpha_j} < p^{\beta_j}$ for at least one j. Let $\langle x \rangle = C_{p^{\beta_j}}$ and note x is a maximum root of P. Now, for some i, we have $x \in A_i$. By the structure of A_i, x must be equal to some power of an element, say y, of A_i, where y has order larger than that of x. This contradicts the fact that x is a maximum root of P. Thus P is not a CIA-group. $\square$

It should come as no surprise that the structure of a finite abelian CIA-group depends on the group's Sylow structure. Here we see that a finite abelian group is a CIA-group if and only if it possesses a Sylow CIA-subgroup.

THEOREM 3.4. *Let G be a finite abelian group with order $p_1^{\alpha_1} \cdots p_n^{\alpha_n}$ where each p_i is prime and $p_i \neq p_j$ for $i \neq j$. Then G is a CIA-group if and only if some Sylow subgroup of G is a CIA-group.*

PROOF. Suppose some Sylow subgroup, say P, of G is a CIA-group. Let H be a Sylow p-complement to P in G. Then $G = P \times H$ and we see G is a CIA-group after applying Lemma 2.7.

Suppose no Sylow subgroup of G is a CIA-group. Applying part (ii) of Theorem 3.3, we see that for each $P_i \in \mathrm{Syl}_{p_i}(G)$,

$$P_i = C_{p_i^{\beta_{i_1}}} \times \cdots \times C_{p_i^{\beta_{i_{t(i)}}}}$$

where $t(i)$ is the number of factors in P_i, $\beta_{i_k} \neq \beta_{i_j}$ for $k \neq j$, and $\sum_{k=1}^{t(i)} \beta_{i_k} = \alpha_i$. Assume G is a CIA-group. Then G is the union of proper subgroups isomorphic to $A_1 \times \cdots \times A_n$ where for each A_i,

$$A_i \simeq C_{p_i^{\gamma_{i_1}}} \times \cdots \times C_{p_i^{\gamma_{i_{t(i)}}}}$$

with $1 \leq p_i^{\gamma_{ij}} \leq p_i^{\beta_{ij}}$ for all i and j. Also, we must have $p_l^{\gamma_{lm}} < p_l^{\beta_{lm}}$ for at least one l and m. Let x be an element generating $C_{p_l^{\beta_{lm}}}$. Then x is a maximum root of P_l. There must exist a subgroup H of G isomorphic to $A_1 \times \cdots \times A_n$ containing x, and x must be an element of the Sylow p_l-subgroup of H, say H_l, which is isomorphic to A_l. By the structure of H_l, x must be equal to some power of an element, say y, of H_l, where y has order larger than that of x. This contradicts the fact that x is a maximum root of P_l. Thus G is not a CIA-group. □

COROLLARY 3.5. *A finite abelian group G is a CIA-group if and only if there exists a direct factor of G of the form $C_{p^n} \times C_{p^n}$ for some prime p and some integer n.*

NOTATION 3.6. For a set S of integers, D_S will be used to denote the direct product $D_S = \times_{n \in S} C_n$.

COROLLARY 3.7. *Let G be a finite abelian group. Then G is a CIA-group if and only if $P \not\simeq D_{\pi_{me}(P)}$ for some Sylow subgroup P of G.*

COROLLARY 3.8. *A finite cyclic group is not a CIA-group.*

COROLLARY 3.9. *Let G be a finite abelian CIA-group with $\{A_i\}_{i=1}^n$ a collection of isomorphic abelian subgroups of G. Then $\{A_i\}_{i=1}^n$ is a covering of G if and only if $A_i \simeq \left(\times_{P \in \mathrm{Syl}(G)} D_{\pi_{me}(P)}\right) \times H$ where H is any subgroup of G trivially intersecting and not complementing $\left(\times_{P \in \mathrm{Syl}(G)} D_{\pi_{me}(P)}\right)$ in G.*

We will end this section with two results on nilpotent CIA-groups. The next theorem will tell us that, ultimately, to characterize the nilpotent CIA-groups, one needs to characterize the CIA-groups of prime power order.

THEOREM 3.10. *Let G be a finite nilpotent group. Write $G = P_1 \times \cdots \times P_n$ where $P_i \in \mathrm{Syl}_{p_i}(G)$. Then G is a CIA-group if and only if G satisfies the following two conditions:*

(1) *Every Sylow subgroup of G is abelian or is a CIA-group;*
(2) *At least one Sylow subgroup of G is a CIA-group.*

PROOF. Suppose every Sylow subgroup of G is abelian or is a CIA-group and further suppose G possesses at least one Sylow subgroup which is a CIA-group. Then we can decompose G into $G = H \times K$, where H is abelian and K is the direct product of CIA-groups. We observe that K is a CIA-group by Lemma 2.7 and then, after applying Lemma 2.7 once more, we see that G is a CIA-group.

Suppose G is a CIA-group. Write $G = \bigcup_{i=1}^m A^i$ where the A^i's are isomorphic abelian groups. Since G is nilpotent, we can write each A^i as $A^i = A_1^i \times \cdots \times A_n^i$ where each $A_j^i \leq P_j$. Let $x \in P_j$. Then $x \in A^i$ for some i and, since x is of p_j-power order, it follows that $x \in A_j^i$. Hence $P_j = \bigcup_{i=1}^m A_j^i$. If A_j^i is proper in P_j, then P_j is a CIA-group. If $A_j^i = P_j$, then P_j is abelian. Thus (1) holds.

To show (2) holds, we can suppose P_j is abelian for all j. Then G is an abelian CIA-group and hence possesses a Sylow subgroup which is a CIA-group by Theorem 3.4. □

COROLLARY 3.11. *Every finite Hamiltonian group is a CIA-group.*

PROOF. This follows from Theorem 3.10, Lemma 2.7, and the fact that the quaternion group of order 8 is a CIA-group, □

4. Non-CIA-groups

In this section we will give several examples of groups which are not CIA-groups. First, we will look at a theorem and a corollary which will aid in showing certain groups are not CIA-groups.

THEOREM 4.1. *Let G be a finite CIA-group and write $G = \bigcup_{i=1}^{n} H_i$, where the H_i's are proper isomorphic abelian groups. Then, given any $x \in G$, there exists an $a \in G$ with $|a| = \exp(G)$ and $a \in C_{H_i}(x)$ for some i. In particular, $\exp(G) = \exp(H_i)$ for all i.*

PROOF. Given any $x \in G$, then x is in at least one H_i. So $|x|$ divides $|H_i|$ for all i and thus $\exp(G)$ divides $|H_i|$ for all i. So for each H_i there is an $a_i \in H_i$ with $|a_i| = \exp(G)$. Given any $x \in G$, then x is in at least one H_i and $a_i \in C_{H_i}(x)$. □

COROLLARY 4.2. *Let G be a CIA-group. If $x \in G$ and p is a prime divisor of $|G|$, then p is a divisor of $|C_G(x)|$. Moreover, $\exp(G)$ is a divisor of $|C_G(x)|$.*

The converse of Theorem 4.1 does not hold. Let $G = C_4 \times C_2$. Then G is not a CIA-group, but certainly if $x \in G$ then $\exp(G)$ divides $|C_G(x)|$, since G is abelian. Now we will look at several classes of groups which are not CIA-groups.

PROPOSITION 4.3. *There are no CIA-groups of square-free order.*

PROOF. Assume that G is CIA-group of square-free order. Let $G = \bigcup_{i=1}^{m} A_i$ with the A_i's isomorphic abelian subgroups of G. By Theorem 4.1, $\exp(G) = \exp(A_i)$. But G is of square-free order and so $|G| = \exp(G)$ giving us $G = A_i$. Hence G is not a CIA-group. □

PROPOSITION 4.4. *If $G = D_n$, the dihedral group of order $2n$, where $n > 2$, then G is not a CIA-group.*

PROOF. First note that D_4 is not a CIA-group. Let x and y be the generators of D_n, where $|x| = 2$ and $|y| = n \geq 3$. Then $|C_{D_n}(x)| \leq 4 \leq \exp(D_n)$. Supposing G is a CIA-group, we see from Theorem 4.1 that $|C_{D_n}(x)|$ is n or $2n$. If $|C_{D_n}(x)| = 2n$, then x is central in D_n which is not the case. So $|C_{D_n}(x)| = n$. We can assume $\exp(G) = n$ as well. Hence $n = 4$ and we have a contradiction. So D_n is not a CIA-group. □

PROPOSITION 4.5. *If G is a Frobenius group, then G is not a CIA-group.*

PROOF. Assume G is a Frobenius CIA-group. Write $G = HN$ with $H \cap N = 1$ where N is the Frobenius kernel of G. Let p be a prime dividing $|H|$ and let x be a nontrivial element of N. By Exercise 8.5.5 of [**18**], we have that p does not divide $|N|$. By Exercise 8.5.6 of [**18**], we have that $C_G(x) \leq N$. However, by Corollary 4.2, we have that p divides $|C_G(x)|$ in which case p divides $|N|$. Hence there are no Frobenius groups which are CIA-groups. □

PROPOSITION 4.6. *If $G = S_n$, the symmetric group of degree n, where $n \geq 2$, then G is not a CIA-group.*

PROOF. By Corollary 3.8, S_2 is not a CIA-group. Suppose $n > 2$. If n is odd, let x be an n-cycle so that $|C_{S_n}(x)| = n$. If n is even, let x be an $(n-1)$-cycle so that $|C_{S_n}(x)| = n - 1$. It is now evident that S_n is not a CIA-group. $\quad\square$

PROPOSITION 4.7. *If $G = A_n$ the alternating group of degree n where $n \geq 3$, then G is not a CIA-group.*

PROOF. By Corollary 3.8, A_3 is not a CIA-group. Suppose $n > 3$. If n is odd, let x be an n-cycle so that $|C_{A_n}(x)| = n$. If n is even, let x be an $(n-1)$-cycle so that $|C_{A_n}(x)| = n - 1$. It is now evident that A_n is not a CIA-group. $\quad\square$

NOTATION 4.8. For a finite group G, let us agree to denote by $\pi_e(G)$ the set of all orders of elements in G, that is

$$\pi_e(G) = \{|a| : a \in G\}.$$

PROPOSITION 4.9. *If G is a simple group, then G is not a CIA-group.*

PROOF. By Theorem 1.3, there are no infinite simple CIA-groups, since CIA-groups are central-by-finite. Assume G is a finite simple CIA-group. Note that by Theorem 4.1, we have $\exp(G) \in \pi_e(G)$. Then G must be abelian by Theorem 6 of [20]. Hence G is cyclic and Corollary 3.8 gives us a contradiction showing us that there are no simple CIA-groups. $\quad\square$

5. CIA-groups

All the groups that we will look at in this section are finite. We now have several examples of groups which are not CIA-groups. However, the list of CIA-groups is still large, for in this section we will see that every group is a direct factor of a CIA-group. We we will end the section with several examples of CIA-groups (mostly found with the use of GAP [10]) and concern ourselves with the question, "When is the center of a CIA-group 'small'?"

THEOREM 5.1. *If G is a finite group, then $G \times D_{\pi_e(G)}$ is a CIA-group covered by abelian subgroups isomorphic to $D_{\pi_e(G)}$.*

PROOF. Let $H = G \times D_{\pi_e(G)}$ and $\Pi_1 : H \to G$ be the projection homomorphism. Given $h \in H$, let $g = \Pi_1(h)$, $k = |g|$, Π_t be the projection homomorphism from H to the kth-order factor of $D_{\pi_e(G)}$, $l = \Pi_t(h)$, and $A = D_{\pi_e(G)-\{k\}} \times \langle gl \rangle$. Then $h \in A$ and $A \simeq D_{\pi_e(G)}$. $\quad\square$

COROLLARY 5.2. *Every finite group is a direct factor of a CIA-group.*

COROLLARY 5.3. *If G is a finite group and $\exp(G) = \prod_{n \in \pi_e(G)} n$, then G is a direct factor of a CIA-group which is covered by copies of the cyclic group $C_{\exp(G)}$.*

PROOF. $D_{\pi_e(G)} = C_{\exp(G)}$. $\quad\square$

EXAMPLE 5.4. Let A_5 be the alternating group on 5 elements and let $G = A_5 \times C_{30}$. Using Theorem 5.1 we see that G is a non-solvable CIA-group and is covered by copies of C_{30}.

For the remainder of this section, we will concern ourselves with the center of a CIA-group. In particular, we are interested in finding examples of CIA-groups with a "small" center.

DEFINITION 5.5. Let us say that a group G has a *small center* if $\exp(Z(G)) < \exp(G)$.

REMARK 5.6. If $\exp(G) > |Z(G)|$, then G has a small center.

EXAMPLE 5.7. The quaternion group Q_8 is a CIA-group with a small center.

The next example shows that there exist p-groups with a small center, for any odd prime p, which are CIA-groups.

EXAMPLE 5.8. Let $G = \langle x, y \mid x^{p^2} = y^{p^2} = 1, x^y = x^{p+1} \rangle$ with p an odd prime. Then the center of G is $\langle x^p \rangle \times \langle y^p \rangle$ and the exponent of G is p^2. Also, G is a CIA-group with small center and G is covered by copies of C_{p^2}.

PROOF. Note that G is a an extension of C_{p^2} by itself and thus has order p^4.

Let us first show G has $\langle x^p \rangle \times \langle y^p \rangle$ as its center. We have $(x^p)^y = (x^y)^p = (x^{p+1})^p = x^{p^2+p} = x^p$ and hence $x^p \in Z(G)$. Note $x^y = x^{p+1}$ says that $[x, y] = x^p$. Hence $[y, x] = x^{-p}$ and one can deduce that $y^x = yx^{-p}$. Thus one has $(y^p)^x = (y^x)^p = (yx^{-p})^p = y^p x^{-p^2} = y^p$ so that $y^p \in Z(G)$. Now $|Z(G)| \neq p^3$ or else $G/Z(G)$ is cyclic in which case G is abelian. So the center of G must be $\langle x^p \rangle \times \langle y^p \rangle$.

Note that $G/Z(G)$ must be elementary abelian of order p^2 so that G has class 2. By Lemma 3.9 in [**11**], we have that $(gh)^p = g^p h^p$ for all g and h in G. In particular, we see that G has exponent p^2.

To show that G is a CIA-group covered by copies of C_{p^2}, we only need to verify that each element $g \in G$ of order p is in a cyclic group of order p^2. Let g be of order p in G. If g is a power of x or a power of y, then we have nothing to show since g would be in $\langle x \rangle$ or $\langle y \rangle$. So write $g = x^i y^j$ with both i and j not divisible by p^2. Note $1 = g^p = x^{ip} y^{jp}$. So p divides both i and j. Write $i = tp$ and $j = kp$ and note that p does not divide t nor k. Hence $h = x^t y^k$ has order p^2 and $g \in \langle h \rangle$. □

The following theorems will allow for some interesting examples of CIA-groups.

THEOREM 5.9. *Let H be a group of exponent p for some prime p. Let Q_8 denote the quaternion group of order 8. If $G = HQ_8$ where each element of H commutes with an element of order 4 in Q_8, each element of Q_8 commutes with an element of order p in H, and the center of G has order divisible by 2, then G is a CIA-group covered by copies of C_{4p}.*

PROOF. Firstly, note that since the center of G has order divisible by 2, we have that G contains a unique element of order 2 which is found in each conjugate of Q_8. Secondly, note that the hypotheses imply that every element of order p commutes with one of order 4 and that every element of order 4 commutes with one of order p. These facts are easily verified using Sylow's Theorem.

Each element of G is of order 1, 2, 4, p, $2p$, or $4p$. It is clear that the elements of order 1, 2, and $4p$ can be found as elements in certain cyclic groups of order $4p$. The identity and the unique element of order 2 will be an any subgroup of order $4p$ and any element of order $4p$ will be in the group generated by itself.

Let g be an element of order 4 and let h be of order p such that g and h commute. Then gh is of order $4p$ and we have $g \in \langle g^p \rangle = \langle (gh)^p \rangle \le \langle gh \rangle$.

Let g be an element of order $2p$. Then g^2 has order p and g^p has order 2. Since g^2 has order p, we know g^4 has order p as well. So $g^4 g^p$ must be of order $2p$ and so $g \in \langle g^4 g^p \rangle$. Now, let k be an element of order 4 commuting with g^2. So $g^2 k$

is of order $4p$ and $k^2 = g^p$. We have $g \in \langle g^4 g^p \rangle = \langle g^4 k^2 \rangle = \langle (g^2 k)^2 \rangle \leq \langle g^2 k \rangle$. This completes the proof. ☐

THEOREM 5.10. *Let H and K be groups such that $\exp(H) = p$ and $\exp(K) = q$ where p and q are distinct primes. If $G = HK$ where every element in K commutes with an element of order p in H and every element in H commutes with an element of order q in K, then G is a CIA-group covered by copies of C_{pq}.*

PROOF. As in the proof of Theorem 5.9, Sylow's Theorem and the hypotheses imply that every element of order p commutes with one of order q and every element of order q commutes with one of order p. Since G is a product of a Sylow p-subgroup of $\exp(H) = p$ and a Sylow q-subgroup of $\exp(K) = q$, every $g \in G$ has order 1, p, q, or pq. Let g be an element of order p and let k be an element of order q such that g and k commute. Then gk has order pq and $g \in \langle g^q \rangle = \langle (gk)^q \rangle \leq \langle gk \rangle$. Likewise, if g is an element of order q commuting with h an element of order p, then $g \in \langle gh \rangle$. ☐

The next example is a nonabelian CIA-group of odd order having a small center.

EXAMPLE 5.11. Let

$$H = \langle a, b, c \mid a^7 = b^7 = c^7 = 1, a^c = a, b^c = b, a^b = ac \rangle$$

be the nonabelian group of order 7^3 and exponent 7 and let

$$K = \langle d, e, f \mid d^3 = e^3 = f^3 = 1, d^f = d, e^f = e, d^e = df \rangle$$

be the nonabelian group of order 3^3 and exponent 3. Define the group G as a semidirect product $H \rtimes K$ where the action of K on H is defined as follows; e and f centralize H, $a^d = a^2$, $b^d = b$, and $c^d = c^2$. So, in short, we have G is the following:

$$\left\langle a,b,c,d,e,f \;\middle|\; \begin{array}{l} a^7=b^7=c^7=d^3=e^3=f^3=1, \ a^c=a, \ b^c=b, \ a^b=ac, \\ d^f=d, \ e^f=e, \ d^e=df, \ a^e=a, \ b^e=b, \ c^e=c, \\ a^d=a^2, \ b^d=b, \ c^d=c^2, \ a^f=a, \ b^f=b, \ c^f=c \end{array} \right\rangle.$$

It is apparent from the relations that the center of K is $\langle f \rangle$ and, moreover, $\langle f \rangle$ is in the center of G. Note from the relations that the center of H must be $\langle c \rangle$. If the center of G contains a subgroup of order 7 then it must contain $\langle c \rangle$. However c does not commute with d. So the center of G is $\langle f \rangle$ and is of order 3.

Clearly f commutes with any $h \in H$. Also, it is apparent from the relations that b commutes with any $k \in K$. So by Theorem 5.10 we see G is a CIA-group covered by copies of C_{21} with a center of order 3. Thus G is a non-nilpotent CIA-group of odd order with a small center.

EXAMPLE 5.12. Let

$$K = \langle a, b, c \mid a^p = b^p = c^p = 1, a^c = a, b^c = b, a^b = ac \rangle$$

be the nonabelian group of order p^3 and exponent p and let

$$Q_8 = \langle x, y \mid y^4 = 1, x^2 = y^2, x^y = x^{-1} \rangle$$

be the quaternion group of order 8. Let us define two groups, G and H, as semidirect products $K \rtimes Q_8$ where the action of Q_8 on K is defined as follows; $a^x = a$, $b^x = b^{-1}$, and $c^x = c^{-1}$ for G, $a^x = a^{-1}$, $b^x = b^{-1}$, and $c^x = c$ for H, and y centralizes K for both G and H. So, in short, we have the following two groups:

$$G = \left\langle a,b,c,x,y \middle| \begin{array}{l} a^p=b^p=c^p=x^4=y^4=1,\ x^2=y^2,\ a^c=a, \\ b^c=b,\ a^b=ac,\ x^y=x^{-1},\ a^x=a,\ b^x=b^{-1}, \\ c^x=c^{-1},\ a^y=a,\ b^y=b,\ c^y=c \end{array} \right\rangle,$$

$$H = \left\langle a,b,c,x,y \middle| \begin{array}{l} a^p=b^p=c^p=x^4=y^4=1,\ x^2=y^2,\ a^c=a, \\ b^c=b,\ a^b=ac,\ x^y=x^{-1},\ a^x=a^{-1},\ b^x=b^{-1}, \\ c^x=c,\ a^y=a,\ b^y=b,\ c^y=c \end{array} \right\rangle.$$

From the relations used to define both G and H, one sees that a, b, and c each commute with y. Hence each element of K commutes with an element of order 4 in Q_8. Also, a and x commute in G and c and x commute in H. Hence each element of Q_8 commutes with an element of order p in K. Also the centers of G and H have orders divisible by 2, since they contain $\langle x^2 \rangle$. So the hypotheses of Theorem 5.9 are satisfied. Hence we see that both G and H are CIA-groups covered by copies of C_{4p}.

Also, the relations make it clear that $\langle x^2 \rangle \leq Z(G)$ and $\langle x^2, c \rangle \leq Z(H)$. Since Q_8 has no element of 4 in its center, neither does G nor H. Suppose $g \in Z(G)$ and $|g| = p$. Then $g \in K$ and so $g \in Z(K) = \langle c \rangle$. So g is c or a p'-power of c. However, c nor any p'-power of c commutes with x as the relation $c^x = c^{-1}$ shows. Now suppose $g \in Z(H)$ and $|g| = p$. Then $g \in K$ and so $g \in Z(K) = \langle c \rangle$. We can conclude that $Z(G) = \langle x^2 \rangle$ and $Z(H) = \langle x^2, c \rangle$. So, not only are G and H CIA-groups, they are CIA-groups with small centers.

It seems likely that by using Theorem 5.10 and a product of two "large" groups, one of exponent p and another of exponent q with q dividing $p - 1$, one will be able to construct a centerless CIA-group. However, we have been unable to construct such a group. It is still an open question as to whether centerless CIA-groups exist.

6. GAP Documentation

Let us briefly elaborate on the elementary, yet crucial, role GAP [10] played in this research. Without knowing much about CIA-groups, we looked to examples to help us understand how CIA-groups arise and to understand their structure. GAP was used to determine which groups in GAP's "SmallGroups" library were CIA-groups. The following function was defined in GAP:

```
# Determines if G is the union of its subgroups isomorphic to S.
#
IsUnionOfIsomorphicSubgroups:=
    function(G,S)
        local c;        # All subgroups of G isomorphic to S.
        c:=Filtered(ConjugacyClassesSubgroups(G),
            C->IdGroup(S)=IdGroup(Representative(C)));
        return IsSubset(Concatenation(
            List(Concatenation(List(c,Elements)),Elements)),G);
end;
```

This function allows one to determine if a given group G can be written as the union of subgroups all isomorphic to S. Note that no attempt was made to ensure this was done in the quickest manner possible. We hoped that several small examples would be sufficient for us to gain a better understanding of CIA-groups and hence memory issues would not be a problem. The following GAP function allows one to determine which small groups, in a certain range of orders, are CIA-groups:

```
# Computes CIA groups for all groups whose order
# is between low to high.
#
CIAGroups:=
  function(low,high)
      local i,j,              # index variables
            G,Agrps,          # Group in question and possible abelian
                              # subgroups of G
            A;                # Abelian groups
      for i in [low..high] do
          for j in [1..NumberSmallGroups(i)] do
              G:=SmallGroup(i,j);
              Agrps:=Filtered(AllGroups([1..Order(g)-1]),
                      H->(IsAbelian(H) and
                      exponent(H)=Exponent(G)));
              for A in Agrps do
                  if IsUnionOfIsomorphicSubgroups(G,A) then
                      Print(IdGroup(G)," ",IdGroup(A)," ",
                          StructureDescription(G),
                          " is covered by copies of ",
                          StructureDescription(A),"\n");
                  fi;
              od;
          od;
      od;
  end;
```

It should be mentioned that this code will not run if 1024 is between low and high because groups of this order are excluded from the "SmallGroups" library. In the above, Agrps is a list of all groups H with order less than the order of G where H is abelian and exp(H)=exp(G). The code checks if G is a CIA-group by making use of the previously defined function IsUnionOfIsomorphicSubgroups. The output makes use of StructureDescription which tells us the structure of G and the subgroups covering G. Not only do wee see which groups are CIA-groups, but we see the different possible coverings used.

The results in Section 3 were discovered by examining the data that GAP provided us. Also, it became apparent that several classes of groups were missing from our lists of CIA-groups. This lead us to the results in Section 4. We also noticed that even though certain groups, like the dihedral, symmetric, and alternating groups, weren't CIA-groups, they kept appearing as direct factors of certain CIA-groups. This lead us to the result in Theorem 5.1. At one point in our research we conjectured that there were no centerless CIA-groups. After all, GAP had not found any. This lead us to search for CIA-groups with small centers. Several groups fitting the hypotheses in Example 5.8 and Theorems 5.9 and 5.10 were found by GAP. After examining these examples, we generalized and discovered the results found in Example 5.8 and Theorems 5.9 and 5.10. All of this was done essentially with the two GAP functions defined in this section.

It should be mentioned that the group G found in Example 5.11 was constructed in GAP as a free group modulo the relations given. We then used GAP to verify that this group was a CIA-group. Of course, Theorem 5.10 does this work for us now.

Acknowledgment. The authors would like to thank the referee for the careful reading of the initial submission and for the suggestions which have lead to a much better presentation of the material.

References

[1] H. Bell, A. Klein and L.C. Kappe, An analogue for rings of a group problem of P. Erdös and B.H. Neumann, *Acta Math. Hungar.* **77** (1997), 57-67.

[2] H. U. Besche, B. Eick and E. A. OBrien, The groups of order at most 2000, *Electron. Res. Announc. Amer. Math. Soc.* **7** (2001), 14.

[3] M. Brodie, Finite n-coverings of groups, *Arch. Math.* **63** (1994), 385-392.

[4] M. Brodie, R. Chamberlain and L.C. Kappe, Finite coverings by normal subgroups, *Proceedings AMS* **104** (1988), 669-674.

[5] M. Brodie and L.C. Kappe, Finite coverings by subgroups with a given property, *Glasgow Math. J.* **35** (1993), 179-188.

[6] M. Bruckheimer, A.C. Bryan and A. Muir, Groups which are the union of proper subgroups, *MAA Monthly* **77** (1970), 52-57.

[7] T. Foguel and L.C. Kappe, On loops covered by subloops, *Expositiones Mathematicae* **23**, (2005), 255-270.

[8] T. Foguel, Simple Power Associative Loops with Exactly One Covering, *Results in Mathematics* **45**, (2004), 241-245.

[9] T. Foguel, Amalgam of a loop over an idempotent quasigroup , *Quasigroups and Related Systems* **13**, (2005)no. 1, 99-104.

[10] The GAP Group, GAP – Groups, Algorithms, and Programming, Version 4.4, 2002, (http://www.gap-system.org)

[11] D. Gorenstein, *Finite Groups*, Chelsea Publishing Co., New York, 1980.

[12] S. Haber and A. Rosenfeld, Groups as unions of proper subgroups, *MAA Monthly* **66** (1959), 491-494.

[13] L.C. Kappe, Finite coverings by 2-Engel groups, *Bull. Austral. Math. Soc.* **38**, (1988), 141-150.

[14] L.C. Kappe, J.C. Lennox, and J. Wiegold, An analogue for semigroups of a group problem of P. Erdös and B.H. Neumann, *Bull. Austral. Math. Soc.* **63** (2001), 59-66.

[15] B.H. Neumann, Groups covered by finitely many cosets, *Publ. Math. Debrecen* **3** (1954), 227-242.

[16] B.H. Neumann, Groups covered by permutable subsets, *J.London Math. Soc.* **29** (1954), 236-248.

[17] D.J.S. Robinson, *Finiteness conditions and generalized soluble groups, Part I*, Springer Verlag, New York, 1972.

[18] D.J.S. Robinson, *A course in the theory of groups,* Springer-Verlag New York, 1996.

[19] G. Scorza, Gruppi che possone pensarsi come somma di tre sottogruppi, *Boll. Un. Mat. Ital.* **5** (1926), 216-218.

[20] W. Shi, Finite groups defined by the sets of their element orders, *Xinan Shifan Daxue Xuebao Ziran Kexue Ban* **22** (1997), no. 5, 481-486.

TUVAL S. FOGUEL, DEPARTMENT OF MATHEMATICS, AUBURN UNIVERSITY AT MONTGOMERY, P.O. BOX 244023, MONTGOMERY, AL 36124-4023
E-mail address: tfoguel@mail.aum.edu

MATTHEW F. RAGLAND, DEPARTMENT OF MATHEMATICS, AUBURN UNIVERSITY AT MONT-GOMERY, P.O. BOX 244023, MONTGOMERY, AL 36124-4023
E-mail address: mragland@mail.aum.edu

Contemporary Mathematics
Volume **470**, 2008

On Some Subnormality Conditions in Metabelian Groups

David Garrison and Luise-Charlotte Kappe

Dedicated to Derek J.S. Robinson on the occasion of his seventieth birthday.

ABSTRACT. We consider several classes of groups, B_n, U_n, and $U_{n,m}$, defined with respect to certain subnormality conditions. A group is in B_n if all its cyclic subgroups are n-subnormal, in U_n if all subgroups are n-subnormal and in $U_{n,m}$ if all subgroups that are nilpotent of class at most m are n-subnormal. We investigate the interrelations between these classes in general for $n = 2$ and for metabelian groups in the case when $n \geq 3$. We show that $U_{2,2} = U_2$ for non-torsion groups and with the help of GAP we give an example of a torsion group in $U_{2,2}$ which is not in U_2. We now turn to the case $n > 2$. Restricting ourselves to metabelian non-torsion groups without elements of prime orders $p \leq n-1$, we show that $B_n = U_{n,n-1}$ when $n+1$ is a prime and that $B_n = U_n$ otherwise. For the former case we give an example of a group that is in $U_{n,n-1}$ but not in U_n. For metabelian torsion groups we show that $B_n = U_{n,n-1}$, if the group does not contain elements of order $p \leq n$. With the help of GAP we provide various examples showing that under our assumptions U_n is a proper subclass of $U_{n,n-1}$ and the restrictions on the element orders cannot be dispensed with.

1. Introduction

Let G be a group. A subgroup H in G is said to be subnormal, if there exists a finite series $H = H_0, H_1, \ldots, H_{k-1}, H_k = G$ such that

$$H = H_0 \lhd H_1 \lhd \ldots \lhd H_{k-1} \lhd H_k = G.$$

Following [14], we say H is subnormal of defect n, if the length of the shortest such series is n. We say that H is n-subnormal, denoted by $H \lhd^n G$, if H is subnormal of defect at most n. A group in which every cyclic subgroup is n-subnormal is called an n-Baer group. The class of n-Baer groups is denoted by B_n and the class of groups with all subgroups n-subnormal is denoted by U_n. By a result of Roseblade [15], groups in U_n are nilpotent, the bound for the nilpotency class depending on n. The exact value for the class is only known for $n = 1$ and 2. For $n = 1$, we have $B_1 = U_1$ and the groups in this class coincide with the Dedekind groups (see [1], [3] or [14, Theorem 6.1.1] for easier reference), hence the nilpotency class is at most 2.

2000 *Mathematics Subject Classification.* Primary 20D35, 20E15, Secondary 20F18, 20F45.
Key words and phrases. n-Baer groups, n-subnormal, metabelian groups.

As shown in [**16**] and [**13**], groups in B_2 have niplotency class not exceeding 3. However not all B_2-groups are U_2-groups [**16**]. For $n \geq 3$, groups in B_n are not necessarily nilpotent, and therefore the class U_n is a proper subclass of B_n.

There are classes lying between B_n and U_n, such as the class of groups where all subgroups that are of nilpotency class not exceeding m are n-subnormal. We denote this class by $U_{n,m}$. The topic of this paper is the investigation of the interdependencies of the properties that define the classes U_n, $U_{n,m}$, and B_n within the class of metabelian groups and for $n = 2$ in general.

In [**11**] and [**12**] the following four properties were considered: (i) $G \in B_2$; (ii) $G \in U_{2,1}$; (iii) $G \in U_{2,2}$; (iv) $G \in U_2$. Obviously, (iv) $\rightarrow$ (iii) $\rightarrow$ (ii) $\rightarrow$ (i). It was shown that for 2-generator groups the four conditions are equivalent. Furthermore, if G is a non-torsion group or a torsion group without involutions, then (i) and (ii) are equivalent and there exists a finite 2-group in B_2 which is not in $U_{2,1}$. An example of a non-torsion group with elements of order 3, which already can be found in [**2**], shows that (ii) not always implies (iii). Finally, it was shown that (iii) and (iv) are equivalent for non-torsion groups, raising the question whether this is also true for torsion groups. Here we show that this is not the case by exhibiting a finite group in $U_{2,2}$ which is not in U_2 (Theorem 3.1). This group, exhibited in Example 3.2, is a 5-group of order 5^{11} and nilpotency class 3. It was constructed using GAP [**4**]. The properties of being in B_2 and $U_{2,2}$ were verified first for a special subset of elements with the help of GAP and then extended to all elements using the fact that the group has nilpotency class 3. Although it would have been desirable to have GAP validate all of the required properties, our group is too large to have GAP do this in a reasonable amount of time. We suspect that a construction of a p-group, $p > 5$, similar to the 5-group of Example 3.2, leads to a p-group in $U_{2,2}$ but not in U_2. But verification of the relevant properties even for the select subset in a reasonable amount of time poses a problem.

Turning to n-subnormality, $n \geq 3$, the situation is more complex. For one thing, groups in B_n need no longer be nilpotent and no explicit bounds for the nilpotency class of the groups in U_n are known. It can be easily seen that every n-Baer group is $(n+1)$-Engel. In [**10**] it was shown that a non-torsion group in U_3 without involution has class at most 4, and that for metabelian groups, every non-torsion n-Baer group is n-Engel. Metabelian groups turn out to be more accessible when it comes to verifying subnormality. In [**11**, Corollary 3] it was shown that a group G has all subgroups 2-subnormal if and only if $[x, h_1, h_2] \in \langle h_1, h_2 \rangle$ for all $x, h_1, h_2 \in G$. We will generalize this criterion for 2-subnormality to n-subnormality in the case of metabelian groups (Corollary 2.6). Having class restrictions for metabelian Engel groups leads to class bounds for the groups in B_n in the absence of elements of prime order for small primes. In [**5**, Lemma 2.6] it was shown that every metabelian n-Engel group is an n-Baer group. This together with restrictions on the nilpotency class of metabelian n-Engel groups is given in [**9**, Theorem 3] and [**5**, Corollary 2.7], lead to the following restrictions on the nilpotency class $cl(G)$ of metabelian n-Baer groups.

THEOREM 1.1. *Let $n \geq 3$ and G a metabelian n-Baer group without elements of order $p \leq n - 1$.*

(i) *If G is a non-torsion group, then $cl(G) \leq n$, if $n + 1$ is not a prime, and $cl(G) \leq n + 1$, if $n + 1$ is prime.*

(ii) *If G is a torsion group, then $cl(G) \leq n+1$, if $n+1$ is not a prime or if $n+1$ is prime and G contains no elements of order $n+1$, and $cl(G) \leq n+2$, if $n+1$ is prime and G contains elements of order $n+1$.*

We mention here that no upper bound for the nilpotency class of the groups in B_n exists, if they contain elements of order $p \leq n-1$ (see [6]).

In Theorem 4.2 we completely characterize the interdependencies of the properties that determine the classes B_n, U_n, and $U_{n,m}$ in the case of metabelian non-torsion groups without elements of order $p \leq n-1$ and $n \geq 3$, namely $B_n = U_n$ for $n+1$ not a prime, and $B_n = U_{n,n-1}$, if $n+1$ is prime, and this result is sharp. By Example 5.6, there exists a non-torsion group in B_n, hence in $U_{n,n-1}$, but not in U_n when $n+1$ is prime. Example 5.6 is an adaptation of Example 3 in [9].

For the torsion case, our results are summarized in Theorem 4.3. If $n \geq 3$ and in the absence of elements of order $p \leq n$, we have $B_n = U_{n,n-1}$. The torsion version of Example 5.6 shows that there exists a p-group, $p = n+1$, which is in B_n, hence in $U_{n,n-1}$, but not in U_n. Example 5.5 provides a 5-group in B_3, but not in U_3. Similar to Example 3.2, this group is constructed using GAP and the relevant properties are verified with the help of GAP and then extended to the rest of the elements using the class restriction of the group and the fact that it is metabelian.

The exclusion of elements of order $p \leq n-1$ comes as no surprise, since there are metabelian p-groups, $p \leq n-1$, which are n-Engel, hence n-Baer by [5, Lemma 2.6], with no bound on the nilpotency class. A well known example by Gruenberg [6] is a centerless metabelian p-group which is $(p+1)$-Engel and it has an elementary abelian commutator subgroup and commutator factor group. Suitable sections of this group should provide examples showing that the absence of elements of order p, where $p \leq n-1$, is needed in the assumptions of Theorem 4.3. Adding an infinite direct factor to any of these groups leads to a metabelian non-torsion group which is again n-Engel and hence n-Baer by [5, Lemma 2.6]. Such a group provides us with examples showing that the same assumption is needed in Theorem 4.2. However it is not easy to derive explicit examples from this approach. For the cases $n = 3$ and $p = 2$, as well as $n = 4$ and $p = 2$ and 3, we provide explicit groups which are in B_3 and B_4, respectively, but have abelian subgroups which are not n-subnormal, $n = 3$ and $n = 4$, respectively (Examples 5.1, 5.3, and 5.4). Lastly, Example 5.2 is a 3-group in B_3 having an abelian subgroup which is not 3-subnormal. This shows that for $n = 3$ we need to require the absence of elements of order 3 in the assumption of Theorem 4.3. We suspect that for all $n = p$, groups with the same properties exist, but they are too big for this to be verified with the help of GAP in a reasonable amount of time. Examples 5.1 through 5.4 have all been constructed by GAP and the verification of their properties has been done with the help of GAP in a similar manner as described for Example 3.2 and 5.5.

2. Preparatory results

For the convenience of the reader, this section contains several formulas and preparatory results to be applied in the rest of the paper. We make use of the following standard notation for commutators: $x^{-1}y^{-1}xy = [x, y]$, and recursively, $[x, _ny] = [[x, _{n-1}y], y]$ for $x, y \in G$ and $n \in \mathbb{N}$, where $[x, _0y] = x$. In addition, we observe that for a metabelian group G the Hall-Witt identity reduces to $1 = [a, b, c][c, a, b][b, c, a]$ for $a, b, c \in G$ and we always have $[a, b, c] = [a, c, b]$, if $a \in G'$ and $b, c \in G$.

The first lemma provides two expansion formulas for metabelian groups. The proofs can be found in [8].

LEMMA 2.1. *Let G be a metabelian group, $n \in \mathbb{N}$ and $a, b \in G$. Then*

(i)
$$(ab^{-1})^n = a^n \left(\prod_{0 < i+j < n} [a, {}_ib, {}_ja]^{\binom{n}{i+j+1}} \right) b^{-n},$$

(ii)
$$[a, b^n] = \prod_{i=1}^{n} [a, {}_ib]^{\binom{n}{i}}.$$

Our next lemma is a commutator expansion formula. Since we are working in metabelian groups, the lemma will use additive notation. The proof is omitted, since it is simply an application of induction on n and the usual commutator identities for metabelian groups.

LEMMA 2.2. *Let G be a metabelian group, $u \in G'$, $v, w \in G$ and $n \in \mathbb{N}$. Then*

$$[u, {}_nvw] = \sum_{i=0}^{n} \sum_{j=0}^{i} \binom{n}{i} \binom{i}{j} [u, {}_iv, {}_{n-j}w] = \sum_{i=0}^{n} \sum_{j=0}^{i} \binom{n}{i} \binom{i}{j} [u, {}_{n-j}v, {}_iw].$$

The following familiar argument is often called the Vandermonde argument, since the determinant of the system in question is a Vandermonde determinant. To facilitate its application, we shall state it here as a lemma without proof.

LEMMA 2.3. *Let W be a module over the integers without elements of order at most m. Then the system of equations*

$$0 = \sum_{i=1}^{k} j^i w_i, \quad j = 1, \ldots, k,$$

has only the trivial solution $w_i = 0$, for $i = 1, \ldots, k$, provided $k \leq m$.

Let H be an arbitrary subgroup of a group G. We define the series, $(H^{G,i})_{i=1}^{\infty}$, of successive normal closures by induction as follows:

$$H^{G,0} = G, \quad H^{G,i+1} = H^{H^{G,i}},$$

where H^K denotes as usual the normal closure of H in K.

By [14, 13.1.3] we have $H^{G,i} = H[G, {}_iH]$ for the successive normal closures of H in G. If $H \lhd^n G$, then $H = H^{G,n}$ with n minimal. This together with the above leads to the following subnormality criterion.

LEMMA 2.4. *Let H be a subgroup of the group G, $n \in \mathbb{Z}$, $n \geq 0$. Then the subgroup H is n-subnormal in G if and only if $[G, {}_nH] \leq H$.*

The next lemma gives us an efficient way to determine $[G, {}_iH]$ in the case of metabelian groups.

LEMMA 2.5. *Let G be a metabelian group and H a subgroup of G. Then $[G, {}_iH] = \langle [g, h_1, \ldots, h_i] \mid g \in G, h_1, \ldots, h_i \in H \rangle$.*

PROOF. The cases $i = 0$ and $i = 1$ are trivial. For $i \geq 2$ we use the fact that for a metabelian group we have $[uv, g] = [u, g][v, g]$ for $u, v \in G'$ and $g \in G$. The inclusion $[G, {}_iH] \leq \langle [g, h_1, \ldots, h_i] \mid g \in G, h_1, \ldots, h_i \in H \rangle$ then follows by induction on i. The reverse inclusion is trivial. $\square$

The following corollary is an immediate consequence of the preceding two lemmas.

COROLLARY 2.6. *Let G be a metabelian group and H a subgroup of G. Then H is n-subnormal in G if and only if $[x, h_1, \ldots, h_n] \in H$ for all $x \in G$ and all $h_1, \ldots, h_n \in H$.*

The last lemma in this section is used repeatedly in Section 4 to show that subgroups of nilpotency class $n - 1$ are n-subnormal under certain conditions.

LEMMA 2.7. *Let p be prime and G be a metabelian group without elements of order $p \leq n$, where n is an integer with $n \geq 3$, and H a subgroup of G satisfying the following conditions:*

(i) $[x, {}_n a] \in H$;

(ii) $[x, {}_i a, {}_j b] = [x, {}_j b, {}_i a]$, $0 \leq i, j \leq n$, $i + j = n$;

(iii) $[x, a_1, \ldots, a_m] \in H$, $m \geq n + 1$,

for all $x \in G$, and all $a, b, a_1, \ldots, a_m \in H$. Then $[x, a_1, \ldots, a_n] \in H$ for all $x \in G$ and all $a_1, \ldots, a_n \in H$.

PROOF. Let G and H be as given in our claim. We first show $[x, a, {}_{n-1} b] \in H$ for all $x \in G$ and all $a, b \in H$. Using additive notation in G', direct expansion yields

$$[x, {}_n ab] = [x, a, {}_{n-1} ab] + [x, b, {}_{n-1} ab] + [x, a, b, {}_{n-1} ab].$$

Observing that $[x, a, b, {}_{n-1} ab] \in H \cap G'$ by (iii), leads to

$$(2.7.1) \qquad [x, {}_n ab] \equiv [x, a, {}_{n-1} ab] + [x, b, {}_{n-1} ab] \pmod{H \cap G'}.$$

Applying Lemma 2.2 to $[x, a, {}_{n-1} ab]$ and $[x, b, {}_{n-1} ab]$ and observing $[x, {}_s a, {}_t b] \in H$ for $s + t \geq n + 1$ by (iii), as well as $[x, {}_n a], [x, {}_n b] \in H$ by (i), we obtain from (2.7.1)

$$(2.7.2) \qquad \begin{aligned} [x, {}_n ab] &\equiv \sum_{i=0}^{n-2} \binom{n-1}{i} [x, {}_{i+1} a, {}_{(n-1)-i} b] \\ &+ \sum_{j=0}^{n-2} \binom{n-1}{j} [x, {}_{j+1} b, {}_{(n-1)-j} a] \pmod{H \cap G'}. \end{aligned}$$

By (ii), we have $[x, {}_{i+1} a, {}_{(n-1)-i} b] = [x, {}_{j+1} b, {}_{(n-1)-j} a]$, when $i + 1 = (n-1) - j$ and $(n-1) - i = j + 1$. Observing that $j = (n-2) - i$, and thus $\binom{n-1}{i} + \binom{n-1}{j} = \binom{n}{i+1}$, (2.7.2) reduces to

$$[x, {}_n ab] \equiv \sum_{i=0}^{n-2} \binom{n}{i+1} [x, {}_{i+1} a, {}_{(n-1)-i} b] \pmod{H \cap G'}.$$

Changing the indices of our summation in the above leads to

$$(2.7.3) \qquad \sum_{i=1}^{n-1} \binom{n}{i} [x, {}_i a, {}_{n-i} b] \in H \cap G'.$$

Next we substitute a^l, $l = 2, \ldots, n - 1$, for a in (2.7.3). Observing (iii), we can expand linearly mod $H \cap G'$ and arrive at

$$(2.7.4) \qquad \sum_{i=1}^{n-1} l^i \binom{n}{i} [x, {}_i a, {}_{n-i} b] \equiv 0 \pmod{H \cap G'} \text{ for } l = 1, \ldots, n - 1.$$

Since G has no elements of order $p \leq n$, we may apply Lemma 2.3 to the system of equations in (2.7.4) and get $[x, _ia, _{n-i}b] \in H$ for $i = 1, \ldots, n - 1$. In particular, we have $[x, a, _{n-1}b] \in H$, which is the first step of our induction proof.

Assume now that for $1 \leq k \leq n - 1$ we have $[x, c_1, \ldots, c_k, _{n-k}b] \in H$ for all $c_1, \ldots, c_k, b \in H$ and all $x \in G$. We will show that $[x, c_1, \ldots, c_{k+1}, _{n-(k+1)}b] \in H$ for all $c_1, \ldots, c_{k+1}, b \in H$. Direct expansion of $[x, c_1, \ldots, c_k, _{n-k}ab]$, using Lemma 2.2, (iii), and the induction hypothesis, leads to

$$\sum_{i=1}^{(n-k)-1} \binom{n-k}{i} [x, c_1, \ldots, c_k, _ia, _{(n-k)-i}b] \equiv 0 \pmod{H \cap G'}.$$

Substituting a^l for a and observing that G has no elements of order $p \leq n$, we obtain by Lemma 2.3 that $[x, c_1, \ldots, c_k, _ia, _{(n-k)-i}b] \in H$ for $i = 1, \ldots, (n-k)-1$. Setting $i = 1$ and $a = c_{k+1}$, yields $[x, c_1, \ldots, c_{k+1}, _{n-(k+1)}b] \in H$, the desired result.   □

3. Groups with all class 2 subgroups 2-subnormal

In this section we show that the classes $U_{2,2}$ and U_2 are equal for non-torsion groups and give an example of a torsion group in $U_{2,2}$ having a subgroup of nilpotency class 3 which is not 2-subnormal. Thus $U_{2,2}$ is not contained in U_2 in the case of torsion groups.

THEOREM 3.1. *Let G be a group where all subgroups of class at most 2 are 2-subnormal. If G is a non-torsion group, then all subgroups of G are 2-subnormal and there exists a torsion group K with all subgroups of class at most 2 being 2-subnormal, but not having all subgroups 2-subnormal.*

PROOF. Let G be a non-torsion group in which all subgroups of class at most 2 are 2-subnormal. Obviously, $G \in B_2$. By a result of Heineken [7], this is equivalent to being 2-Engel. Since G is metabelian, it suffices by Corollary 2.6 to show that $[x, a, b] \in H$ for $x \in G$ and $a, b \in H$, where $H \leq G$. Observing that any 2-generator 2-Engel group has class 2, our assumption that $G \in U_{2,2}$ yields $[x, a, b] \in \langle a, b \rangle$ for all $a, b \in G$. Thus we obtain $[x, a, b] \in H$ for $x \in G$ and all $a, b \in H$, proving our claim in case of a non-torsion group.

Now turning to the case of torsion groups, it is shown that the 5-group K of Example 3.2 has all class 2 subgroups 2-subnormal. But there exists a proper subgroup of class 3 which is not 2-subnormal.   □

EXAMPLE 3.2. *Let $K = G/M^G$, where*

$$G = \left\langle x, y, z \; \middle| \; \begin{array}{l} x^{125} = y^{125} = z^{125} = [x_1, x_2]^{25} = [x_1, x_2, x_3]^5 = 1, \\ x_1, x_2, x_3 \in \{x, y, z\}, \; G_4 = 1 \end{array} \right\rangle,$$

and M^G is the normal closure of

$$M = \langle x^5, [y, _2x], [z, _2x], [x, _2y]y^{-25}, [z, _2y]y^{25}, [x, _2z]z^{-25}, [y, _2z]z^{-25}, [y, _2xyz],$$
$$[x, _2xyz](xyz)^{-50} \rangle.$$

Then all subgroups of K, that are of class at most 2, are 2-subnormal, however not all subgroups are 2-subnormal.

PROOF. We will use GAP (see [4]) to construct the group and to validate some of its properties. Although it would be convenient to use GAP for verifying all of the required properties, our group is too large to have GAP do this in a reasonable amount of time. To validate some of the properties, we will have GAP do it for a small subset of elements, and then prove that the remaining elements of the group have the desired properties using what GAP has shown for the subset and other characteristics of the group.

We observe using GAP that G and K are 3-generator groups of class 3, hence metabelian, and have order 5^{23} and 5^{11}, respectively. Furthermore, $K = \langle a, b, c \rangle$ with $|a| = 5$, and $|b| = |c| = 125$, as well as $exp(K) = 125$ and $exp(K') = 25$.

First we will show that $K \in B_2$. Observing that $a^5, b^{25}, c^{25} \in K' \cap Z(K)$, every $k \in K$ can be represented as $k = gd$, where $d \in K'$ and $g = a^i b^j c^l$ for $1 \leq i \leq 5$, $1 \leq j, l \leq 25$. Using the GAP function $\mathtt{SubnormalSeries}$, we verify that $\langle g \rangle \vartriangleleft^2 K$, and in addition we show that if $|g| \leq 25$, we have $[w, {}_2 g] = 1$ for $w \in \{a, b, c\}$.

Now let $k = gd$, with $d \neq 1$. Observing that K has class 3 and $d \in K'$, we obtain by linear expansion $[w, {}_2 gd] = [w, {}_2 g]$ for $w \in \{a, b, c\}$. If $|g| \leq 25$ we have $[w, {}_2 k] = [w, {}_2 g] = 1$ and hence $\langle k \rangle \vartriangleleft^2 K$. If $|g| = 125$, then $[w, {}_2 g] \in \langle g \rangle \cap K'$ implies $[w, {}_2 g] \in \langle g^{25} \rangle$, since $\langle g \rangle \cap K' \leq \langle g^{25} \rangle$. By Lemma 2.1 and the facts that $d \in K'$ and $exp(K') = 25$, it follows $(gd)^{25} = g^{25} [g, d^{-1}]^{\binom{25}{2}} d^{25}$, hence $k^{25} = g^{25}$. We conclude $[w, {}_2 k] \in \langle k^{25} \rangle$. It follows $[h, {}_2 k] \in \langle k^{25} \rangle$ for all h and $k \in K$. Hence $K \in B_2$.

Next we will show that K has all class 2 subgroups 2-subnormal. By Corollary 2.6 it suffices to show that $[x, u, v] \in H$ for all $x \in K$ and $u, v \in H$, where H is a proper subgroup of G and the class of H is 2. If $u, v \in H$ and $cl(H) = 2$, then $\langle u, v \rangle$ has class 2 which is equivalent to being 2-Engel for the 2-generator case. Thus proving our claim reduces to showing $\langle u, v \rangle \vartriangleleft^2 K$, whenever $\langle u, v \rangle$ is 2-Engel.

Let $u, v \in K$. Then $u = gd_1$ and $v = hd_2$ with $d_1, d_2 \in K'$ and $g = a^i b^j c^l$, $h = a^r b^s c^t$ for $1 \leq i, r \leq 5$, and $1 \leq j, s, l, t \leq 25$. We proceed as follows: we will have GAP verify that if $\langle g, h \rangle$ is 2-Engel, then $\langle g, h \rangle \vartriangleleft^2 K$. In addition we have GAP check that $\langle [w, g, h], [w, h, g] \rangle \leq \langle g^{25}, h^{25} \rangle$ for $w \in \{a, b, c\}$.

It remains to be shown that if $\langle u, v \rangle$ is 2-Engel and $u = gd_1$, $v = hd_2$ with $d_1, d_2 \in K'$ not both equal to 1, we have $\langle u, v \rangle \vartriangleleft^2 K$. The class restriction on K and $d_1, d_2 \in K'$ yield $\langle u, v \rangle$ is 2-Engel if and only if $\langle g, h \rangle$ is 2-Engel. It can easily be seen that $[w, u, v] = [w, g, h]$ for $u, v \in K$ and $w \in \{a, b, c\}$. This together with the fact that $u^{25} = g^{25}$ and $v^{25} = h^{25}$ yields $\langle [w, u, v], [w, v, u] \rangle = \langle [w, g, h], [w, h, g] \rangle \leq K' \cap \langle g, h \rangle \leq \langle g^{25}, h^{25} \rangle = \langle u^{25}, v^{25} \rangle$. We conclude $\langle u, v \rangle \vartriangleleft^2 K$ and hence all subgroups of class at most 2 are 2-subnormal in K.

The verification that $\langle a, b \rangle$ has nilpotency class 3 and is not 2-subnormal in K is easily shown using GAP and the functions $\mathtt{LowerCentralSeries}$ and $\mathtt{SubnormalSeries}$. This completes the construction and validation of Example 3.2. $\square$

4. Groups with class n-1 subgroups n-subnormal

In this section we discuss the interrelations among B_n, U_n, and $U_{n,n-1}$ for $n \geq 3$ for metabelian groups in the case of non-torsion groups as well as torsion groups.

PROPOSITION 4.1. *Let n be an integer, $n \geq 3$, and G a metabelian group of class $n + 1$ without elements of order $p \leq n$, where p is prime. If $G \in B_n$, then $H \vartriangleleft^n G$ for all $H \leq G$ with $cl(H) \leq n - 1$.*

PROOF. Let G be a group as given in the statement and $n \geq 3$. Let H be a subgroup of G, such that $cl(H) \leq n - 1$. We want to verify now that H satisfies the hypothesis of Lemma 2.7. Since G has all cyclic subgroups n-subnormal, it follows by Corollary 2.6 that $[x, {}_n a] \in H$ for all $x \in G$ and all $a \in H$. Thus (i) holds. By the Hall-Witt identity for metabelian groups we have

$$[x, {}_ia, {}_jb][b, x, {}_ia, {}_{j-1}b][a, b, x, {}_{i-1}a, {}_{j-1}b] = 1$$

for $x \in G$, $a, b \in H$ and $i, j \in \mathbb{N}$. If $i + j = n$, then $1 = [a, b, {}_{i-1}a, {}_{j-1}b, x] = [a, b, x, {}_{i-1}a, {}_{j-1}b]$, since $cl(H) \leq n - 1$. Thus (ii) holds.

Because of the class restriction on G we have $[x, h_1, \ldots, h_{n+1}] = 1$ for all $x \in G$ and all $h_1, \ldots, h_{n+1} \in H$. Thus (iii) holds. It follows by Lemma 2.7 that $[x, h_1, \ldots, h_n] \in H$ for all $x \in G$ and all $h_1, \ldots, h_n \in H$. By Corollary 2.6 this yields $H \lhd^n G$. $\square$

The preceding proposition together with Theorem 1.1 completely settles the case for metabelian non-torsion groups.

THEOREM 4.2. *Let n be an integer, $n \geq 3$, and G a metabelian non-torsion group in B_n without elements of order $p \leq n - 1$, where p is prime. If $n + 1$ is not prime, then $G \in U_n$ if and only if $cl(G) \leq n$. If $n + 1$ is prime, then $H \lhd^n G$ for all $H \leq G$ with $cl(H) \leq n - 1$, and there exists a group $G \in B_n$ with $H \leq G$ and $cl(H) = n$, such that H is not n-subnormal in G.*

PROOF. If $n + 1$ is not a prime, then $cl(G) \leq n$ by Theorem 1.1. Let $H \leq G$. Then $[G, {}_nH] = 1$ and hence $[G, {}_nH] \leq H$ and thus $H \lhd^n G$ by Lemma 2.4. Therefore $G \in U_n$. Conversely, suppose $G \in U_n$. Then $G \in B_n$ and by Theorem 1.1 we have $cl(G) \leq n$.

Now let $n + 1$ be a prime and $G \in B_n$. By Theorem 1.1 we have $cl(G) \leq n + 1$. If G has class $n + 1$, then G satifies the assumptions of Proposition 4.1, since $n \geq 3$ and n is not prime. We conclude $H \lhd^n G$ for all $H \leq G$ and $cl(H) \leq n - 1$. To prove the last part of our claim, we consider the non-torsion group $G = \langle a, b, c \rangle$ of Example 5.6. There it is shown that G is an n-Engel group of class $n + 1$. It follows that $G \in B_n$, thus G satisfies our assumptions. However, as shown in Example 5.6, the subgroup $H = \langle b, c \rangle$ has class n and is not n-subnormal in G. $\square$

As is to be expected, our result for metabelian torsion groups is somewhat weaker. In the last section we will provide various examples showing that we cannot dispense with the order restrictions on elements and class restrictions on subgroups in the assumptions of Theorem 4.3.

THEOREM 4.3. *Let n be an integer, $n \geq 3$, and G a metabelian torsion group without elements of order $p \leq n$, where p is prime. If $G \in B_n$, then $H \lhd^n G$ for all $H \leq G$ with H having class less than or equal to $n - 1$.*

PROOF. Let G be a group as in the assumptions. If $n + 1$ is not prime, then G has class not exceeding $n + 1$ by Theorem 4.1 of [5]. Thus G satisfies the assumptions of Proposition 4.1 and we conclude that $H \lhd^n G$ for all $H \leq G$ with $cl(H) \leq n - 1$.

Now let $n + 1$ be a prime. Then G has class not exceeding $n + 2$ by Theorem 1.1. In this case we will show that G satisfies the assumptions of Lemma 2.7. The verification of (i) and (ii) is as in the proof of Proposition 4.1, since this did not depend on the class restriction on G.

To verify (iii), we observe that $[x, h_1, \ldots, h_{n+2}] = 1$ for all $x \in G$ and all $h_1, \ldots, h_{n+2} \in H$, where $H \leq G$ and $cl(H) \leq n - 1$. It remains to be shown that $[x, h_1, \ldots, h_{n+1}] \in H$ for all $x \in G$ and all $h_1, \ldots, h_{n+1} \in H$. Let $x \in G$ and $a, b, c \in H$. Then $[x, c, {}_nab] \in H \cap Z(G)$ by our assumptions. Thus we can expand linearly and obtain by Lemma 2.2

$$(4.3.1) \qquad 0 \equiv [x, c, {}_n ab] = \sum_{i=1}^{n-1} \binom{n}{i} [x, c, {}_i a, {}_{n-i} b] \pmod{H \cap Z(G)}.$$

Substituting a^l, $l = 1, \ldots, n-1$ for a and applying Lemma 2.3 yields $[x, c, {}_i a, {}_{n-i} b]$ in H for $i = 1, \ldots, n-1$. In particular, $[x, c, a, {}_{n-1} b] \in H$. Setting $c = h_1$ and $a = h_2$ we obtain $[x, h_1, h_2, {}_{n-1} b] \in H$. Continuing in the same manner as in Lemma 2.7 yields $[x, h_1, \ldots, h_{n+1}] \in H$ for all $x \in G$ and $h_1, \ldots, h_{n+1} \in H$. Thus (iii) holds. By Lemma 2.7 we conclude that $[x, h_1, \ldots, h_n] \in H$. This implies $H \lhd^n G$ by Corollary 2.6. $\qquad\square$

5. Examples

In this section we provide various examples showing that certain assumptions about the subnormality defect n and the absence of elements of certain prime orders cannot be dispensed with. All examples, with the exception of the last one, have been constructed using GAP, and the verification of their properties was done with the help of GAP.

The following metabelian 2-group has all cyclic subgroups 3-subnormal, but has an abelian subgroup that is not 3-subnormal. This example shows that the restriction of not having any elements of order 2 for $n = 3$ is required in Theorem 4.3.

EXAMPLE 5.1. *Let $K = G/M^G$, where*

$$G = \left\langle x, y, z \ \middle| \ \begin{array}{l} x^{16} = y^{16} = z^{16} = [x_1, x_2]^8 = [x_1, x_2, x_3]^8 = [x_1, \ldots, x_4]^4 = 1, \\ x_1, \ldots, x_4 \in \{x, y, z\}, \ G_5 = 1, \ G'' = 1 \end{array} \right\rangle,$$

and M^G is the normal closure of

$$M = \langle [x, y], [x, {}_3 xyz](xyz)^{16}, [y, {}_3 xyz](xyz)^{16}, [z, {}_3 xyz^2](xyz^2)^{16},$$
$$[x, {}_3 xy^2 z](xy^2 z)^{16}, [y, {}_3 xy^2 z](xy^2 z)^{16}, [x, {}_3 x^2 yz](x^2 yz)^{32}, [y, {}_3 x^2 yz](x^2 yz)^{32},$$
$$[z, {}_3 xy^2 z^2](xy^2 z^2)^{16}\rangle.$$

Then K is a metabelian 2-group and has all cyclic subgroups 3-subnormal, but does not have all subgroups 3-subnormal. In particular, K has an abelian subgroup which is not 3-subnormal.

PROOF. We begin with the construction of the group G. This is accomplished with GAP, using the functions FreeGroup, PcGroupFpGroup and RefinedPcGroup. Once G has been constructed, we construct the subgroup M, the normal closure of M in G, and finally the quotient group $K = G/M^G$. The groups G and K are metabelian, have nilpotency class 4, and have order 2^{75} and 2^{37}, respectively.

Among the generators of K there are three generators, namely xM^G, yM^G, and zM^G, call them a, b, and c, such that $K = \langle a, b, c \rangle$, and $|a| = |b| = |c| = 16$.

First we will show that K is a 3-Engel group from which it follows that $K \in B_3$. Observing that $a^{16} = b^{16} = c^{16} = 1$, every element $k \in K$ can be represented as $k = gd$, where $d \in K'$ and $g = a^i b^j c^l$ for $1 \leq i, j, l \leq 16$. With the help of GAP we proceed to verify that $[u, {}_3 g] = 1$ for $u \in \{a, b, c\}$. Observing that K has class 4 and $d \in K'$ yields by linear expansion that $[u, {}_3 k] = [u, {}_3 gd] = [u, {}_3 g] = 1$ for $u \in \{a, b, c\}$, and hence $[h, {}_3 k] = 1$ for all $h, k \in K$. Thus $\langle k \rangle \lhd^3 K$ for all $k \in K$.

Since $[x, y] \in M^G$, it follows that $\langle a, b \rangle$ is an abelian subgroup of K. It can easily be verified using the GAP function SubnormalSeries that $\langle a, b \rangle$ is not 3-subnormal in K. $\qquad \square$

The following 3-group has all cyclic subgroups 3-subnormal, but has an abelian subgroup that is not 3-subnormal. This group shows that the restriction of not having any elements of order 3 for $n = 3$ is required for Theorem 4.3.

EXAMPLE 5.2. *Let $K = G/M^G$, where*

$$G = \left\langle x, y, z \,\middle|\, \begin{array}{l} x^{27} = y^{27} = z^{27} = [x_1, x_2]^9 = [x_1, x_2, x_3]^9 = [x_1, \ldots, x_4]^3 = 1, \\ x_1, \ldots, x_4 \in \{x, y, z\}, \; G_5 = 1, \; G'' = 1 \end{array} \right\rangle,$$

and M^G is the normal closure of

$$\begin{aligned}
M = \big\langle & [x, y], [y, {}_2x], [x, {}_3xyz](xyz)^9, [y, {}_3xyz](xyz)^9, [x, {}_3xyz^2](xyz^2)^9, \\
& [y, {}_3xyz^2](xyz^2)^{18}, [x, {}_3xyz^3](xyz^3)^9, [z, {}_3xyz^3](xyz^3)^{27}, [x, {}_3xy^2z](xy^2z)^{18}, \\
& [y, {}_3xy^2z](xy^2z)^{27}, [x, {}_3xy^3z](xy^3z)^{18}, [y, {}_3xy^3z](xy^3z)^{27}, [x, {}_3x^2yz](x^2yz)^{27}, \\
& [y, {}_3x^2yz](x^2yz)^{27}, [x, {}_3x^3yz](x^3yz)^{27}, [y, {}_3x^3yz](x^3yz)^{27}, \\
& [x, {}_3x^2y^2z^2](x^2y^2z^2)^{18}, [x, {}_3x^3y^2z^3](x^3y^2z^3)^{18}, [x, {}_3x^3y^3z^2](x^3y^3z^2)^9, \\
& [x, {}_3x^9y^8z](x^9y^8z)^9, [y, {}_3x^9y^8z](x^9y^8z)^9, [x, {}_3x^8y^9z](x^8y^9z)^9, \\
& [x, {}_3x^2yz^3](x^2yz^3)^9 \big\rangle.
\end{aligned}$$

Then K is a metabelian 3-group in which all cyclic subgroups are 3-subnormal, but does not have all subgroups 3-subnormal. In particular, K has an abelian subgroup which is not 3-subnormal.

PROOF. We begin with the construction of the group G. This is accomplished with GAP, using the functions FreeGroup, PcGroupFpGroup and RefinedPcGroup. Once G has been constructed, we construct the subgroup M, the normal closure of M in G, and finally the quotient group $K = G/M^G$. The group G is metabelian, has order 3^{46}, and is nilpotent of class 4.

Among the generators of K there are three generators, call them a, b, and c, such that $K = \langle a, b, c \rangle$, and $|a| = |b| = 9$, $|c| = 27$. The quotient group K has order 3^{16}, $cl(K) = 4$, $K'' = 1$, $exp(K') = 9$, and $exp(K_4) = 3$.

First we will show that $K \in B_3$. Observing that $a^9, b^9, c^9 \in K'$, every $k \in K$ can be represented as $k = gd$ where $d \in K'$ and $g = a^i b^j c^l$ for $1 \le i, j, l \le 9$. As in Example 3.2, we proceed by verifying with the help of GAP that all cyclic subgroups $\langle g \rangle$ are 3-subnormal. We do this by using the GAP function SubnormalSeries. Additionally, while showing this, we also verified that if $|g| \le 9$, then $[u, {}_3g] = 1$, for $u \in \{a, b, c\}$.

Next we show that for arbitrary $k \in K$ we have $\langle k \rangle \lhd^3 K$ by reducing it to the case $\langle g \rangle \lhd^3 K$, where $k = gd$ with $d \in K'$. Observing $d \in K'$ and K having class 4, we obtain by linear expansion that $[u, {}_3k] = [u, {}_3gd] = [u, {}_3g]$ for $u \in \{a, b, c\}$. If $|g| = 9$, we have $[u, {}_3k] = [u, {}_3g] = 1$, hence $\langle k \rangle \lhd^3 K$. If $|g| > 9$, then $[u, {}_3g] \in \langle g \rangle \cap K'$ implies $[u, {}_3g] \in \langle g^9 \rangle$, since $\langle g \rangle \cap K' \le \langle g^9 \rangle$. By Lemma 2.1 and the facts that $d \in K'$, $exp(K') = 9$, $exp(K_3) = 3$, and $cl(K) = 4$, we obtain

$$k^9 = (gd)^9 = g^9[g, d^{-1}]^{\binom{9}{2}}[g, d^{-1}, g]^{\binom{9}{3}}d^9 = g^9.$$

Thus $[u, {}_3k] = [u, {}_3g] \in \langle g^9 \rangle = \langle k^9 \rangle$ for $u \in \{a, b, c\}$. From the class restriction on K it follows that $[h, {}_3k] \in \langle k \rangle$ for all $h, k \in K$ and hence $K \in B_3$.

The verification that $\langle a, b \rangle$ is abelian and not 3-subnormal in K is easily shown using the GAP functions Comm and SubnormalSeries. □

The following 2-group has all cyclic subgroups 4-subnormal, but has an abelian subgroup that is not 4-subnormal. This group shows that the restriction of not having any elements of order 2 for $n = 4$ is required in Theorem 4.3.

EXAMPLE 5.3. *Let* $K = G/M^G$, *where*

$$G = \left\langle x, y, z \;\middle|\; \begin{array}{l} x^{16} = y^{16} = z^{16} = [x_1, x_2]^8 = [x_1, x_2, x_3]^8 = [x_1, \ldots, x_4]^8 = 1, \\ [x_1, \ldots, x_5]^2 = 1, \; x_1, \ldots, x_5 \in \{x, y, z\}, \; G_6 = 1, \; G'' = 1 \end{array} \right\rangle,$$

and M^G *is the normal closure of*

$$M = \big\langle [x, y], [z, {}_3x], [z, {}_3y], [x, z]^8, [y, z]^8, [x, {}_2z, x, y], [x, {}_3z, x], [x, {}_4z], [y, {}_4z],$$
$$[x, {}_4xyz](xyz)^{16}, [y, {}_4xyz](xyz)^{16}, [z, {}_4xyz](xyz)^{16}, [x, {}_4xy^2z](xy^2z)^{16},$$
$$[y, {}_4xy^2z](xy^2z)^{16}, [x, {}_4x^2yz](x^2yz)^{16} \big\rangle.$$

Then K *is a metabelian 2-group and has all cyclic subgroups 4-subnormal, but does not have all subgroups 4-subnormal. In particular, K has an abelian subgroup which is not 4-subnormal.*

PROOF. We begin with the construction of the group G. This is accomplished with GAP, using the functions FreeGroup, PcGroupFpGroup and RefinedPcGroup. Once G has been constructed, we construct the subgroup M, the normal closure of M in G, and finally the quotient group $K = G/M^G$. The group G has order 2^{114} and is nilpotent of class 5.

Among the generators of K there are three generators, call them a, b, and c, such that $K = \langle a, b, c \rangle$, and $|a| = |b| = |c| = 16$. The quotient group K has order 2^{52} and is also nilpotent of class 5. Furthermore, $exp(K') = 8$ and $exp(K_5) = 2$.

First we will verify that K has all cyclic subgroups 4-subnormal. Observing that $a^{16} = b^{16} = c^{16} = 1$, all elements $k \in K$ can be represented in the form $k = gd$, $d \in K'$ and $g = a^i b^j c^l$, $1 \le i, j, l \le 16$. As in the preceding examples, we proceed by verifying that all cyclic subgroups $\langle g \rangle$ are 4-subnormal, using the GAP function SubnormalSeries. Additionally, while showing this, we also verified that if $|g| \le 16$, then $[u, {}_4g] = 1$, for $u \in \{a, b, c\}$.

Next we show that for arbitrary $k \in K$ we have $\langle k \rangle \lhd^4 K$ by reducing it to the case $\langle g \rangle \lhd^4 K$, where $k = gd$ with $d \in K'$. Observing that K has class 5 and $d \in K'$, we obtain by linear expansion that $[u, {}_4k] = [u, {}_4gd] = [u, {}_4g]$ for $u \in \{a, b, c\}$. If $|g| \le 16$, we have $[u, {}_4k] = [u, {}_4g] = 1$, hence $\langle k \rangle \lhd^4 K$. If $|g| > 16$, then $[u, {}_4g] \in \langle g \rangle \cap K'$ implies $[u, {}_4g] \in \langle g^{16} \rangle$, since $\langle g \rangle \cap K' \le \langle g^{16} \rangle$. By Lemma 2.1 and the facts that $d \in K'$, $exp(K') = 8$, and $exp(K_5) = 2$, we obtain

$$k^{16} = (gd)^{16} = g^{16}[g, d^{-1}]^{\binom{16}{2}}[g, d^{-1}, g]^{\binom{16}{3}}[g, d^{-1}, {}_2g]^{\binom{16}{4}} d^{16} = g^{16}.$$

Thus $[u, {}_4k] = [u, {}_4g] \in \langle g^{16} \rangle = \langle k^{16} \rangle$ for $u \in \{a, b, c\}$. Since K has class 5, it follows that $[h, {}_4k] \in \langle k \rangle$ for all $h, k \in K$ and hence $K \in B_4$.

The verification that $\langle a, b \rangle$ is abelian and is not 4-subnormal in K is easily shown using GAP and the functions Comm and SubnormalSeries. □

The following 3-group has all cyclic subgroups 4-subnormal, but has an abelian subgroup that is not 4-subnormal. This example shows that the restriction of not having any elements of order 3 for $n = 4$ is required for Theorem 4.3.

EXAMPLE 5.4. *Let $K = G/M^G$, where*

$$G = \langle x, y, z \mid x^9 = y^9 = z^9 = [x_1, x_2]^3 = [x_1, x_2, x_3]^3 = [x_1, \dots, x_4]^3 = 1,$$

$$[x_1, \dots, x_5]^3 = [x_1, \dots, x_6]^3 = 1, \ x_1, \dots, x_6 \in \{x, y, z\}, \ G_7 = 1, \ G'' = 1 \rangle,$$

and M^G is the normal closure of $M = \langle x^3, y^3, z^3, [x, y], [x, {}_3 z], [y, {}_3 z] \rangle$.

Then K is a metabelian 3-group and has all cyclic subgroups 4-subnormal, but does not have all subgroups 4-subnormal. In particular, K has an abelian subgroup which is not 4-subnormal.

PROOF. We begin with the construction of the group G. This is accomplished with GAP, using the functions FreeGroup, PcGroupFpGroup and RefinedPcGroup. Once G has been constructed, we construct the subgroup M, the normal closure of M in G, and finally the quotient group $K = G/M^G$. The group G has order 3^{91}, is nilpotent of class 6, and is metabelian.

Among the generators of K there are three generators, call them a, b, and c, such that $K = \langle a, b, c \rangle$, and $|a| = |b| = |c| = 3$. The quotient group K has order 3^{19} and is also nilpotent of class 6.

First we will show that K is a 4-Engel group from which it follows that $K \in B_4$. Observing that $a^3 = b^3 = c^3 = 1$, every element $k \in K$ can be represented as $k = gd$, where $d \in K_3$ and $g = a^i b^j c^l [a, c]^s [b, c]^t$ for $1 \leq i, j, l, s, t \leq 3$. With the help of GAP we proceed to verify that $[u, {}_4 g] = 1$ for $u \in \{a, b, c\}$. Observing that $d \in K_3$ and K having class 6, it follows $[u, {}_4 k] = [u, {}_4 gd] = [u, {}_4 g] = 1$ for $u \in \{a, b, c\}$. We conclude $[h, {}_4 k] = 1$ for all $h, k \in K$. This can be seen as follows. If $[v, {}_4 k] = [w, {}_4 k] = 1$ for $v, w, k \in K$, we obtain by straightforward expansion that $[vw, {}_4 k] = [v, {}_4 k]^w [w, {}_4 k] = 1$. We conclude that K is a 4-Engel group.

The verification that $\langle a, b \rangle$ is abelian and is not 4-subnormal in K is easily shown using GAP and the functions Comm and SubnormalSeries. □

The following 5-group has all cyclic subgroups 3-subnormal, but has a subgroup of class 3 that is not 3-subnormal. This group shows that the restriction to subgroups of class ≤ 2 is required in Theorem 4.3.

EXAMPLE 5.5. *Let $K = G/M^G$, where*

$$G = \left\langle x, y, z \ \middle| \ \begin{array}{l} x^{25} = y^{25} = z^{25} = [x_1, x_2]^5 = [x_1, x_2, x_3]^5 = [x_1, \dots, x_4]^5 = 1, \\ x_1, \dots, x_4 \in \{x, y, z\}, \ G_5 = 1, \ G'' = 1 \end{array} \right\rangle,$$

and M^G is the normal closure of

$$M = \langle x^5, y^5, [y, {}_3 x], [z, {}_3 x], [x, {}_3 y], [z, {}_3 y], [x, {}_3 z] z^5, [y, {}_3 z] z^5, [x, {}_3 xy^2](xy^2)^5,$$

$$[x, {}_3 x^2 y](x^2 y)^5, [x, {}_3 xyz](xyz)^5, [y, {}_3 xyz](xyz)^5, [x, {}_3 xyz^2](xyz^2)^5,$$

$$[y, {}_3 xyz^2](xyz^2)^{10}, [x, {}_3 xy^2 z](xy^2 z)^5, [y, {}_3 xy^2 z](xy^2 z)^5, [x, {}_3 xy^2 z^2](xy^2 z^2)^5,$$

$$[y, {}_3 xy^2 z^2](xy^2 z^2)^{15} \rangle.$$

Then K is a metabelian 5-group and has all cyclic subgroups 3-subnormal, but does not have all subgroups 3-subnormal. In particular, K has a subgroup of nilpotency class 3 which is not 3-subnormal.

PROOF. We begin with the construction of the group G. This is accomplished with GAP, using the functions FreeGroup, PcGroupFpGroup and RefinedPcGroup. Once G has been constructed, we construct the subgroup M, the normal closure of M in G, and finally the quotient group $K = G/M^G$. The group G is metabelian, has order 5^{32}, and is nilpotent of class 4.

Among the generators of K there are three generators, call them a, b, and c, such that $K = \langle a, b, c \rangle$, and $|a| = |b| = 5$, $|c| = 25$. The quotient group K has order 5^{15} and is also nilpotent of class 4. Furthermore, $exp(K') = 5$ and $K'' = 1$.

We begin by showing that K has all cyclic subgroups 3-subnormal. Noting that $a^5, b^5, c^5 \in K'$, all elements $k \in K$ can be represented in the form $k = gd$ with $d \in K'$ and $g = a^i b^j c^l$, $1 \le i, j, l \le 5$. We proceed by verifying that all cyclic subgroups $\langle g \rangle$ are 3-subnormal using the GAP function SubnormalSeries. Additionally, while showing this, we also verified that if $|g| \le 5$, then $[u, {}_3 g] = 1$, for $u \in \{a, b, c\}$.

Next we show that for arbitrary $k \in K$ we have $\langle k \rangle \lhd^3 K$ by reducing it to the case $\langle g \rangle \lhd^3 K$, where $k = gd$ with $d \in K'$. Observing that K has class 4 and $d \in K'$, we obtain by linear expansion that $[u, {}_3 k] = [u, {}_3 gd] = [u, {}_3 g]$ for $u \in \{a, b, c\}$. If $|g| = 5$, we have $[u, {}_3 k] = [u, {}_3 g] = 1$, hence $\langle k \rangle \lhd^3 K$. If $|g| > 5$, then $[u, {}_3 g] \in \langle g \rangle \cap K'$ implies $[u, {}_3 g] \in \langle g^5 \rangle$, since $\langle g \rangle \cap K' \le \langle g^5 \rangle$. By Lemma 2.1 and the facts that $d \in K'$, $exp(K') = 5$, and $cl(K) = 4$, we obtain

$$k^5 = (gd)^5 = g^5 [g, d^{-1}]^{\binom{5}{2}} [g, d^{-1}, g]^{\binom{5}{3}} d^5 = g^5.$$

Thus $[u, {}_3 k] = [u, {}_3 g] \in \langle g^5 \rangle = \langle k^5 \rangle$ for $u \in \{a, b, c\}$. From the class restriction on K it follows that $[h, {}_3 k] \in \langle k \rangle$ for all $h, k \in K$ and hence $K \in B_3$.

By Theorem 4.3 it follows that all subgroups of K that are of class at most 2 are 3-subnormal. However not all subgroups of K are 3-subnormal as the subgroup $\langle a, b \rangle$ is not 3-subnormal. The verification that $\langle a, b \rangle$ is nilpotent of class 3 and is not 3-subnormal in K is easily shown using GAP and the functions LowerCentralSeries and SubnormalSeries.   □

Our last example is a non-torsion group which has all cyclic subgroups n-subnormal, where $n \ge 4$ and $n + 1$ is prime, but it has a class n subgroup that is not n-subnormal. This shows that the restriction to subgroups of class $n - 1$ is required for Theorem 4.2 in case $n + 1$ is prime. A slight modification of the group leads to a p-group, $p = n + 1$, with the same properties, showing that the restriction to subgroups of class $n - 1$ is required in Theorem 4.3. For details of the latter example we refer to Example 3 of [9].

EXAMPLE 5.6. Let $n \ge 4$ be an integer with $n + 1 = p$, a prime. We obtain the desired group G by three split extensions, starting with an elementary abelian p-group. Consider the elementary abelian p-groups $X = \langle x_1 \rangle \times \langle x_2 \rangle \times \langle x_3 \rangle$, $Z = \langle z \rangle$, and $Y_i = \langle y_{1i} \rangle \times \langle y_{2i} \rangle \times \langle y_{3i} \rangle \times \langle y_{4i} \rangle$ for $i = 1, \ldots, p - 3$. We start with the group $U = X \times Y_1 \times \cdots \times Y_{p-3} \times Z$ which is an elementary abelian p-group of order $p^{4(p-2)}$. The group V is the semidirect product of U with an infinite cyclic group $\langle a \rangle$ inducing an automorphism of order p on U. The relations of $V = U \rtimes \langle a \rangle$ are those of U and in addition $[x_1, a] = y_{11}$, $[x_2, a] = [x_3, a] = 1$, $[y_{ji}, a] = 1$ for $j = 1, 2, 3$ and $i = 1, \ldots, p - 3$, $[y_{4i}, a] = y_{1,i+1}$ for $i = 1, \ldots, p - 4$, $[y_{4,p-3}, a] = z$, and $[z, a] = 1$.

Again, the group W is the semidirect product of V with an infinite cyclic group $\langle b \rangle$ inducing an automorphism of order p on V. The relations of $W = V \rtimes \langle b \rangle$ are those of V and in addition $[a, b] = x_2$, $[x_1, b] = [x_2, b] = 1$, $[x_3, b] = y_{11}y_{21}$, $[y_{ji}, b] = 1$ for $j = 1, 2, 4$ and $i = 1, \ldots, p-3$, $[y_{3i}, b] = y_{1,i+1}y_{2,i+1}$ for $i = 1, \ldots, p-4$, $[y_{3,p-3}, b] = z^{-1}$, and $[z, b] = 1$.

Finally, the group G is obtained in the same way as the semidirect product of W with an infinite cyclic group $\langle c \rangle$ inducing an automorphism of order p on W. The relations of $G = W \rtimes \langle c \rangle$ are those of W and in addition $[a, c] = x_3$, $[b, c] = x_1$, $[x_1, c] = y_{41}$, $[x_2, c] = y_{21}$, $[x_3, c] = y_{31}$, $[y_{ji}, c] = y_{j,i+1}$, for $j = 1, 2, 3, 4$ and $i = 1, \ldots, p-4$, $[y_{1,p-3}, c] = z$, $[y_{2,p-3}, c] = z^{-2}$, $[y_{3,p-3}, c] = [y_{4,p-3}, c] = 1$, and $[z, c] = 1$. If we choose $\langle a \rangle$, $\langle b \rangle$, $\langle c \rangle$ to be cyclic groups of order p, then G is a p-group of order p^{4p-5} (see [9]). In either case G has the following properties:

(i) $G = \langle a, b, c \rangle$ is metabelian, since $G' = U$;
(ii) G is nilpotent of class p. For the proof we refer to (ii) in Example 3 of [9];
(iii) G has all class $n - 1$ subgroups n-subnormal;
(iv) The subgroup $H = \langle b, c \rangle$ has class n and is not n-subnormal in G.

To prove (iii), it suffices to show that G is an n-Engel group. Our claim then follows by Proposition 4.1. We first will show that $[u, {}_n g] = 1$ for $u \in \{a, b, c\}$ and $g = a^i b^j c^l$ with i, j, l integers. Let $u = a$. Because of the class restriction on G we can expand linearly and, using the relations of G, we obtain

$$[a, {}_n g] = ([a, {}_{n-1}c, b]^{n-1}[a, b, {}_{n-1}c])^{jl^{n-1}} = (z^{-n+1}z^{-2})^{jl^{n-1}} = (z^{-p})^{jl^{n-1}} = 1$$

for all integers i, j, l. Similarly, if $u = b$, we obtain

$$[b, {}_n g] = ([b, {}_{n-1}c, a]^{n-1}[b, a, {}_{n-1}c])^{il^{n-1}} = (z^{n-1}z^2)^{il^{n-1}} = (z^p)^{il^{n-1}} = 1$$

for all integers i, j, l. Finally, let $u = c$. Then

$$[c, {}_n g] = ([c, b, {}_{n-2}c, a]^{n-1}[c, a, {}_{n-2}c, b]^{n-1})^{ijl^{n-2}} = (z^{-n+1}z^{n-1})^{ijl^{n-2}} = 1$$

for all integers i, j, l. We observe that any $x \in G$ can be represented as $x = gd$ with $d \in G'$ and $g = a^i b^j c^l$, i, j, l integers. Because of the class restriction on G we can expand linearly and obtain $[u, {}_n x] = [u, {}_n gd] = [u, {}_n g] = 1$ for $u \in \{a, b, c\}$. Similarly, it follows $[y, {}_n x] = 1$ for all $y, x \in G$. Thus G is n-Engel as claimed.

To prove (iv), we observe that $[b, {}_{n-1}c]$ is a commutator of weight n in H and $[b, {}_{n-1}c] = y_{4,n-2} \neq 1$. Thus $cl(H) \geq n$. On the other hand, all commutators of total weight $n+1$ in H are trivial. Hence $cl(H) \leq n$. We conclude that H is nilpotent of class n. To show that H is not n-subnormal, we observe that $[a, {}_{n-1}c, b] = z \notin H$. Thus our claim follows by Corollary 2.6.

Acknowledgements. The authors want to thank Wolfgang P. Kappe and the anonymous referee for many important comments, helpful suggestions and careful reading of the paper.

References

1. R. Baer, Situation der Untergruppen und Struktur der Gruppe, *S. B. Heidelberg Akad. Math. Nat. Klasse* **2** (1933), 12–17.

2. D. Cappit, On groups with every subgroup 2-subnormal, *J. London Math. Soc.* **2** (1973), 17–18.

3. R. Dedekind, Über Gruppen, deren sämtliche Teiler Normalteiler sind, *Math. Ann.* **48** (1897), 548–561.

4. The GAP Group, GAP – Groups, Algorithms, and Programming, Version 4.4.9; 2006, (http://www.gap-system.org).

5. D. J. Garrison and L. C. Kappe, Metabelian groups with all cyclic subgroups subnormal of bounded defect, *Proceedings of the Conference on Infinite Groups 1994, Walter de Gruyter, Berlin* (1995), 73–85.

6. K. W. Gruenberg, The Engel Elements of a Soluble Group, *Illinois J. Math.* **3** (1959), 151–168.

7. H. Heineken, A Class of Three-Engel Groups, *J. Algebra* **17** (1971), 341–345.

8. G. T. Hogan and W. P. Kappe, On the H_p-problem for finite p-groups, *Proc. AMS* **20** (1969), 450–454.

9. L. C. Kappe, Engel Margins in Metabelian Groups, *Comm. Algebra* **11** (1983), 1965–1987.

10. L. C. Kappe and G. Traustason, Subnormality conditions in non-torsion groups, *Bull. Austral. Math. Soc.* **59** (1999), 459–465.

11. S. K. Mahdavi, On groups with every subgroup 2-subnormal, *Arch. Math.* **47** (1986), 282–289.

12. S. K. Mahdavianary, Groups with many subgroups 2-subnormal, *Dissertation, State University of New York at Binghamton* (1983).

13. S. K. Mahdavianary, A special class of three-Engel groups, *Arch. Math.* **40** (1983), 193–199.

14. D. J. S. Robinson, *A Course in the Theory of Groups*, Springer-Verlag, New York-Berlin-Heidelberg (1982).

15. J. E. Roseblade, On groups in which every subgroup is subnormal, *J. Algebra* **2** (1965), 402–412.

16. M. Stadelmann, Gruppen deren Untergruppen subnormal vom Defekt zwei sind, *Arch. Math.* **30** (1978), 364–371.

LOCKHEED MARTIN SYSTEMS INTEGRATION-OWEGO, OWEGO, NY 13827, USA
E-mail address: `dave.garrison@lmco.com`

DEPARTMENT OF MATHEMATICAL SCIENCES, STATE UNIVERSITY OF NEW YORK AT BINGHAMTON, BINGHAMTON, NY 13902-6000, USA
E-mail address: `menger@math.binghamton.edu`

Contemporary Mathematics
Volume **470**, 2008

Normalizer calculation using automorphisms

Alexander Hulpke

ABSTRACT. We study how to reduce the calculation of the normalizer of a
subgroup U in G to the calculation of centralizers and element conjugacy in G
and calculations in the automorphism group $\mathrm{Aut}(U)$. Experimental run times
show that this can be substantially faster than existing backtrack algorithms.

1. Introduction

A fundamental technique for the computation of subgroups in a permutation
group, such as centralizer, normalizer, subgroup intersection or set stabilizer, is
backtrack search. The idea for this stems from Sims's work on stabilizer chains
[**Sim70**]. A description can be found in [**Ser03**] or [**HEO05**]. More recent devel-
opments using partition backtrack algorithms are described in [**Leo97**] or [**The97**].

In general, backtrack algorithms are of exponential complexity, though in practice the performance is often good. For example the computation of centralizers
is considered to be easy in practice ([**Ser03**, p.205]). Furthermore, [**Luk93**] shows
that centralizer, set stabilizer and subgroup intersection are polynomial time equivalent (and therefore also could be considered as "easy in practice", even though
neither of them is known to be polynomial time).

Luks also shows that the computation of normalizers is at least as hard as
computing centralizers, though it is not known whether the algorithmic complexity
actually differs.

Recently, Luks and Miyazaki [**LM02**] have shown that the calculation of normalizers in permutation groups is polynomial time for groups with bounded composition factors. The complexity of the general case is unknown.

In practice – contrary to centralizer and its equivalents – normalizer computations are often hard with current backtrack-based algorithms. (A footnote on p. 121
in [**HEO05**] describes them as "among the most difficult problems in CGT".) This
motivates the search for a reduction to easier problems.

To see why normalizer computations are hard consider the reductions possible
in a backtrack search for $N = N_G(U)$ with $U \le G \le S_n$. First, for each coset of

2000 *Mathematics Subject Classification.* Primary 20B40; Secondary 20B05.
Key words and phrases. normalizer, computation, automorphism group, backtrack, permu-
tation equivalent.
Supported in part by NSF Grant # 0633333.

the normalizer (or a subgroup of the normalizer) only one element has to be tested. If the index $[G : N]$ is large this method alone will not yield sufficient reductions.

In many cases it is possible to reduce this index by considering combinatorial or geometric invariants of the group U that is to be normalized (see for example [**RD04**], [**Hul05**, section 11], [**Miy06**]): the normalizer N must preserve the set of such invariants, thus the group G can be replaced by the stabilizer S of this invariant set in G and thus $N_G(U) = N_S(U)$. This is most fruitful if G is the full symmetric group.

This approach fails, however, if the subgroup U is very "symmetric" and has only little distinct permutational structure.

The second type of reduction employed is the actual problem-specific pruning of the backtrack tree: Sims's original work already suggests using the fact that $N_G(U)$ permutes the orbits of U on pairs of points. If U is elementary abelian or regular, however this approach often yields little improvement.

Another improvement, due to Holt [**Hol91**], uses the fact that elements of $N_G(U)$ induce automorphisms of U. In the case of a regular subgroup U this implies that any element of $N_G(U)$ is determined uniquely by specifying the images of two points. Furthermore [**Hol91**] suggests a reduction if G and U have a faithful permutation representation on a subset of the permutation domain: in this case the images of points in this domain determine the automorphism and thus determine the images of all other points.

However, if $U \leq S_n$ is elementary abelian but neither regular nor the intransitive direct product C_p^k, neither of these reductions might be applicable. In the author's recent work on classifying transitive permutation groups [**Hul05**] these problems frequently occurred.

2. Determining the normalizer from its action

As the computation of element centralizers (and the corresponding task of finding conjugating elements) seems to be so much easier in practice, the following approach for normalizing "small" or "known" subgroups looks promising:

Let Ω be a finite set, $G \leq S_\Omega$ be a permutation group and $U \leq G$. Then the normalizer $N_G(U)$ induces (by conjugation) automorphisms of U, and the kernel of this action is $C_G(U)$. Thus we can consider the quotient $N_G(U)/C_G(U)$ as a subgroup of $\mathrm{Aut}(U)$. We will assume that we know $\mathrm{Aut}(U)$ or can get it very cheaply. (For example this is the case if U is comparatively small or if it is abelian.)

Next, suppose we know the subgroup $A \leq \mathrm{Aut}(U)$ which is induced by $N_G(U)$. We then can compute, for every α in a generating set X of A, a conjugating element $g_\alpha \in G$ such that $u^\alpha = u^{g_\alpha}$ (see Section 3.3 for details). With this we get that

$$(1) \qquad\qquad N_G(U) = \langle C_G(U), g_\alpha \rangle_{\alpha \in X}.$$

Assuming we know the subgroup A, this would reduce the normalizer computations to computations of centralizers and element conjugacies. Unfortunately, the exact determination of A seems to be hard in general as well. We therefore employ the following approximation:

Conjugacy in a permutation group preserves the cycle structure of elements. We thus partition the elements of U into classes $\{C_i\}$ based on cycle structure. Let $B \leq \mathrm{Aut}(U)$ be the stabilizer of this partitioning. Clearly $A \leq B$.

If we consider $\mathrm{Aut}(U)$ as a permutation group on the (non-identity) elements of U, we can calculate B as an iterated set stabilizer. This again is a backtrack calculation, albeit of a different degree.

Using the conjugacy test from Section 3.3 for finding a conjugating element g_α, we can also check whether an element $\beta \in B$ is in fact contained in A. We thus can use a backtrack search over B (considering it as a permutation group and using [**HEO05**, §4.6.2]) to determine A. Doing so with the constructive test for normalizer induced automorphisms will automatically find conjugating permutations g_α for all the generators α that are needed in equation (1).

Essentially we are trading a normalizer backtrack search in G for a backtrack search in $\mathrm{Aut}(U)$ and an element conjugacy backtrack in G. If U (and thus $\mathrm{Aut}(U)$) is substantially smaller than G, this can be expected to yield shorter overall run time. The example runtimes in Section 6 support this claim.

We will describe the individual steps of such a calculation, including the choice of classes $\{\mathcal{C}_i\}$ in detail in the next section. However even from the short description the following question arises:

> Can one guarantee that $B = A$ or at least limit the index of A in B? (If so, one could bound the backtrack search for A in B which is potentially the worst-behaving part of the algorithm.)

We will study this question in Section 4.

3. Details of the algorithm

3.1. Normalizer classes. To obtain B we want to partition the elements of U into classes according to the action of $N_G(U)$. For this we observe the following:

- Conjugacy by $N_G(U)$ will join U-conjugacy classes of the same cardinality as $U \lhd N_G(U)$.
- Conjugacy by $N_G(U)$ preserves the cycle structure of elements.

In the first approximation we therefore consider the following classes $\{\mathcal{B}_i\}$ as the union of U-conjugacy classes: two elements $u_1, u_2 \in U$ are in the same $\mathcal{B}$-class if $\left|u_1^U\right| = \left|u_2^U\right|$ and u_1 has the same cycle structure as u_2. (If G is not transitive on Ω one can even separate by cycle structures when restricted to G-orbits of a particular length.)

To refine this partition, we observe that the class sums for the $N_G(U)$-classes in U form a subalgebra of the center of the group algebra for $N_G(U)$. We therefore calculate "structure constants" for the multiplication of the classes $\{\mathcal{B}_i\}$:

For $x \in \mathcal{B}_i$, let $z_{j,k}$ be the number of elements $y \in \mathcal{B}_j$ such that $xy \in \mathcal{B}_k$. We define the *signature* of x as the collection of values $z_{j,k}$ for all j, k.

Since the action of $N_G(U)$ induces automorphisms of U, elements with different signatures cannot be in the same class. This offers the possibility for a refinement by splitting up the class $\mathcal{B}_i$ according to element signatures; the same argument can be used to split up the corresponding class $\mathcal{B}_j$ using the same signatures. As $(xy)^{-1} = y^{-1}x^{-1}$ and elements and their inverses stay in the same class, it is sufficient to consider only ordered pairs i, j. We thus obtain (often finer) classes $\{\mathcal{C}_i\}_j$.

In principle this process can be iterated. In practice, however it turns out that further iterations do not typically yield improvements (that is, the classes turn out to be maximally refined after one iteration).

The calculations required in this step involve the multiplication of group elements and the identification of the class in which the result lies. Instead of multiplying permutations, it is faster to work with base images and to do class identifications based on hash values of base images.

3.2. Partition stabilizer. The next step is to calculate the automorphism group of U, for example following [**Sho28**] for abelian groups, [**ELGO02**] for p-groups, and [**CH03**] for non-solvable groups. Since we assume that U is small this calculation will be fast.

We then construct the permutation action of $\mathrm{Aut}(U)$ on the set $U^{\#}$ of non-identity elements of U as a subgroup of $S_{|U|-1}$. The classes $\mathcal{C}_i$ of elements obtained in the previous step thus correspond to sets of numbers. We arrange these sets in increasing order (typically the calculation of a set stabilizer is faster if the corresponding set is smaller) and iteratively compute the stabilizers of these sets using a partition backtrack algorithm [**Leo97**]. The last stabilizer then stabilizes all sets and therefore the (ordered) partition of U into classes; it is therefore equal to the group B.

In the special case that U is elementary abelian two improvements are possible: first, if any of the classes of U spans a subspace, the automorphism group can be reduced from $\mathrm{GL}_n(p)$ to a subspace stabilizer consisting of block matrices. Furthermore if the characteristic is different from 2, one can first consider the stabilizer under the projective action.

3.3. Normalizing elements. We now assume that we have obtained a subgroup $B \le \mathrm{Aut}(U)$ which contains the group A of all automorphisms that are induced by $N_G(U)$, but might be larger. Assume that $U = \langle u_1, \ldots, u_n \rangle$. We calculate iteratively $C_G(u_1), C_G(\langle u_1, u_2 \rangle) = C_{C_G(u_1)}(u_2), \ldots$ etc. to obtain $C_G(U)$. The following lemma describes an element test for A which simultaneously produces elements $g_\alpha \in G$ inducing a particular automorphism α.

LEMMA 1. *For $\alpha \in \mathrm{Aut}(U)$ we have that α is induced by $N_G(U)$ if and only if there exists*

1. $g_1 \in G$ such that $u_1^\alpha = u_1^{g_1}$
2. $g_2 \in C_G(u_1^{g_1})$ such that $u_2^\alpha = (u_2^{g_1})^{g_2}$.

   $\vdots$

m. $g_m \in C_G(u_1^{g_1}, u_2^{g_1 g_2}, \ldots, u_{m-1}^{g_1 g_2 \cdots g_{m-1}})$ such that $u_m^\alpha = (u_m^{g_1 g_2 \cdots g_{m-1}})^{g_m}$.

In this case the element $x = g_1 g_2 \cdots g_m$ is an element that induces α. Any other element inducing the same automorphism will differ from x only by an element of $C_G(U)$.

PROOF. Assume that x is as given. Then $u_i^x = u_i^{g_1 \cdots g_i}$. As $u_i^\alpha \in U$, the element x maps every generator of U into U and therefore normalizes the finite group U. Furthermore for any $u = u_{i_1} \cdots u_{i_k} \in U$, we have that

$$u^x = (u_{i_1} \cdots u_{i_k})^x = u_{i_1}^x \cdots u_{i_k}^x = u_{i_1}^\alpha \cdots u_{i_k}^\alpha = (u_{i_1} \cdots u_{i_k})^\alpha = u^\alpha.$$

Conversely, if α is induced by $y \in N_G(U)$ we can set $g_1 = y$, $g_i = 1$ for $i > 1$. Also in this case we have that $x \cdot y^{-1} \in C_G(U)$ as it fixes all generators of U. $\square$

For performance reasons it can be advantageous to arrange the generators u_i in order of decreasing support (number of points moved). While this does not usually result in a notable increase in the cost of a conjugacy test, it tends to decrease the size of the first centralizer, thus making the subsequent conjugacy tests easier. This is particularly relevant when computing normalizers in the full symmetric group.

4. Are all automorphisms induced?

An obvious question that arises at this point is whether indeed all automorphisms that stabilize the partition of U into classes are induced by the normalizer of U in G or, if not, what we can say about the index $[B : A]$.

One way to study this is via representation theory:

LEMMA 2. *Let G be a permutation group acting on Ω with the natural permutation matrix representation $\nu : G \to GL_{|\Omega|}(\mathbb{C})$ and $\alpha \in \mathrm{Aut}(G)$. The following are equivalent:*

a) α *preserves the cycle shape of every element.*
b) α *preserves the number of fixed points of every element*
c) *The (complex) representations ν and $\alpha\nu$ afford the same permutation character.*
d) *The representations ν and $\alpha\nu$ are equivalent (as complex representations).*

In this case we say that α is equidistributing.

PROOF. As complex representations are equivalent if and only if they afford the same character we only need to show that b) implies a). This follows, because for a prime p dividing an integer n every n-cycle of an element $g \in G$ becomes a set of p disjoint $\frac{n}{p}$-cycles in g^p. This gives rise to a recursive formula for the number of n-cycles in terms of the number of fixed points of powers of g.   □

This lemma incidentally shows that instead of cycle structure it would have been sufficient to group elements according to their number of fixed points. However, this would have given initially larger classes $\mathcal{B}_i$ and led to longer runtime since the backtrack algorithm for set stabilizers runs faster if the set to be stabilized is smaller.

On the other hand, α is induced by $N_{S_n}(U)$ if the representations ν and $\alpha\nu$ are *permutation equivalent*, i.e., if they can be transformed into each other by a permutation of the basis vectors, or equivalently if they can be conjugated into each other by permutation matrices.

As the following examples show, these two concepts are not equivalent.

EXAMPLE 1. Let $U = \mathrm{PSL}_3(2)$, acting 2-transitively on 7 points. Then the rational classes of U (and thus also the cycle structures) are determined solely by the orders of elements. Thus the outer automorphism of order 2 is equidistributing. This automorphism can be considered as a duality between points and lines in the underlying projective geometry, so it cannot be induced by S_7.

EXAMPLE 2. For our second example let $G = S_8$ and let

$$U = \langle u_1 = (2,3)(6,8), u_2 = (2,6)(3,8)(5,7), u_3 = (1,4)(5,7)\rangle \leq G.$$

Then $U \cong C_2^3$, $|C_{S_8}(U)| = 16$, $|N_{S_8}(U)| = 64$. There exists an automorphism α of U that maps the generators of U as $u_1 \mapsto u_3 = (1,4)(5,7)$, $u_2 \mapsto u_3 u_2 u_1 = (2,6)(3,8)(5,7)$, and $u_3 \mapsto u_1 = (2,3)(6,8)$. This automorphism is equidistributing, but is not induced by $N_{S_8}(U)$.

As the first example shows, such behavior is often associated with the existence of interesting combinatorial structures. However – typically there are few interesting structures associated to a permutation group – one can expect this behavior to be infrequent.

For transitive groups the existence of an equidistributing automorphism α which is not induced by a permutation means that $\mathrm{Stab}_G(1)^\alpha$ is not a point stabilizer, i.e. G must have (at least) two classes of subgroups isomorphic to $\mathrm{Stab}_G(1)$. Thus the index $[B : A]$ counts classes of subgroups isomorphic to, but not conjugate to $\mathrm{Stab}_G(1)$. In general there are just a few classes of such subgroups, indicating that $[B : A]$ ought to be small.

We can prove equality $A = B$ in some cases, detailed below. Some of these cases are of practical relevance, as the groups are small in comparison to the degree and have comparatively little permutational structure to aid a backtrack search for the normalizer.

LEMMA 3. *Suppose that $U \leq S_n$ acts transitively with a cyclic point stabilizer (for example, if U is regular), and let $\alpha \in \mathrm{Aut}(U)$ be equidistributing. Then α is induced by $N_{S_n}(U)$.*

PROOF. Let $S = \mathrm{Stab}_U(1) = \langle g \rangle$. Then $S^\alpha = \langle g^\alpha \rangle$. As α is equidistributing, g^α and thus S^α has a fixed point and thus $S^\alpha \leq \mathrm{Stab}_U(\omega)$ for some point ω. As the sizes coincide we must have equality. Thus α maps a point stabilizer to a point stabilizer. By [**DM96**, Theorem 4.2B], α is therefore induced by conjugation in the symmetric group.   □

A related question that has been studied in the literature, for example in [**PSZ78**] and [**Cam05**], is whether two permutation groups which have the same number of elements for any cycle structure of elements must be (permutation) isomorphic. (This has applications to the recognition of Galois groups, see e.g. [**Hul99**]). In this context Woltermann and Sehgal [**WS79**] obtained a result about uniqueness of such groups; the proof can be translated easily to yield the following result:

LEMMA 4. *Let U be a solvable $\frac{3}{2}$-transitive permutation group. Then every automorphism $\alpha \in \mathrm{Aut}(U)$ is induced by $N_{S_n}(U)$.*

PROOF. By [**Wie64**, Theorem 10.4] a solvable $\frac{3}{2}$-transitive group is either primitive or a Frobenius group. In either case it has a characteristic, regular normal subgroup N with $S = \mathrm{Stab}_U(1)$ a complement to N and all complements of N are conjugate [**Hup67**, Satz II.3.2, Satz V.8.3, Satz I.18.3]. As N is characteristic $N = N^\alpha$. Therefore S^α is a complement to N and thus conjugate to S. Thus S^α is a point stabilizer and by [**DM96**, Theorem 4.2B] α is induced by $N_{S_n}(U)$.   □

The third important special case we consider is that of intransitive elementary abelian groups that occur as base groups for imprimitive permutation groups. (This is essentially the relevant case for the construction of transitive groups in [**Hul05**].)

LEMMA 5. *Let p be a prime and $m = \lfloor \frac{n}{p} \rfloor$. Then S_n has a subgroup $V \cong Z_p^m$ whose m orbits are $\{1, \ldots, p\}$, $\{p+1, \ldots, 2p\}, \ldots, \{(m-1)p+1, \ldots, mp\}$. Suppose that $U \leq V$ and that $\alpha \in \mathrm{Aut}(U)$ is equidistributing. Then α is induced by $N_{S_n}(U)$.*

PROOF. The action of V on each orbit is the regular action of Z_p. Because of this the cycle structure of each element of V is determined by the *number of orbits* on which it acts nontrivially.

We consider V as an m-dimensional vector space over $\mathbb{F}_p$ with basis

$$\{(1, 2, \ldots, p), (p+1, \ldots, 2p), \ldots\}.$$

The weight (defined as in coding theory to be the number of nonzero entries in a vector) of each element of V corresponds to its cycle structure, considered as a permutation: the weight equals the number of p-cycles.

Now suppose that $U \leq V$. Then B consists of the automorphisms that preserve the weight of each vector, i.e., it consists of weight preserving linear transformations.

By the theorem of MacWilliams [**Mac62**], [**HP03**, Theorem 7.9.4] this implies that every automorphism in B is a monomial transformation, that is an element of $\mathbb{F}_p^* \wr S_m$. Such elements however can be represented as elements of $S_p \wr S_m$ and thus as elements of S_n. $\qquad\square$

5. Generalizations

The results of the previous section concentrating on elementary abelian groups might seem to be very weak. If we consider a general permutation group U however the following argument applies:

If U has a trivial radical the structure of U is very restricted and $[\text{Aut}(U) : U]$ is small. We can therefore simply set $B := \text{Aut}(U)$ and just run the backtrack search for $A \leq \text{Aut}(U)$. (One can obtain the structure of $\text{Aut}(U)$ easily by embedding into a direct product of wreath products [**CH03**].)

Otherwise, U has a nontrivial radical and therefore contains a characteristic elementary abelian subgroup $V \leq U$ and thus $N_G(U) \leq N_G(V)$. Setting $M := N_G(V)$ we have that $N_G(U) = N_M(U)$. As $V \triangleleft M$ we can compute this second normalizer in the factor group as $N_{M/V}(U/V)$, which provides a further reduction.

While we have considered permutation groups so far, the same same strategy can also be applied to the case of normalizing subgroups of $\text{GL}_n(q)$. The best permutation representation for this group has degree at least $(q^n - 1)/(q-1)$ which very quickly makes it infeasible to use such a permutation representation for a backtrack-type calculation. In effect therefore there exists no practical algorithm for the computation of subgroup normalizers.

On the other hand, the computation of normal forms of matrices yields an effective conjugacy test (as well as a determination of conjugating permutations). The calculation of module automorphisms [**Smi94**] provides an algorithm to compute element centralizers in $\text{GL}_n(q)$.

Furthermore, even a single element centralizer will be comparatively small. This makes it feasible to then use a stabilizer-chain based approach [**But82**] for the calculation of conjugating elements within the centralizer and to determine iterated centralizers.

Therefore $\text{GL}_n(q)$ fulfills all prerequisites for the algorithm proposed in this paper. Instead of cycle structure of elements, one can use normal forms for matrices. With this modification, the proposed algorithm makes it possible to compute the normalizer of somewhat small matrix groups in $\text{GL}_n(q)$. Again it might be preferable to start by normalizing small characteristic subgroups for which the automorphism group can be determined easily.

6. Examples

As stated above, the algorithm proposed does not offer better complexity than the ordinary backtrack for a normalizer calculation. On the other hand, for the case of elementary abelian subgroups sometimes a dramatically better practical performance has been observed. This section will present some evidence for such a claim.

In the following description the "old" algorithm is the partition backtrack algorithm for the normalizer as implemented in GAP 4.4 [**GAP04**], following [**The97**]. It incorporates an initial reduction from S_n to a direct product of wreath products, following [**Hul05**, section 11].

The "new" algorithm is the author's GAP implementation of the automorphism-based approach of this paper for the case of elementary abelian subgroups.

We shall consider randomly generated elementary abelian subgroups of S_{30}. They were generated by picking a random p-element of the group and then repeatedly selecting random p elements that centralize the elements chosen so far until a group of the desired order was generated. It was possible that the process stopped before the desired order was reached if there were no further p-elements in the centralizer. In this case the attempt was abandoned and the construction started anew from scratch.

Table 1 summarizes the results of these experiments. The column entries are:

> $|U|$: Order of the subgroups to be normalized.
>
> **Runs:** Number of experimental runs. (The attempt was made to run up to 100 examples but some runs were interrupted by hand after spending substantially more time, as long as the results obtained up to that point appeared consistent.)
>
> **AvgOld:** The average runtime for the "old" (backtrack-based) algorithm (in milliseconds).
>
> **AvgNew:** The average runtime for the "new" (automorphism-based) algorithm (in milliseconds).
>
> **AvgRatio:** The average ratio "old" to "new".
>
> **MinRatio:** The minimal ratio "old" to "new".
>
> **MaxRatio:** The maximal ratio "old" to "new".

Runtimes were calculated by GAP 4.4 on a 2.4GHz Pentium 4 under Linux and are given in milliseconds.

The cases in which runs were aborted by hand typically affected situations in which the "old" algorithm was performing particularly badly. As the terminated runs did not complete they were not included in the table but would have increased substantially the "worst case" factors.

Again, the results show that the new algorithm not only is faster in most cases, it also performs much less badly in the cases in which the old algorithm is superior than vice versa. In particular for larger subgroups that are still away from the theoretical maximum size for subgroups of S_{30} (e.g. 2^7, 3^6) the performance of the old algorithm is spectacularly worse.

The only cases in which the old algorithm is consistently better are subgroups of order 5^5, 5^6, 7^4 and 11^2. The reason for this is that the structure of the p-Sylow subgroups of S_{30} (($5 \wr 5$) × 5, 7^4, 11^2) very much restricts the possibilities for random subgroups of the given order: they will be almost always direct products

$|U|$	Runs	AvgOld	AvgNew	AvgRatio	MinRatio	MaxRatio
2^2	100	261	152	2.37	0.23	8.12
2^3	100	419	142	3.98	0.51	77.0
2^4	100	737	342	5.87	0.14	90.9
2^5	100	17546	174	206	0.66	15707
2^6	100	48698	236	491	0.67	12656
2^7	29	956058	448	3245	0.50	87103
3^2	100	426	182	3.73	0.31	11.8
3^3	100	3489	579	21.6	0.12	381
3^4	23	316277	834	2235	1.67	41826
3^5	13	376365	381	1202	1.86	12528
3^6	4	935833	1758	378	0.91	1103
5^2	100	2120	106	26.8	0.13	318
5^3	31	94583	195	477	0.70	4642
5^4	21	85266	1018	114	1.52	477
5^5	70	2176	23059	0.41	0.028	2.34
5^6	100	312	10770	0.029	0.019	0.047
7^2	100	575	109	5.81	0.51	27.7
7^3	100	457	328	1.43	0.30	3.09
7^4	100	139	1305	0.10	0.081	0.18
11^2	100	106	197	0.61	0.24	1.61

TABLE 1. Runtime Comparisons

$|U|$	Runs	AvgOld	AvgNew	AvgRatio	MinRatio	MaxRatio
5^5	20	737772	15425	186	1	648
7^3	20	11729	428	22	0.7	64

TABLE 2. Runtime Comparisons in S_{35}

of disjoint cycles. The stabilizer of this orbit partition, obtained as the first step of the normalizer algorithm is then very close to the normalizer, which leaves very little work for the backtrack search.

As soon as G is increased this second effect vanishes, and the "new" algorithm again performs better, as the times for normalizers in S_{35}, given in table 2 shows.

References

[But82]   Gregory Butler, *Computing in permutation and matrix groups II: backtrack algorithms*, Math. Comp. **39** (1982), no. 160, 671–670.

[Cam05]   Peter J. Cameron, *Partitions and permutations*, Discrete Math. **291** (2005), 45–54.

[CH03]   John Cannon and Derek Holt, *Automorphism group computation and isomorphism testing in finite groups*, J. Symbolic Comput. **35** (2003), no. 3, 241–267.

[DM96]   John D. Dixon and Brian Mortimer, *Permutation groups*, Graduate Texts in Mathematics, vol. 163, Springer, 1996.

[ELGO02] Bettina Eick, C.R. Leedham-Green, and E.A. O'Brien, *Constructing automorphism groups of p-groups*, Comm. Algebra **30** (2002), no. 5, 2271–2295.

[GAP04]   The GAP Group, http://www.gap-system.org, GAP – *Groups, Algorithms, and Programming*, Version *4.4*, 2004.

[HEO05]   Derek F. Holt, Bettina Eick, and Eamonn A. O'Brien, *Handbook of Computational Group Theory*, Discrete Mathematics and its Applications, Chapman & Hall/CRC, Boca Raton, FL, 2005.

[Hol91]   D. F. Holt, *The computation of normalizers in permutation groups*, J. Symbolic Comput. **12** (1991), no. 4-5, 499–516, Computational group theory, Part 2.

[HP03]   W. Cary Huffman and Vera Pless, *Fundamentals of error-correcting codes*, Cambridge University Press, 2003.

[Hul99]   Alexander Hulpke, *Techniques for the computation of Galois groups*, Algorithmic Algebra and Number Theory (B. H. Matzat, G.-M. Greuel, and G. Hiss, eds.), Springer, 1999, pp. 65–77.

[Hul05]   _____, *Constructing transitive permutation groups*, J. Symbolic Comput. **39** (2005), no. 1, 1–30.

[Hup67]   Bertram Huppert, *Endliche Gruppen I*, Grundlehren der mathematischen Wissenschaften, vol. 134, Springer, 1967.

[Leo97]   Jeffrey S. Leon, *Partitions, refinements, and permutation group computation*, Proceedings of the 2nd DIMACS Workshop held at Rutgers University, New Brunswick, NJ, June 7–10, 1995 (Larry Finkelstein and William M. Kantor, eds.), DIMACS: Series in Discrete Mathematics and Theoretical Computer Science, vol. 28, American Mathematical Society, Providence, RI, 1997, pp. 123–158.

[LM02]   Eugene M. Luks and Takunari Miyazaki, *Polynomial-time normalizers for permutation groups with restricted composition factors*, Proceedings of the 2002 International Symposium on Symbolic and Algebraic Computation (Teo Mora, ed.), The Association for Computing Machinery, ACM Press, 2002, pp. 176–183.

[Luk93]   Eugene M. Luks, *Permutation groups and polynomial-time computation*, Groups and Computation (Providence, RI) (Larry Finkelstein and William M. Kantor, eds.), DIMACS: Series in Discrete Mathematics and Theoretical Computer Science, vol. 11, American Mathematical Society, 1993, pp. 139–175.

[Mac62]   F. J. MacWilliams, *Combinatorial problems of elementary abelian groups*, Ph.d. thesis, Harvard University, 1962.

[Miy06]   Izumi Miyamoto, *An improvement of* GAP *normalizer function for permutation groups*, Proceedings of the 31st International Symposium on Symbolic and Algebraic Computation held in Genova, July 9–12, 2006 (Jean-Guillaume Dumas, ed.), ACM Press, New York, 2006.

[PSZ78]   Ann Scrandis Playtis, Surinder Sehgal, and Hans Zassenhaus, *Equidistributed permutation groups*, Comm. Algebra **6** (1978), no. 1, 35–57.

[RD04]   Colva M. Roney-Dougal, *Conjugacy of subgroups of the general linear group*, Experimental Mathematics **13** (2004), no. 2, 151–163.

[Ser03]   Ákos Seress, *Permutation group algorithms*, Cambridge University Press, 2003.

[Sho28]   Kenjiro Shoda, *Über die Automorphismen einer endlichen Abelschen Gruppe*, Math. Ann. **100** (1928), 674–686.

[Sim70]   Charles C. Sims, *Computational methods in the study of permutation groups*, Computational Problems in Abstract Algebra (John Leech, ed.), Pergamon press, 1970, pp. 169–183.

[Smi94]   Michael J. Smith, *Computing automorphisms of finite soluble groups*, Ph.D. thesis, Australian National University, Canberra, 1994.

[The97]   Heiko Theißen, *Eine Methode zur Normalisatorberechnung in Permutationsgruppen mit Anwendungen in der Konstruktion primitiver Gruppen*, Dissertation, Rheinisch-Westfälische Technische Hochschule, Aachen, Germany, 1997.

[Wie64]   Helmut Wielandt, *Finite permutation groups*, Academic Press, 1964.

[WS79]   Michael Woltermann and Surinder Sehgal, *Equidistributed $\frac{3}{2}$-transitive solvable permutation groups. I, II*, Comm. Algebra **7** (1979), no. 15, 1599–1643, 1645–1672.

DEPARTMENT OF MATHEMATICS, COLORADO STATE UNIVERSITY, FORT COLLINS, CO 80523
E-mail address: hulpke@math.colostate.edu

Contemporary Mathematics
Volume **470**, 2008

Group theory in **SAGE**

David Joyner and David Kohel

ABSTRACT. SAGE is an open source computer algebra system implemented using an object-oriented categorical framework, with methods for objects, methods for their elements, and methods for their morphisms. Currently, SAGE
has the ability to deal with abelian groups, permutation groups, and matrix
groups over a finite field. This paper will present an overview of the implementations of the group-theoretical algorithms in SAGE, with some examples.
We conclude with some possible future directions.

SAGE [**S**] is a general purpose computer algebra system started in 2005, built
on top of existing open source packages, including GAP for group theory, Maxima
for symbolic computation, Pari for number theory, and Singular for multivariate
polynomial computations and commutative algebra. In design, SAGE uses the best
of the ideas in MAGMA [**M**], Mathematica [**Ma**] and other systems, but uses the
popular mainstream language Python as its interpreter. This paper will restrict
itself to presenting an overview of the implementations of the group-theoretical
algorithms in SAGE.

According to the GAP website [**G**], each year between 50 and 100 papers are
published which use GAP in an essential way. SAGE is far too young to have such
an impressive research record. Still, it has several people for published research in
coding theory, number theory, and modular forms, as well as being used in teaching
both graduate and undergraduate math classes.

Currently, SAGE has the ability to deal with abelian groups (finitely generated
multiplicative abelian groups, groups of Dirichlet characters, and dual groups of
finite abelian groups), permutation groups, matrix groups over a finite field, and
congruence subgroups. The sections below will deal with these classes of groups
separately. As a recreational aside, some algorithms enabling one to model the
Rubik's cube using group theory are also included; for example, three fast optimized
solvers are included with SAGE.

In rough design, group theory in SAGE is implemented in an object-oriented
categorical framework, with methods for group objects G and H, methods for
their elements g and h, methods for their set of morphisms $\mathrm{Hom}(G, H)$ and methods for the element homomorphisms $\phi : G \rightarrow H$. For instance, a permutation
group G would be an object which has, for example, a method called **order**, which

2000 *Mathematics Subject Classification.* Primary: 20-01. Secondary: 20-04, 20B40.
Key words and phrases. SAGE, GAP, finite group computations.

returns $|G|$. This class with its methods are collected into a permutation group
Python "module". Likewise the class of permutations $g \in G$ also has a method
called **order**, which computes the smallest $n > 0$ for which $g^n = 1$. The class
of permutations with associated methods, are collected into a permutation group
element "module". Similarly, there exist a class for the sets $\mathrm{Hom}(G, H)$ of group
homomorphisms between groups G and H, with associated methods such as **domain**
and **codomain**. When both G and H are both permutation groups, the elements
of $\mathrm{Hom}(G, H)$ belong to a class of permutation group homomorphisms, and its associated member functions such as **kernel** are collected into a permutation group
morphism "module". The SAGE source code contains comprehensive documentation on these modules.

We conclude with a section on possible future directions of SAGE and group
theory.

1. The GAP interface

For the most part, SAGE's high-level group-theoretic capabilities are derived
from the extensive functionality of the computer algebra system GAP. Fast low-
level arithmetic is achieved by native code in C or Python. SAGE communicates
with GAP and the other components using pseudotty's and a Python package called
pexpect [**P**]. The SAGE/Python functions which call GAP using **pexpect** are called
"wrappers".

- **Pexpect:** *makes Python a better tool for controlling other applications.*
  (**pexpect.sourceforge.net**)
  Pexpect is a pure Python module for spawning child applications;
  controlling them; and responding to expected patterns in their output.
  gap.eval('gapcommand') sends 'gapcommand' to GAP
  For example[1],

```
───────────────────────── SAGE ─────────────────────────

sage: gap.eval('2+3')
'5'
```

  evaluates $2 + 3$ in GAP and returns the answer as a string. Some such
  GAP "string" output can be used in SAGE using the **eval** command.
  For example, integers and lists can (**eval('5')+1** returns 6), but also
  polynomials in some cases. Except for these simple data structures, in
  most cases, GAP output cannot be used directly in SAGE and conversely.
- **Pseudotty:** A device which appears to an application program as an
  ordinary terminal but which is *in fact* connected to a different process.
  Pseudo-ttys have a slave half and a control half.
  The command **gap_console()** brings up a GAP prompt in SAGE.
  For example,

---

[1]Note that usually GAP requires a semi-colon at the end, but single semi-colons are not
needed inside a **gap.eval** command. If you want to suppress the output then you do need to use
a double semi-colon though.

```
─────────────────────── SAGE ───────────────────────

sage: gap_console()
GAP4, Version: 4.4.10 of 02-Oct-2007, x86_64-unknown-linux-gnu-gcc
gap> 2+3;
5
gap>
```

brings up GAP inside SAGE. There is no preparsing.

A functional understanding of pseudotty's and pexpect is not needed to use SAGE, however, it helps to understand the functioning of SAGE commands.

For objects and morphisms, but not elements, SAGE uses Python wrappers of GAP functions, which it communicates to using pexpect. For many high level algorithms, for example the computation of derived series, there is no noticeable overhead. For some low level operations, such as computations with permutation group elements, SAGE has a native optimized compiled implementation (still not as fast as GAP's implementation, written in C code). However, for almost all matrix group and abelian group operations, SAGE currently passes the computation to GAP and returns the result via pexpect.

2. Abelian groups

Finitely generated abelian groups are supported, both finite and infinite, with a multiplicative notation for elements. For the finite abelian groups, we simply wrap the appropriate GAP functions. However, for the infinite abelian groups, the corresponding GAP functions are located in a GAP package which had ambiguous licensing, so the code could not be used. Some SAGE functions for infinite abelian groups are 100% pure Python.

To be concrete, we present some background in order to introduce notation. A finitely generated abelian group is a group A for which there exists a finite presentation defined by an exact sequence

$$\mathbb{Z}^k \to \mathbb{Z}^\ell \to A \to 1,$$

for positive integers k, ℓ with $k \le \ell$.

For example, a finite abelian group has a decomposition

$$A = \langle a_1 \rangle \times \langle a_2 \rangle \times \cdots \times \langle a_\ell \rangle,$$

where $ord(a_i) = p_i^{c_i}$, for some primes p_i and some positive integers c_i, $i = 1, 2, \ldots, \ell$. GAP calls the list (ordered by size) of the $p_i^{c_i}$ the *abelian invariants*. In SAGE they will be called *invariants*. In this situation, $k = \ell$ and $\phi : \mathbb{Z}^\ell \to A$ is the map $\phi(x_1, \ldots, x_\ell) = a_1^{x_1} \cdots a_\ell^{x_\ell}$, for $x_i \in \mathbb{Z}$. The matrix of relations $M : \mathbb{Z}^k \to \mathbb{Z}^\ell$ is the matrix whose rows generate the kernel of ϕ as a $\mathbb{Z}$-module. In other words, $M = (M_{ij})$ is an $k \times \ell$ diagonal matrix with $M_{ii} = p_i^{c_i}$. Consider now the subgroup $B \subset A$ generated by $b_1 = a_1^{f_{1,1}} \cdots a_\ell^{f_{\ell,1}}, \ldots, b_m = a_1^{f_{1,m}} \cdots a_\ell^{f_{\ell,m}}$. The kernel of the map $\phi_B : \mathbb{Z}^m \to B$ defined by $\phi_B(x_1, \ldots, x_m) = b_1^{x_1} \cdots b_m^{x_m}$, for $x_i \in \mathbb{Z}$, is the kernel of the matrix $F = (f_{i,j})$ regarded as a map $\mathbb{Z}^m \to (\mathbb{Z}/p_1^{c_1}\mathbb{Z}) \times \cdots \times (\mathbb{Z}/p_\ell^{c_\ell}\mathbb{Z})$. In particular, $B \cong \mathbb{Z}^m / \ker(F)$. If $B = A$ then the Smith normal form (SNF) of a generator matrix of $\ker(F)$ and the SNF of M are the same. The diagonal entries s_i of the SNF $S = \text{diag}(s_1, s_2, \ldots, s_r, 0, \ldots, 0)$, are called *determinantal divisors* of F, where r is the rank. The *invariant factors* of A are:

$$s_1, s_2/s_1, \ldots, s_r/s_{r-1}.$$

The elementary divisors use the highest (non-trivial) prime powers occuring in the factorizations of the numbers $s_1, s_2, \ldots, s_r$. The definition of elementary divisors of an abelian group used by SAGE is that of Rotman [**R**][2].

SAGE supports multiplicative abelian groups on any prescribed finite number of generators.

Here's a simple example:

──────────────── SAGE ────────────────

```
sage: F.<a,b,c,d,e> = AbelianGroup(5, [5,5,7,8,9])
sage: F(1)
1
sage: prod([ a, b, a, c, b, d, c, d ])
a^2*b^2*c^2*d^2
sage: d * b**2 * c**3
b^2*c^3*d
sage: G = AbelianGroup(3,[2,2,2]); G
Multiplicative Abelian Group isomorphic to C2 x C2 x C2
sage: H = AbelianGroup([2,3], names="xy"); H
Multiplicative Abelian Group isomorphic to C2 x C3
sage: AbelianGroup(5)
Multiplicative Abelian Group isomorphic to
  Z x Z x Z x Z x Z
sage: AbelianGroup(5).order()
+Infinity
```

What is created above is, first, a group F with 5 generators a, b, c, d, e, of orders $5, 5, 7, 8, 9$, respectively. This group could also be created using the two lines:

──────────────── SAGE ────────────────

```
sage: F = AbelianGroup(5, [5,5,7,8,9], names='abcde')
sage: (a, b, c, d, e) = F.gens()
```

Secondly, a group G with 3 (unnamed) generators of orders $2, 2, 2$. Next, a group H is created having 2 generators x, y, of orders $2, 3$. Finally, a free abelian group of rank 5 is created.

EXAMPLE 1. Abelian groups arise naturally throughout mathematics. Inheritance in Python means that an object can inherit from the class of abelian groups to achieve functionality as an abelian group, or in this example, construct an isomorphism with a group object.

──────────

[2]Since different texts have different definitions of this, this usage may be "non-standard", depending on your background.

```
—————————————————— SAGE ——————————————————
sage: k = FiniteField(101)
sage: E = EllipticCurve([k(1),k(3)])
sage: G, gens = E.abelian_group()
sage: G
Multiplicative Abelian Group isomorphic to C87
sage: gens
((32 : 68 : 1),)
```

We note that the second return value is a sequence of group generators which implicitly determines an isomorphism. In order to return instead an isomorphism, it remains to develop a generic framework for morphisms between arbitrary groups. In this instance the abelian group homomorphism must handle the translation between multiplicative and additive notation.

The elements themselves of an abelian group A have some convenient methods implemented. For example, you can determine the permutation which a group element is associated to when you regard A as a permutation group. Also, you can determine the order of an element and read off the powers of the generators occurring in the element's expression.

Here's a simple example:

```
—————————————————— SAGE ——————————————————
sage: A = AbelianGroup(3,[2,3,4],names="abc"); A
Multiplicative Abelian Group isomorphic to C2 x C3 x C4
sage: a,b,c=A.gens()
sage: (c^3*b).list()
[0, 1, 3]
sage: G = A.permutation_group(); G
Permutation Group with generators \
[(1,13)(2,14)(3,15)(4,16)(5,17)(6,18)(7,19)\
(8,20)(9,21)(10,22)(11,23)(12,24),\
 (1,5,9)(2,6,10)(3,7,11)(4,8,12)(13,17,21)\
(14,18,22)(15,19,23)(16,20,24),\
 (1,3,2,4)(5,7,6,8)(9,11,10,12)(13,15,14,16)\
(17,19,18,20)(21,23,22,24)]
sage: g = a.as_permutation(); g
(1,13)(2,14)(3,15)(4,16)(5,17)(6,18)(7,19)(8,20)\
(9,21)(10,22)(11,23)(12,24)
sage: g in G
True
```

Also implemented are homomorphisms. With them, you can compute images and kernels. Here's an example:

```
—————————————————— SAGE ——————————————————
sage: H = AbelianGroup(3,[2,3,4],names="abc"); H
Multiplicative Abelian Group isomorphic to C2 x C3 x C4
sage: a,b,c = H.gens()
sage: G = AbelianGroup(2,[2,3],names="xy"); G
Multiplicative Abelian Group isomorphic to C2 x C3
```

```
sage: x,y = G.gens()
sage: phi = AbelianGroupMorphism(G,H,[x,y],[a,b])
sage: phi(y*x)
a*b
sage: phi(y^2)
b^2
sage: phi.parent() == Hom(G,H)
True
sage: phi.parent()
Set of Morphisms from Multiplicative Abelian Group
isomorphic to C2 x C3 to Multiplicative Abelian Group
isomorphic to C2 x C3 x C4 in Category of groups
```

The dual group (the group of complex characters) of a finite abelian group is also implemented. GAP does not have such dual groups functionality, so for this implementation it is not simply a matter of wrapping GAP functions.

Here's an example:

---------------------------------- SAGE ----------------------------------

```
sage: F = AbelianGroup(5, [2,3,5,7,8], names="abcde")
sage: a,b,c,d,e = F.gens()
sage: Fd = DualAbelianGroup(F, names="ABCDE")
sage: A,B,C,D,E = Fd.gens()
sage: A*B^2*D^7
A*B^2
sage: A(a)      ## random last few digits
-1.0000000000000000 + 0.00000000000000013834419720915037*I
sage: B(b)      ## random last few digits
-0.49999999999999983 + 0.86602540378443871*I
sage: A(a*b)     ## random last few digits
-1.0000000000000000 + 0.00000000000000013834419720915037*I
```

Note that since the field of complex numbers is represented using floating point numbers, inaccuracies may enter into the least significant digits. If desired, a higher precision or even a different base ring may be specified (though characteristic 0 is currently assumed). Similarly, a **DirichletCharacter** is the extension of a homomorphism $(\mathbb{Z}/N\mathbb{Z})^* \to R^*$, for some ring R, to the map $\mathbb{Z}/N\mathbb{Z} \to R$ obtained by sending nonunits to 0:

---------------------------------- SAGE ----------------------------------

```
sage: G = DirichletGroup(35)
sage: G.gens()
([zeta12^3, 1], [1, zeta12^2])
sage: g0,g1 = G.gens()
sage: g = g0*g1; g
[zeta12^3, zeta12^2]
sage: g.order()
12
```

Dirichlet group elements in SAGE support, among others, special methods for computing Galois orbits, Gauss sums, and generalized Bernoulli numbers

3. Permutation groups

A *permutation group* is a finite group G whose elements are permutations of a given finite set X (i.e., bijections $X \to X$) and whose group operation is the composition of permutations. The number of elements of X is called the *degree* of G.

3.1. Some permutation group methods. A permutation is inputted into SAGE as either a string that defines a permutation using disjoint cycle notation, or a list of tuples which represent disjoint cycles:

$$(1) \qquad \begin{array}{ccc} (a,\ldots,b)(c,\ldots,d)\cdots(e,\ldots,f) & \leftrightarrow & [(a,\ldots,b),(c,\ldots,d),\ldots,(e,\ldots,f)] \\ () = \text{identity} & \leftrightarrow & [()]. \end{array}$$

Warning: There is a SAGE command `Permutation` which does *not* currently make a permutation group element. This command forms part of the combinatorics package, and is not to be confused with permutation group constructors.

You can construct the following standard groups as permutation groups. The finite matrix groups over $GF(q)$ in the list use standard permutation representations to construct them as permutation groups.

- `SymmetricGroup(n)`, S_n of order $n!$, and `AlternatingGroup(n)`, A_n of order $n!/2$ (n can also be a list X of distinct positive integers)
- `DihedralGroup(n)`, D_n of order $2n$, and `CyclicPermutationGroup(n)`, C_n of order n
- `TransitiveGroup(i,n)`f, the ith transitive group of degree n from the GAP tables of transitive groups (this command requires the "optional" package `database_gap`)
- `MathieuGroup(d)`, of degrees 9, 10, 11, 12, 21, 22, 23, or 24
- `KleinFourGroup`, the subgroup of S_4 of order 4 isomorphic to $C_2 \times C_2$
- `PGL(n,q)`, `PSL(n,q)`, `PSp(2n,q)`, projective (general, special, symplectic, resp.) linear group of $n \times n$ matrices over the finite field $GF(q)$
- `PSU(n,q)`, projective special unitary group of $n \times n$ matrices having coefficients in the finite field $GF(q^2)$ that respect a fixed nondegenerate sesquilinear form of determinant 1.
- `PGU(n,q)`, projective general unitary group of $n \times n$ matrices having coefficients in the finite field $GF(q^2)$ that respect a fixed nondegenerate sesquilinear form, modulo the centre.
- `Suzuki(q)`, Suzuki group over $GF(q)$, $^2B_2(2^{2k+1}) = Sz(2^{2k+1})$.
- `direct_product_permgroups`, which takes a list of permutation groups and returns their direct product.

Permutation groups include methods for computing composition series, lower and upper central series, multiplication table, character table, quotient group by a normal subgroup, Sylow subgroups, the number of groups of a given order, and many others. The are made available via wrappers for corresponding GAP functions.

There are a number of functions which interface with GAP's `SmallGroups` database, so it is a good idea to have that installed before trying out the examples below[3].

Here are some examples which illustrate various ways to construct permutations in S_4:

```
                          ── SAGE ──

sage: G = SymmetricGroup(4)
sage: G((1,2,3,4))
(1,2,3,4)
sage: G([(1,2),(3,4)])
(1,2)(3,4)
sage: G('(1,2)(3,4)')
(1,2)(3,4)
sage: G([1,2,4,3])
(3,4)
sage: G([2,3,4,1])
(1,2,3,4)
sage: G(G((1,2,3,4)))
(1,2,3,4)
sage: G(1)
()
```

The constructor `PermutationGroupElement` creates an element of some the symmetric group S_n for n minimal. Automatic coercion using $S_n \subset S_{n+m}$ allows the multiplication of elements in different groups, with the result in a minimal group containing both elements. Moreover, such coercion permits equality testing between elements in different permutation groups.

```
                          ── SAGE ──

sage: g1 = PermutationGroupElement([(1,2),(3,4,5)])
sage: g1.parent()
Symmetric group of order 5! as a permutation group
sage: g2 = PermutationGroupElement([(1,2)])
sage: g2.parent()
Symmetric group of order 2! as a permutation group
sage: g1*g2
(3,4,5)
sage: g2*g1
(3,4,5)
sage: g1 == g2
False
```

A permutation group can be constructed in SAGE either by standard constructors or by a given list of generators. SAGE recognises mathematically identical groups as equal even if they are constructed with different sets of generators.

---

[3]The `SmallGroups` database is not GPL'd and so must be installed into SAGE separately. See the optional packages page http://www.sagemath.org/packages.html for instructions.

———————————————————— SAGE ————————————————————

```
sage: S5 = SymmetricGroup(5)
sage: A5 = AlternatingGroup(5)
sage: A5
Alternating group of order 5!/2 as a permutation group
sage: A5.gens()
((1,2,3,4,5), (3,4,5))
sage: A5.is_subgroup(S5)
True
sage: G = PermutationGroup([(1,2,3,4,5),(1,2,3)])
sage: G
Permutation Group with generators [(1,2,3,4,5), (1,2,3)]
sage: G == A5
True
sage: G.group_id()    # requires database_gap* package
[60, 5]
```

Similarly, Galois groups computed with PARI can be imported into SAGE using the GAP small groups database:

———————————————————— SAGE ————————————————————

```
sage: H = pari('x^4 - 2*x^3 - 2*x + 1').polgalois()
sage: G = PariGroup(H, 4); G
PARI group [8, -1, 3, "D(4)"] of degree 4
sage: H = PermutationGroup(G); H  # requires database_gap
Transitive group number 3 of degree 4
sage: H.gens()                         # requires database_gap
((1,2,3,4), (1,3))
```

There is an underlying GAP object that implements each permutation group. One can apply GAP commands (such as DerivedSeries) to this GAP object.

———————————————————— SAGE ————————————————————

```
sage: G = PermutationGroup([[(1,2,3,4)]])
sage: H = gap(G); H
Group([ (1,2,3,4) ])
sage: G = PermutationGroup([[(1,2,3),(4,5)],[(3,4)]])
sage: H.DerivedSeries()    # output somewhat random
[ Group([ (1,2,3)(4,5), (3,4) ]),
  Group([ (1,5)(3,4), (1,5)(2,4), (1,4,5) ]) ]
sage: G.derived_series()
[Permutation Group with generators
 [(1,2,3)(4,5), (3,4)],
 Permutation Group with generators
 [(1,5)(3,4), (1,5)(2,4), (1,3,5)]]
```

Here G is a "SAGE group", to which SAGE's methods (such as **order**) apply. On the other hand, H is a "GAP group", to which (any and all of) GAP's group-theoretical methods (such as **Size**) apply. The command **G.derived_series()** returns the groups as SAGE objects, whereas the output of **H.DerivedSeries()** are GAP objects.

A *composition series* of a group G is a normal series

$$1 = H_0 \subset H_1 \subset \cdots \subset H_n = G,$$

with strict inclusions, such that each H_i is a maximal normal subgroup of H_{i+1}. The SAGE method **composition_series** wraps the corresponding GAP function **CompositionSeries**:

──────────────── SAGE ────────────────

```
sage: G = PermutationGroup([[(1,2,3),(4,5)],[(3,4)]])
sage: G.composition_series()  # random output order
[Permutation Group with generators [(1,2,3)(4,5), (3,4)],
 Permutation Group with generators
 [(1,5)(3,4), (1,5)(2,4), (1,3,5)],
 Permutation Group with generators [()]]
```

SAGE has similar commands **lower_central_series** and **upper_central_series**.

You can also save and reload objects created in a SAGE session:

──────────────── SAGE ────────────────

```
sage: G = DihedralGroup(6)
sage: Z = G.center()
sage: save(G, 'G')
sage: save(Z, 'Z')
sage:
Exiting SAGE (CPU time 0m2.08s, Wall time 615m27.05s).
Exiting spawned Gap process.
wdj@wooster:~/wdj/sagefiles/sage-2.9.alpha5$ ./sage
----------------------------------------------------------------------
| SAGE Version 2.9.alpha5, Release Date: 2007-12-10                  |
| Type notebook() for the GUI, and license() for information.        |
----------------------------------------------------------------------

sage: G = load('G')
sage: Z = load('Z')
sage: Z.is_subgroup(G)
True
```

SAGE does not remember Z as the center of G, but rather that Z and G are given by certain generators as a permutation group. From that, it can determine that Z is a subgroup of G. If you use **save_session** and **load_session** instead, the behaviour is similar, except that it automatically saves the variables for you:

──────────────── SAGE ────────────────

```
----------------------------------------------------------------------
| SAGE Version 2.9.alpha5, Release Date: 2007-12-10                  |
| Type notebook() for the GUI, and license() for information.        |
----------------------------------------------------------------------

sage: G = DihedralGroup(6)
sage: Z = G.center()
```

```
sage: Z.is_subgroup(G)
True
sage: save_session('dihedral6')
sage:
Exiting SAGE (CPU time 0m0.23s, Wall time 1m3.16s).
Exiting spawned Gap process.
wdj@wooster:~/wdj/sagefiles/sage-2.9.alpha5$ ./sage
--------------------------------------------------------------------
| SAGE Version 2.9.alpha5, Release Date: 2007-12-10                 |
| Type notebook() for the GUI, and license() for information.       |
--------------------------------------------------------------------

sage: load_session('dihedral6')
sage: G; Z
Dihedral group of order 12 as a permutation group
Permutation Group with generators [(1,4)(2,5)(3,6)]
sage: Z.is_subgroup(G)
True
```

SAGE's `direct_product` wraps the GAP functions `DirectProduct`, `Embedding`, and `Projection` functions. The direct product of permutation groups will be a permutation group again. As input one gives two permutation groups G_1, G_2, and the output produced is a 5-tuple $(D, \iota_1, \iota_2, \mathrm{pr}_1, \mathrm{pr}_2)$, where:

- D is a direct product of G_1, G_2, returned as a permutation group;
- ι_1 is an embedding of G_1 into D;
- ι_2 is an embedding of G_2 into D;
- pr_1 is the projection of D onto G_1 (satisfying $\iota_1 \circ \mathrm{pr}_1 = 1$); and
- pr_2 is the projection of D onto G_2 (satisfying $\iota_2 \circ \mathrm{pr}_2 = 1$).

Some examples:

```
———————————————— SAGE ————————————————
sage: G = CyclicPermutationGroup(4)
sage: D = G.direct_product(G,False)
sage: D
Permutation Group with generators [(1,2,3,4), (5,6,7,8)]
sage: D,iota1,iota2,pr1,pr2 = G.direct_product(G)
sage: D; iota1; iota2; pr1; pr2
Permutation Group with generators [(1,2,3,4), (5,6,7,8)]
Homomorphism : Cyclic group of order 4 as a permutation group
   --> Permutation Group with generators [(1,2,3,4), (5,6,7,8)]
Homomorphism : Cyclic group of order 4 as a permutation group
   --> Permutation Group with generators [(1,2,3,4), (5,6,7,8)]
Homomorphism : Permutation Group with generators
               [(1,2,3,4), (5,6,7,8)]
   --> Cyclic group of order 4 as a permutation group
Homomorphism : Permutation Group with generators
               [(1,2,3,4), (5,6,7,8)]
   --> Cyclic group of order 4 as a permutation group
sage: g=D([[(1,3),(2,4)]]); g
(1,3)(2,4)
sage: d=D([[(1,4,3,2),(5,7),(6,8)]]); d
(1,4,3,2)(5,7)(6,8)
sage: iota1(g); iota2(g); pr1(d); pr2(d)
(1,3)(2,4)
(5,7)(6,8)
(1,4,3,2)
(1,3)(2,4)
```

SAGE can also display the matrix of values of the irreducible characters of a permutation group G at the conjugacy classes of G. The columns represent the conjugacy classes of G and the rows represent the different irreducible characters in the ordering given by GAP.

Some examples[4]:

─────────────────────────── SAGE ───────────────────────────

```
sage: # Alternating group A_4 of order 12:
sage: G = PermutationGroup([[(1,2),(3,4)], [(1,2,3)]])
sage: G.order()
12
sage: G.character_table()
[   1     1       1               1    ]
[   1     1   -zeta3 - 1      zeta3 ]
[   1     1   zeta3 -zeta3     - 1   ]
[   3    -1       0               0   ]
sage: # Dihedral group D_4 of order 8:
sage: G = PermutationGroup([[(1,2),(3,4)], [(1,2,3,4)]])
sage: G.order()
8
sage: G.character_table()
[ 1  1  1  1  1]
[ 1 -1 -1  1  1]
[ 1 -1  1 -1  1]
[ 1  1 -1 -1  1]
[ 2  0  0  0 -2]
```

SAGE can compute the Sylow p-subgroups of a permutation group G. Here's an example:

─────────────────────────── SAGE ───────────────────────────

```
sage: G = AlternatingGroup(5)
sage: G.order()
60
sage: G2 = G.sylow_subgroup(2)
sage: G2.order()
4
sage: G.sylow_subgroup(3)
Permutation Group with generators [(1,2,3)]
sage: G.sylow_subgroup(5)
Permutation Group with generators [(1,2,3,4,5)]
sage: G.sylow_subgroup(7)
Permutation Group with generators [()]
```

Now might be a good time to recall what J. Neubüser said in his invitation to Computational Group Theory in 1995:

───────────

[4]Here, zeta3 refers to a primitive cubic root of unity.

Nobody has ever paid a license fee for the proof that Sylow subgroups exist in every finite group, Nobody should ever pay a license
fee for computing Sylow subgroups in a given finite group.

Joachim Neubüser [**N**]

SAGE also computes the quotient group G/N, where N is a normal subgroup
of a permutation group. The method `quotient_group` wraps the GAP operator /.

```
──────────────────────── SAGE ────────────────────────

sage: G = PermutationGroup([[(1,2,3), (2,3)]])
sage: N = PermutationGroup([[(1,2,3)]])
sage: G.quotient_group(N)
Permutation Group with generators [(1,2)]
```

HAP ("Homology, Algebra, Programming") is a GAP package for group homology and cohomology written by Graham Ellis [**H**]. Thanks to SAGE's wrappers of
several HAP functions, homology and cohomology of permutation groups can be
computed in SAGE. To compute the homology groups $H^5(S_3, \mathbb{Z})$ and $H^5(S_3, \mathrm{GF}(2))$,
use the following commands.

```
──────────────────────── SAGE ────────────────────────

sage: G = SymmetricGroup(3)
sage: G.cohomology(5)        # needs optional gap_packages
  Trivial Abelian Group
sage: G.cohomology(5,2)      # needs optional gap_packages
  Multiplicative Abelian Group isomorphic to C2
```

The GAP package HAP is not distributed with SAGE, so an additional SAGE
package must be loaded for these commands to work. For further details, we refer
the interested reader to the expository paper [**J1**].

3.2. Element methods. Currently, elements of a permutation group can act
on some simple objects, such as strings, lists, and multivariate polynomials.

```
──────────────────────── SAGE ────────────────────────

sage: G = SymmetricGroup(4)
sage: g = G((1,2,3,4))
sage: g('abcd')
'bcda'
sage: g([0,6,-2,11])
[6, -2, 11, 0]
sage: h = G((1,2))
sage: g*h == h*g
False
sage: (g*h)(1); g(h(1)); h(g(1))
1
3
1
```

Unfortunately the current notation, which suggests g(i) is a left action, is not compatible with the current definition of multiplication (g*h)(i) is h(g(i)) rather than g(h(i)). Moreover, methods implementing more sophisticated actions are currently lacking (see §5.4 below for further discussion).

3.3. Recreational aside: Rubik's cube. The Rubik's cube[5] is a puzzle with $9 \times 6 = 56$ facets, but only 48 of them move. (You can think of the 6 center facets as being fixed, since we assume you have fixed an orientation in space of your cube once and for all.) Labeling the facets 1, 2, ..., 48 in any fixed way you like, each move of the Rubik's cube amounts to permuting the symbols in $\{1, 2, \ldots, 48\}$.

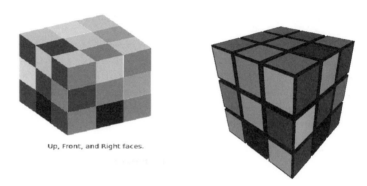

Up, Front, and Right faces.

FIGURE 1. Different SAGE plots of the "superflip" in 3d.

A "basic move" consists of a quarter turn in the clockwise direction of one of the 6 faces. These moves are usually denoted R(ight), L(eft), U(p), D(own), F(ront), B(ack). Each quarter face turn of a Rubik's cube has order 4. Let us compute the order of the Rubik's cube group and the order of a face turn:

```
───────────────── SAGE ─────────────────

sage: f= [(17,19,24,22),(18,21,23,20),(6,25,43,16),\
(7,28,42,13),(8,30,41,11)]
sage: b=[(33,35,40,38),(34,37,39,36),( 3, 9,46,32),\
( 2,12,47,29),( 1,14,48,27)]
sage: l=[( 9,11,16,14),(10,13,15,12),( 1,17,41,40),\
( 4,20,44,37),( 6,22,46,35)]
sage: r=[(25,27,32,30),(26,29,31,28),( 3,38,43,19),\
( 5,36,45,21),( 8,33,48,24)]
sage: u=[( 1, 3, 8, 6),( 2, 5, 7, 4),( 9,33,25,17),\
(10,34,26,18),(11,35,27,19)]
sage: d=[(41,43,48,46),(42,45,47,44),(14,22,30,38),\
(15,23,31,39),(16,24,32,40)]
sage: cube = PermutationGroup([f,b,l,r,u,d])
sage: F,B,L,R,U,D = cube.gens()
sage: cube.order()
43252003274489856000
```

---

[5]The reader wishing more background may consult [**J2**] for details.

```
sage: F.order()
4
```

The face turns applied to the cube with the following labeling

```
 ───────────── Rubik's cube labeling ─────────────

                    +--------------+
                    |  1    2    3 |
                    |  4   top    5 |
                    |  6    7    8 |
    +--------------+--------------+--------------+--------------+
    |  9   10   11 | 17   18   19 | 25   26   27 | 33   34   35 |
    | 12  left  13 | 20 front  21 | 28 right  29 | 36  rear  37 |
    | 14   15   16 | 22   23   24 | 30   31   32 | 38   39   40 |
    +--------------+--------------+--------------+--------------+
                    | 41   42   43 |
                    | 44 bottom 45 |
                    | 46   47   48 |
                    +--------------+
```

gives rise to an embedding $G \hookrightarrow S_{48}$. It turns out that SAGE already has the Rubik's cube group G "pre-programmed":

```
 ───────────────────── SAGE ─────────────────────

sage: rubik = CubeGroup()
sage: rubik
The PermutationGroup of all legal moves of the Rubik's cube.
```

Next, we shall construct the "superflip" (every edge is flipped, but otherwise the cube is in the solved state) using a known shortest manuever in the face-turn metric.

```
 ───────────────────── SAGE ─────────────────────

sage: rubik = CubeGroup()
sage: superflip = "R*L*D^2*B^3*L^2*F^2*R^2*U^3*D*R^3*D^2*\
                   F^3*B^3*D^3*F^2*D^3*R^2*U^3*F^2*D^3"
sage: P = rubik.plot3d_cube(superflip)
sage: show(P)
sage: G = rubik.group()
sage: rubik.move(superflip)[0]
(2,34)(4,10)(5,26)(7,18)(12,37)(13,20)(15,44)(21,28)(23,42)(29,36)(31,45)(39,47)
```

This last line tells us what the superflip move is as an element of our permutation group representation of the Rubik's cube group. Now we shall construct this move group theoretically, using the fact that it is the unique non-trivial element in the center of the Rubik's cube group.

```
 ───────────────────── SAGE ─────────────────────

sage: Z = G.center()
sage: s = Z.gens()[0]
```

```
sage: s == rubik.move(superflip)[0]
True
sage: S = RubiksCube(s)
sage: S.solve()
"L2 B2 D2 F L2 R2 B U2 B D2 R2 F' L' R D' B' F R2 D2 R' U' D'"
```

This uses a program written by Dik T. Winter implementing Kociemba's algorithm. Other options are S.solve("dietz"), which uses Eric Dietz's cubex program; S.solve("optimal"), which uses Michael Reid's optimal program; and also S.solve("gap"), which uses GAP to solve the "word problem". In fact, Kociemba's algorithm is not bad since it returns a move which is 22 moves in the face-turn metric. (The move given above is only 20 moves.)

3.4. Permutation group homomorphisms. SAGE can also compute kernels and images. We shall give a simple example:

———————————————————— SAGE ————————————————————
```
sage: G = CyclicPermutationGroup(4)
sage: gens = G.gens()
sage: H = DihedralGroup(4)
sage: g = G([(1,2,3,4)]); g
(1,2,3,4)
sage: phi = PermutationGroupMorphism_im_gens( G, H, gens, gens)
sage: phi.image(G)
'Group([ (1,2,3,4) ])'
sage: phi.kernel()
Group(())
sage: phi.image(g)
'(1,2,3,4)'
sage: phi(g)
'(1,2,3,4)'
sage: phi.range()
Dihedral group of order 8 as a permutation group
sage: phi.codomain()
Dihedral group of order 8 as a permutation group
sage: phi.domain()
Cyclic group of order 4 as a permutation group
```

4. Matrix groups

The projective groups $\mathrm{PSp}_{2n}(K)$ and $\mathrm{PSL}_n(K)$ are mathematically the quotients of the matrix groups $\mathrm{Sp}_{2n}(K)$ and $\mathrm{SL}_n(K)$. However, over a finite field, GAP (and, at least currently, SAGE) implements these completely differently. The groups $\mathrm{PSL}(n, \mathrm{GF}(q))$, $\mathrm{PSp}(2n, \mathrm{GF}(q))$, and the other projective classical groups, are realized as *permutation* groups and all the methods in the above section apply[6].

On the other hand, the MatrixGroup class is designed for computing in the standard matrix groups GL_n, SL_n, GO_n, SO_n, GU_n and SU_n, and Sp_{2n}, and their subgroups defined by a (relatively small) finite set of generators.

---

[6]In fact, for some $\mathrm{PSL}(n, \mathrm{GF}(q))$ ($n > 5$ must be a prime), extra methods have been implemented to help with the computation in [**JK**], but details would take us too far afield.

──────────────── SAGE ────────────────

```
sage: F = GF(3)
sage: gens = [matrix(F,2, [1,0, -1,1]), matrix(F, 2, [1,1,0,1])]
sage: G = MatrixGroup(gens)
sage: G.conjugacy_class_representatives()
[
[1 0]
[0 1],
[0 1]
[2 1],
[0 1]
[2 2],
[0 2]
[1 1],
[0 2]
[1 2],
[0 1]
[2 0],
[2 0]
[0 2]
]
```

4.1. General and special linear groups. In SAGE, general linear groups can be constructed over rings other than finite fields. However, at the present time, not many methods are implemented unless the base ring is finite.

──────────────── SAGE ────────────────

```
sage: GL(4,QQ)
General Linear Group of degree 4 over Rational Field
sage: GL(1,ZZ)
General Linear Group of degree 1 over Integer Ring
sage: GL(100,RR)
General Linear Group of degree 100 over Real Field
 with 53 bits of precision
sage: GL(3,GF(49,'a'))
General Linear Group of degree 3 over Finite Field
 in a of size 7^2
```

For all the above groups but the last one, SAGE has very few methods implemented so far. However, when the ground field is finite, various methods are already implemented for computing with them.

──────────────── SAGE ────────────────

```
sage: F = GF(3); MS = MatrixSpace(F,2,2)
sage: gens = [MS([[0,1],[1,0]]),MS([[1,1],[0,1]])]
sage: G = MatrixGroup(gens)
sage: G.order()
48
sage: H = GL(2,F)
sage: H.order()
48
```

```
sage: H == G
True
sage: H.as_matrix_group() == G
True
```

Similarly, the special linear groups have special methods implemented over finite fields:

――――――――――――――――― SAGE ―――――――――――――――――

```
sage: FF = GF(3)
sage: for n in range(1,9):
....:       SL(n,FF).order()
....:
1
24
5616
12130560
237783237120
42064805779476480
6703422210133904166912 0
961721214905722855895197286400
sage: G = SL(2,GF(3)); G
Special Linear Group of degree 2 over Finite Field of size 3
sage: G.conjugacy_class_representatives()
[
 [1 0]
 [0 1],
 [0 2]
 [1 1],
 [0 1]
 [2 1],
 [2 0]
 [0 2],
 [0 2]
 [1 2],
 [0 1]
 [2 2],
 [0 2]
 [1 0]
]
```

In addition, SAGE has implementations of methods for the special linear group over the integers, SL(2,ZZ), and its congruence subgroups $\Gamma_0(N)$, $\Gamma_1(N)$, and $\Gamma(N)$ motivated by their associated spaces of modular forms.

――――――――――――――――― SAGE ―――――――――――――――――

```
sage: G = Gamma0(5)
sage: G
Congruence Subgroup Gamma0(5)
```

For the number theoretic methods for congruence subgroups we refer the interested reader to the SAGE reference manual for details.

4.2. Orthogonal groups. The general orthogonal group $GO(e, d, q)$ consists of those $d \times d$ matrices over the field $GF(q)$ that respect a non-singular quadratic form specified by $e \in \{-1, 0, 1\}$. The value of e must be 0 for odd d (and can optionally be omitted in this case), and one of 1 or -1 for even d.

`SpecialOrthogonalGroup` returns a group isomorphic to the special orthogonal group $SO(e, d, q)$, which is the subgroup of all those matrices in the general orthogonal group that have determinant one. (The index of $SO(e, d, q)$ in $GO(e, d, q)$ is 2 if q is odd, but $SO(e, d, q) = GO(e, d, q)$ if q is even.)

Warning: GAP's notation for the finite orthogonal groups differs from that of SAGE. (This is forced by the Python constraint that optional arguments must follow all required arguments.) Whereas GAP uses the notation `GO([e,]d,q)` and `SO([e,]d,q)` where [...] denotes an optional value, SAGE uses the notation `GO(d,GF(q),e=0)` and `SO(d,GF(q),e=0)` where `e=0` denotes an optional argument that takes value 0 if omitted.

─────────────────────── SAGE ───────────────────────

```
sage: G = SO(3,GF(5))
sage: G.gens()
[
[2 0 0]
[0 3 0]
[0 0 1],
[3 2 3]
[0 2 0]
[0 3 1],
[1 4 4]
[4 0 0]
[2 0 4]
]
sage: G = SO(3,GF(5))
sage: G.as_matrix_group()
Matrix group over Finite Field of size 5 with 3 generators:
[[[2, 0, 0]
sage: GO( 3, GF(7), 0)
General Orthogonal Group of degree 3, form parameter 0,
 over the Finite Field of size 7
sage: GO( 3, GF(7), 0).order()
672
```

Special classes also exist for the general and special unitary groups GU_n and SU_n, and the symplectic groups Sp_{2n}. Similarly for the orthogonal groups there exist special methods for these groups over finite fields.

As with homomorphisms between permutation groups, SAGE can compute with homomorphisms between matrix groups over finite fields. The `hom` code wraps GAP's `GroupHomomorphismByImages` function but only for matrix groups.

─────────────────────── SAGE ───────────────────────

```
sage: F = GF(5); MS = MatrixSpace(F,2,2)
sage: G = MatrixGroup([MS([1,1,0,1])])
sage: H = MatrixGroup([MS([1,0,1,1])])
sage: phi = G.hom(H.gens())
sage: phi
```

```
Homomorphism : Matrix group over Finite Field
 of size 5 with 1 generators:
 [[[1, 1], [0, 1]]] --> Matrix group over Finite Field
 of size 5 with 1 generators:
 [[[1, 0], [1, 1]]]
sage: phi(MS([1,1,0,1]))
[1 0]
[1 1]
sage: F = GF(7); MS = MatrixSpace(F,2,2)
xsage: F.multiplicative_generator()
3
sage: G = MatrixGroup([MS([3,0,0,1])])
sage: a = G.gens()[0]^2
sage: phi = G.hom([a])
```

5. Future developments

The main advantages of SAGE over a stand-alone package like GAP (from which SAGE draws much of its group-theoretic functionality) is the combination of

(1) a standard, well-supported, and modern programming language Python with user-friendly interfaces;

(2) a comprehensive libary of the best open source algorithms covering most all areas of mathematics;

(3) a well thought out categorical design of classes of objects and their elements together with the sets of morphisms $\mathrm{Hom}(X, Y)$ and their element homomorphisms $\phi : X \to Y$.

These characteristics allow anyone with a basic understanding of group theory and Python/SAGE to contribute to the further development of SAGE. Below we mention some of the areas where SAGE is currently being further developed, or needs to be developed and expanded in the near future.

5.1. Categories and morphisms. The categorical framework allows one create objects X and homsets $\mathrm{Hom}(X, Y)$ in SAGE and manipulate them. This framework needs to be extended to include more functionality on the homsets and their subsets $\mathrm{Iso}(X, Y)$ or $\mathrm{Aut}(X)$. Additional work is needed on efficient integrity checking to ensure, for given groups G and H that a map created in $\mathrm{Hom}(G, H)$ specifies a valid homomorphism. Constructors for homomorphisms between the different classes of abelian groups, permutation groups, and matrix groups will aid in translating between classes on which different methods may apply or be more efficient.

5.2. Efficiency of algorithms. In order to optimize performance, more of the basic arithmetic and needs to be coded in Cython/Pyrex [**Cy**], so that calls to GAP are necessary only for sophisticated high-level algorithms. We envision that more SAGE/Python code will emerge as native applications are developed to make use of the diverse range of SAGE code for rings, algebras, and combinatorics, or to develop applications to to number theory and algebraic geometry.

5.3. Abelian groups. Currently, abelian groups are implemented with a multiplicative notation providing consistency of group operations $*$ and $\char94$ (as opposed to $+$ and $*$) within the category of groups. An additive interface for abelian groups is envisioned to provide an alternative representation, and a class from which many naturally occurring additive groups could inherit (like the group of points on an elliptic curve of Example 1 or the additive group of a finite ring). A metastructure for groups would handle translation between the additive and multiplicative notation.

Moreover, it would be desirable to support more (potentially) infinite abelian groups like unit groups of rings like $\mathbb{Q}^\times$, $\mathbb{Z}_{(p)}^\times$, $\mathbb{Z}_p^\times$, $\mathbb{Q}_p^\times$, $\mathbb{R}^\times$ and $\mathbb{C}^\times$ (unit groups of the rationals, the valuation ring $\mathbb{Z}_{(p)} = \{n/m \mid (m,p) = 1\}$, the p-adics, reals and complexes) and additive groups of $GF(q)$, $\mathbb{Q}$, and $\mathbb{Z}$, and permitting the construction of quotients such as $\mathbb{Q}/\mathbb{Z}$ and of dual groups $\operatorname{Hom}(A, \mathbb{Q}^\times)$, $\operatorname{Hom}(A, \mathbb{Z}_p^\times)$, $\operatorname{Hom}(A, \mathbb{C}^\times)$, and $\operatorname{Hom}(A, \mathbb{Q}/\mathbb{Z})$. There are an infinity of possible structures, but these groups arise naturally in mathematics, and the framework for category theory gives a clean interface in which to define them.

5.4. Permutation groups. Permutation groups arise naturally in the representation of group actions on finite sets. We envision enhancing and extending the permutation group class to allow left or right actions on arbitrary finite sets, with additional Python classes in SAGE for the G-sets on which they act. This would constitute a new category in SAGE that would include G-set morphisms, consisting of a homomorphism $G \to H$ with compatible map of sets.

Implementing permutations groups which act naturally (on the right or left) on more general sets than $\{1, 2, \ldots, n\}$ would simplify actions on the index sets $\{0, 1, \ldots, n-1\}$. In terms of efficiency, one would like to do something like:

---------- SAGE? ----------

```
sage: G = SymmetricGroup(n,0,n-1)
sage: for i in range(n):
....:     g = G.random_element()
....:     x[g(i)]
```

Obviously, for large n, one does not want to create the entire list `g([ i for i in range(n) ])` in order to extract one element.

The example of the Rubik's cube is a less trivial example of a G-set. The Rubik's code group acts naturally on the set S of states of the Rubik's cube. Moreover, this group action $G \times S \to S$ is compatible with the map $\pi_1 : S \to C$ to the corner states C, and $\phi_2 : S \to E$ to the set of edge states E, compatible in the sense that the natural actions

$$G \times C \to C \text{ and } G \times E \to E,$$

satisfy $\phi_i(g(x)) = g(\phi_i(x))$. A category of G-sets would give a natural framework for investigating the kernel of the actions of G on C and E (or on equivalence classes of corners and edges), giving a mathematical structure to the classes of reduction moves.

In the mathematical "real world", a category of G-sets would provide a natural framework for investigating Galois actions on sets (of class groups, torsion points on abelian varieties, or a set of special points on curves or algebraic varieties). The infrastructure for G-sets would give a mechanism for representing the action directly on a finite set, and handle the induced maps under morphisms applied to the underlying sets.

5.5. Matrix groups. Matrix groups are the primary objects of study of representation theory, and arise naturally in geometric group actions (e.g. on a vector space equipped with the structure of a bilinear form). Permutation group representations give rise a powerful algorithms for studying finite groups but linear representations provide other tools. Standard packages for group theory – GAP and MAGMA – give preference to the permutation group representation for $PSL(n, GF(q))$, ignoring the geometric structure encoded in the matrix representation.

A future implementation in SAGE could differentiate the permutation group and matrix group representations as follows:

```
—————————————————————————— SAGE? ——————————————————————————

sage: PSL(2,7)
Permutation Group with generators [(3,7,5)(4,8,6),
  (1,2,6)(3,4,8)]
sage: G = PSL(2,GF(7))
sage: G
Projective special linear group of degree 2 over
Finite field of size 7
sage: G([1,1,0,1])
[1 1]
[0 1]
```

Over a finite field, `PSL(n,R)` would have augmented features, like the ability to determine the order and generators (based on the permutation representation), while explicit homomorphisms could translate between the permutation representation when R is a finite field.

```
—————————————————————————— SAGE? ——————————————————————————

sage: G = PSL(2, GF(7))
sage: H = G.permutation_group()
sage: H
Permutation Group with generators [(3,7,5)(4,8,6),
  (1,2,6)(3,4,8)]
sage: f = G.permutation_representation()
sage: f
Homomorphism from Permutation Group ... to Projective
special linear group of degree 2 over Finite field of size 7
```

These matrix repsentations also provide a tool for analysis of infinite matrix groups. For instance, a coset of $\Gamma(p)$ in $SL_2(\mathbb{Z})$, has a representative in $SL_2(GF(p))$ using the natual exact sequence:

$$1 \to \Gamma(p) \to SL_(\mathbb{Z}) \to SL_2(GF(p)) \to 1$$

One could then pass to an element of $PSL_2(GF(p))$, via the exact sequence of matrix groups:

$$1 \to GF(p)^* \to SL_2(GF(p)) \to PSL_2(GF(p)) \to 1$$

and use homomorphisms to access the efficient permutation group algorithms for $PSL(2, p)$. The role of a computer algebra system like SAGE is to facilitate such translation of a mathematical structures.

5.6. Other group families. Besides increased functionality in matrix groups and permutation groups, a wider class of groups, and their methods, remain to be implemented. These include Lie groups and Lie algebras, Coxeter groups, p-groups, crystallographic groups, polycyclic groups, solvable groups, free groups, and braid groups. At the time of this writing, SAGE's functionality with these groups has room for a great deal of improvement (to put it euphemistically). However, there exist a wide range of open source programs (particularly in GAP) for dealing with such groups.

To illustrate how straightforward adding this functionality to SAGE really is (anyone with time and a basic understanding of group theory and Python/SAGE can do this), we give an example of a simple wrapper. Wrapping a GAP function in SAGE is a matter of writing a program in Python which uses the pexpect interface to pipe various commands to GAP and read back the input into SAGE.

For example[7], suppose we want to make a wrapper for computation of the Cartan matrix of a simple Lie algebra. The Cartan matrix of G_2 is available in GAP using the commands

———————————— GAP ————————————

```
gap> L:= SimpleLieAlgebra( "G", 2, Rationals );
<Lie algebra of dimension 14 over Rationals>
gap> R:= RootSystem( L );
<root system of rank 2>
gap> CartanMatrix( R );
[ [ 2, -1 ], [ -3, 2 ] ]
```

(Incidentally, most of the GAP Lie algebra implementation was written by Thomas Breuer, Willem de Graaf and Craig Struble.) In SAGE, one can simply type

———————————— GAP ————————————

```
sage: L = gap.SimpleLieAlgebra('"G"', 2, 'Rationals'); L
<Lie algebra of dimension 14 over Rationals>
sage: R = L.RootSystem(); R
<root system of rank 2>
sage: R.CartanMatrix()
[ [ 2, -1 ], [ -3, 2 ] ]
```

Note the '"G"' which is evaluated in GAP as the string "G".

Using this example, we show how one might write a Python/SAGE program whose input is, say, ('G',2) and whose output is the matrix above (but as a SAGE matrix).

First, the input must be converted into strings consisting of legal GAP commands. Then the GAP output, which is also a string, must be parsed and converted if possible to a corresponding SAGE/Python class object.

---

[7]We emphasize that this is just an example to illustrating wrappers. In fact, the module `sage/combinat/cartan_matrix.py`, written by Mike Hansen, implements the method `cartan_matrix` directly in Python, without calling GAP, so this particular wrapper is not needed.

```
———————————————— Python ——————————————————
def cartan_matrix(type, rank):
    """
    Return the Cartain matrix of given Chevalley type and rank.

    INPUT:
        type -- a Chevalley letter name, as a string, for
                 a family type of simple Lie algebras
        rank -- an integer (legal for that type).

    EXAMPLES:
        sage: cartan_matrix("A",5)
        [ 2 -1  0  0  0]
        [-1  2 -1  0  0]
        [ 0 -1  2 -1  0]
        [ 0  0 -1  2 -1]
        [ 0  0  0 -1  2]
        sage: cartan_matrix("G",2)
        [ 2 -1]
        [-3  2]
    """

    L = gap.SimpleLieAlgebra('"%s"'%type, rank, 'Rationals')
    R = L.RootSystem()
    sM  = R.CartanMatrix()
    ans = eval(str(sM))
    MS  = MatrixSpace(ZZ, rank)
    return MS(ans)
```

The output **ans** is a Python list. The last two lines convert that list to a SAGE class object Matrix instance.

If you are interested in contributing code to SAGE, please subscribe to the **sage-devel** list at http://www.sagemath.org/list.html and post your idea. All contributions are welcome!

5.7. Open source groups database. On the internet, you can find various online databases of mathematical objects:

- Sloane's online database of integer sequences,
  http://www.research.att.com/~njas/sequences/,
- Linear error-correcting codes, http://www.codetables.de and
  http://www.win.tue.nl/%7Eaeb/voorlincod.html,
- Atlas character tables for certain group families,
  http://brauer.maths.qmul.ac.uk/Atlas/v3/,

and so on. However, to our knowledge, no such an online database exists for finite groups. We envision a peer-review system, much in the same way that Sloane has set up for the OEIS, which allows anyone to contribute valid group-theoretic data to the database. It would also be available for download in a suitable format, and licensed in such a way that the downloaded data could be redistributed. Furthermore, we envision a certificate system that SAGE can issue which can help with the following scenario. Suppose ten different people want to contribute methematical data to this peer-reviewed open source database, and each of these contributions was obtained from an hour of computer calculation using SAGE. It would be useful for the referees if SAGE could implement a "trusted certificate of computation" that verified that a specific computation was actually performed (on a specific computer which a specific version of SAGE at a specific time).

Acknowledgements: We thank Mike Hansen and the referee for their careful reading and many helpful suggestions.

References

[C1] H. Cohen, **Advanced topics in computational number theory**, Springer, 2000.

[C2] H. Cohen, **A course in computational algebraic number theory**, Springer, 1996.

[Cy] Cython, `http://www.cython.org/`.

[H] G. Ellis, The GAP package HAP 1.8.4, `http://www.gap-system.org/Packages/hap.html`.

[G] The GAP Group, GAP – *Groups, Algorithms, and Programming, Version 4.4*, 2005. `http://www.gap-system.org`.

[J1] D. Joyner, *A primer on computational group homology and cohomology*, to appear in **Aspects of Infinite Groups**, (ed. Ben Fine), World Scientific. Available at `http://arxiv.org/abs/0706.0549`

[J2] D. Joyner, **Adventures with group theory: Rubik's cube, Merlin's machine, and other mathematical toys**, 2nd edition, The Johns Hopkins Univ. Press, 2008.

[JK] D. Joyner and A. Ksir, *Modular representations on some Riemann-Roch spaces of modular curves X(N)*, in **Computational Aspects of Algebraic Curves**, (Editor: T. Shaska) Lecture Notes in Computing, WorldScientific, 2005. Available at `http://front.math.ucdavis.edu/math.AG/0502586`

[M] MAGMA, `http://magma.maths.usyd.edu.au/magma/`.

[Ma] Mathematica, `http://www.wolfram.com/`.

[N] J. Neubüser, "An invitation to Computational Group Theory," in **Groups St Andrews, Galway, 1993**, (ed. C. M. Campbell), Cambridge University Press, 1995. Available at: `http://www.gap-system.org/Doc/Talks/talks.html`.

[P] Pexpect, `http://pexpect.sourceforge.net/`.

[R] J. Rotman, **An introduction to the theory of groups**, 4th ed, Springer, 1995.

[S] W. Stein, *SAGE Mathematics Software (Version 2.10)*, The SAGE Group, 2008, `http://www.sagemath.org`.

[St] W. Stein, **Modular Forms: A Computational Approach**, with an appendix by Paul Gunnells, AMS Graduate Studies in Mathematics, Vol. 79, 2007.

MATHEMATICS DEPARTMENT, U. S. NAVAL ACADEMY, ANNAPOLIS, MD 21402, USA
E-mail address: `wdj@usna.edu`

INSTITUT DE MATHÉMATIQUES DE MUNINY, 163 AVENUE DE LUMINY, CASE 907, 13288 MARSEILLE CEDEX 9, FRANCE
E-mail address: `kohel@iml.univ-mrs.fr`

Contemporary Mathematics
Volume **470**, 2008

Simultaneous Constructions of the Sporadic Groups Co_2 and Fi_{22}

Hyun Kyu Kim and Gerhard O. Michler

ABSTRACT. In this article we give self-contained existence proofs for the sporadic simple groups Co_2 and Fi_{22} using the second author's algorithm [**10**] constructing finite simple groups from irreducible subgroups of $GL_n(2)$. These two sporadic groups were originally discovered by J. Conway [**4**] and B. Fischer [**7**], respectively, by means of completely different and unrelated methods. In this article $n = 10$ and the irreducible subgroups are the Mathieu group $\mathcal{M}_{22}$ and its automorphism group $\mathrm{Aut}(\mathcal{M}_{22})$. We construct their five non-isomorphic extensions E_i by the two 10-dimensional non-isomorphic simple modules of $\mathcal{M}_{22}$ and by the two 10-dimensional simple modules of $A_{22} = \mathrm{Aut}(\mathcal{M}_{22})$ over $F = GF(2)$. In two cases we construct the centralizer $H_i = C_{G_i}(z_i)$ of a 2-central involution z_i of E_i in any target simple group $\mathfrak{G}_i$. Then we prove that all the conditions of Algorithm 7.4.8 of [**11**] are satisfied. This allows us to construct $\mathfrak{G}_3 \cong Co_2$ inside $GL_{23}(13)$ and $\mathfrak{G}_2 \cong Fi_{22}$ inside $GL_{78}(13)$. We also calculate their character tables and presentations.

1. Introduction

In 1969 J.H. Conway [**4**] discovered three sporadic simple groups which he defined in terms of the automorphism group $A = \mathrm{Aut}(\Lambda)$ of the 24-dimensional Leech lattice Λ, see also [**5**]. The center $Z(A)$ of A has order 2, and $Co_1 = A/Z(A)$ is the largest of these three simple groups. He obtained his groups Co_2 and Co_3 as stabilizers in A of suitable vectors of the Leech lattice.

Two years later B. Fischer found three sporadic simple groups by characterizing the finite groups G that can be generated by a conjugacy class $D = z^G$ of p-transpositions, which means that the product of two elements of D has order 1, 2, or p, where p is an odd prime. He proved that besides the symmetric groups S_n, the symplectic groups $Sp_n(2)$, the projective unitary groups $U_n(2)$ over the field with 4 elements and certain orthogonal groups, his three sporadic simple groups Fi_{22}, Fi_{23}, and Fi'_{24} describe all 3-transposition groups, see [**7**]. Whereas the 24-dimensional Leech lattice Λ provides a natural action of $A = \mathrm{Aut}(\Lambda)$ and has two

---

2000 *Mathematics Subject Classification.* 20D08, 20D05, 20C40.

Key words and phrases. construction of simple groups, sporadic simples groups, Conway sporadic simple group Co_2, Fischer sporadic simply group Fi_{22}.

The first author kindly acknowledges financial support by the Hunter R. Rawlings III Cornell Presidential Research Scholars Program for participation in this research project. This collaboration has also been supported by the grant NSF/SCREMS DMS-0532/06. Both authors thank the referee for helpful suggestions and corrections.

well defined 23-dimensional sublattices on which the Conway groups Co_2 and Co_3 act, Fischer had to construct a graph $\mathcal{G}$ and an action on it for each 3-transposition group $G = \langle D \rangle$. As its vertices he took the 3-transpositions x of D. Two distinct elements $x, y \in D$ are called to be *connected* and joined by an edge (x, y) in $\mathcal{G}$ if they commute in G. He showed that each of the groups considered in his theorem has a natural representation as an automorphism group of its graph $\mathcal{G}$. See also [1] for a coherent account on Fischer's theorem.

It is the purpose of this paper to provide uniform existence proofs for the simple sporadic groups Co_2 and Fi_{22} by means of Algorithm 2.5 of [10] constructing finite simple groups from irreducible subgroups T of $GL_n(2)$. Here we deal with the case $n = 10$ and the irreducible subgroups $\mathcal{M}_{22}$ and $A_{22} = \operatorname{Aut}(\mathcal{M}_{22})$. Our methods are purely algebraic and do not require any geometric insight. Furthermore, they are not restricted to the study of sporadic groups. Since every finite simple group G has finitely many irreducible representations V, Algorithm 2.5 of [10] can be applied again to the extensions E of G by V. In particular, $\mathcal{M}_{22}$ has been constructed this way from the irreducible subgroup S_5 of $GL_4(2)$, see Proposition 8.2.4 of [11]. Here the algorithm is applied again to the simple modules of $\mathcal{M}_{22}$ and its automorphism group A_{22}.

In Section 2 we state some known facts about $\mathcal{M}_{22}$, its automorphism group A_{22} and their irreducible representations over $F = GF(2)$. As shown in [9], the groups $\mathcal{M}_{22}$ and A_{22} each have two simple modules V_1, V_2 and V_3, V_4 of dimension 10, respectively. We show that the second cohomological dimensions $\dim_F[H^2(\mathcal{M}_{22}, V_i)]$ and $\dim_F[H^2(A_{22}, V_i)]$ are 0 for $i = 1, 2$ and $i = 4$, but $\dim_F[H^2(A_{22}, V_3)] = 1$. Therefore there are four split extensions $E_i = V_i \rtimes \mathcal{M}_{22}$, $i = 1, 2$, $E_i = V_i \rtimes A_{22}$, $i = 3, 4$ and one non-split extension E_5 of V_3 by A_{22}. For each of these five groups we state a presentation in terms of generators and relations, see Lemmas 2.4 and 2.5. We also construct for each of them a faithful permutation representation, see Lemma 2.6. In all five cases it has been checked that the elementary abelian normal subgroup V_i is a maximal elementary abelian subgroup of a Sylow 2-subgroup S_i of E_i as required in Step 4 of Algorithm 2.5 of [10].

In Section 3 we apply Step 5 of that algorithm to the extension groups E_3 and E_2. Let E be any of these two groups. Then E has a unique conjugacy class z^G of 2-central involutions. Let $D = C_E(z)$. Using a faithful permutation representation of E we find in both cases a uniquely determined nonabelian normal subgroup Q of D such that $V = Q/Z(Q)$ is elementary abelian of order 2^8, where $Z(Q)$ denotes the center of Q. Furthermore, Q has a complement W in D. In both cases W has an elementary abelian normal Fitting subgroup B which has a complement L in W.

In the first case, $E = E_3$, it follows that W has a center $Z(W) = \langle u \rangle$ of order 2 and that $L \cong S_6$. Using Yamaki's Theorem 8.6.6 of [11] we see that W is isomorphic to the centralizer of a 2-central involution of the symplectic group $Sp_6(2)$. Applying now Algorithm 7.4.8 of [11] we construct a simple subgroup K of $GL_8(2)$ such that $C_K(u) \cong W$, $|K : W|$ is odd, and $K \cong Sp_6(2)$. In order to perform all these steps we start with the factor group $D/Z(Q)$. Let U_1 be the split extension of K by V. Then we construct all central extensions H_3 of U_1 by $Z(Q)$ and check which ones have a Sylow 2-subgroup S_3 which is isomorphic to the ones of D. We prove that this happens exactly once. Thus the group H_3 with center $Z(H_3) = \langle z \rangle$ is uniquely determined, see Propositions 3.1. Furthermore, we prove in Lemma 3.2

that S_3 has a maximal elementary abelian normal subgroup A of order 2^{10} such that $N_{H_3}(A) \cong D$ as it is required by Algorithm 2.5 of [10].

In the second case, $E = E_2$, it follows that Q has an elementary abelian center $Z(W) = \langle z_1, z_2 \rangle$ of order 4. This time the Fitting subgroup B of the complement W of Q has order 16 and its complement L in W is isomorphic to the symmetric group S_5. Since the center of W is trivial we apply Algorithm 2.5 of [10] to construct a subgroup K of $GL_8(2)$ such that $N_K(A) \cong W$, $|K : W|$ is odd, and $K \cong \mathrm{Aut}(U_4(2))$. Again we start from the factor group $D/Z(Q)$. Let U_1 be the split extension of K by V. Since $Z(Q)$ is elementary abelian of order 4 we have to construct two consecutive central extensions of U_1. As above it follows that there is a uniquely determined extension group H_2 which satisfies all the required conditions of Algorithm 2.5 of [10], see Proposition 3.3 and Lemma 3.4.

In Section 4 we take the constructed presentation of H_3 as the input of Algorithm 7.4.8 of [11]. It returns a finite simple group $\mathfrak{G}_3$ inside $GL_{23}(13)$ of order $2^{18} \cdot 3^6 \cdot 5^3 \cdot 7 \cdot 11 \cdot 23$ having a 2-central involution $\mathfrak{z}$ such that $C_{\mathfrak{G}_3}(\mathfrak{z}) \cong H_3$, see Theorem 4.2. We also calculate the character table of $\mathfrak{G}_3$. It is equivalent to the one of Conway's second sporadic group Co_2. For further applications we also state a finite presentation of $\mathfrak{G}_3$ in Corollary 4.3.

In Section 5 we apply the same methods to the presentation of the involution centralizer H_2 and we obtain a finite simple group $\mathfrak{G}_2$ inside $GL_{78}(13)$ of order $2^{17} \cdot 3^9 \cdot 5^2 \cdot 7 \cdot 11 \cdot 13$ having a 2-central involution $\mathfrak{z}$ such that $C_{\mathfrak{G}_2}(\mathfrak{z}) \cong H_2$, see Theorem 5.1. We calculated the character table of $\mathfrak{G}_2$ and verified that it is equivalent to the one of Fischer's sporadic group Fi_{22}. We also obtained a presentation of $\mathfrak{G}_2$, see Corollary 5.2.

In the final section we show that Algorithm 2.5 of [10] constructs the automorphism group of the Fischer group Fi_{22} when it is applied to the extension E_4, see Corollary 6.1. However, it fails to construct a centralizer H_i from E_i in the cases $i = 1, 5$. The following diagram summarizes our experiments with Algorithm 2.5 of [10] described in this article.

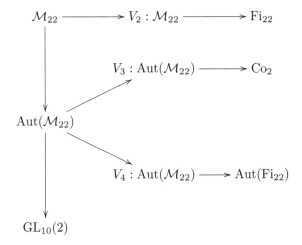

In the Appendices we collect all the systems of representatives of conjugacy classes in terms of the given generators of the local subgroups E_i, H_i and D_i for $i = 3, 2$. We also state the character tables of these groups and the four generating matrices of the matrix group $\mathfrak{G}_2 \cong Fi_{22}$.

Concerning our notation and terminology we refer to the books [3] and [11]. The computer algebra system MAGMA is described in [2].

2. Extensions of Mathieu group $\mathcal{M}_{22}$ and $\text{Aut}(\mathcal{M}_{22})$

The Mathieu group $\mathcal{M}_{22}$ is defined in Definition 8.2.1 of [11] by means of generators and relations. This beautiful presentation is due to J.A. Todd. The irreducible 2-modular representations of the Mathieu group $\mathcal{M}_{22}$ were determined by G. James [9]. Here only the two non-isomorphic simple modules V_i, $i = 1, 2$, of dimension 10 over $F = \text{GF}(2)$ will be considered. Todd's permutation representations of the Mathieu groups are stated in Lemma 8.2.2 of [11]. All conditions of Holt's Algorithm [8] implemented in MAGMA are satisfied, and it is applied here. Thus we show that for each simple module V_i there is exactly one extension E_i of $\mathcal{M}_{22}$ by V_i, the split extension. The presentations of E_1 and E_2 are given in Lemma 2.4.

Then we show that the automorphism group $A_{22} = \text{Aut}(\mathcal{M}_{22})$ has two 10-dimensional irreducible representations V_i, $i = 3, 4$. In Lemma 2.5 we also prove that A_{22} has a non-split extension E_5 by V_3 besides the two split extensions by V_3 and V_4. All extension groups are described here by generators and relations. We also give faithful permutation representations of all five extension groups E_i, see Lemma 2.6. They yield the basic information about these groups that will be used in the following sections.

In order to avoid any ambiguities we specify our concrete definition of a semidirect product. This is necessary to document the close relation between the irreducible representations V_i of $\mathcal{M}_{22}$ and A_{22} and the documented presentations of the extension groups E_i, $1 \leq i \leq 5$. Furthermore, these presentations work for all split extensions of linear groups over prime fields by their natural vector space.

DEFINITION 2.1. *Let F be a prime field of characteristic $p > 0$, G be a subgroup of $\text{GL}_n(F)$, and $V = F^n$ be the canonical n-dimensional vector space. Then V is a right FG-module of the group algebra FG of G over F with respect to the multiplication $v \cdot g$ defined by the product of the row vector $v \in V$ times the matrix $g \in G$. For each matrix $g \in G$ let $g^* = [g^{-1}]^T$ be the transpose of the inverse matrix g^{-1} of g. Then by Definition 3.4.4 of [11] V becomes a right FG-module under the multiplication $[v, g] := [g^* \cdot (v^T)]^T$ where $\cdot$ denotes the product of the matrix g^* times the column vector v^T. This right FG-module V^* is called the dual FG-module of V.*

The semidirect product $V \rtimes G$ of V and G consists of all pairs (v, g), with $v \in V$ and $g \in G$. Its multiplication $*$ is defined by:

$$(v_1, g_1) * (v_2, g_2) := (v_1 + [v_2, g_1], g_1 g_2) \text{ for all } v_1, v_2 \in V \text{ and all } g_1, g_2 \in G.$$

If we identify each $v \in V$ with $(v, 1)$ and each $g \in G$ with $(0, g)$, then we have that $v * g = (v + [0, 1], 1g) = (v, g)$ for each $v \in V$ and $g \in G$.

The proof of the following well-known result is routine and therefore omitted.

LEMMA 2.2. *Let $G = \langle x_i \mid 1 \leq i \leq r \rangle$ be a finitely generated subgroup of $\text{GL}_n(F)$ with set of defining relations $\mathcal{R}(G)$ in the given generators x_i. Let $V = F^n$ be the canonical n-dimensional vector space with standard basis $\{e_j \mid 1 \leq j \leq n\}$. Let*

$\mathcal{R}_1(V \rtimes G)$ be the set of all relations:

$$e_j^p = 1 \qquad \text{for all } 1 \le j \le n,$$
$$e_k * e_j = e_j * e_k \quad \text{for all } 1 \le j, k \le n.$$

Let $\mathcal{R}_2(V \rtimes G)$ be the set of all relations:

$$x_i * e_j * x_i^{-1} = e_1{}^{a(i)_{1,j}} * \cdots * e_n{}^{a(i)_{n,j}} \quad \text{for all } 1 \le i \le r,\ 1 \le j \le n,$$

where $(a(i)_{j1}, a(i)_{j2}, \ldots, a(i)_{jn}) = [(x_i)^* \cdot ((e_j)^T)]^T$. Then

$$\mathcal{R}(V \rtimes G) = \mathcal{R}_1(V \rtimes G) \cup \mathcal{R}_2(V \rtimes G) \cup \mathcal{R}(G).$$

REMARK 2.3. (a) For a given finitely presented group G the set $\mathcal{R}(G)$ of defining relations is equal for all its split extensions E by elementary abelian p-groups V. Furthermore, the set $\mathcal{R}_1(V \rtimes G)$ of relations describing the multiplication in the elementary abelian group is uniquely determined by the dimension of V over F. Therefore this set of relations is abbreviated by $\mathcal{R}_1(V \rtimes G)$ in the rest of this article. In particular, it suffices to state explicitly only the set of essential relations $\mathcal{R}_2(V \rtimes G)$ of the split extension $V \rtimes G$.

(b) The semidirect products $V \rtimes G$ constructed in this article by means of Lemma 2.2 are isomorphic to the semidirect products of G by V calculated by means of MAGMA. However, sometimes the sets $\mathcal{R}(V \rtimes G)$ of defining relations can be different. Furthermore, Lemma 2.2 can be applied to large groups and it is independent of MAGMA.

In the following $F = \mathrm{GF}(2)$ and $p = 2$.

LEMMA 2.4. Let $M_{22} = \langle a, b, c, d, t, g, h, i \rangle$ be the finitely presented group of Definition 8.2.1 of [11]. Then the following statements hold:

(a) A faithful permutation representation of degree 24 of M_{22} is stated in Lemma 8.2.2 of [11].

(b) The first irreducible representation V_1 of M_{22} is described by the following matrices:

$$Ma_1 = \begin{pmatrix} 1&0&1&0&0&0&0&0&0&0 \\ 0&0&0&1&0&0&0&0&0&0 \\ 0&0&1&0&0&0&0&0&0&0 \\ 0&1&0&0&0&0&0&0&0&0 \\ 0&0&1&0&0&0&0&1&0&0 \\ 0&0&0&0&0&0&0&0&1&0 \\ 0&0&1&0&0&0&1&0&0&0 \\ 0&0&1&0&1&0&0&0&0&0 \\ 0&0&0&0&0&1&0&0&0&0 \\ 0&0&1&0&0&0&0&0&0&1 \end{pmatrix}, \quad Mb_1 = \begin{pmatrix} 1&0&1&0&1&0&0&1&0&0 \\ 0&0&0&0&1&1&0&0&0&0 \\ 0&0&1&0&0&0&0&0&0&0 \\ 0&0&1&0&0&0&0&1&1&0 \\ 0&0&0&0&0&0&0&1&0&0 \\ 0&1&0&0&0&0&0&1&0&0 \\ 0&0&1&0&0&1&0&0&0&0 \\ 0&0&0&0&1&0&0&0&0&0 \\ 0&0&1&1&1&0&0&0&0&0 \\ 0&0&1&0&1&0&0&1&0&1 \end{pmatrix},$$

$$Mc_1 = \begin{pmatrix} 1&0&0&0&0&1&0&0&1&0 \\ 0&0&1&0&1&1&0&0&0&0 \\ 0&0&1&0&0&0&0&0&0&0 \\ 0&0&0&0&0&0&0&1&1&0 \\ 0&1&0&0&0&0&0&0&1&0 \\ 0&0&1&0&0&0&0&0&1&0 \\ 0&0&1&0&0&0&1&0&0&0 \\ 0&0&1&1&0&1&0&0&0&0 \\ 0&0&1&0&0&1&0&0&0&0 \\ 0&0&1&0&0&1&0&0&1&1 \end{pmatrix}, \quad Md_1 = \begin{pmatrix} 1&0&0&1&0&1&0&1&0&0 \\ 0&0&1&0&1&0&0&0&1&0 \\ 0&1&0&1&1&1&0&1&1&0 \\ 0&0&0&0&0&1&0&1&0&0 \\ 0&0&0&1&0&0&0&0&1&0 \\ 0&0&0&0&1&0&0&1&1&0 \\ 0&0&0&1&1&1&1&0&0&0 \\ 0&0&0&1&1&0&0&1&1&0 \\ 0&0&0&0&1&1&0&1&0&0 \\ 0&0&0&1&1&1&0&0&0&1 \end{pmatrix},$$

$$Mt_1 = \begin{pmatrix} 1&0&1&1&0&0&0&0&1&0 \\ 0&1&1&0&1&1&0&0&1&0 \\ 0&0&1&1&1&1&0&0&0&0 \\ 0&0&1&1&1&0&0&0&0&0 \\ 0&0&1&0&1&1&0&0&0&0 \\ 0&0&1&1&0&1&0&0&0&0 \\ 0&0&1&0&0&1&0&0&0&0 \\ 0&0&0&1&0&1&0&0&1&0 \\ 0&0&0&1&1&0&0&1&1&0 \\ 0&0&0&0&1&0&0&0&1&1 \end{pmatrix}, \quad Mg_1 = \begin{pmatrix} 1&0&1&0&0&0&0&0&0&0 \\ 0&0&1&0&0&0&1&1&1&0 \\ 0&0&1&0&0&0&0&0&0&0 \\ 0&0&1&1&0&0&0&1&1&0 \\ 0&0&1&0&1&0&0&1&0&0 \\ 0&0&1&0&0&1&0&0&1&0 \\ 0&1&1&0&0&0&0&1&1&0 \\ 0&0&0&0&0&0&0&1&0&0 \\ 0&0&0&0&0&0&0&0&1&0 \\ 0&0&0&0&0&0&0&1&0&1 \end{pmatrix},$$

$$Mh_1 = \begin{pmatrix} 1000000010 \\ 0100010000 \\ 0010010010 \\ 0001000010 \\ 0000010100 \\ 0000010000 \\ 0000000011 \\ 0000110000 \\ 0000000010 \\ 0000001010 \end{pmatrix} \quad and \quad Mi_1 = \begin{pmatrix} 0000110001 \\ 0111100000 \\ 0011110000 \\ 0010100000 \\ 0010110000 \\ 0001100000 \\ 0001111000 \\ 0001010010 \\ 0011000100 \\ 1011010000 \end{pmatrix}.$$

(c) *The second irreducible representation V_2 of $\mathcal{M}_{22}$ is described by the transpose inverse matrices of the generating matrices of $\mathcal{M}_{22}$ defining V_1:*
$Ma_2 = [Ma_1^{-1}]^T$, $Mb_2 = [Mb_1^{-1}]^T$, $Mc_2 = [Mc_1^{-1}]^T$, $Md_2 = [Md_1^{-1}]^T$, $Mt_2 = [Mt_1^{-1}]^T$, $Mg_2 = [Mg_1^{-1}]^T$, $Mh_2 = [Mh_1^{-1}]^T$ and $Mi_2 = [Mi_1^{-1}]^T$.

(d) $\dim_F[H^2(\mathcal{M}_{22}, V_i)] = 0$ for $i = 1, 2$.

(e) $E_1 = V_1 \rtimes \mathcal{M}_{22} = \langle a, b, c, d, t, g, h, i, v_1, v_2, v_3, v_4, v_5, v_6, v_8, v_8, v_9, v_{10} \rangle$ has set $\mathcal{R}(E_1)$ of defining relations consisting of $\mathcal{R}(\mathcal{M}_{22})$, $\mathcal{R}_1(V_1 \rtimes \mathcal{M}_{22})$ and the following set $\mathcal{R}_2(V_1 \rtimes \mathcal{M}_{22})$ of essential relations:

$av_1a^{-1}v_1v_3 = av_2a^{-1}v_4 = av_3a^{-1}v_3 = av_4a^{-1}v_2 = av_5a^{-1}v_3v_8 = av_6a^{-1}v_9 = 1,$

$av_7a^{-1}v_3v_7 = av_8a^{-1}v_3v_5 = av_9a^{-1}v_6 = av_{10}a^{-1}v_3v_{10} = 1,$

$bv_1b^{-1}v_1v_3v_5v_8 = bv_2b^{-1}v_5v_6 = bv_3b^{-1}v_3 = bv_4b^{-1}v_3v_8v_9 = bv_5b^{-1}v_8 = 1,$

$bv_6b^{-1}v_2v_8 = bv_7b^{-1}v_3v_7 = bv_8b^{-1}v_5 = bv_9b^{-1}v_3v_4v_5 = bv_{10}b^{-1}v_3v_5v_8v_{10} = 1,$

$cv_1c^{-1}v_1v_6v_9 = cv_2c^{-1}v_3v_5v_6 = cv_3c^{-1}v_3 = cv_4c^{-1}v_8v_9 = cv_5c^{-1}v_2v_9 = 1,$

$cv_6c^{-1}v_3v_9 = cv_7c^{-1}v_3v_7 = cv_8c^{-1}v_3v_4v_6 = cv_9c^{-1}v_3v_6 = cv_{10}c^{-1}v_3v_6v_9v_{10} = 1,$

$dv_1d^{-1}v_1v_4v_6v_8 = dv_2d^{-1}v_3v_5v_9 = dv_3d^{-1}v_2v_4v_5v_6v_8v_9 = dv_4d^{-1}v_6v_8 = 1,$

$dv_5d^{-1}v_4v_9 = dv_6d^{-1}v_5v_8v_9 = dv_7d^{-1}v_4v_5v_6v_7 = dv_8d^{-1}v_4v_5v_8v_9 = 1,$

$dv_9d^{-1}v_5v_6v_8 = dv_{10}d^{-1}v_4v_5v_6v_{10} = tv_1t^{-1}v_1v_3v_5v_6v_8 = tv_2t^{-1}v_2v_4v_8 = 1,$

$tv_3t^{-1}v_4v_5v_6 = tv_4t^{-1}v_3v_5 = tv_5t^{-1}v_3v_6 = tv_6t^{-1}v_3v_4 = tv_7t^{-1}v_4v_5v_6v_7 = 1,$

$tv_8t^{-1}v_4v_6v_8v_9 = tv_9t^{-1}v_4v_5v_8 = tv_{10}t^{-1}v_3v_4v_5v_6v_8v_{10} = 1,$

$gv_1g^{-1}v_1v_3 = gv_2g^{-1}v_3v_7v_8v_9 = gv_3g^{-1}v_3 = gv_4g^{-1}v_3v_4v_8v_9 = 1,$

$gv_5g^{-1}v_3v_5v_8 = gv_6g^{-1}v_3v_6v_9 = gv_7g^{-1}v_2v_3v_8v_9 = gv_8g^{-1}v_8 = 1,$

$gv_9g^{-1}v_9 = gv_{10}g^{-1}v_8v_{10} = hv_1h^{-1}v_1v_9 = hv_2h^{-1}v_2v_6 = hv_3h^{-1}v_3v_6v_9 = 1,$

$hv_4h^{-1}v_4v_9 = hv_5h^{-1}v_6v_8 = hv_6h^{-1}v_6 = hv_7h^{-1}v_9v_{10} = hv_8h^{-1}v_5v_6 = 1,$

$hv_9h^{-1}v_9 = hv_{10}h^{-1}v_7v_9 = iv_1i^{-1}v_5v_6v_{10} = iv_2i^{-1}v_2v_3v_4v_5 = 1,$

$iv_3i^{-1}v_3v_4v_5v_6 = iv_4i^{-1}v_3v_5 = iv_5i^{-1}v_3v_5v_6 = iv_6i^{-1}v_4v_5 = 1,$

$iv_7i^{-1}v_4v_5v_6v_7 = iv_8i^{-1}v_4v_6v_9 = iv_9i^{-1}v_3v_4v_8 = iv_{10}i^{-1}v_1v_3v_4v_6 = 1.$

(f) $E_2 = V_2 \rtimes \mathcal{M}_{22} = \langle a, b, c, d, t, g, h, i, v_1, v_2, v_3, v_4, v_5, v_6, v_8, v_8, v_9, v_{10} \rangle$ has set $\mathcal{R}(E_2)$ of defining relations consisting of $\mathcal{R}(\mathcal{M}_{22})$, $\mathcal{R}_1(V_2 \rtimes \mathcal{M}_{22})$ and the following set $\mathcal{R}_2(V_2 \rtimes \mathcal{M}_{22})$ of essential relations:

$av_1a^{-1}v_1 = av_2a^{-1}v_4 = av_3a^{-1}v_1v_3v_5v_7v_8v_{10} = av_4a^{-1}v_2 = av_5a^{-1}v_8 = 1,$

$av_6a^{-1}v_9 = av_7a^{-1}v_7 = av_8a^{-1}v_5 = av_9a^{-1}v_6 = av_{10}a^{-1}v_{10} = bv_1b^{-1}v_1 = 1,$

$bv_2b^{-1}v_6 = bv_3b^{-1}v_1v_3v_4v_7v_9v_{10} = bv_4b^{-1}v_9 = bv_5b^{-1}v_1v_2v_8v_9v_{10} = 1,$

$bv_6b^{-1}v_2 = bv_7b^{-1}v_7 = bv_8b^{-1}v_1v_4v_5v_6v_{10} = bv_9b^{-1}v_4 = bv_{10}b^{-1}v_{10} = 1,$

$cv_1c^{-1}v_1 = cv_2c^{-1}v_5 = cv_3c^{-1}v_2v_3v_6v_7v_8v_9v_{10} = cv_4c^{-1}v_8 = cv_5c^{-1}v_2 = 1,$

$$cv_6c^{-1}v_1v_2v_8v_9v_{10} = cv_7c^{-1}v_7 = cv_8c^{-1}v_4 = cv_9c^{-1}v_1v_4v_5v_6v_{10} = 1,$$

$$cv_{10}c^{-1}v_{10} = dv_1d^{-1}v_1 = dv_2d^{-1}v_3 = dv_3d^{-1}v_2 = dv_4d^{-1}v_1v_3v_5v_7v_8v_{10} = 1,$$

$$dv_5d^{-1}v_2v_3v_6v_7v_8v_9v_{10} = dv_6d^{-1}v_1v_3v_4v_7v_9v_{10} = dv_7d^{-1}v_7 = 1,$$

$$dv_8d^{-1}v_1v_3v_4v_6v_8v_9 = dv_9d^{-1}v_2v_3v_5v_6v_8 = dv_{10}d^{-1}v_{10} = tv_1t^{-1}v_1 = 1,$$

$$tv_2t^{-1}v_2 = tv_3t^{-1}v_1v_2v_3v_4v_5v_6v_7 = tv_4t^{-1}v_1v_3v_4v_6v_8v_9 = 1,$$

$$tv_5t^{-1}v_2v_3v_4v_5v_9v_{10} = tv_6t^{-1}v_2v_3v_5v_6v_8 = tv_7t^{-1}v_7 = tv_8t^{-1}v_9 = 1,$$

$$tv_9t^{-1}v_1v_2v_8v_9v_{10} = tv_{10}t^{-1}v_{10} = gv_1g^{-1}v_1 = gv_2g^{-1}v_7 = 1,$$

$$gv_3g^{-1}v_1v_2v_3v_4v_5v_6v_7 = gv_4g^{-1}v_4 = gv_5g^{-1}v_5 = gv_6g^{-1}v_6 = gv_7g^{-1}v_2 = 1,$$

$$gv_8g^{-1}v_2v_4v_5v_7v_8v_{10} = gv_9g^{-1}v_2v_4v_6v_7v_9 = gv_{10}g^{-1}v_{10} = hv_1h^{-1}v_1 = 1,$$

$$hv_2h^{-1}v_2 = hv_3h^{-1}v_3 = hv_4h^{-1}v_4 = hv_5h^{-1}v_8 = hv_6h^{-1}v_2v_3v_5v_6v_8 = 1,$$

$$hv_7h^{-1}v_{10} = hv_8h^{-1}v_5 = hv_9h^{-1}v_1v_3v_4v_7v_9v_{10} = hv_{10}h^{-1}v_7 = 1,$$

$$iv_1i^{-1}v_{10} = iv_2i^{-1}v_2 = iv_3i^{-1}v_2v_3v_4v_5v_9v_{10} = iv_4i^{-1}v_2v_3v_6v_7v_8v_9v_{10} = 1,$$

$$iv_5i^{-1}v_1v_2v_3v_4v_5v_6v_7 = iv_6i^{-1}v_1v_3v_5v_7v_8v_{10} = iv_7i^{-1}v_7 = 1,$$

$$iv_8i^{-1}v_9 = iv_9i^{-1}v_8 = iv_{10}i^{-1}v_1 = 1.$$

PROOF. The two irreducible FM_{22}-modules V_i, $i = 1, 2$, occur as composition factors with multiplicity 1 in the permutation module $(1_{M_{21}})^{M_{22}}$ of degree 22 where $M_{21} = \langle a, b, c, d, t, g, h, i \rangle$. They are dual to each other. Using the faithful permutation representation of M_{22} stated in (a) and the Meataxe Algorithm implemented in MAGMA we obtain the generating matrices of M_{22} given in (b) that define V_1. Their dual matrices define V_2, yielding (c).

(d) The cohomological dimensions $d_i = \dim_F[H^2(M_{22}, V_i)]$, $i = 1, 2$, have been calculated by means of MAGMA using Holt's Algorithm 7.4.5 of [11]. Its hypothesis is given by the presentation of M_{22} in Definition 8.2.1 of [11] and all the data stated in (a), (b) and (c). It follows that $d_1 = 0 = d_2$.

(e) Since $d_1 = 0$ there exists only the split extension E_1 of M_{22} by V_1. The presentation of E_1 has been obtained automatically by application of Lemma 2.2 and MAGMA.

(f) Since $d_2 = 0$ the split extension E_2 of M_{22} by V_2 is well defined. Its presentation has been obtained by application of Lemma 2.2.   □

LEMMA 2.5. Let $M_{22} = \langle a, b, c, d, t, g, h, i \rangle$ be the finitely presented group of Definition 8.2.1 of [11]. Let $A_{22} = \mathrm{Aut}(M_{22})$. Then the following statements hold:

(a) A_{22} has a faithful permutation representation of degree 44 and is isomorphic to the subgroup $\langle Pp, Pq \rangle$ of the symmetric group S_{44} generated by the following permutations:

$$Pp = (1,2,7,4)(3,21,44,25)(5,24,22,26)(6,8,39,36)(9,35)(10,37,12,13)$$
$$\cdot(11,20,14,17)(18,19)(23,42,41,32)(27,29)(28,33)(30,31,43,34),$$

$$Pq = (1,3,16,44,13,5)(2,15)(4,28,36,22,17,29)(6,24,12,43,39,38)(7,30)$$
$$\cdot(8,33,11,25,19,41)(9,31,18,21,10,42)(14,27,20,23,35,32)(26,37)(34,40).$$

(b) $A_{22} = \langle p, q \rangle$ has the following set $\mathcal{R}(A_{22})$ of defining relations:

$$p^4 = q^6 = 1,$$
$$q^{-1}p^{-2}q^{-1}p^{-1}q^{-1}p^{-2}q^{-1}p^{-1}q^{-1}p^2q^{-1}p^{-1} = 1,$$
$$pq^{-3}p^{-1}q^{-1}p^{-2}q^{-1}pqp^{-1}qpq^{-1} = (p^{-1}q^{-2}p^{-1}qpq^{-1}p^{-1}q^{-1})^2 = 1,$$
$$q^{-1}p^{-2}q^{-1}pq^{-1}p^{-2}q^{-1}pq^{-1}p^2q^{-1}p = (q^{-1}p^{-1}q^{-1})^5 = 1,$$
$$qpq^2pqp^{-1}q^{-1}p^{-2}q^{-2}p^2q^{-1}p^{-1}q^{-2}p^{-1} = 1.$$

(c) $A_{22} = \langle p, q \rangle$ has (up to isomorphism) two irreducible modules V_3 and V_4 of degree 10 over $F = \mathrm{GF}(2)$. V_3 is described by the following matrices:

$$Mp = \begin{pmatrix} 1&0&0&0&0&0&0&0&0&0 \\ 0&0&0&0&0&0&1&0&0&1 \\ 1&1&0&0&1&0&1&0&0&1 \\ 0&1&1&1&0&0&1&0&0&0 \\ 1&1&0&0&0&0&1&0&0&0 \\ 0&1&0&0&0&0&0&0&1&1 \\ 1&0&0&1&0&1&1&0&0&0 \\ 1&0&0&1&0&0&0&0&0&1 \\ 1&1&0&1&0&0&0&0&0&1 \\ 1&1&0&1&0&0&1&1&0&1 \end{pmatrix} \quad \text{and} \quad Mq = \begin{pmatrix} 0&0&1&0&1&0&0&1&0&0 \\ 0&0&1&0&0&0&0&0&0&0 \\ 0&0&0&0&1&0&0&0&0&0 \\ 1&0&1&0&0&1&0&0&0&0 \\ 0&0&0&0&1&0&0&0&1&0 \\ 0&0&1&0&0&0&1&0&0&0 \\ 0&0&0&0&1&1&0&0&0&0 \\ 0&0&0&1&1&1&0&0&0&0 \\ 0&0&0&0&1&1&0&0&0&1 \\ 0&1&1&0&1&1&0&0&0&0 \end{pmatrix}.$$

(d) V_4 is described by the matrices $Mp_1 = Mp^* = [Mp^{-1}]^T$, $Mq_1 = Mq^* = [Mq^{-1}]^T$ in $\mathrm{GL}_{10}(2)$.

(e) $\dim_F[H^2(A_{22}, V_3)] = 1$ and $\dim_F[H^2(A_{22}, V_4)] = 0$.

(f) $E_3 = V_3 \rtimes A_{22} = \langle p, q, v_1, v_2, v_3, v_4, v_5, v_6, v_8, v_8, v_9, v_{10} \rangle$ has set $\mathcal{R}(E_3)$ of defining relations consisting of $\mathcal{R}(A_{22})$, $\mathcal{R}_1(V_3 \rtimes A_{22})$ and the following set $\mathcal{R}_2(V_3 \rtimes A_{22})$ of essential relations:

$$pv_1p^{-1}v_1 = pv_2p^{-1}v_8v_9 = pv_3p^{-1}v_1v_2v_4v_9 = pv_4p^{-1}v_2v_5v_9 = 1,$$
$$pv_5p^{-1}v_1v_2v_3v_8v_9 = pv_6p^{-1}v_2v_7v_8 = pv_7p^{-1}v_1v_5v_8v_9 = 1,$$
$$pv_8p^{-1}v_1v_5v_8v_{10} = pv_9p^{-1}v_1v_2v_5v_6 = pv_{10}p^{-1}v_1v_2v_5v_8v_9 = 1,$$
$$qv_1q^{-1}v_2v_3v_4v_7 = qv_2q^{-1}v_2v_7v_{10} = qv_3q^{-1}v_2 = 1,$$
$$qv_4q^{-1}v_7v_8 = qv_5q^{-1}v_3 = qv_6q^{-1}v_3v_7 = qv_7q^{-1}v_2v_6 = 1,$$
$$qv_8q^{-1}v_1v_2v_3 = qv_9q^{-1}v_3v_5 = qv_{10}q^{-1}v_7v_9 = 1.$$

(g) $E_4 = V_4 \rtimes A_{22} = \langle p_1, q_1, v_1, v_2, v_3, v_4, v_5, v_6, v_8, v_8, v_9, v_{10} \rangle$ has set $\mathcal{R}(E_4)$ of defining relations consisting of $\mathcal{R}(A_{22})$, $\mathcal{R}_1(V_4 \rtimes A_{22})$ and the following set $\mathcal{R}_2(V_4 \rtimes A_{22})$ of essential relations:

$$p_1v_1p_1^{-1}v_1v_3v_5v_7v_8v_9v_{10} = p_1v_2p_1^{-1}v_3v_4v_5v_6v_9v_{10} = 1,$$
$$p_1v_3p_1^{-1}v_4 = p_1v_4p_1^{-1}v_4v_7v_8v_9v_{10} = 1,$$
$$p_1v_5p_1^{-1}v_3 = p_1v_6p_1^{-1}v_7 = p_1v_7p_1^{-1}v_2v_3v_4v_5v_7v_{10} = 1,$$
$$p_1v_8p_1^{-1}v_{10} = p_1v_9p_1^{-1}v_6 = p_1v_{10}p_1^{-1}v_2v_3v_6v_8v_9v_{10} = 1,$$
$$q_1v_1q_1^{-1}v_4 = q_1v_2q_1^{-1}v_{10} = q_1v_3q_1^{-1}v_1v_2v_4v_6v_{10} = 1,$$
$$q_1v_4q_1^{-1}v_8 = q_1v_5q_1^{-1}v_1v_3v_5v_7v_8v_9v_{10} = q_1v_6q_1^{-1}v_4v_7v_8v_9v_{10} = 1,$$
$$q_1v_7q_1^{-1}v_6 = q_1v_8q_1^{-1}v_1 = q_1v_9q_1^{-1}v_5 = q_1v_{10}q_1^{-1}v_9 = 1.$$

(h) The non-split extension $E_5 = \langle p_2, q_2, v_1, v_2, v_3, v_4, v_5, v_6, v_7, v_8, v_9, v_{10} \rangle$ of A_{22} by V_3 has a set $\mathcal{R}(E_5)$ of defining relations consisting of $\mathcal{R}_1(V_3 \rtimes A_{22})$ and the following relations:

$$p_2^8 = q_2^{12} = 1,$$
$$(p_2, v_1^{-1}) = p_2^{-1}v_2p_2v_7^{-1}v_{10}^{-1} = p_2^{-1}v_3p_2v_1^{-1}v_2^{-1}v_5^{-1}v_7^{-1}v_{10}^{-1} = 1,$$
$$p_2^{-1}v_4p_2v_2^{-1}v_3^{-1}v_4^{-1}v_7^{-1} = p_2^{-1}v_5p_2v_1^{-1}v_2^{-1}v_7^{-1} = 1,$$

$$p_2^{-1}v_6p_2v_2^{-1}v_9^{-1}v_{10}^{-1} = p_2^{-1}v_7p_2v_1^{-1}v_4^{-1}v_6^{-1}v_7^{-1} = 1,$$

$$p_2^{-1}v_8p_2v_1^{-1}v_4^{-1}v_{10}^{-1} = p_2^{-1}v_9p_2v_1^{-1}v_2^{-1}v_4^{-1}v_{10}^{-1} = 1,$$

$$p_2^{-1}v_{10}p_2v_1^{-1}v_2^{-1}v_4^{-1}v_7^{-1}v_8^{-1}v_{10}^{-1} = 1,$$

$$q_2^{-1}v_1q_2v_3^{-1}v_5^{-1}v_8^{-1} = q_2^{-1}v_2q_2v_3^{-1} = q_2^{-1}v_3q_2v_5^{-1} = q_2^{-1}v_4q_2v_1^{-1}v_3^{-1}v_6^{-1} = 1,$$

$$q_2^{-1}v_5q_2v_5^{-1}v_9^{-1} = q_2^{-1}v_6q_2v_3^{-1}v_7^{-1} = q_2^{-1}v_7q_2v_5^{-1}v_6^{-1} = 1,$$

$$q_2^{-1}v_8q_2v_4^{-1}v_5^{-1}v_6^{-1} = q_2^{-1}v_9q_2v_5^{-1}v_6^{-1}v_{10}^{-1} = q_2^{-1}v_{10}q_2v_2^{-1}v_3^{-1}v_5^{-1}v_6^{-1} = 1,$$

$$p_2^4v_3^{-1}v_4^{-1}v_5^{-1}v_6^{-1}v_7^{-1}v_8^{-1}v_9^{-1}v_{10}^{-1} = q_2^6v_1^{-1}v_4^{-1}v_8^{-1} = 1,$$

$$q_2^{-1}p_2^{-2}q_2^{-1}p_2^{-1}q_2^{-1}p_2^{-2}q_2^{-1}p_2^{-1}q_2^{-1}p_2^2q_2^{-1}p_2^{-1}v_1^{-1}v_2^{-1}v_4^{-1} = 1,$$

$$p_2q_2^{-3}p_2^{-1}q_2^{-1}p_2^{-2}q_2^{-1}p_2q_2p_2^{-1}q_2p_2q_2^{-1}v_1^{-1}v_2^{-1}v_5^{-1}v_6^{-1}v_8^{-1}v_{10}^{-1} = 1,$$

$$q_2^{-1}p_2^{-2}q_2^{-1}p_2q_2^{-1}p_2^{-2}q_2^{-1}p_2q_2^{-1}p_2^2q_2^{-1}p_2v_5^{-1}v_6^{-1}v_7^{-1}v_9^{-1} = 1,$$

$$q_2^{-1}p_2^{-1}q_2^{-2}p_2^{-1}q_2^{-2}p_2^{-1}q_2^{-2}p_2^{-1}q_2^{-1}p_2^{-1}q_2^{-1}v_1^{-1}v_4^{-1}v_5^{-1}v_6^{-1}v_7^{-1}v_8^{-1}v_9^{-1} = 1,$$

$$p_2^{-1}q_2^{-2}p_2^{-1}q_2p_2q_2^{-1}p_2^{-1}q_2^{-1}p_2^{-1}q_2^{-2}p_2^{-1}q_2p_2q_2^{-1}p_2^{-1}q_2^{-1}$$
$$\cdot v_1^{-1}v_2^{-1}v_3^{-1}v_4^{-1}v_5^{-1}v_6^{-1}v_8^{-1}v_9^{-1} = 1,$$

$$q_2p_2q_2^2p_2q_2p_2^{-1}q_2^{-1}p_2^{-2}q_2^{-2}p_2^2q_2^{-1}p_2^{-1}q_2^{-2}p_2^{-1}v_1^{-1}v_4^{-1}v_5^{-1}v_9^{-1} = 1.$$

PROOF. (a) Using Todd's permutation representation PM_{22} of M_{22} stated in Lemma 8.2.2 of [11] and the MAGMA command `AutomorphismGroup(PM_{22})` we calculated the automorphism group $A_{22} = \mathrm{Aut}(M_{22})$ of M_{22}. MAGMA provided eleven generators and a faithful permutation representation PA_{22} of degree 44 by means of the command `PermutationGroup(A_{22})`. Using this permutation representation we were able to show that PA_{22} is even generated by the two permutations $Pp, Pq \in S_{44}$ given in the statement. Also it has been checked computationally that the derived subgroup A'_{22} is a simple group which is isomorphic to M_{22} and that $|A_{22} : A'_{22}| = 2$.

(b) The given presentation of $A_{22} = \langle p, q \rangle$ in the above generators has been obtained by MAGMA and the faithful permutation representation of degree 44.

(c) Decomposing the permutation module PA_{22} of degree 44 into irreducible composition factors by means of the Meataxe Algorithm we obtained the first irreducible representation V_3 of A_{22} over $F = \mathrm{GF}(2)$ as given in the statement.

(d) The second irreducible representation V_4 of A_{22} is the dual of V_3, which has been checked to be non-isomorphic to V_3.

(e) The cohomological dimensions $d_i = \dim_F[H^2(A_{22}, V_i)]$, $i = 3, 4$, have been calculated by means of MAGMA using Holt's Algorithm 7.4.5 of [11], the presentation of A_{22} given in (b), the faithful permutation representation PA_{22} constructed in (a) and the two irreducible representations of A_{22} given in (c) and (d). As a result of the computation, we get $d_3 = 1$ and $d_4 = 0$.

(f) and (g). The two presentations of the split extensions E_3 and E_4 have been obtained by means of Lemma 2.2.

(h) The presentation of the non-split extension E_5 has been obtained by means of Holt's Algorithm 7.4.5 of [11] implemented in MAGMA [8]. This completes the proof.

$\square$

LEMMA 2.6. *Keep the notation of Lemmas 2.4 and 2.5. Then the following statements hold:*

(a) *Each of the four split extensions E_i, $i \in \{1, 2, 3, 4\}$ has a faithful permutation representation of degree 1024. Its stabilizer is the complement of V_i in E_i, which generates E_i together with $v_1 \in V_i$ for all 4 groups.*

(b) *The non-split extension $E_5 = \langle p_2, q_2, v_1, v_2, v_3, v_4, v_5, v_6, v_7, v_8, v_9, v_{10} \rangle = \langle p_2, q_2 \rangle$ has a faithful permutation representation of degree 88 with stabilizer $U = \langle q_2^2, (p_2 q_2^2)^2 \rangle$.*

(c) *The split extension E_1 of $\mathcal{M}_{22}$ by V_1 has 47 conjugacy classes. The element $z_1 = (tv_1)^3$ represents the unique conjugacy class of 2-central involutions of E_1 and $|C_{E_1}(z_1)| = 2^{17} \cdot 3^2 \cdot 5$.*

(d) *The split extension E_2 of $\mathcal{M}_{22}$ by V_2 has 43 conjugacy classes. The element $z_2 = (iv_1)^2$ represents the unique conjugacy class of 2-central involutions of E_2 and $|C_{E_2}(z_2)| = 2^{17} \cdot 3 \cdot 5$.*

(e) *The split extension E_3 of A_{22} by V_3 has 79 conjugacy classes. The element $z_3 = (pq^2 v_1)^5$ represents the unique conjugacy class of 2-central involutions of E_3 and $|C_{E_3}(z_3)| = 2^{18} \cdot 3^2 \cdot 5$.*

(f) *The split extension E_4 of A_{22} by V_4 has 77 conjugacy classes. The element $z_4 = (p_1^2 q_1 v_1)^{10}$ represents the unique conjugacy class of 2-central involutions of E_4 and $|C_{E_4}(z_4)| = 2^{18} \cdot 3 \cdot 5$.*

(g) *The non-split extension E_5 of A_{22} by V_3 has 79 conjugacy classes. The element $z_5 = p_2^4$ represents the unique conjugacy class of 2-central involutions of E_5 and $|C_{E_5}(z_5)| = 2^{18} \cdot 3^2 \cdot 5$.*

PROOF. (a) The presentations of the 4 split extensions E_i are given in Lemmas 2.4 and 2.5. Taking their complements $\mathcal{M}_{22}$ or A_{22} of V_i as stabilizers MAGMA provides the faithful permutation representation PE_i of E_i of degree 1024 for $i = 1, 2, 3, 4$. The final assertions have been proved by means of these permutation representations and MAGMA.

(b) The presentation of the non-split extension E_5 is given in Lemma 2.4(h). Let $U = \langle q_2^2, (p_2 q_2^2)^2 \rangle$. Using the MAGMA command CosetAction(E_5,U) we obtained a faithful permutation representation PE_5 of E_5 with stabilizer U and degree 88. It has been used to show that $E_5 = \langle p_2, q_2 \rangle$.

All remaining assertion were proved as follows. For each extension group E_i, $i \in \{1, 2, 3, 4, 5\}$ we used its faithful permutation representation PE_i, MAGMA and Kratzer's Algorithm 5.3.18 of [11] to determine a system of representatives of all conjugacy classes of E_i in terms of the given generators of the presentation of E_i. In view of the limited space they are not stated in this article, except for the given words of the 2-central involutions z_i of E_i which in each case were uniquely determined up to conjugation in E_i. This completes the proof. $\square$

3. Construction of the 2-central involution centralizers

In this Section we apply Algorithm 2.5 of [10] to the extension groups E_3 and E_2 in order to construct 2 groups H_3 and H_2 which are isomorphic to the centralizers of a 2-central involution of the simple groups Co_2 and Fi_{22}, respectively.

PROPOSITION 3.1. *Keep the notation of Lemmas 2.4 and 2.5. Let $E_3 = \langle p, q, v_1 \rangle$ be the split extension of $A_{22} = \mathrm{Aut}(\mathcal{M}_{22})$ by its simple module V_3 of dimension 10 over $F = \mathrm{GF}(2)$. Then the following statements hold:*

(a) $z = (pq^2v_1)^5$ is a 2-central involution of $E_3 = \langle p, q, v_1 \rangle$ with centralizer $D = C_{E_3}(z) = \langle n_i, f_j \mid 1 \le i \le 12, \, 1 \le j \le 5 \rangle$, where

$$n_1 = (p^2qp^2q^2v_1)^6, \qquad n_2 = (p^2qpv_1qpv_1qv_1)^6, \qquad n_3 = (v_1qp^3qpv_1qp)^4,$$
$$n_4 = (pqpq^4pqv_1q)^7, \qquad n_5 = (pqpv_1qp)^3, \qquad n_6 = (pqp^2q^2v_1)^6,$$
$$n_7 = (pq^2pv_1qp)^6, \qquad n_8 = (qpqp^2qv_1)^6, \qquad n_9 = (p^3q^3p^2q^2)^4,$$
$$n_{10} = (pqpq^2p^2qpq)^6, \qquad n_{11} = (pq^2p^3q^3p)^4, \qquad n_{12} = (qp^2q^4pqp)^3,$$
$$f_1 = (p^3q^2v_1qpv_1q)^5, \qquad f_2 = (q^2pqv_1qpqpq)^7, \qquad f_3 = (pqp^2q^3v_1qpq)^6,$$
$$f_4 = (p^3q^2pqpv_1q)^7, \qquad f_5 = (p^2q^3v_1qv_1q^2)^2.$$

(b) D has a unique extra-special normal subgroup Q of order 512. It is generated by the following involutions:

$$q_1 = n_8 n_{10} n_{11} n_8 n_{12}, \qquad q_2 = n_5 n_6 n_7 n_{12},$$
$$q_3 = n_5 n_7 n_9 n_6, \qquad q_4 = n_6 n_8 n_9 n_{10},$$
$$q_5 = n_5, \qquad q_6 = n_5 n_6 n_7,$$
$$q_7 = n_5 n_6 n_7 n_8, \qquad q_8 = n_5 n_7 n_8.$$

(c) Q has a complement W in D of order $|W| = 2^9 \cdot 3^2 \cdot 5$. The Fitting subgroup $B = \langle f_j \mid 1 \le j \le 5 \rangle$ of W is elementary abelian of order 2^5 and has a complement $L = \langle s_k \mid 1 \le k \le 3 \rangle \cong S_6$ in W, where

$$s_1 = n_2 n_3 n_2 n_1 n_3 n_4, \quad s_2 = n_1 n_2 n_3 n_1 n_2 n_3, \quad \text{and} \quad s_3 = (n_3 n_1 n_4)^2.$$

(d) $D = \langle q_i, f_j, s_k \mid 1 \le i \le 8, 1 \le j \le 5, 1 \le k \le 3 \rangle$ is a finitely presented group with center $Z(D) = \langle z \rangle$ of order 2, where $z = (q_4 q_8)^2$, and the following set $\mathcal{R}(D)$ of defining relations:

$$s_1^4 = s_2^5 = s_3^3 = 1,$$
$$(s_2 s_1)^2 = s_1 s_3 s_2^{-1} s_1^{-1} s_3^{-1} s_2 = (s_2 s_3)^3 = s_2^2 s_3 s_1 s_2^{-1} s_1 s_3 = 1,$$
$$s_2^{-1} s_1 s_3^{-1} s_2^{-1} s_1 s_3^{-1} s_2 s_3 = 1,$$
$$f_1^2 = f_2^2 = f_3^2 = f_4^2 = f_5^2 = 1,$$
$$f_1^{s_1}(f_1)^{-1} = f_1^{s_2}(f_1)^{-1} = f_1^{s_3}(f_1)^{-1} = 1,$$
$$f_2^{s_1}(f_1 f_4 f_5)^{-1} = f_2^{s_2}(f_1 f_2 f_3 f_5)^{-1} = f_2^{s_3}(f_3 f_5)^{-1} = 1,$$
$$f_3^{s_1}(f_3 f_5)^{-1} = f_3^{s_2}(f_2 f_3)^{-1} = f_3^{s_3}(f_2 f_5)^{-1} = 1,$$
$$f_4^{s_1}(f_1 f_3 f_4 f_5)^{-1} = f_4^{s_2}(f_1 f_3 f_4)^{-1} = f_4^{s_3}(f_1 f_3 f_4)^{-1} = 1,$$
$$f_5^{s_1}(f_2 f_3 f_4)^{-1} = f_5^{s_2}(f_2 f_3 f_4)^{-1} = f_5^{s_3}(f_1 f_2)^{-1} = 1,$$
$$[f_1, f_2] = [f_1, f_3] = [f_1, f_4] = [f_1, f_5] = [f_2, f_3] = [f_2, f_4] = [f_2, f_5] = 1,$$
$$[f_3, f_4] = [f_3, f_5] = [f_4, f_5] = 1,$$
$$q_1^2 = q_2^2 = q_3^2 = q_4^2 = q_5^2 = q_6^2 = q_7^2 = q_8^2 = 1,$$
$$[q_1, q_2] = [q_1, q_3] = [q_1, q_4] = [q_1, q_6] = [q_1, q_8] = 1, \quad [q_1, q_5] = [q_1, q_7] = (q_4 q_8)^2,$$
$$[q_2, q_3] = [q_2, q_4] = [q_2, q_5] = [q_2, q_6] = [q_2, q_7] = 1, \quad [q_2, q_8] = (q_4 q_8)^2,$$
$$[q_3, q_4] = [q_3, q_6] = [q_3, q_7] = 1, \quad [q_3, q_5] = [q_3, q_8] = (q_4 q_8)^2,$$
$$[q_4, q_5] = [q_4, q_6] = [q_4, q_7] = [q_4, q_8] = (q_4 q_8)^2,$$

$[q_5, q_6] = [q_5, q_7] = [q_5, q_8] = [q_6, q_7] = [q_6, q_8] = [q_7, q_8] = 1,$

$q_1^{s_1}(q_3 q_4 q_6 q_7 q_8)^{-1} = (q_4 q_8)^2, \quad q_1^{s_2}(q_1 q_2 q_3 q_8)^{-1} = q_1^{s_3}(q_1 q_3 q_5 q_7 q_8)^{-1} = 1,$

$q_1^{f_1}(q_1 q_5 q_7)^{-1} = q_1^{f_2}(q_1)^{-1} = q_1^{f_3}(q_1 q_5 q_6 q_7)^{-1} = q_1^{f_4}(q_1)^{-1} = 1,$

$q_1^{f_5}(q_1 q_6 q_8)^{-1} = q_2^{s_1}(q_1 q_4 q_5)^{-1} = (q_4 q_8)^2,$

$q_2^{s_2}(q_2 q_4 q_5 q_6)^{-1} = q_2^{s_3}(q_1 q_2 q_4 q_5)^{-1} = 1,$

$q_2^{f_1}(q_2 q_6)^{-1} = (q_4 q_8)^2, \quad q_2^{f_2}(q_2)^{-1} = q_2^{f_3}(q_2 q_5 q_6 q_7)^{-1} = 1,$

$q_2^{f_4}(q_2 q_5 q_7)^{-1} = q_2^{f_5}(q_2 q_6 q_7)^{-1} = (q_4 q_8)^2, \quad q_3^{s_1}(q_3 q_5 q_7 q_8)^{-1} = (q_4 q_8)^2,$

$q_3^{s_2}(q_2 q_3 q_6 q_8)^{-1} = (q_4 q_8)^2, \quad q_3^{s_3}(q_1 q_5 q_6 q_7 q_8)^{-1} = 1, \quad q_3^{f_1}(q_3 q_7)^{-1} = (q_4 q_8)^2,$

$q_3^{f_2}(q_3 q_6)^{-1} = q_3^{f_3}(q_3 q_5 q_8)^{-1} = 1, \quad q_3^{f_4}(q_3 q_5 q_6 q_7 q_8)^{-1} = q_3^{f_5}(q_3 q_7)^{-1} = (q_4 q_8)^2,$

$q_4^{s_1}(q_2 q_4 q_6 q_7)^{-1} = q_4^{s_2}(q_1 q_3 q_5 q_7 q_8)^{-1} = 1, \quad q_4^{s_3}(q_1 q_2 q_3 q_5 q_8)^{-1} = (q_4 q_8)^2,$

$q_4^{f_1}(q_4 q_6 q_8)^{-1} = 1, \quad q_4^{f_2}(q_4 q_5 q_7)^{-1} = q_4^{f_5}(q_4 q_5 q_7)^{-1} = (q_4 q_8)^2,$

$q_4^{f_3}(q_4 q_5 q_8)^{-1} = q_4^{f_4}(q_4)^{-1} = 1,$

$q_5^{s_1}(q_6 q_8)^{-1} = q_5^{s_2}(q_5 q_7)^{-1} = 1, \quad q_5^{s_3}(q_7)^{-1} = (q_4 q_8)^2, \quad q_5^{f_1}(q_5)^{-1} = 1,$

$q_5^{f_2}(q_5)^{-1} = q_5^{f_3}(q_5)^{-1} = q_5^{f_4}(q_5)^{-1} = q_5^{f_5}(q_5)^{-1} = 1,$

$q_6^{s_1}(q_5 q_6 q_7 q_8)^{-1} = q_6^{s_2}(q_8)^{-1} = q_6^{s_3}(q_5 q_7 q_8)^{-1} = (q_4 q_8)^2,$

$q_6^{f_1}(q_6)^{-1} = 1, \quad q_6^{f_2}(q_6)^{-1} = q_6^{f_3}(q_6)^{-1} = q_6^{f_4}(q_6)^{-1} = q_6^{f_5}(q_6)^{-1} = 1,$

$q_7^{s_1}(q_7)^{-1} = 1, \quad q_7^{s_2}(q_6 q_7)^{-1} = q_7^{s_3}(q_5 q_7)^{-1} = (q_4 q_8)^2,$

$q_7^{f_1}(q_7)^{-1} = q_7^{f_2}(q_7)^{-1} = q_7^{f_3}(q_7)^{-1} = q_7^{f_4}(q_7)^{-1} = q_7^{f_5}(q_7)^{-1} = 1,$

$q_8^{s_1}(q_5 q_6 q_7)^{-1} = q_8^{s_2}(q_5 q_8)^{-1} = (q_4 q_8)^2, \quad q_8^{s_3}(q_6 q_7 q_8)^{-1} = 1,$

$q_8^{f_1}(q_8)^{-1} = q_8^{f_2}(q_8)^{-1} = q_8^{f_3}(q_8)^{-1} = q_8^{f_4}(q_8)^{-1} = q_8^{f_5}(q_8)^{-1} = 1,$

$[(q_4 q_8)^2, s_k] = 1 \quad for \quad k = 1, 2, 3,$

$[(q_4 q_8)^2, f_j] = 1 \quad for \quad 1 \le j \le 5,$

$[(q_4 q_8)^2, q_i] = 1 \quad for \quad 1 \le i \le 8.$

(e) $D_1 = D/Z(D)$ has a faithful permutation representation of degree 256 with stabilizer W. Furthermore, $V = Q/Z(D)$ is the unique elementary abelian normal subgroup $V = Q/Z(D)$ of order 256 in D_1, and D_1 is a semidirect product of W by V.

(f) There is a basis $\mathcal{B} = \{v_i = q_i Z(D) \mid 1 \le i \le 8\}$ of V such that the three generators s_k of L are represented by the following matrices with respect to $\mathcal{B}$:

$$Ms_1 = \begin{pmatrix} 0 1 0 0 0 0 0 0 \\ 0 0 0 1 0 0 0 0 \\ 1 0 1 0 0 0 0 0 \\ 1 1 0 1 0 0 0 0 \\ 0 1 0 1 0 0 1 0 1 \\ 1 0 0 1 1 1 0 1 \\ 1 0 1 1 0 1 1 1 \\ 1 0 1 0 1 1 0 0 \end{pmatrix}, \quad Ms_2 = \begin{pmatrix} 1 0 0 1 0 0 0 0 \\ 1 1 1 0 0 0 0 0 \\ 1 0 1 1 0 0 0 0 \\ 0 1 0 0 0 0 0 0 \\ 0 1 0 1 1 0 1 1 \\ 0 1 1 0 0 0 1 0 \\ 0 0 0 1 1 0 1 0 \\ 1 0 1 1 0 1 0 1 \end{pmatrix}, \quad and \quad Ms_3 = \begin{pmatrix} 1 1 1 1 0 0 0 0 \\ 0 1 0 1 0 0 0 0 \\ 1 0 0 1 0 0 0 0 \\ 0 1 0 0 0 0 0 0 \\ 1 1 1 1 0 1 1 0 \\ 0 0 1 0 0 0 0 1 \\ 1 0 1 0 1 1 1 1 \\ 1 0 1 1 0 1 0 1 \end{pmatrix}.$$

(g) $W = \langle u, r, s_1, s_2, s_3 \rangle$ where $u = f_1$ is the central involution of W and $r = f_2 f_4$.

(h) The subgroup $S = \langle f_j, a_s, w \mid 1 \le j \le 5, 1 \le s \le 6 \rangle$ is a Sylow 2-subgroup of W which has a unique maximal elementary abelian normal 2-subgroup

$A = \langle a_1, a_2, a_3, a_4, a_5 = u, a_6 \rangle$ of order 64 where

$$w = (s_1 s_2^2 s_3 s_2)^3, \quad a_1 = s_2^2 f_3 s_1 s_3 s_2, \quad a_2 = (f_5 s_2 s_1^3)^3,$$
$$a_3 = (s_1 f_3)^2, \qquad a_4 = (s_1)^2, \qquad a_6 = (f_2 s_1)^4.$$

(i) Let MA be the subgroup of $\mathrm{GL}_8(2)$ generated by the matrices Ma_1, Ma_2, Ma_3, Ma_4, Mu, Ma_6 of the generators of A with respect to $\mathcal{B}$. Then $NA = N_{\mathrm{GL}_8(2)}(MA)$ has a cyclic Sylow 7-subgroup generated by the matrix:

$$Ms = \begin{pmatrix} 1 & 1 & 0 & 0 & 0 & 0 & 0 & 0 \\ 1 & 1 & 0 & 0 & 1 & 0 & 0 & 1 \\ 1 & 1 & 0 & 0 & 0 & 1 & 0 & 0 \\ 0 & 1 & 0 & 1 & 0 & 0 & 0 & 0 \\ 0 & 0 & 1 & 1 & 0 & 0 & 0 & 0 \\ 0 & 1 & 0 & 0 & 1 & 1 & 0 & 0 \\ 0 & 1 & 0 & 1 & 0 & 1 & 1 & 1 \\ 1 & 1 & 1 & 0 & 0 & 0 & 0 & 0 \end{pmatrix}.$$

(j) Let $MW = \langle Mr, Mu, Ms_1, Ms_2, Ms_3 \rangle$. Let MK be the subgroup of $\mathrm{GL}_8(2)$ generated by MW and the matrix Ms. Then MK is a simple group of order $|MK| = 2^9 \cdot 3^4 \cdot 5 \cdot 7$ which is isomorphic to $\mathrm{Sp}_6(2)$.

(k) MK is an irreducible subgroup of $\mathrm{GL}_8(2)$ generated by the five matrices $m_1 = Ms$, $m_2 = Mr$, $m_3 = Ms_1$, $m_4 = Ms_2$ and $m_5 = Ms_3$ of respective orders 7, 2, 4, 5, and 3. With respect to this set of generators MK has the following set $\mathcal{R}(K)$ of defining relations:

$$m_1^7 = m_2^2 = m_3^4 = m_4^5 = m_5^3 = 1,$$
$$m_2 m_4^{-1} m_3^{-1} m_2 m_3 m_4 = (m_3^{-1} m_4^{-1})^2 = 1,$$
$$m_3 m_5 m_4^{-1} m_3^{-1} m_5^{-1} m_4 = (m_5^{-1} m_4^{-1})^3 = 1,$$
$$(m_1^{-1} m_2)^4 = (m_2 m_3^{-1})^4 = 1, \quad (m_2 m_3 m_2 m_3^{-1})^2 = 1,$$
$$m_5^{-1} m_3^{-1} m_4 m_3^{-1} m_5^{-1} m_4^{-2} = 1,$$
$$m_2 m_3 m_5^{-1} m_3^{-1} m_2 m_3 m_5 m_3^{-1} = 1,$$
$$m_4 m_1^{-1} m_4 m_3 m_1 m_4^{-1} m_3 m_4 = 1,$$
$$m_4^{-1} m_3 m_5^{-1} m_4^{-1} m_3 m_5^{-1} m_4 m_5 = 1,$$
$$[m_5, m_1^{-1}, m_5] = 1, \quad (m_1^{-2} m_2)^3 = 1,$$
$$m_5 m_3 m_1 m_2 m_1^{-1} m_5^{-1} m_2 m_5^{-1} m_4^{-1} = 1,$$
$$(m_3^{-1} m_1^{-1} m_2)^3 = 1, \quad (m_3 m_1^{-1} m_2)^3 = 1,$$
$$m_1 m_3^{-1} m_2 m_3 m_1^{-1} m_4^{-2} m_3 m_5 = 1,$$
$$m_1^{-1} m_3 m_2 m_1^3 m_3 m_1^2 m_3^{-1} = 1,$$
$$m_2 m_1 m_3 m_1^3 m_2 m_1 m_2 m_3^{-1} = 1.$$

(l) Let K be the finitely presented group constructed in (k). Then V is an irreducible 8-dimensional representation of K over $F = \mathrm{GF}(2)$. Furthermore, K has a faithful permutation representation PK of degree 63.

(m) Let $H_1 = \langle m_i, v_j \mid 1 \le i \le 5, 1 \le j \le 8 \rangle$ be the split extension of K by V. Then H_1 has a set $\mathcal{R}(H_1)$ of defining relations consisting of $\mathcal{R}(K)$, $\mathcal{R}_1(V \rtimes K)$ and the following set $\mathcal{R}_2(V \rtimes K)$ of essential relations:

$$m_1 v_1 m_1^{-1} v_7 v_8 = m_1 v_2 m_1^{-1} v_1 = m_1 v_3 m_1^{-1} v_2 v_8 = m_1 v_4 m_1^{-1} v_3 = 1,$$
$$m_1 v_5 m_1^{-1} v_4 v_8 = m_1 v_6 m_1^{-1} v_5 v_8 = m_1 v_7 m_1^{-1} v_6 = m_1 v_8 m_1^{-1} v_8 = 1,$$
$$m_2 v_1 m_2^{-1} v_4 v_5 v_6 = m_2 v_2 m_2^{-1} v_1 v_2 v_3 v_4 v_7 v_8 = m_2 v_3 m_2^{-1} v_5 v_6 v_7 v_8 = 1,$$
$$m_2 v_4 m_2^{-1} v_3 v_4 v_5 v_6 v_7 v_8 = m_2 v_5 m_2^{-1} v_1 v_4 v_6 = m_2 v_6 m_2^{-1} v_3 v_5 v_7 v_8 = 1,$$

$$m_2v_7m_2^{-1}v_1v_4v_5v_6v_7 = m_2v_8m_2^{-1}v_8 = 1,$$

$$m_3v_1m_3^{-1}v_1v_2v_7v_8 = m_3v_2m_3^{-1}v_1v_3v_4v_5v_7 = m_3v_3m_3^{-1}v_1v_2v_3 = 1,$$

$$m_3v_4m_3^{-1}v_1v_4v_5v_6v_7 = m_3v_5m_3^{-1}v_2v_3v_4 = m_3v_6m_3^{-1}v_1v_3v_4 = 1,$$

$$m_3v_7m_3^{-1}v_1v_3v_5 = m_3v_8m_3^{-1}v_8 = 1,$$

$$m_4v_1m_4^{-1}v_1v_2v_5v_7 = m_4v_2m_4^{-1}v_1v_3v_7v_8 = m_4v_3m_4^{-1}v_4v_6v_7v_8 = 1,$$

$$m_4v_4m_4^{-1}v_1v_3v_4v_5v_6v_7v_8 = m_4v_5m_4^{-1}v_1v_2v_3v_6 = m_4v_6m_4^{-1}v_4v_5v_8 = 1,$$

$$m_4v_7m_4^{-1}v_1v_2v_6v_8 = m_4v_8m_4^{-1}v_3v_4v_5v_7 = 1,$$

$$m_5v_1m_5^{-1}v_1v_4v_8 = m_5v_2m_5^{-1}v_1v_3v_8 = m_5v_3m_5^{-1}v_2v_4v_5v_6v_7v_8 = 1,$$

$$m_5v_4m_5^{-1}v_1v_3v_5v_6v_7v_8 = m_5v_5m_5^{-1}v_1v_6 = m_5v_6m_5^{-1}v_4v_5v_6v_8 = 1,$$

$$m_5v_7m_5^{-1}v_1v_2v_6v_7v_8 = m_5v_8m_5^{-1}v_3v_5v_6v_7v_8 = 1.$$

(n) $\dim_F[H^2(H_1, F)] = 2$ and there exists a unique central extension H of H_1 whose Sylow 2-subgroups are isomorphic to the ones of $D = C_{E_3}(z)$. H is a split extension of K by its Fitting subgroup Q and H is isomorphic to the finitely presented group $H = \langle h_i \mid 1 \le i \le 14 \rangle$ with the following set $\mathcal{R}(H)$ of defining relations:

$$h_1^7 = h_2^2 = h_3^4 = h_4^5 = h_5^3 = h_6^2 = h_7^2 = h_{10}^2 = h_{13}^2 = h_{14}^2 = 1,$$

$$h_8^2 = h_9^2 = h_{11}^2 = h_{12}^2 = h_{14}, \quad [h_1, h_{14}] = 1,$$

$$[h_2, h_{14}] = [h_3, h_{14}] = [h_4, h_{14}] = [h_5, h_{14}] = [h_6, h_{14}] = [h_7, h_{14}] = 1,$$

$$[h_8, h_{14}] = [h_9, h_{14}] = [h_{10}, h_{14}] = [h_{11}, h_{14}] = [h_{12}, h_{14}] = [h_{13}, h_{14}] = 1,$$

$$h_2h_4^{-1}h_3^{-1}h_2h_3h_4 = (h_3^{-1}h_4^{-1})^2 = h_3h_5h_4^{-1}h_3^{-1}h_5^{-1}h_4 = 1,$$

$$(h_5^{-1}h_4^{-1})^3 = (h_1^{-1}h_2)^4 = (h_2h_3^{-1})^4 = (h_2h_3h_2h_3^{-1})^2 = 1,$$

$$h_5^{-1}h_3^{-1}h_4h_3^{-1}h_5^{-1}h_4^{-2} = h_2h_3h_5^{-1}h_3^{-1}h_2h_3h_5h_3^{-1} = 1,$$

$$h_4h_1^{-1}h_4h_3h_1h_4^{-1}h_3h_4 = h_4^{-1}h_3h_5^{-1}h_4^{-1}h_3h_5^{-1}h_4h_5 = 1,$$

$$[h_5, h_1^{-1}, h_5] = (h_1^{-2}h_2)^3 = h_5h_3h_1h_2h_1^{-1}h_5^{-1}h_2h_5^{-1}h_4^{-1} = 1,$$

$$(h_3^{-1}h_1^{-1}h_2)^3 = (h_3h_1^{-1}h_2)^3 = h_1h_3^{-1}h_2h_3h_1^{-1}h_4^{-2}h_3h_5 = 1,$$

$$h_1^{-1}h_3h_2h_1^3h_3h_1^2h_3^{-1} = h_2h_1h_3h_1^3h_2h_1h_2h_3^{-1} = 1,$$

$$h_1h_6h_1^{-1}h_{12}h_{13} = h_1h_7h_1^{-1}h_6 = h_1h_8h_1^{-1}h_7h_{13} = 1,$$

$$h_1h_9h_1^{-1}h_8h_{14} = h_1h_{10}h_1^{-1}h_9h_{13} = h_1h_{11}h_1^{-1}h_{10}h_{13} = 1,$$

$$h_1h_{12}h_1^{-1}h_{11}h_{14} = h_1h_{13}h_1^{-1}h_{13} = h_2h_6h_2^{-1}h_9h_{10}h_{11} = 1,$$

$$h_2h_7h_2^{-1}h_6h_7h_8h_9h_{12}h_{13}h_{14} = h_2h_8h_2^{-1}h_{10}h_{11}h_{12}h_{13} = 1,$$

$$h_2h_9h_2^{-1}h_8h_9h_{10}h_{11}h_{12}h_{13} = h_2h_{10}h_2^{-1}h_6h_9h_{11} = 1,$$

$$h_2h_{11}h_2^{-1}h_8h_{10}h_{12}h_{13} = h_2h_{12}h_2^{-1}h_6h_9h_{10}h_{11}h_{12}h_{14} = 1,$$

$$h_2h_{13}h_2^{-1}h_{13} = h_3h_6h_3^{-1}h_6h_7h_{12}h_{13}h_{14} = 1,$$

$$h_3h_7h_3^{-1}h_6h_8h_9h_{10}h_{12}h_{14} = h_3h_8h_3^{-1}h_6h_7h_8h_{14} = 1,$$

$$h_3h_9h_3^{-1}h_6h_9h_{10}h_{11}h_{12} = h_3h_{10}h_3^{-1}h_7h_8h_9 = 1,$$

$$h_3h_{11}h_3^{-1}h_6h_8h_9 = h_3h_{12}h_3^{-1}h_6h_8h_{10}h_{14} = 1,$$

$$h_3 h_{13} h_3^{-1} h_{13} = h_4 h_6 h_4^{-1} h_6 h_7 h_{10} h_{12} = 1,$$

$$h_4 h_7 h_4^{-1} h_6 h_8 h_{12} h_{13} = h_4 h_8 h_4^{-1} h_9 h_{11} h_{12} h_{13} h_{14} = 1,$$

$$h_4 h_9 h_4^{-1} h_6 h_8 h_9 h_{10} h_{11} h_{12} h_{13} = h_4 h_{10} h_4^{-1} h_6 h_7 h_8 h_{11} h_{14} = 1,$$

$$h_4 h_{11} h_4^{-1} h_9 h_{10} h_{13} h_{14} = h_4 h_{12} h_4^{-1} h_6 h_7 h_{11} h_{13} = 1,$$

$$h_4 h_{13} h_4^{-1} h_8 h_9 h_{10} h_{12} h_{14} = h_5 h_6 h_5^{-1} h_6 h_9 h_{13} h_{14} = 1,$$

$$h_5 h_7 h_5^{-1} h_6 h_8 h_{13} = h_5 h_8 h_5^{-1} h_7 h_9 h_{10} h_{11} h_{12} h_{13} h_{14} = 1,$$

$$h_5 h_9 h_5^{-1} h_6 h_8 h_{10} h_{11} h_{12} h_{13} = h_5 h_{10} h_5^{-1} h_6 h_{11} = 1,$$

$$h_5 h_{11} h_5^{-1} h_9 h_{10} h_{11} h_{13} h_{14} = h_5 h_{12} h_5^{-1} h_6 h_7 h_{11} h_{12} h_{13} h_{14} = 1,$$

$$h_5 h_{13} h_5^{-1} h_8 h_{10} h_{11} h_{12} h_{13} = h_6^{-1} h_7^{-1} h_6 h_7 h_{14} = 1,$$

$$h_6^{-1} h_8^{-1} h_6 h_8 h_{14} = h_6^{-1} h_9^{-1} h_6 h_9 h_{14} = [h_6, h_{10}] = 1,$$

$$h_6^{-1} h_{11}^{-1} h_6 h_{11} h_{14} = [h_6, h_{12}] = h_6^{-1} h_{13}^{-1} h_6 h_{13} h_{14} = 1,$$

$$[h_7, h_8] = h_7^{-1} h_9^{-1} h_7 h_9 h_{14} = [h_7, h_{10}] = h_7^{-1} h_{11}^{-1} h_7 h_{11} h_{14} = 1,$$

$$h_7^{-1} h_{12}^{-1} h_7 h_{12} h_{14} = h_7^{-1} h_{13}^{-1} h_7 h_{13} h_{14} = 1,$$

$$h_8^{-1} h_9^{-1} h_8 h_9 h_{14} = h_8^{-1} h_{10}^{-1} h_8 h_{10} h_{14} = [h_8, h_{11}] = 1,$$

$$[h_8, h_{12}] = h_8^{-1} h_{13}^{-1} h_8 h_{13} h_{14} = [h_9, h_{10}] = [h_9, h_{11}] = [h_9, h_{12}] = 1,$$

$$h_9^{-1} h_{13}^{-1} h_9 h_{13} h_{14} = [h_{10}, h_{11}] = h_{10}^{-1} h_{12}^{-1} h_{10} h_{12} h_{14} = 1,$$

$$h_{10}^{-1} h_{13}^{-1} h_{10} h_{13} h_{14} = h_{11}^{-1} h_{12}^{-1} h_{11} h_{12} h_{14} = 1,$$

$$h_{11}^{-1} h_{13}^{-1} h_{11} h_{13} h_{14} = h_{12}^{-1} h_{13}^{-1} h_{12} h_{13} h_{14} = 1.$$

PROOF. (a) By Lemma 2.6 the split extension $E_3 = V_3 \rtimes A_{22} = \langle p, q, v_1 \rangle$ has a unique conjugacy class of involutions of highest defect. It is represented by $z = (pq^2 v_1)^5$ and its centralizer $D = C_{E_3}(z)$ has order $2^{18} \cdot 3^2 \cdot 5$.

Using the faithful permutation representation of E_3 with stabilizer A_{22} and MAGMA it has been checked that D has a unique normal subgroup Q of order 512 with center $Z(Q) = \langle z \rangle$. Furthermore, Q is extra-special and it has a complement W of order $2^9 \cdot 3^2 \cdot 5$. Another application of MAGMA yields that the Fitting subgroup B of W is elementary abelian of order 2^5, and that B has a complement $L \cong S_6$ in W. Using then a stand alone program written by the first author we found the generators of the subgroups $B = \langle f_j \mid 1 \leq j \leq 5$, $L = \langle n_l \mid 1 \leq l \leq 4 \rangle$ and $Q = \langle n_l \mid 5 \leq l \leq 12 \rangle$ of D. Hence (a) holds.

(b) Since Q is extra-special of order 2^9 and its center $Z(Q) = Z(D) = \langle z \rangle$ we determined by means of MAGMA eight involutions q_i generating Q. Using the stand alone program mentioned before the words in the generators n_l of Q of the eight involutions were calculated. Since their residue classes generate the elementary abelian normal subgroup $V = Q/Z(Q)$ of $D_1 = D/Z(Q)$ we obtained the following set $\mathcal{R}(Q)$ of defining relations of $Q = \langle q_i \mid 1 \leq i \leq 8 \rangle$:

$$q_1^2 = q_2^2 = q_3^2 = q_4^2 = q_5^2 = q_6^2 = q_7^2 = q_8^2 = 1,$$

$$[q_1, q_2] = [q_1, q_3] = [q_1, q_4] = [q_1, q_6] = [q_1, q_8] = 1, \quad [q_1, q_5] = [q_1, q_7] = z,$$

$$[q_2, q_3] = [q_2, q_4] = [q_2, q_5] = [q_2, q_6] = [q_2, q_7] = 1, \quad [q_2, q_8] = z,$$

$$[q_3, q_4] = [q_3, q_6] = [q_3, q_7] = 1, \quad [q_3, q_5] = [q_3, q_8] = z,$$

$$[q_4, q_5] = [q_4, q_6] = [q_4, q_7] = [q_4, q_8] = z,$$
$$[q_5, q_6] = [q_5, q_7] = [q_5, q_8] = [q_6, q_7] = [q_6, q_8] = [q_7, q_8] = 1.$$

In particular, $z = (q_4 q_8)^2$, hence (b) holds.

(c) Another application of MAGMA and the permutation representation PE_3 yields that the permutation group PL of the complement $L = \langle n_k \mid 1 \leq k \leq 4 \rangle$ can be generated by the three elements s_1, s_2 and s_3. For these generators s_i MAGMA provided the following set $\mathcal{R}(L)$ of defining relations:

$$s_1^4 = s_2^5 = s_3^3 = 1,$$
$$(s_2 s_1)^2 = s_1 s_3 s_2^{-1} s_1^{-1} s_3^{-1} s_2 = (s_2 s_3)^3 = s_2^2 s_3 s_1 s_2^{-1} s_1 s_3 = 1,$$
$$s_2^{-1} s_1 s_3^{-1} s_2^{-1} s_1 s_3^{-1} s_2 s_3 = 1.$$

Using MAGMA again it has been checked that $L \cong S_6$. Thus (c) holds.

(d) Since the elementary abelian subgroup B is normal in the semidirect product W of L by B the presentation of $W = \langle f_j, s_k \rangle$ can easily be calculated by means of $\mathcal{R}(L)$ and the conjugation action of the $s_k \in L$ on B. Hence W has a defining set $\mathcal{R}(W)$ of relations consisting of $\mathcal{R}(L)$ and the following relations:

$$f_1^2 = f_2^2 = f_3^2 = f_4^2 = f_5^2 = 1,$$
$$f_1^{s_1}(f_1)^{-1} = f_1^{s_2}(f_1)^{-1} = f_1^{s_3}(f_1)^{-1} = 1,$$
$$f_2^{s_1}(f_1 f_4 f_5)^{-1} = f_2^{s_2}(f_1 f_2 f_3 f_5)^{-1} = f_2^{s_3}(f_3 f_5)^{-1} = 1,$$
$$f_3^{s_1}(f_3 f_5)^{-1} = f_3^{s_2}(f_2 f_3)^{-1} = f_3^{s_3}(f_2 f_5)^{-1} = 1,$$
$$f_4^{s_1}(f_1 f_3 f_4 f_5)^{-1} = f_4^{s_2}(f_1 f_3 f_4)^{-1} = f_4^{s_3}(f_1 f_3 f_4)^{-1} = 1,$$
$$f_5^{s_1}(f_2 f_3 f_4)^{-1} = f_5^{s_2}(f_2 f_3 f_4)^{-1} = f_5^{s_3}(f_1 f_2)^{-1} = 1,$$
$$[f_1, f_2] = [f_1, f_3] = [f_1, f_4] = [f_1, f_5] = [f_2, f_3] = [f_2, f_4] = [f_2, f_5] = 1,$$
$$[f_3, f_4] = [f_3, f_5] = [f_4, f_5] = 1.$$

As Q is normal in the semidirect product D of W by Q the conjugate action of the f_j and s_k on Q provides the following set $\mathcal{R}(Q)$ of relations:

$$q_1^{s_1}(q_3 q_4 q_6 q_7 q_8)^{-1} = z, \quad q_1^{s_2}(q_1 q_2 q_3 q_8)^{-1} = q_1^{s_3}(q_1 q_3 q_5 q_7 q_8)^{-1} = 1,$$
$$q_1^{f_1}(q_1 q_5 q_7)^{-1} = q_1^{f_2}(q_1)^{-1} = q_1^{f_3}(q_1 q_5 q_6 q_7)^{-1} = q_1^{f_4}(q_1)^{-1} = 1,$$
$$q_1^{f_5}(q_1 q_6 q_8)^{-1} = q_2^{s_1}(q_1 q_4 q_5)^{-1} = (q_4 q_8)^2,$$
$$q_2^{s_2}(q_2 q_4 q_5 q_6)^{-1} = q_2^{s_3}(q_1 q_2 q_4 q_5)^{-1} = 1,$$
$$q_2^{f_1}(q_2 q_6)^{-1} = (q_4 q_8)^2, \quad q_2^{f_2}(q_2)^{-1} = q_2^{f_3}(q_2 q_5 q_6 q_7)^{-1} = 1,$$
$$q_2^{f_4}(q_2 q_5 q_7)^{-1} = q_2^{f_5}(q_2 q_6 q_7)^{-1} = z, \quad q_3^{s_1}(q_3 q_5 q_7 q_8)^{-1} = z,$$
$$q_3^{s_2}(q_2 q_3 q_6 q_8)^{-1} = (q_4 q_8)^2, \quad q_3^{s_3}(q_1 q_5 q_6 q_7 q_8)^{-1} = 1, \quad q_3^{f_1}(q_3 q_7)^{-1} = z,$$
$$q_3^{f_2}(q_3 q_6)^{-1} = q_3^{f_3}(q_3 q_5 q_8)^{-1} = 1, \quad q_3^{f_4}(q_3 q_5 q_6 q_7 q_8)^{-1} = q_3^{f_5}(q_3 q_7)^{-1} = (q_4 q_8)^2,$$
$$q_4^{s_1}(q_2 q_4 q_6 q_7)^{-1} = q_4^{s_2}(q_1 q_3 q_5 q_7 q_8)^{-1} = 1, \quad q_4^{s_3}(q_1 q_2 q_3 q_5 q_8)^{-1} = z,$$
$$q_4^{f_1}(q_4 q_6 q_8)^{-1} = 1, \quad q_4^{f_2}(q_4 q_5 q_7)^{-1} = q_4^{f_5}(q_4 q_5 q_7)^{-1} = z,$$
$$q_4^{f_3}(q_4 q_5 q_8)^{-1} = q_4^{f_4}(q_4)^{-1} = 1,$$
$$q_5^{s_1}(q_6 q_8)^{-1} = q_5^{s_2}(q_5 q_7)^{-1} = 1, \quad q_5^{s_3}(q_7)^{-1} = (q_4 q_8)^2, \quad q_5^{f_1}(q_5)^{-1} = 1,$$

$$q_5^{f_2}(q_5)^{-1} = q_5^{f_3}(q_5)^{-1} = q_5^{f_4}(q_5)^{-1} = q_5^{f_5}(q_5)^{-1} = 1,$$

$$q_6^{s_1}(q_5q_6q_7q_8)^{-1} = q_6^{s_2}(q_8)^{-1} = q_6^{s_3}(q_5q_7q_8)^{-1} = z,$$

$$q_6^{f_1}(q_6)^{-1} = 1, \quad q_6^{f_2}(q_6)^{-1} = q_6^{f_3}(q_6)^{-1} = q_6^{f_4}(q_6)^{-1} = q_6^{f_5}(q_6)^{-1} = 1,$$

$$q_7^{s_1}(q_7)^{-1} = 1, \quad q_7^{s_2}(q_6q_7)^{-1} = q_7^{s_3}(q_5q_7)^{-1} = (q_4q_8)^2,$$

$$q_7^{f_1}(q_7)^{-1} = q_7^{f_2}(q_7)^{-1} = q_7^{f_3}(q_7)^{-1} = q_7^{f_4}(q_7)^{-1} = q_7^{f_5}(q_7)^{-1} = 1,$$

$$q_8^{s_1}(q_5q_6q_7)^{-1} = q_8^{s_2}(q_5q_8)^{-1} = (q_4q_8)^2, \quad q_8^{s_3}(q_6q_7q_8)^{-1} = 1,$$

$$q_8^{f_1}(q_8)^{-1} = q_8^{f_2}(q_8)^{-1} = q_8^{f_3}(q_8)^{-1} = q_8^{f_4}(q_8)^{-1} = q_8^{f_5}(q_8)^{-1} = 1,$$

Since z generates the center $Z(D)$ of D, the additional relations of the set $\mathcal{R}(D)$ of defining relations of D given in (d) are clear by the proof of (b). Using MAGMA again we have checked that this is a presentation of D.

(e) All the statements follow immediately from the presentation of D given in (d).

(f) From (c) and (d) follows that the Fitting subgroup B of W has order 2^5 and a complement $L = \langle s_1, s_2, s_3 \rangle \cong S_6$ in W. Hence W is isomorphic to the centralizer $H(\mathrm{Sp}_6(2))$ of a 2-central involution of $\mathrm{Sp}_6(2)$ by Proposition 8.6.5 of [11]. Thus we may apply Algorithm 7.4.8 of [11] to construct a simple overgroup K of W such that $|K : W|$ is odd and K normalizes V. For that we choose the basis $\mathcal{B} = \{v_i = q_i Z(Q) \mid 1 \leq i \leq 8\}$ of V. With respect to $\mathcal{B}$ the conjugate action of the s_k on V is described by the three matrices Ms_k of the statement.

(g) From the presentation of D follows that $u = f_1$ is an involution of $Z(W)$. Another application of MAGMA yields that $W = \langle u, r, y_1, y_2, y_3 \rangle$ for $r = f_2 f_4$ and that the conjugate actions of r and u on V have the following matrices with respect to $\mathcal{B}$:

$$Mu = \begin{pmatrix} 1&0&0&0&0&0&0&0 \\ 0&1&0&0&0&0&0&0 \\ 0&0&1&0&0&0&0&0 \\ 0&0&0&1&0&0&0&0 \\ 1&0&0&0&1&0&0&0 \\ 0&1&0&1&0&1&0&0 \\ 1&0&1&0&0&0&1&0 \\ 0&0&0&1&0&0&0&1 \end{pmatrix} \quad \text{and} \quad Mr = \begin{pmatrix} 1&0&0&0&0&0&0&0 \\ 0&1&0&0&0&0&0&0 \\ 0&0&1&0&0&0&0&0 \\ 0&0&0&1&0&0&0&0 \\ 0&1&1&1&1&0&0&0 \\ 0&0&0&0&0&1&0&0 \\ 0&1&1&1&0&0&1&0 \\ 0&0&1&0&0&0&0&1 \end{pmatrix}.$$

(h) Using the faithful permutation representation PD_1 of D_1 given in (e) and MAGMA we find the generators of the given Sylow 2-subgroup S of W. It has a unique maximal elementary abelian normal subgroup $A = \langle a_1, a_2, a_3, a_4, a_5 = u, a_6 \rangle$ of order 64. Its generators a_i are represented by the following matrices with respect to $\mathcal{B}$:

$$Ma_1 = \begin{pmatrix} 1&0&0&0&0&0&0&0 \\ 0&1&0&0&0&0&0&0 \\ 1&0&1&1&0&0&0&0 \\ 0&0&0&1&0&0&0&0 \\ 0&0&0&0&1&0&0&0 \\ 0&0&0&0&0&1&0&0 \\ 0&0&0&0&1&0&1&1 \\ 0&0&0&0&0&0&0&1 \end{pmatrix}, \quad Ma_2 = \begin{pmatrix} 0&0&0&1&0&0&0&0 \\ 0&0&0&1&0&0&0&0 \\ 0&1&1&0&0&0&0&0 \\ 1&0&0&0&0&0&0&0 \\ 0&0&0&0&0&0&0&1 \\ 0&0&0&0&1&1&0&1 \\ 0&0&0&0&1&1&1&0 \\ 0&0&0&0&1&0&0&0 \end{pmatrix}, \quad Ma_3 = \begin{pmatrix} 0&0&0&1&0&0&0&0 \\ 1&1&0&1&0&0&0&0 \\ 1&1&1&0&0&0&0&0 \\ 1&0&0&0&0&0&0&0 \\ 0&0&0&0&0&0&0&1 \\ 0&0&0&0&0&1&0&0 \\ 0&0&0&0&0&1&1&0 \\ 0&0&0&0&1&0&0&0 \end{pmatrix},$$

$$Ma_4 = \begin{pmatrix} 0&0&0&1&0&0&0&0 \\ 1&1&0&1&0&0&0&0 \\ 1&1&1&0&0&0&0&0 \\ 1&0&0&0&0&0&0&0 \\ 1&0&0&0&0&0&0&1 \\ 1&1&0&0&0&1&0&0 \\ 1&0&1&1&0&1&1&0 \\ 0&0&0&1&1&0&0&0 \end{pmatrix}, \quad \text{and } Ma_6 = \begin{pmatrix} 1&0&0&0&0&0&0&0 \\ 0&1&0&0&0&0&0&0 \\ 0&0&1&0&0&0&0&0 \\ 0&0&0&1&0&0&0&0 \\ 1&0&0&1&1&0&0&0 \\ 0&0&0&0&0&1&0&0 \\ 1&1&0&0&0&0&1&0 \\ 1&0&0&1&0&0&0&1 \end{pmatrix}.$$

Observe that the map $\varphi \colon Px \to Mx$ sending the permutation Px of PD_1 to its matrix Mx with respect to the basis $\mathcal{B}$ yields an anti-epimorphism from PD_1 onto $MW = \langle Mr, Mu, Ms_1, Ms_2, Ms_3 \rangle \leq \mathrm{GL}_8(2)$ with kernel PV.

(i) $MA = \langle Ma_1, Ma_2, Ma_3, Ma_4, Ma_5 = Mu, Ma_6 \rangle$ is an elementary abelian subgroup of $\Gamma = \mathrm{GL}_8(2)$ of order 2^6. In view of Proposition 8.6.5(i) of [11] we now calculate $NA = N_\Gamma(MA)$. The matrix Ms of the statement generates a Sylow 7-subgroup of order 7 of NA.

(j) Let $MK = \langle MW, Ms \rangle$. Then an application of MAGMA yields that $MU = NA \cap MK = N_{MK}(MA)$ has order $2^9 \cdot 3 \cdot 7$. It is a split extension of $GL_3(2)$ by MA. Furthermore, $MK = \langle MW, MU \rangle$ is isomorphic to the simple group $Sp_6(2)$.

(k) The group MK has a faithful permutation representation of degree 63 with stabilizer MW. Using it and MAGMA the set $\mathcal{R}(K)$ of the defining relations of $K = \langle m_1, m_2, m_3, m_4, m_5 \rangle$ has been calculated where $m_1 = Ms$, $m_2 = Mr$, $m_3 = Ms_1$, $m_4 = My_4$ and $m_5 = Ms_3$.

(l) By (k) and (j), K has an irreducible 8-dimensional matrix representation V. The given presentation of the split extension $H_1 = V \rtimes K$ has been calculated by means of MAGMA using the presentation of K given in (k).

(m) Assertion (c) implies that $D_1 = D/Z(D)$ splits over $V = Q/Z(D)$ with complement W. By Algorithm 2.5 of [10] the centralizer $H = C_G(z)$ of z in the (at this time unknown) target simple group G must have odd index $|H : D| = |K : W|$. Therefore Theorem 1.4.15 of [11] implies that H is a central extension of H_1 by a cyclic subgroup $Z(H) = \langle z \rangle$ having a normal subgroup Q containing z such that $V = Q/Z(H)$ has a complement in $H/Z(H)$ isomorphic to K.

By construction $H_1 = \langle m_1, m_2, m_3, m_4, m_5, v_1, v_2, v_3, v_4, v_5, v_6, v_7, v_8 \rangle$ has a faithful permutation representation PH_1 of degree 256 with stabilizer K. Let FPH_1 be the presentation of H_1 given in (m). Then we can apply Holt's Algorithm implemented in MAGMA [8] to the trivial matrix representation of H_1 over $F = GF(2)$. It yields that the second cohomological dimension $\dim_F[H^2(H_1, F)] = 2$. Thus there are three non-split central extensions $E_{0,1}, E_{1,0}, E_{1,1}$ of H_1. As D does not split over $Z(D)$ H can only be isomorphic to a non-split extension.

Let $TH_1 := $ `GModule(PH_1, FEalg)` be the trivial module of the matrix algebra FEalg generated by the thirteen identity matrices corresponding to the thirteen generators of H_1. Using the MAGMA command

$$\texttt{P\_H :=ExtensionProcess(PH\_1,TH\_1,FPH\_1)}$$

we construct a presentation for each of the three non-split central extensions

$$
\begin{aligned}
E_{0,1} &:= \texttt{Extension(P\_H, [0,1])}, \\
E_{1,0} &:= \texttt{Extension(P\_H, [1,0])}, \text{ and} \\
E_{1,1} &:= \texttt{Extension(P\_H, [1,1])},
\end{aligned}
$$

corresponding to the three linearly independent 2-cocycles $[a, b] \in F^2$. Each of these three presentations has 14 generators where the 14th generator corresponds to the new central element z. The first five generators correspond to m_1, m_2, m_3, m_4, m_5 in the factor group H_1. Taking the corresponding generators in $E_{1,0}$ as the stabilizer of a permutation representation $PE_{1,0}$ of $E_{1,0}$ it follows by another application of MAGMA that $PE_{1,0}$ is a faithful permutation of $E_{1,0}$ of degree 512, and that $E_{1,0}$ has an extra-special normal subgroup Q of order 2^9 with complement $K = \langle m_1, m_2, m_3, m_4, m_5 \rangle$. We also have checked that the Sylow 2-subgroups of $E_{1,0}$ and D are isomorphic, and that this is not the case for the the two other extensions $E_{0,1}$ and $E_{1,1}$. Therefore only $E_{1,0}$ can be isomorphic to the centralizer $H = C_G(z)$. The presentation of $E_{1,0}$ is given in the statement. This completes the proof. $\qquad\square$

LEMMA 3.2. *Keep the notation of Lemmas 2.4, 2.5 and Proposition 3.1. Let $H = \langle h_i \,|\, 1 \leq i \leq 14 \rangle$ be the finitely presented group constructed in Proposition 3.1. Then the following statements hold:*

(a) *H has a faithful permutation representation of degree 512 with stabilizer $\langle h_1, h_4 \rangle$.*

(b) *Each Sylow 2-subgroup S of H has a unique maximal elementary abelian normal subgroup A of order 2^{10} and $N_H(A) \cong D = C_{E_3}(z)$.*

(c) *There is a Sylow 2-subgroup S such that $D = N_H(A) = \langle x, y \rangle$, where $x = h_2 h_{16} h_1$ and $y = h_9 h_{14} h_{15}$ have orders 12 and 6, respectively. Furthermore, $H = \langle x, y, h \rangle$ where $h = h_1$ has order 7.*

(d) *The Goldschmidt index of the amalgam $H \leftarrow D \rightarrow E_3$ is 2.*

(e) *A system of representatives r_i of the 100 conjugacy classes of H and the corresponding centralizers orders $|C_H(r_i)|$ are given in Table A.1.*

(f) *A system of representatives d_i of the 148 conjugacy classes of D and the corresponding centralizers orders $|C_D(d_i)|$ are given in Table A.2.*

(g) *Let $\sigma : N_H(A) \rightarrow D = C_{E_3}(z)$ be the isomorphism given in (b). Then there is an element $e \in E_3$ of order 4 such that $E_3 = \langle \sigma(D), e \rangle$. A system of representatives e_i of the 79 conjugacy classes of E_3 and the corresponding centralizers orders $|C_{E_3}(e_i)|$ are given in Table A.3.*

(h) *The character tables of H, D and E_3 are given in Tables B.1, B.2, and B.3, respectively.*

PROOF. (a) This assertion follows at once from Proposition 3.1(n).

In particular, H has a faithful permutation representation PH of degree 512. Using it and MAGMA it is straightforward to verify statements (b) and (c).

(d) The Goldschmidt index has been calculated by means of Kratzer's Algorithm 7.1.10 of [11].

The systems of representatives of the conjugacy classes of H and D have been calculated by means of PH, MAGMA and Kratzer's Algorithm 5.3.18 of [11].

(g) It has been checked with MAGMA that $e = p$ satisfies $E_3 = \langle \sigma(D), e \rangle$.

The character tables mentioned in (h) have been calculated by means of PH and MAGMA. $\qquad\square$

PROPOSITION 3.3. *Keep the notation of Lemma 2.4. Let*
$$E_2 = \langle a, b, c, d, t, g, h, i, v_1 \rangle$$
be the split extension of $\mathcal{M}_{22}$ by its simple module V_2 of dimension 10 over $F = \mathrm{GF}(2)$. Then the following statements hold:

(a) *$z = (iv_1)^2$ is a 2-central involution of E_2 with centralizer $D = C_{E_2}(z)$.*

(b) *D is a finitely presented group $D = \langle y_i \,|\, 1 \leq i \leq 11 \rangle$ having the following set $\mathcal{R}(D)$ of defining relations:*

$$y_2^2 = y_3^2 = y_5^2 = y_6^2 = y_7^2 = y_8^2 = y_9^2 = y_{11}^2 = (y_5 y_9)^2 = (y_5 y_{11})^2 = (y_9 y_{11})^2 = 1,$$

$$y_{11} y_3 y_5 y_4 y_{10} y_3 y_5 y_4 y_{10} y_5^{-1} y_{10}^{-1} y_4^{-1} y_5^{-1} y_3^{-1} y_{10}^{-1} y_4^{-1} y_5^{-1} y_3^{-1} = 1,$$

$$y_{11} y_3 y_5 y_4 y_{10} y_3 y_5 y_4 y_{10} y_9^{-1} y_{10}^{-1} y_4^{-1} y_5^{-1} y_3^{-1} y_{10}^{-1} y_4^{-1} y_5^{-1} y_3^{-1} y_5 = 1,$$

$$y_{11} y_{10}^{-1} y_4^{-1} y_5^{-1} y_3^{-1} y_{10}^{-1} y_4^{-1} y_5^{-1} y_3^{-1} y_9^{-1} y_5^{-1} y_3 y_5 y_4 y_{10} y_3 y_5 y_4 y_{10} = 1,$$

$$y_{10} y_5^{-1} y_{10}^{-1} y_4^{-1} y_5^{-1} y_3^{-1} y_5 y_9 y_3 y_5 y_4 = y_{11} y_3 y_5 y_4 y_{10} y_9^{-1} y_{10}^{-1} y_4^{-1} y_5^{-1} y_3^{-1} y_5 y_9 = 1,$$

$$y_{11} y_{10}^{-1} y_4^{-1} y_5^{-1} y_3^{-1} y_5^{-1} y_3 y_5 y_4 y_{10} = (y_{10} y_3 y_5 y_4)^3 = y_9 y_4^{-1} y_5^{-1} y_4 = y_9 y_4^{-1} y_9^{-1} y_4 y_5 = 1,$$

$$y_{11} y_4^{-1} y_{11}^{-1} y_4 y_5 = y_{11} y_3 y_5 y_4 y_{10} y_4^{-1} y_3 y_5 y_4 y_{10} y_4^{-1} y_9 = 1,$$

$$y_{10}y_4^{-1}y_{10}^{-1}y_4^{-1}y_5^{-1}y_3^{-1}y_4^{-1}y_3y_5y_4y_{10}y_4y_3y_5y_4 = 1,$$

$$y_{11}y_3y_5y_4y_{10}y_3y_5y_4y_{10}y_4^{-2}y_{10}^{-1}y_4^{-1}y_5^{-1}y_3^{-1}y_4y_9 = 1,$$

$$y_{11}y_{10}y_5^{-1}y_{10}^{-1}y_{11}^{-1}y_3^{-1}y_5y_9y_3y_5y_4y_{10}y_3y_5y_4y_{10}y_4^{-1}y_{10}^{-1}y_4^{-1}y_5^{-1} = 1,$$

$$y_{11}y_{10}y_9^{-1}y_{10}^{-1}y_{11}^{-1}y_3^{-1}y_5y_3y_5y_4y_{10}y_4^{-1}y_3 = 1,$$

$$y_{11}y_{10}^{-1}y_{11}^{-1}y_3^{-1}y_4y_{10}^{-1}y_4^{-1}y_5^{-1}y_3^{-1}y_{11}^{-1}y_9^{-1}y_5^{-1}y_3y_{11}y_{10} = 1,$$

$$y_{11}y_4^{-1}y_5^{-1}y_3^{-1}y_{10}y_4^{-1}y_5^{-1}y_3^{-1}y_{10}^{-1}y_{11}^{-1}y_3^{-1}y_5y_9y_{11}y_3y_5y_4y_{10}y_3y_5y_4y_{10}y_3 = 1,$$

$$y_{11}y_4^{-1}y_5^{-1}y_3^{-1}y_4y_{10}^{-1}y_4^{-1}y_5^{-1}y_3^{-1}y_{11}^{-1}y_9^{-1}y_{10}^{-1}y_{11}^{-1}y_3^{-1}$$
$$\cdot y_5y_3y_5y_4y_{10}y_3y_5y_4y_{10}y_4^{-1}y_3 = 1,$$

$$y_{11}y_{10}y_3y_{11}y_{10}y_5y_4^{-1}y_3 = y_{11}y_{10}y_3y_5y_9y_{10}^{-1}y_5y_3y_5y_4y_{10}y_4^{-1}y_{10}^{-1}y_4^{-1}y_5^{-1} = 1,$$

$$y_{11}y_3y_5y_4y_{10}y_3y_5y_4y_{10}y_4^{-1}y_3y_5y_{10}^{-1}y_4^{-1}y_5^{-1}y_3^{-1}y_{10}y_4^{-1}y_5^{-1}y_3^{-1}y_5^{-1}y_3^{-1}y_5 = 1,$$

$$y_{11}y_4^{-1}y_{10}^{-1}y_4^{-1}y_5^{-1}y_3^{-1}y_5^{-1}y_3^{-1}y_{10}^{-1}y_{11}y_3^{-1}y_{11}^{-1}y_9^{-1}y_5^{-1}y_3y_5^2 = 1,$$

$$y_{11}y_{10}y_5^{-1}y_3^{-1}y_{10}^{-1}y_{11}^{-1}y_3^{-1}y_4y_9^{-1}y_3y_5y_9y_3 = 1,$$

$$y_{11}y_5^{-1}y_3^{-1}y_{10}^{-1}y_{11}^{-1}y_5y_4y_{10}y_4y_{11}^{-1}y_3y_5y_9 = (y_5y_3y_5)^2 = 1,$$

$$y_{11}y_3y_5y_4y_{10}y_4^{-1}y_3y_5y_4y_{10}y_3y_5y_4y_3y_5y_4^{-1}y_5^{-1}y_3^{-1}y_5 = 1,$$

$$y_9y_3y_5y_4^{-1}y_2y_3^{-1}y_2^{-1}y_5 = y_9y_4^{-1}y_2y_4^{-1}y_2^{-1} = y_9y_2y_5^{-1}y_2^{-1} = 1,$$

$$y_9y_2^{-1}y_5^{-1}y_2 = y_{10}y_2^{-1}y_5^{-1}y_4^{-1}y_{10}^{-1}y_2 = y_{11}y_2^{-1}y_{11}^{-1}y_9^{-1}y_5^{-1}y_2 = 1,$$

$$y_9y_2y_1y_2y_1y_2^{-1}y_1^{-1}y_2^{-1}y_1^{-1}y_5 = y_9y_4^{-1}y_1y_2y_1y_3^{-1}y_1^{-1}y_2^{-1}y_1^{-1}y_5y_3 = 1,$$

$$y_9y_4^{-1}y_1y_2y_1y_4^{-1}y_1^{-1}y_2^{-1}y_1^{-1}y_5 = y_9y_1y_2y_1y_5^{-1}y_1^{-1}y_2^{-1}y_1^{-1} = 1,$$

$$y_9y_1^{-1}y_2^{-1}y_1^{-1}y_5^{-1}y_1y_2y_1 = y_{10}y_1^{-1}y_2^{-1}y_1^{-1}y_4^{-1}y_9^{-1}y_{10}y_9^{-1}y_1y_2y_1 = 1,$$

$$y_{11}y_1^{-1}y_2^{-1}y_1^{-1}y_{11}^{-1}y_9^{-1}y_5^{-1}y_1y_2y_1 = y_9y_1y_2y_1^2y_2y_1y_5 = y_9y_4y_1y_2y_1^2y_2^{-1}y_1^{-1} = 1,$$

$$y_{11}y_4^{-1}y_1y_3^{-1}y_1^{-1}y_9y_{11}y_3y_9 = [y_4^{-1}, y_1] = y_9y_1y_5^{-1}y_1^{-1} = y_9y_1^{-1}y_9^{-1}y_5^{-1}y_1 = 1,$$

$$y_{10}y_1^{-1}y_4y_{10}y_1 = y_{11}y_1^{-1}y_{11}^{-1}y_5^{-1}y_1 = y_9y_1y_4^{-1}y_5^{-1}y_1y_5 = y_4y_1^{-2} = 1,$$

$$y_{11}y_2y_1y_6y_3y_{11}y_{10}y_7y_1^{-1}y_2^{-1}y_4^{-1}y_{10}^{-1}y_3^{-1}y_5^{-1}y_7^{-1}y_{10}^{-1}y_{11}^{-1}y_3^{-1}y_6^{-1}y_5y_9y_{11}y_{10}^2 = 1,$$

$$y_{11}y_{10}^{-1}y_9y_2y_1y_2y_1y_6y_3y_{11}y_{10}y_7y_1^{-1}y_2^{-1}y_1^{-1}y_2^{-1}y_4y_9^{-1}y_{10}y_4^{-1}$$
$$\cdot y_{11}^{-1}y_5^{-1}y_7^{-1}y_{10}^{-1}y_{11}^{-1}y_3^{-1}y_6^{-1} = 1,$$

$$y_{11}y_{10}y_7y_4y_{10}^{-1}y_4y_5^{-1}y_7^{-1}y_{10}^{-1}y_{11}^{-1}y_3^{-1}y_6^{-1}y_5y_9y_{11}y_{10}^2y_6y_3 = 1,$$

$$y_{11}y_4^{-1}y_{10}^{-1}y_5y_2y_6y_3y_{11}y_{10}y_7y_2^{-1}y_3^{-1}y_{10}y_3^{-1}y_7^{-1}y_{10}^{-1}y_{11}^{-1}y_3^{-1}y_6^{-1}y_9 = 1,$$

$$y_{11}y_3y_4y_9y_{10}y_4^{-1}y_6y_3y_{11}y_{10}y_7y_4^{-1}y_{10}^{-1}y_9^{-1}y_3^{-1}y_{11}^{-1}y_9^{-1}y_5^{-1}y_7^{-1}y_{10}^{-1}y_{11}^{-1}y_3^{-1}y_6^{-1} = 1,$$

$$y_{11}y_4^{-1}y_{10}y_9y_7^{-1}y_{10}^{-1}y_{11}^{-1}y_3^{-1}y_6^{-1}y_4^{-1}y_{10}^{-1}y_4^{-1}y_{11}^{-1}y_9^{-1}y_6y_3y_{11}y_{10}y_7 = 1,$$

$$y_{11}y_7^{-1}y_{10}^{-1}y_{11}^{-1}y_3^{-1}y_6^{-1}y_{11}^{-1}y_9^{-1}y_6y_3y_{11}y_{10}y_7y_9 = 1,$$

$$y_{11}y_{10}y_7y_{10}^{-1}y_6y_3y_{11}y_{10}y_7y_5y_9y_{10}^{-1}y_4^{-1}y_{10}y_6y_3 = 1,$$

$$y_{11}y_3y_2y_1y_2y_1y_6y_3y_{11}y_{10}y_7y_3y_{10}y_6y_3y_{11}y_{10}y_7y_4y_7^{-1}y_{10}^{-1}y_{11}^{-1}y_3^{-1}y_6^{-1}y_5y_3y_4 = 1,$$

$$y_{11}y_{10}y_7y_4^{-1}y_2^{-1}y_7^{-1}y_{10}^{-1}y_{11}^{-1}y_3^{-1}y_6^{-1}y_5y_9y_{10}y_3y_{10}y_1y_2y_1^2y_6y_3y_{11}y_{10}y_7y_9y_6y_3 = 1,$$

$$y_{11}y_3y_{11}y_{10}y_6y_3y_{11}y_{10}y_7y_5y_6y_3y_{11}y_{10}y_7y_9^{-1}y_4^{-1}y_7^{-1}y_{10}^{-1}y_{11}^{-1}y_3^{-1}y_6^{-1}y_9 = 1,$$

$$y_7y_4^{-1}y_2^{-1}y_6^{-1}y_2^{-1}y_4y_5^{-1} = y_9y_6y_2^2y_8^{-1}y_5 = y_9y_6y_2y_5^{-1}y_2^{-1}y_6^{-1} = 1,$$

$$y_9y_2^{-1}y_6^{-1}y_5^{-1}y_6y_2 = y_{11}y_2^{-1}y_6^{-1}y_{11}^{-1}y_9^{-1}y_5^{-1}y_6y_2 = 1,$$

$$y_{11}y_{10}y_7y_3y_4y_2^{-1}y_6^{-1}y_3^{-1}y_7^{-1}y_{10}^{-1}y_{11}^{-1}y_3^{-1}y_6^{-1}y_1^{-1}y_2^{-1}y_1^{-1}y_2^{-1}y_4^{-1}y_{10}^{-1}y_3^{-1}$$
$$\cdot y_9^{-1}y_6y_2y_5y_3y_4y_{10}^{-1}y_6y_3 = 1,$$

$$y_{11}y_4y_{10}y_2y_1y_2y_1y_6y_3y_{11}y_{10}y_7y_2y_3y_2^{-1}y_6^{-1}y_{10}y_7^{-1}y_{10}^{-1}y_{11}^{-1}y_3^{-1}y_6^{-1}y_1^{-1}y_2^{-1}y_1^{-1}y_9^{-1}$$
$$\cdot y_{10}y_4^{-1}y_6y_2y_5y_9 = 1,$$

$$y_{11}y_{10}y_7y_2^{-1}y_6^{-1}y_9^{-1}y_7^{-1}y_{10}^{-1}y_{11}^{-1}y_3^{-1}y_6^{-1}y_1^{-2}y_2^{-1}y_1^{-1}y_2^{-1}$$
$$\cdot y_9^{-1}y_{10}^{-1}y_{11}^{-1}y_3^{-1}y_9^{-1}y_5^{-1}y_6y_2y_5y_3y_{11}y_{10}y_9y_2y_1y_2y_1^2y_6y_3 = 1,$$

$$y_{11}y_{10}y_7y_4y_9y_6y_2y_1^{-1}y_2^{-1}y_1^{-1}y_2^{-1}y_{10}^{-1}y_5^{-1}y_{10}^{-1}y_2^{-1}y_6^{-1}y_2y_1y_2y_1y_6y_3 = 1,$$

$$y_{11}y_3y_4y_{10}y_4y_2y_1y_2y_1y_6y_3y_{11}y_{10}y_7y_3y_4y_6y_2y_1^{-1}y_3^{-1}y_9^{-1}y_{10}$$
$$\cdot y_9^{-1}y_5^{-1}y_2^{-1}y_6^{-1}y_5y_9 = 1,$$

$$y_{11}y_3y_{11}y_{10}y_6^{-1}y_{10}y_7^{-1}y_{10}^{-1}y_{11}^{-1}y_3^{-1}y_6^{-1}y_2^{-1}y_3^{-1}y_{10}^{-1}y_{11}^{-1}y_9^{-1}y_6y_2 = 1,$$

$$y_{11}y_{10}y_2^{-1}y_6^{-1}y_{10}y_7^{-1}y_{10}^{-1}y_{11}^{-1}y_3^{-1}y_6^{-1}y_4^{-1}y_9^{-1}y_{10}^{-1}y_5^{-1}y_6y_2y_5y_3y_4^{-1} = 1,$$

$$(y_6y_2^2)^2 = y_9y_4^{-1}y_2y_6y_2y_6y_2y_4^{-1}y_2^{-1}y_6^{-1} = 1,$$

$$y_{11}y_3y_5y_2y_1y_2y_1^2y_6y_3y_{11}y_{10}y_7y_2y_6y_2y_4y_{10}^{-1}y_6y_2y_{10}y_2^{-1}y_6^{-1}y_9 = 1,$$

$$y_{11}y_{10}y_7y_2y_{10}^{-1}y_6y_2^2y_{10}^{-1}y_6y_2y_5^{-1}y_3^{-1}y_2^{-1}y_6^{-1}y_5y_9y_4y_{10}y_9y_2y_6y_3 = 1.$$

(c) D has a faithful permutation representation PD of degree 1024 with stabilizer $U = \langle y_1, y_4, y_{10}\rangle$.

(d) D has a unique normal nonabelian subgroup Q of order 1024 with center $Z(Q) = \langle z_1, z_2\rangle$ of order 4, where $z_1 = y_9y_1y_2y_1y_2$, $z_2 = y_9y_{11}y_4y_1$. Furthermore, Q has a complement W in D of order $|W| = 2^7 \cdot 3 \cdot 5$, where $W = \langle w_1, w_2, w_3\rangle$ and

$$w_1 = y_{11}y_{10}^2y_4^{-1}y_1y_6y_{10}^{-1}y_8y_{10}y_8y_2,$$
$$w_2 = y_{11}y_{10}^{-1}y_1y_6y_{10}y_4y_8y_{10}y_7, \quad \text{and}$$
$$w_3 = y_{11}y_{10}y_3y_4y_1y_6y_{10}y_4y_8y_{10}y_7.$$

(e) Let $\alpha : D \to D_1 = D/Z(Q)$ be the canonical epimorphism with kernel $\ker(\alpha) = Z(Q)$. Let $V = \alpha(Q)$ and let $u_i = \alpha(y_i) \in D_1$ for $i = 1, 2, \ldots, 11$. Then V is an elementary abelian normal subgroup of order 2^8 of D_1 having a complement $W_1 = \langle k_1, k_2, k_3\rangle \cong W$, where $k_j = \alpha(w_j)$ for $j = 1, 2, 3$. Furthermore, D_1 has a faithful permutation representation of degree 256 with stabilizer W_1, and $V = \langle q_l \,|\, 1 \le l \le 8\rangle$, where

$$q_1 = u_5u_9u_4u_{10}^{-1}u_4^{-1}u_1u_6u_3u_1u_{10},$$
$$q_2 = u_3u_{11}u_4u_1u_7u_3, \quad q_3 = u_9u_7, \quad q_4 = u_{11}u_1u_6u_4,$$
$$q_5 = u_9u_{11}u_3u_5u_1u_7u_{10}u_8u_3u_6,$$
$$q_6 = u_9u_{11}u_3u_{11}u_{10}u_1u_6u_3u_1^{-1}u_3u_7,$$
$$q_7 = u_5u_{11}u_3u_9u_{11}u_4^{-1}u_6u_3u_8u_{10}^{-1}u_7,$$
$$q_8 = u_9u_3u_4^{-1}u_1u_6u_3u_8u_{10}^{-1}u_7u_1.$$

(f) The conjugate action of the three generators k_j of W_1 on V with respect to the basis $\mathcal{B} = \{q_l \,|\, 1 \le l \le 8\}$ of V is given by the following matrices:

$$Mk_1 = \begin{pmatrix} 1&1&0&0&0&0&0&0 \\ 0&1&0&0&0&0&0&0 \\ 0&0&1&1&0&0&0&0 \\ 0&0&0&1&0&0&0&0 \\ 0&0&0&0&0&0&1&0 \\ 0&1&0&0&1&1&1&1 \\ 0&0&0&0&1&0&0&0 \\ 0&0&0&0&0&0&0&1 \end{pmatrix}, \quad Mk_2 = \begin{pmatrix} 1&0&1&0&0&0&0&0 \\ 1&1&1&1&0&0&0&0 \\ 1&1&0&1&0&0&0&0 \\ 1&0&0&1&0&0&0&0 \\ 0&1&0&0&1&0&0&0 \\ 0&0&0&1&0&0&1&1 \\ 0&1&0&0&1&0&1&1 \\ 1&0&1&0&0&0&1&0 \end{pmatrix}, \quad \text{and} \quad Mk_3 = \begin{pmatrix} 1&1&0&0&0&0&0&0 \\ 1&0&0&0&0&0&0&0 \\ 1&0&1&1&0&0&0&0 \\ 1&1&1&0&0&0&0&0 \\ 1&0&0&0&0&1&0&0 \\ 0&1&0&0&1&1&0&0 \\ 0&0&0&0&1&0&1&1 \\ 1&0&0&0&0&1&1&0 \end{pmatrix}.$$

(g) *The Fitting subgroup A of W_1 is elementary abelian of order 16 and generated by $a_1 = k_1 k_3 k_2 k_1 k_2^3$, $a_2 = k_2 k_3^2 k_2 k_3^2$, $a_3 = k_2 k_1 k_3^3 k_1 k_3$, and $a_4 = k_1 k_3 k_2 k_3 k_2 k_1 k_2$. It has a complement $L = \langle k_1, k_2 \rangle$ in W_1 which is isomorphic to the symmetric group S_5.*

(h) *The Fitting subgroup A is a maximal elementary abelian normal subgroup of the Sylow 2-subgroup $S = \langle A, k_1, r \rangle$ of W_1 with center $Z(S) = \langle u \rangle$, where $r = k_2 k_1 k_3 k_2^2 k_1 k_2 k_3$ and $u = k_1 (k_3 k_2)^2 k_1 k_2$. The centralizer $C_{W_1}(u)$ of u has order $2^7 \cdot 3$. It is generated by S and the element $d = (k_1 k_2)^2 (k_2 k_1)^2$ of order 3.*

(i) *Let MW_1 be the subgroup of $\mathrm{GL}_8(2)$ generated by the matrices of the conjugate action of the generators of W_1 on V with respect to the basis $\mathcal{B}$. Let $M a_i$, $M u$ and $M d$ be the corresponding matrix of a_i, u and d, respectively. Let $MX = C_{\mathrm{GL}_8(2)}(Mu) \cap C_{\mathrm{GL}_8(2)}(Md)$. Then MX has an abelian Sylow 3-subgroup MT of order 3^3 and MT contains the matrix*

$$
Mx = \begin{pmatrix}
0 & 1 & 0 & 0 & 0 & 0 & 0 & 0 \\
1 & 1 & 0 & 0 & 0 & 0 & 0 & 0 \\
1 & 1 & 0 & 1 & 0 & 0 & 0 & 1 \\
1 & 0 & 1 & 1 & 0 & 1 & 1 & 1 \\
0 & 0 & 0 & 1 & 1 & 1 & 0 & 0 \\
0 & 1 & 1 & 0 & 1 & 0 & 0 & 0 \\
1 & 1 & 1 & 1 & 1 & 0 & 0 & 0 \\
0 & 1 & 1 & 1 & 0 & 0 & 0 & 0
\end{pmatrix}
$$

of order 3 such that $MC = \langle C_{MW_1}(Mu), Mx \rangle$ has order $|MC| = 2^7 \cdot 3^2$.

(j) *Let $MK = \langle MC, MW_1 \rangle$. Then the derived subgroup MK' of MK is a simple group of order $|MK'| = 2^6 \cdot 3^4 \cdot 5$, which is isomorphic to the unitary group $U_4(2)$.*

(k) *MK is an irreducible subgroup of $\mathrm{GL}_8(2)$ generated by the matrices $m_1 = Mk_1$, $m_2 = Mk_2$, $m_3 = Mk_3$, $m_4 = Mx$ of respective orders 2, 5, 3, and 3. With respect to this set of generators MK has the following set $\mathcal{R}(K)$ of defining relations:*

$$m_1^2 = m_2^5 = m_3^3 = m_4^3 = 1, \quad [m_3, m_4] = 1, \quad m_3^{-1} m_2 m_1 m_2^{-1} m_3 m_1 = 1,$$
$$(m_2^{-1} m_1)^4 = 1, \quad m_3^{-1} m_2 m_3^{-1} m_1 m_2 m_3 m_1 m_2^{-1} = (m_2 m_3^{-1})^4 = 1,$$
$$[m_4, m_2^{-1}, m_4] = 1, \quad m_1 m_4^{-1} m_1 m_4^{-1} m_1 m_4 m_1 m_4 = 1,$$
$$m_2^{-1} m_3^{-1} m_1 m_2^{-2} m_3 m_4 m_1 m_4^{-1} = 1, \quad m_2 m_3^{-1} m_2 m_4 m_2^2 m_3^{-1} m_2^{-1} m_3 m_4 = 1.$$

(l) *Let K be the finitely presented group constructed in (k). Then V is an irreducible 8-dimensional representation of K over $F = \mathrm{GF}(2)$ of second cohomological dimension $\dim_F[H^2(K, V)] = 0$. Furthermore, K has a faithful permutation representation PK of degree 640 having a stabilizer which is the Sylow 3-subgroup $\langle (m_2 m_4)^2, (m_1 m_3^2 m_4)^2 \rangle$ of K.*

(m) *Let $H_1 = \langle m_i, q_j \mid 1 \le i \le 4, 1 \le j \le 8 \rangle$ be the split extension of K by V. Then H_1 has a set $\mathcal{R}(H_1)$ of defining relations consisting of $\mathcal{R}(K)$, $\mathcal{R}_1(V \rtimes K)$ and the following set $\mathcal{R}_2(V \rtimes K)$ of essential relations:*

$$m_1 q_1 m_1^{-1} q_1 q_2 = 1, \quad m_1 q_2 m_1^{-1} q_2 = 1, \quad m_1 q_3 m_1^{-1} q_3 q_4 = 1,$$
$$m_1 q_4 m_1^{-1} q_4 = 1, \quad m_1 q_5 m_1^{-1} q_7 = 1, \quad m_1 q_6 m_1^{-1} q_2 q_5 q_6 q_7 q_8 = 1,$$
$$m_1 q_7 m_1^{-1} q_5 = 1, \quad m_1 q_8 m_1^{-1} q_8 = 1, \quad m_2 q_1 m_2^{-1} q_1 q_2 q_3 = 1,$$
$$m_2 q_2 m_2^{-1} q_3 q_4 = 1, \quad m_2 q_3 m_2^{-1} q_2 q_3 = 1, \quad m_2 q_4 m_2^{-1} q_1 q_2 q_3 q_4 = 1,$$
$$m_2 q_5 m_2^{-1} q_1 q_2 q_6 q_7 = 1, \quad m_2 q_6 m_2^{-1} q_3 q_4 q_5 = 1, \quad m_2 q_7 m_2^{-1} q_1 q_8 = 1,$$
$$m_2 q_8 m_2^{-1} q_2 q_3 q_4 q_6 q_8 = 1, \quad m_3 q_1 m_3^{-1} q_2 = 1, \quad m_3 q_2 m_3^{-1} q_1 q_2 = 1,$$

$$m_3q_3m_3^{-1}q_1q_4 = 1, \quad m_3q_4m_3^{-1}q_1q_2q_3q_4 = 1, \quad m_3q_5m_3^{-1}q_1q_5q_6 = 1,$$

$$m_3q_6m_3^{-1}q_2q_5 = 1, \quad m_3q_7m_3^{-1}q_5q_8 = 1, \quad m_3q_8m_3^{-1}q_1q_6q_7q_8 = 1,$$

$$m_4q_1m_4^{-1}q_1q_2 = 1, \quad m_4q_2m_4^{-1}q_1 = 1, \quad m_4q_3m_4^{-1}q_2q_6q_7q_8 = 1,$$

$$m_4q_4m_4^{-1}q_1q_2q_6q_7 = 1, \quad m_4q_5m_4^{-1}q_1q_2q_7q_8 = 1, \quad m_4q_6m_4^{-1}q_5q_6q_8 = 1,$$

$$m_4q_7m_4^{-1}q_1q_2q_3q_4q_5q_7 = 1, \quad m_4q_8m_4^{-1}q_1q_3q_6q_7 = 1.$$

(n) $\dim_F[H^2(H_1, F)] = 4$ and there exists a unique central extension H of H_1 of order $|H| = 2^{17} \cdot 3^4 \cdot 5$ whose Sylow 2-subgroups are isomorphic to a Sylow 2-subgroup of D. H is a split extension of K by its Fitting subgroup Q, and H is isomorphic to the finitely presented group $H = \langle h_i \mid \le i \le 14 \rangle$ with the following set $\mathcal{R}(H)$ of defining relations:

$$h_1^2 = h_2^5 = h_3^3 = h_4^3 = h_5^2 = h_6^2 = h_7^2 = h_8^2 = h_9^2 = h_{11}^2 = h_{12}^2 = h_{13}^2 = h_{14}^2 = 1,$$

$$h_{10}^2 h_{14}^{-1} = 1, \quad [h_1, h_{14}^{-1}] = [h_2, h_{14}^{-1}] = [h_3, h_{14}^{-1}] = [h_4, h_{14}^{-1}] = [h_5, h_{14}^{-1}] = 1,$$

$$[h_6, h_{14}^{-1}] = [h_7, h_{14}^{-1}] = [h_8, h_{14}^{-1}] = [h_9, h_{14}^{-1}] = [h_{10}, h_{14}^{-1}] = [h_{11}, h_{14}^{-1}] = 1,$$

$$[h_{12}, h_{14}^{-1}] = [h_{13}, h_{14}^{-1}] = h_1^{-1}h_{13}h_1h_{13}^{-1}h_{14}^{-1} = 1,$$

$$[h_2, h_{13}^{-1}] = [h_3, h_{13}^{-1}] = [h_4, h_{13}^{-1}] = [h_5, h_{13}^{-1}] = [h_6, h_{13}^{-1}] = [h_7, h_{13}^{-1}] = 1,$$

$$[h_8, h_{13}^{-1}] = [h_9, h_{13}^{-1}] = [h_{10}, h_{13}^{-1}] = [h_{11}, h_{13}^{-1}] = [h_{12}, h_{13}^{-1}] = [h_3, h_4] = 1,$$

$$h_3^{-1}h_2h_1h_2^{-1}h_3h_1 = (h_2^{-1}h_1)^4 = h_3^{-1}h_2h_3^{-1}h_1h_2h_3h_1h_2^{-1} = (h_2h_3^{-1})^4 = 1,$$

$$[h_4, h_2^{-1}, h_4] = h_1h_4^{-1}h_1h_4^{-1}h_1h_4h_1h_4 = h_2^{-1}h_3^{-1}h_1h_2^{-2}h_3h_4h_1h_4^{-1} = 1,$$

$$h_2h_3^{-1}h_2h_4h_2^2h_3^{-1}h_2^{-1}h_3h_4 = h_1h_5h_1^{-1}h_5h_6h_{13}^{-1} = h_1h_6h_1^{-1}h_6h_{14}^{-1} = 1,$$

$$h_1h_7h_1^{-1}h_7h_8h_{14}^{-1} = h_1h_8h_1^{-1}h_8 = h_1h_9h_1^{-1}h_{11}h_{13}^{-1}h_{14}^{-1} = 1,$$

$$h_1h_{10}h_1^{-1}h_6h_9h_{10}h_{11}h_{12}h_{13}^{-1} = h_1h_{11}h_1^{-1}h_9h_{13}^{-1} = 1,$$

$$h_1h_{12}h_1^{-1}h_{12}h_{14}^{-1} = h_2h_5h_2^{-1}h_5h_6h_7h_{14}^{-1} = 1,$$

$$h_2h_6h_2^{-1}h_7h_8h_{13}^{-1}h_{14}^{-1} = h_2h_7h_2^{-1}h_6h_7h_{14}^{-1} = h_2h_8h_2^{-1}h_5h_6h_7h_8 = 1,$$

$$h_2h_9h_2^{-1}h_5h_6h_{10}h_{11}h_{13}^{-1}h_{14}^{-1} = h_2h_{10}h_2^{-1}h_7h_8h_9h_{13}^{-1} = 1,$$

$$h_2h_{11}h_2^{-1}h_5h_{12}h_{14}^{-1} = h_2h_{12}h_2^{-1}h_6h_7h_8h_{10}h_{12} = 1,$$

$$h_3h_5h_3^{-1}h_6h_{14}^{-1} = h_3h_6h_3^{-1}h_5h_6h_{13}^{-1} = h_3h_7h_3^{-1}h_5h_8 = 1,$$

$$h_3h_8h_3^{-1}h_5h_6h_7h_8 = h_3h_9h_3^{-1}h_5h_9h_{10}h_{14}^{-1} = h_3h_{10}h_3^{-1}h_6h_9 = 1,$$

$$h_3h_{11}h_3^{-1}h_9h_{12}h_{14}^{-1} = h_3h_{12}h_3^{-1}h_5h_{10}h_{11}h_{12}h_{13}^{-1}h_{14}^{-1} = 1,$$

$$h_4h_5h_4^{-1}h_5h_6h_{13}^{-1}h_{14}^{-1} = h_4h_6h_4^{-1}h_5h_{14}^{-1} = 1,$$

$$h_4h_7h_4^{-1}h_6h_{10}h_{11}h_{12}h_{14}^{-1} = h_4h_8h_4^{-1}h_5h_6h_{10}h_{11}h_{13}^{-1}h_{14}^{-1} = 1,$$

$$h_4h_9h_4^{-1}h_5h_6h_{11}h_{12}h_{14}^{-1} = h_4h_{10}h_4^{-1}h_9h_{10}h_{12}h_{13}^{-1}h_{14}^{-1} = 1,$$

$$h_4h_{11}h_4^{-1}h_5h_6h_7h_8h_9h_{11}h_{13}^{-1} = h_4h_{12}h_4^{-1}h_5h_7h_{10}h_{11}h_{14}^{-1} = 1,$$

$$[h_5, h_6] = [h_5, h_7] = [h_5, h_8] = [h_5, h_9] = h_5^{-1}h_{10}^{-1}h_5h_{10}h_{14}^{-1} = 1,$$

$$h_5^{-1}h_{11}^{-1}h_5h_{11}h_{14}^{-1} = [h_5, h_{12}] = [h_6, h_7] = [h_6, h_8] = h_6^{-1}h_9^{-1}h_6h_9h_{14}^{-1} = 1,$$

$$h_6^{-1}h_{10}^{-1}h_6h_{10}h_{14}^{-1} = h_6^{-1}h_{11}^{-1}h_6h_{11}h_{14}^{-1} = [h_6, h_{12}] = [h_7, h_8] = [h_7, h_9] = 1,$$

$$h_7^{-1}h_{10}^{-1}h_7h_{10}h_{14}^{-1} = h_7^{-1}h_{11}^{-1}h_7h_{11}h_{14}^{-1} = h_7^{-1}h_{12}^{-1}h_7h_{12}h_{14}^{-1} = 1,$$

$$h_8^{-1}h_9^{-1}h_8h_9h_{14}^{-1} = [h_8, h_{10}] = h_8^{-1}h_{11}^{-1}h_8h_{11}h_{14}^{-1} = [h_8, h_{12}] = [h_9, h_{10}] = 1,$$

$$[h_9, h_{11}] = [h_9, h_{12}] = h_{10}^{-1}h_{11}^{-1}h_{10}h_{11}h_{14}^{-1} = [h_{10}, h_{12}] = [h_{11}, h_{12}] = 1.$$

PROOF. (a) By Table A.7 the split extension $E_2 = V_2 \rtimes \mathcal{M}_{22}$ has a unique conjugacy class of involutions of highest defect. It is represented by $z = (iv_1)^2$ and its centralizer $D = C_{E_2}(z)$ has order $2^{17} \cdot 3 \cdot 5$.

(b) The presentation of D given in the statement was obtained using the faithful permutation representation of E_2 with stabilizer $\mathcal{M}_{22}$ and the MAGMA command FPGroupStrong(PD).

(c) Another application of MAGMA yields that the finitely presented group given in (b) has a faithful permutation representation PD of degree 1024 with stabilizer $X = \langle y_1, y_4, y_{10} \rangle$.

(d) Using this permutation representation PD of D and the MAGMA command NormalSubgroups(PD) it follows that D has a unique nonabelian normal subgroup Q of order 2^{10} with center $Z(Q) = \langle z_1, z_2 \rangle$ of order 4. The words of z_1, z_2 in the generators y_i of D given in the statement have been obtained computationally as well. Another application of MAGMA determined the generators of the stated complement $W = \langle w_1, w_2, w_3 \rangle$ of Q in D.

(e) Since Q is a characteristic subgroup of D its center $Z(Q)$ is normal in D. Let $\alpha : D \to D_1 = D/Z(Q)$ be the canonical epimorphism with $\ker(\alpha) = Z(Q)$. Let $V = \alpha(Q)$ and let $u_i = \alpha(y_i) \in D_1$ for $i = 1, 2, \ldots, 11$. Then $W_1 = \alpha(W) = \langle k_1, k_2, k_3 \rangle$ is a complement of V, where $k_j = \alpha(w_j)$. Furthermore, D_1 has a faithful permutation representation PD_1 of degree 256 with stabilizer W_1. Another application of MAGMA now yields that the V is elementary abelian and generated by the eight elements q_l given in the statement.

(f) This assertion has been verified computationally.

(g) The faithful permutation representation PD_1 and MAGMA make it straightforward to see that the Fitting subgroup $A = \langle a_1, a_2, a_3, a_4 \rangle$ of W_1 is elementary abelian of order 16, and that $L = \langle k_1, k_2 \rangle$ is a complement of A in W_1. Also the generators a_i of A and the isomorphism $L \cong S_5$ have been obtained by this application of MAGMA.

(h) Another calculation with MAGMA in the permutation group PD_1 shows that $S = \langle A, k_1, r \rangle$ is a Sylow 2-subgroup of W_1 of order 2^7, where $r = k_2 k_1 k_3 k_2^2 k_1 k_2 k_3$. Furthermore, A is the unique maximal elementary abelian normal subgroup of S and the involution $u = k_1(k_3 k_2)^2 k_1 k_2$ generates the center $Z(S)$ of S. Its centralizer $C_{W_1}(u) = \langle S, d \rangle$ has order $2^7 \cdot 3$, and $d = (k_1 k_2)^2 (k_2 k_1)^2$ generates a cyclic Sylow 3-subgroup of $C_{W_1}(u)$.

(i) In order to apply Algorithm 2.5 of [10] to construct a larger centralizer of u we use the anti-isomorphism between PW_1 and its image in $\mathrm{GL}_8(2)$ induced by the conjugate action of W_1 on V. In particular, the matrix d with respect to the basis $\mathcal{B}$ is $Md = (Mk_1 Mk_2)^2 (Mk_2 Mk_1)^2$ and $Mu = Mk_2 Mk_1(Mk_2 Mk_3)^2 Mk_1$. Let MW_1 be the subgroup of $\mathrm{GL}_8(2)$ generated by the matrices Mk_1, Mk_2, and Mk_3 in $\mathrm{GL}_8(2)$.

Another application of MAGMA in $\mathrm{GL}_8(2)$ shows that

$$MX = C_{\mathrm{GL}_8(2)}(Mu) \cap C_{\mathrm{GL}_8(2)}(Md)$$

has an abelian Sylow 3-subgroup MT of order 3^3 and $|MX| = 2^{12} \cdot 3^3 \cdot 5$. Furthermore, it follows that the matrix Mx of the statement belongs to the set of all matrices $My \in MX$ of order 3 such that $|\langle C_{MW_1}(Mu), My \rangle| = 2^7 \cdot 3^2$. In particular, Mu is the unique central involution of $MC = \langle C_{MW_1}(Mu), Mx \rangle$ and $|MC| = 2^7 \cdot 3^2$.

(j) This statement has been verified by means of the previous results and MAGMA.

(k) Let MK be the subgroup of $GL_8(2)$ generated by the matrices $m_1 = Mk_1$, $m_2 = Mk_2$, $m_3 = Mk_3$, and $m_4 = Mx$. Taking a Sylow 3-subgroup of MK as a stabilizer one obtains a faithful permutation representation PK of this matrix group having degree 640. Then the presentation of the finitely generated group $K = \langle m_1, m_2, m_3, m_4 \rangle$ has been calculated by means of the MAGMA command FPGroup(PK).

(l) All assertions of this statement follow easily from (j), (k), and Holt's Algorithm 7.4.5 of [11] implemented in MAGMA.

(m) The presentation of the split extension $H_1 = V \rtimes K$ has been calculated by means of Lemma 2.2 using the presentation of K given in (k).

(n) Part (e) states that $D_1 = D/Z(D)$ splits over $V = Q/Z(D)$ with complement W_1. By Algorithm 2.5 of [10] the desired centralizer $H = C_G(z)$ of $z = z_1$ in the (at this time unknown) target simple group G has to have an odd index $|H : D| = |K : W|$. Therefore Theorem 1.4.15 of [11] implies that H is a central extension of H_1 by a central subgroup $Z(H) = \langle z_1, z_2 \rangle$ having a normal complement Q containing $Z(H)$ such that $V = Q/Z(H)$ has a complement isomorphic to K.

Since $Z(H) = Z(D) = \langle z_1, z_2 \rangle$ is a Klein four group, the central extension H has to be constructed in two steps.

Clearly, $H_1 = \langle m_1, m_2, m_3, m_4, v_1, v_2, v_3, v_4, v_5, v_6, v_7, v_8 \rangle$ has a faithful permutation PH_1 of degree 256 with stabilizer K. Let FPH_1 be the presentation of H_1 given in (m). Then we apply Holt's Algorithm 7.4.5 of [11] implemented in MAGMA [8] to the trivial matrix representation of H_1 over $F = GF(2)$. It yields that the second cohomological dimension $\dim_F[H^2(H_1, F)] = 4$. Thus the first central extension H_2 of H_1 by a central involution $y \in Z(D)$ is one of the sixteen central extensions $E_{a,b,c,d}$, $0 \le a, b, c, d \le 1$. As D does not split over $Z(D)$ the group H_2 can only be isomorphic to a non-split extension. Let $nmodQ := $ GModule(PH_1, FEalg) be the trivial module of the matrix algebra FEalg generated by the twelve identity matrices corresponding to the twelve generators of H_1. Using the MAGMA command P_H :=ExtensionProcess(PH_1,nmodQ,FPH_1) we obtain a presentation for each of the fifteen non-split central extensions $E_{a,b,c,d}$, where $(a, b, c, d) \neq (0, 0, 0, 0)$. Another application of Theorem 1.4.15 of [11] implies that H_2 has a normal complement Q_2 containing $Z_2 = \langle y \rangle$ such that $Q_1 = Q_2/Z_2$. In particular, H_2 has a faithful permutation representation of degree 512, and its Sylow 2-subgroups are isomorphic to the ones of D/Z_2. Constructing the corresponding permutation representations of the groups $E_{a,b,c,d}$ it follows that only the three groups $E_{0,1,0,0}$, $E_{1,0,0,0}$ and $E_{1,1,0,0}$ have a faithful permutation representation of degree 512 whose stabilizer is a subgroup isomorphic to L. As $y \in \{z_1, z_2, z_3 = z_1 z_2\}$ another application of MAGMA yields that only $E_{1,0,0,0}$ has a Sylow 2-subgroup which is isomorphic to those of exactly one factor group $D/\langle z_k \rangle$, where $1 \le k \le 3$. In fact, this $k = 1$.

By construction the extension group $H_2 = E_{1,0,0,0}$ has a complement and therefore a faithful permutation representation of degree 512. Its central involution is the new generator. Applying Holt's Algorithm 7.4.5 of [11] implemented in MAGMA [8] again to the trivial matrix representation of H_2 over $F = GF(2)$ it follows that the second cohomological dimension $\dim_F[H^2(H_1, F)] = 4$. After constructing the corresponding suitable permutation representations of the fifteen non-split extension groups $F_{a,b,c,d}$ we see that only the three groups $F_{0,0,1,0}$, $F_{1,0,0,0}$, and $F_{1,0,1,0}$ have a faithful permutation representation of degree 1024 whose stabilizer is a subgroup isomorphic to K. Now the isomorphism check of MAGMA yields that only $F_{1,0,1,0}$ has a Sylow 2-subgroup which is isomorphic to a Sylow 2-subgroup of D. Hence $H = C_G(z) := F_{1,0,1,0}$. Its presentation is given in the statement. This completes the proof.   □

LEMMA 3.4. *Keep the notation of Lemma 2.4 and Proposition 3.3. Let $H = \langle h_i \mid 1 \le i \le 14 \rangle$ be the finitely presented group constructed in Proposition 3.3. Then the following statements hold:*

(a) *H has a faithful permutation representation of degree 1024 with stabilizer $\langle h_1, h_2, h_3, h_4 \rangle$.*

(b) *Each Sylow 2-subgroup S of H has a unique maximal elementary abelian normal subgroup A of order 2^{10} and $N_H(A) \cong D = C_{E_2}(z)$.*

(c) *There is a Sylow 2-subgroup S such that $D = N_H(A) = \langle x, y \rangle$, where*

$$x = \left(h_2 h_4 h_{12} h_4 (h_2 h_4)^4 \right)^2,$$
$$y = \left((h_2 h_4^2)^2 h_{12} h_4^2 h_2 h_1 h_4^2 h_2 h_4 h_1 h_4 h_2 \right)^2 (h_2 h_4^2 h_2)^3 h_4^2 h_{12} h_4^2 h_2 h_1 h_4 h_{12} h_4$$

have orders 2 and 6, respectively. Furthermore, $H = \langle x, y, h \rangle$, where $h = h_4$ has order 3.

(d) *The amalgam $H \leftarrow D \to E_2$ has Goldschmidt index 1.*

(e) *A system of representatives r_i of the 115 conjugacy classes of H and the corresponding centralizers orders $|C_H(r_i)|$ are given in Table A.5.*

(f) *A system of representatives d_i of the 97 conjugacy classes of D and the corresponding centralizers orders $|C_D(d_i)|$ are given in Table A.6.*

(g) *Let $\sigma : N_H(A) \to D = C_{E_2}$ be the isomorphism given in (b). Then there is an element $e \in E_2$ of order 3 such that $E_2 = \langle \sigma(D), e \rangle$. A system of representatives e_i of the 43 conjugacy classes of E_2 and the corresponding centralizers orders $|C_{E_2}(e_i)|$ are given in Table A.7.*

(h) *The character tables of H, D, and E_2 are given in Tables B.4, B.5, and B.6, respectively.*

PROOF. (a) This assertion follows at once from Proposition 3.3(n).

In particular, H has a faithful permutation representation PH of degree 1024. Using it and MAGMA it is straightforward to verify statements (b) and (c). The words for the generators x and y of D in the generators h_i of H have been found by a stand alone program written by the first author.

(d) The Goldschmidt index has been calculated by means of Kratzer's Algorithm 7.1.10 of [11].

The systems of representatives of the conjugacy classes of H, D, and E_2 have been calculated by means of PH, MAGMA and Kratzer's Algorithm 5.3.18 of [11].

(g) It has been checked with MAGMA that $e = t$ satisfies $E_3 = \langle \sigma(D), e \rangle$.

The character tables of (h) have been obtained by using PH and MAGMA.   □

4. Construction of Conway's simple group Co_2

By Lemma 3.2 the amalgam $H \leftarrow D \to E_3$ constructed in Sections 2 and 3 satisfies the conditions of Step 5 of Algorithm 2.5 of [10]. Therefore we can apply Algorithm 7.4.8 of [11] to give here a new existence proof for Conway's sporadic group Co_2, see [5].

The set of all faithful characters of the finite group U is denoted by $f\mathrm{char}_{\mathbb{C}}(U)$, and $mf\mathrm{char}_{\mathbb{C}}(U)$ denotes the set of all multiplicity-free faithful characters of U.

DEFINITION 4.1. *Let U_1, U_2 be a pair of finite groups intersecting in D. Then*

$$\Sigma = \{(\nu, \omega) \in mf\mathrm{char}_{\mathbb{C}}(U_1) \times mf\mathrm{char}_{\mathbb{C}}(U_2) \mid \nu_{|D} = \omega_{|D}\}$$

is called the set of compatible pairs of multiplicity-free faithful characters *of U_1 and U_2. For each $(\nu, \omega) \in \Sigma$ the integer $n = \nu(1) = \omega(1)$ is called the* degree *of the compatible pair (ν, ω).*

THEOREM 4.2. *Keep the notation of Lemma 3.2 and Proposition 3.1. Using the notation of the three character tables B.1, B.2, and B.3 of the groups H, D, and E_3, respectively, the following statements hold:*

(a) *The smallest degree of a nontrivial pair*

$$(\chi, \tau) \in mf\mathrm{char}_{\mathbb{C}}(H) \times mf\mathrm{char}_{\mathbb{C}}(E_3)$$

   of compatible characters is 23.

(b) *There is exactly one compatible pair $(\chi, \tau) \in mf\mathrm{char}_{\mathbb{C}}(H) \times mf\mathrm{char}_{\mathbb{C}}(E_3)$ of degree 23 of the groups $H = \langle D, h \rangle$ and $E_3 = \langle D, e \rangle$: $(\chi_2 + \chi_4, \tau_2 + \tau_6)$ with common restriction $\tau_{|D} = \chi_{|D} = \psi_2 + \psi_8 + \psi_{\mathbf{26}}$, where irreducible characters with boldface indices denote faithful irreducible characters.*

(c) *Let $\mathfrak{V}$ and $\mathfrak{W}$ be the uniquely determined (up to isomorphism) faithful semisimple multiplicity-free 23-dimensional modules of H and E_3 over $F = \mathrm{GF}(13)$ corresponding to the compatible pair χ, τ, respectively. Let $\kappa_{\mathfrak{V}} : H \to \mathrm{GL}_{23}(13)$ and $\kappa_{\mathfrak{W}} : E_3 \to \mathrm{GL}_{23}(13)$ be the representations of H and E_3 afforded by the modules $\mathfrak{V}$ and $\mathfrak{W}$, respectively. Let $\mathfrak{h} = \kappa_{\mathfrak{V}}(h)$, $\mathfrak{x} = \kappa_{\mathfrak{V}}(x)$, $\mathfrak{y} = \kappa_{\mathfrak{V}}(y)$ in $\kappa_{\mathfrak{V}}(H) \leq \mathrm{GL}_{23}(13)$. Then the following assertions hold:*

   (1) $\mathfrak{V}_{|D} \cong \mathfrak{W}_{|D}$, *and there is a transformation matrix $T \in \mathrm{GL}_{23}(13)$ such that*

$$\mathfrak{x} = T^{-1}\kappa_{\mathfrak{W}}(x_1)T, \mathfrak{y} = T^{-1}\kappa_{\mathfrak{W}}(y_1)T.$$

   Let $\mathfrak{e} = T^{-1}\kappa_{\mathfrak{W}}(e)T \in \mathrm{GL}_{23}(13)$.

   (2) *In $\mathfrak{G}_3 = \langle \mathfrak{h}, \mathfrak{x}, \mathfrak{y}, \mathfrak{e} \rangle$ the subgroup $\mathfrak{E}_3 = \langle \mathfrak{x}, \mathfrak{y}, \mathfrak{e} \rangle$ is the stabilizer of a 1-dimensional subspace $\mathfrak{U}$ of $\mathfrak{V}$ such that the $\mathfrak{G}_3$-orbit $\mathfrak{U}^{\mathfrak{G}}$ has degree 46575.*

   (3) *The generating matrices of $\mathfrak{G}_3$ are:*

$$\mathfrak{h} = \begin{pmatrix}
0 & 12 & 1 & 11 & 3 & 1 & 11 & 0 & 0 & 0 & 0 & 0 & 0 & 0 & 0 & 0 & 0 & 0 & 0 & 0 & 0 & 0 & 0 \\
11 & 1 & 12 & 11 & 1 & 11 & 4 & 0 & 0 & 0 & 0 & 0 & 0 & 0 & 0 & 0 & 0 & 0 & 0 & 0 & 0 & 0 & 0 \\
7 & 10 & 12 & 6 & 11 & 2 & 3 & 0 & 0 & 0 & 0 & 0 & 0 & 0 & 0 & 0 & 0 & 0 & 0 & 0 & 0 & 0 & 0 \\
8 & 3 & 3 & 7 & 10 & 6 & 0 & 0 & 0 & 0 & 0 & 0 & 0 & 0 & 0 & 0 & 0 & 0 & 0 & 0 & 0 & 0 & 0 \\
0 & 2 & 9 & 9 & 12 & 4 & 5 & 0 & 0 & 0 & 0 & 0 & 0 & 0 & 0 & 0 & 0 & 0 & 0 & 0 & 0 & 0 & 0 \\
7 & 7 & 2 & 12 & 11 & 10 & 6 & 0 & 0 & 0 & 0 & 0 & 0 & 0 & 0 & 0 & 0 & 0 & 0 & 0 & 0 & 0 & 0 \\
0 & 1 & 11 & 11 & 6 & 9 & 10 & 0 & 0 & 0 & 0 & 0 & 0 & 0 & 0 & 0 & 0 & 0 & 0 & 0 & 0 & 0 & 0 \\
0 & 0 & 0 & 0 & 0 & 0 & 0 & 12 & 2 & 12 & 12 & 12 & 1 & 1 & 2 & 12 & 0 & 1 & 1 & 0 & 11 & 0 \\
0 & 0 & 0 & 0 & 0 & 0 & 0 & 11 & 2 & 12 & 0 & 12 & 1 & 1 & 1 & 0 & 1 & 1 & 1 & 12 & 11 & 0 \\
0 & 0 & 0 & 0 & 0 & 0 & 1 & 0 & 12 & 1 & 2 & 1 & 0 & 1 & 11 & 1 & 1 & 11 & 12 & 11 & 3 & 0 \\
0 & 0 & 0 & 0 & 0 & 0 & 0 & 12 & 0 & 0 & 0 & 0 & 1 & 0 & 1 & 0 & 12 & 0 & 0 & 0 \\
0 & 0 & 0 & 0 & 0 & 0 & 1 & 11 & 0 & 0 & 2 & 1 & 0 & 1 & 0 & 1 & 2 & 12 & 0 & 11 & 1 & 12 \\
0 & 0 & 0 & 0 & 0 & 0 & 0 & 1 & 12 & 1 & 0 & 0 & 0 & 12 & 0 & 12 & 0 & 12 & 1 & 0 \\
0 & 0 & 0 & 0 & 0 & 0 & 0 & 0 & 0 & 11 & 12 & 0 & 0 & 1 & 12 & 12 & 1 & 0 & 1 & 12 & 0 \\
0 & 0 & 0 & 0 & 0 & 0 & 0 & 11 & 2 & 12 & 0 & 0 & 0 & 1 & 0 & 1 & 1 & 1 & 0 & 11 & 0 \\
0 & 0 & 0 & 0 & 0 & 0 & 0 & 12 & 1 & 0 & 0 & 12 & 1 & 1 & 1 & 0 & 0 & 1 & 12 & 12 & 0 \\
0 & 0 & 0 & 0 & 0 & 0 & 0 & 12 & 1 & 0 & 1 & 0 & 0 & 0 & 1 & 1 & 12 & 1 & 12 & 0 & 12 \\
0 & 0 & 0 & 0 & 0 & 0 & 12 & 12 & 2 & 12 & 10 & 11 & 1 & 0 & 2 & 12 & 12 & 3 & 0 & 2 & 9 & 1 \\
0 & 0 & 0 & 0 & 0 & 0 & 0 & 12 & 0 & 0 & 1 & 0 & 0 & 1 & 0 & 0 & 1 & 12 & 0 & 11 & 1 & 12 \\
0 & 0 & 0 & 0 & 0 & 0 & 12 & 1 & 0 & 0 & 11 & 12 & 0 & 12 & 1 & 12 & 12 & 1 & 0 & 2 & 12 & 0 \\
0 & 0 & 0 & 0 & 0 & 0 & 0 & 12 & 1 & 0 & 12 & 0 & 0 & 0 & 1 & 0 & 0 & 1 & 0 & 1 & 12 & 0 \\
0 & 0 & 0 & 0 & 0 & 0 & 0 & 1 & 12 & 1 & 0 & 0 & 0 & 12 & 0 & 12 & 12 & 12 & 0 & 2 & 0 \\
0 & 0 & 0 & 0 & 0 & 0 & 0 & 1 & 12 & 1 & 0 & 0 & 0 & 12 & 12 & 1 & 0 & 12 & 12 & 0 & 1 & 0
\end{pmatrix},$$

$$\mathfrak{x} =$$

```
/ 12  0  0  0  0  0  0  0  0  0  0  0  0  0  0  0  0  0  0  0  0  0  0  0 \
|  0  2  0 11 12 12  9  0  0  0  0  0  0  0  0  0  0  0  0  0  0  0  0  0 |
|  0  0 10 12  8  4  5  0  0  0  0  0  0  0  0  0  0  0  0  0  0  0  0  0 |
|  0 11  2 12 12  2  3  0  0  0  0  0  0  0  0  0  0  0  0  0  0  0  0  0 |
|  0  9  7 10  6  5  2  0  0  0  0  0  0  0  0  0  0  0  0  0  0  0  0  0 |
|  0  6  6  1  0  8  2  0  0  0  0  0  0  0  0  0  0  0  0  0  0  0  0  0 |
|  0  3  3  7  0 11  0  0  0  0  0  0  0  0  0  0  0  0  0  0  0  0  0  0 |
|  0  0  0  0  0  1  0 12  0  0  0  0  0  0 12 12  1  0  1  0  1 |
|  0  0  0  0  0  0 12  0  2 12 11 11  1  0  2 12 12  2  1  1 10  1 |
|  0  0  0  0  0  0 12  1  0  0 12  0  1  1  0  1  0  1 12 12  0 |
|  0  0  0  0  0  0 12  0  1 12 11 12  0  0  2 12 12  2  1  2 11  0 |
|  0  0  0  0  0 12  1  1 12 10 11  0 12  2 11 11  3  1  3 10  1 |
|  0  0  0  0  0  0  0  1 12  1  1  1  0  0 12  1  0 11  0 12  2 12 |
|  0  0  0  0  0  0  0  2 10  1  2  2 12 12 10  1  0 11 12 12  4 12 |
|  0  0  0  0  0  0  0  1  0 12 12  1  0  0  0  0  0  0  0  0 12  0 |
|  0  0  0  0  0 12  0  2 11 10 11  1  0  3 11 11  4  1  3  8  2 |
|  0  0  0  0  0  0 12  2 12 12 12  1  0  1  0  0  2  0  1 10  1 |
|  0  0  0  0  0 12  0  0  0  0  0  0  0  0  0  0  0  0  0  0 |
|  0  0  0  0  0 12  1  0  0 12 12  0  0  1 12 12  1  0  1  0  0 |
|  0  0  0  0  0 12  0  0  2  1  0  0 12  1  1 12  0 12  1 12 |
|  0  0  0  0  0  0  0  0  0 12  0 12  0  0  0  0  0  0  0  0 |
|  0  0  0  0  0 12  0  0  1  0  0  1  0  0  1  0 12  0  0 |
\  0  0  0  0  0 12  0  1  1  0  0  1  1  0  0  0  0 /
```

$$\mathfrak{y} =$$

```
/ 12  0  0  0  0  0  0  0  0  0  0  0  0  0  0  0  0  0  0  0  0  0  0  0 \
|  0  9  3 12  2  3  9  0  0  0  0  0  0  0  0  0  0  0  0  0  0  0  0  0 |
|  0  1  0 12 10  1  7  0  0  0  0  0  0  0  0  0  0  0  0  0  0  0  0  0 |
|  0 10  4 11  9  3  3  0  0  0  0  0  0  0  0  0  0  0  0  0  0  0  0  0 |
|  0  7  7 12  6 11  2  0  0  0  0  0  0  0  0  0  0  0  0  0  0  0  0  0 |
|  0  9  7 10  0 12  2  0  0  0  0  0  0  0  0  0  0  0  0  0  0  0  0  0 |
|  0 10 10  6 10 12  0  0  0  0  0  0  0  0  0  0  0  0  0  0  0  0  0  0 |
|  0  0  0  0  0  0  1  0 12  0  1  1  0  0 12  1  1 12 12 12  1  0 |
|  0  0  0  0  0  0  1  0 12  1  2  2  0 11  2  1 11 12 11  3 12 |
|  0  0  0  0  0  0  1  1 11  1  2  2 12  0 11  1  0 11 12 12  4  0 |
|  0  0  0  0  0  0  1 12  1  0  0  0  0  1  0  0  0 12  0 |
|  0  0  0  0  0  0 12  2 12 12 12  1  0  1  0  0  2  0  1 10  1 |
|  0  0  0  0  0  0  1  0 12  0  0  0  0 12 12  0  0  1  0  0 |
|  0  0  0  0  0 12 12  2 11 10 11  0  0  3 11 12  4  1  3  8  1 |
|  0  0  0  0  0  0  1  2  9  3  4  3 12 12  8  3  1  8 11 10  7 11 |
|  0  0  0  0  0  0  0  1 12 11 12  1  0  1 12 12  2  0  1 11  1 |
|  0  0  0  0  0  0  1  0 11  1  3  2  0  0 10  2  1 10 12 10  4 12 |
|  0  0  0  0  0  0 12  1  1  1 12  0 12  1  1 12  0 12  2 12 |
|  0  0  0  0  0  0  1 11  1  0  1  0  0  1  0  1  2 12  0 11  0 12 |
|  0  0  0  0  0  0  0 12  0  1  1 12 12 12  1  1 12  0  0  1 12 |
|  0  0  0  0  0  0  1 11  1  1  1 12 12 11  1  0 12 12  0  2  0 |
|  0  0  0  0  0  0  1 12  0  0  0  0  0  0  0  0  0  0  1  0 |
\  0  0  0  0  0  0  0  0  0  0  0  0  0  0  0  0  1  0  0  0 /
```

, *and*

$$\mathfrak{e} =$$

```
/  1  0  0  0  0  0  0  0  0  0  0  0  0  0  0  0  0  0  0  0  0  0  0  0 \
|  0  6  5  5  9  4  1 11 12 10  8  4  6  9  7  5  2  5  3  9  0  9  9 |
|  0  1  9 11  5  1  5 10  9  5 12  3  6 10  4 10  3  1 11  7  0  1  7 |
|  0  8  9  4  0 11 12  2  9  2  2  0 11  5  1  0  3  3  8 12  8  9 12 |
|  0  2  9 10  8  6 10  3  3  0  4  7  3 10  3 10  7  3  0  6  4  0 |
|  0 10  3  7  9  3  7 10  5  8  1 11  1  3  3  1  4  0  8  6 12  6  6 |
|  0  5  8 10 11  8 10  8  5  5  0 11  3  5  8  3  5  0 10 11  0 |
|  0 11 12  4 12  1  1  7  0  6  0  0  6  6  6  0  0  7  0  0 |
|  0  0  0  0  7  7  0  0  6  7  6  6  6  7  7  6  0  0  0  0  0 |
|  0  9  8  8  8 10  1  0 12  7  0  0  6  0  6  0  7  6  6  6  1  6 |
|  0  6  6  1  6  2 11  7  0  5  8  2  1 12  6 11  2  8  4  6 11  3  5 |
|  0  6  6  1  7  2  0  0  6  6  7  1  7  6  6  6  1  1  6  0  6  7 12 |
|  0  6  6  1  6  2 11  6  0  7  6 12 12  0  7  1 12  6  7  7  0 12  7 |
|  0  8  7 10  8  1 12  6  6  2  5 11  5  1  0  2 12 12  9  1  2  3  1 |
|  0 11 12  4  5  8 12  0  7 12  7  7  1  6  0  6  7  7 12  6  0  8  6 |
|  0  0  0  0  7  6  2  6  6  7  7  6  6 12  0  7  1  6  6  1  7  7  5  7 |
|  0  5  6  3  5 12  1  6  7  0  6 12  6  0  0  1 12 12  8  0  2  5  1 |
|  0  2  1  9  8  6 12  7  1 12  7  7  7  0  0  0 12  7  0 12 12  6  2  0 |
|  0  6  6  1  0  9 11  7  0 12  1  8  7  6  0  5  7  7  5  6 12  2  6 |
|  0  0  0  0  6  7 11  7  7  6  6  7  1  0  6 12  7  6  0  6  7  7 |
|  0  0  0  0  0  0  0  1 12  0  0  1  0 12 12  1  0  0 12  1  1  0 |
|  0  9  8  8  8 10  1  0  0  0  6  0  0  7  0  6  0  6  7  6  7  0  7 |
\  0  7  7 12  6 11  0  0  6  7  6  0  6  7  7  7  0  0  7  0  6  6  0 /
```

.

(4) $\mathfrak{G}_3 = \langle \mathfrak{h}, \mathfrak{x}, \mathfrak{y}, \mathfrak{e} \rangle$, *and* $\mathfrak{G}_3$ *has* 60 *conjugacy classes* $\mathfrak{g}_i^{\mathfrak{G}_3}$ *with representatives* $\mathfrak{g}_i$ *and centralizer orders* $|C_{\mathfrak{G}_3}(\mathfrak{g}_i)|$ *as given in Table A.4.*

(5) *The character table of* $\mathfrak{G}_3$ *coincides with that of* Co_2 *in the Atlas* [**6**, pp. 154–155].

(d) $\mathfrak{G}_3$ *is a finite simple group with* 2-*central involution* $\kappa_{\mathfrak{V}}(z) = (\mathfrak{x}\mathfrak{y}\mathfrak{h})^{15}$ *such that* $C_{\mathfrak{G}_3}(\kappa_{\mathfrak{V}}(z)) = \kappa_{\mathfrak{V}}(H)$, *and* $|\mathfrak{G}_3| = 2^{18} \cdot 3^6 \cdot 7 \cdot 11 \cdot 13$.

PROOF. (a) The character tables of the groups H, D, and E_3 are stated in the Appendices. In the following we use their notations. Using MAGMA and the character tables of H, D, and E_3 and the fusion of the classes of $D(H)$ in H and $D(E_3)$ in E_3 an application of Kratzer's Algorithm 7.3.10 of [**11**] yields the compatible pair stated in assertion (a).

(b) Kratzer's Algorithm also shows that the pair (χ, τ) of (b) is the unique compatible pair of degree 23 with respect to the fusion of the D-classes into the H- and E_3-classes.

(c) In order to construct the semisimple faithful representation $\mathfrak{V}$ corresponding to the character $\chi = \chi_2 + \chi_4$ we determine the irreducible constituents of the faithful permutation representation $(1_T)^H$ of H with stabilizer $T = \langle h_1, h_4 \rangle$ given in Lemma 3.2. Calculating inner products with the irreducible characters of H it follows $(1_T)^H$ has five irreducible constituents of degrees 1, 16, 120, 135, and 240. By the character table of H we know that χ_4 is the only irreducible character of H of degree 16. Constructing the permutation matrices of the three generators h, x, and y of H over the field $K = \mathrm{GF}(13)$ and applying the Meataxe Algorithm implemented in MAGMA we obtain the matrices $\mathfrak{J}x, \mathfrak{J}y, \mathfrak{J}h \in \mathrm{GL}_{16}(13)$ of the generators of H in the 16-dimensional representation $\mathfrak{V}_1$. Furthermore, it has been checked that $\mathfrak{V}_1$ restricted to $D(H) = \langle x, y \rangle$ is irreducible.

In order to find the irreducible module $\mathfrak{V}_2$ corresponding to the character χ_2 of degree 7 we applied Theorem 2.5.4 of [11] by calculating the 4th exterior power $\bigwedge^4 \mathfrak{V}_1$ of dimension 1820 over $K = \mathrm{GF}(13)$ and decomposing it into irreducible composition factors. It follows that it has three composition factors $\mathfrak{W}_j$, $1 \leq j \leq 3$, of dimensions 35, 840, and 945. Let $\mathfrak{W}_1$ be the one of dimension 35. Its second exterior power $\bigwedge^2 \mathfrak{W}_1$ of dimension 595 splits into four irreducible constituents $\mathfrak{X}_k$, $1 \leq k \leq 4$, of dimensions 7, 21, 189, and 378. H has exactly one irreducible character of dimension 7 by Table B.1. Hence $\mathfrak{V}_2 = \mathfrak{X}_1$. Choosing a basis in this KH-module the generators of H are represented by the following matrices:

$$\mathfrak{N}x = \begin{pmatrix} 5 & 8 & 8 & 5 & 9 & 11 & 8 \\ 9 & 6 & 11 & 1 & 6 & 3 & 1 \\ 10 & 3 & 2 & 11 & 10 & 7 & 12 \\ 3 & 8 & 10 & 0 & 10 & 12 & 9 \\ 11 & 11 & 6 & 5 & 3 & 7 & 11 \\ 0 & 6 & 6 & 1 & 0 & 8 & 2 \\ 0 & 3 & 3 & 7 & 0 & 11 & 0 \end{pmatrix}, \quad \mathfrak{N}y = \begin{pmatrix} 4 & 10 & 4 & 1 & 10 & 8 & 5 \\ 8 & 1 & 7 & 6 & 1 & 8 & 12 \\ 10 & 4 & 5 & 11 & 12 & 4 & 1 \\ 5 & 5 & 0 & 4 & 10 & 11 & 0 \\ 12 & 8 & 0 & 3 & 11 & 12 & 0 \\ 12 & 10 & 0 & 1 & 5 & 0 & 0 \\ 6 & 4 & 0 & 8 & 6 & 6 & 12 \end{pmatrix}, \quad \text{and} \quad \mathfrak{N}h = \begin{pmatrix} 12 & 10 & 11 & 8 & 5 & 9 & 3 \\ 4 & 0 & 11 & 1 & 11 & 4 & 5 \\ 0 & 3 & 6 & 5 & 6 & 9 & 2 \\ 7 & 5 & 4 & 10 & 0 & 3 & 6 \\ 1 & 4 & 0 & 11 & 11 & 0 & 0 \\ 1 & 3 & 12 & 2 & 2 & 12 & 2 \\ 7 & 2 & 0 & 12 & 12 & 7 & 1 \end{pmatrix}.$$

Using MAGMA and the Meataxe Algorithm again we see that the restriction of $\mathfrak{V}_2$ to $D_H = \langle x, y \rangle$ splits into two irreducible KD_H-modules $\mathfrak{V}_{21}$ and $\mathfrak{V}_{22}$ of dimensions 1 and 6 having respective bases B_1 and B_2. Hence their union B is a basis of $\mathfrak{V}_2$. Let $T_1 \in \mathrm{GL}_7(13)$ be the matrix of the base change from the canonical basis to the basis B of $\mathfrak{V}_2$. Then $\mathfrak{L}x = T_1 \mathfrak{N}x (T_1)^{-1}$, $\mathfrak{L}y = T_1 \mathfrak{N}y (T_1)^{-1}$ and $\mathfrak{L}h = T_1 \mathfrak{N}h (T_1)^{-1}$ represent the three generators x, y, h of H with respect to the basis B of $\mathfrak{V}_2$. It follows that the blocked diagonal 7×7 matrices $\mathfrak{L}x$ and $\mathfrak{L}y$ are the 7×7 matrices in the upper left corner of the matrices $\mathfrak{x}$ and $\mathfrak{y}$ given in the statement. Furthermore, the matrix $\mathfrak{L}h$ of $h \in H$ is the corresponding 7×7 block of the matrix $\mathfrak{h}$ in the statement.

Let $\mathfrak{x}$, $\mathfrak{y}$ and $\mathfrak{h}$ be the diagonal joins of the matrices $\mathfrak{L}x$ and $\mathfrak{J}x$, $\mathfrak{L}y$ and $\mathfrak{J}y$ and $\mathfrak{L}h$ and $\mathfrak{J}h$ in $\mathrm{GL}_{23}(13)$, respectively. Then these three 23×23 matrices are given in the statement. Clearly, they satisfy the equations $\mathfrak{x} = \kappa_{\mathfrak{V}}(x)$, $\mathfrak{y} = \kappa_{\mathfrak{V}}(y)$ and $\mathfrak{h} = \kappa_{\mathfrak{V}}(h)$ where $\kappa_{\mathfrak{V}} : H \to \mathrm{GL}_{23}(13)$ denotes the semisimple representation of $H = \langle x, y, h \rangle$ afforded by $\mathfrak{V}$.

Now we construct the faithful representation $\mathfrak{W}$ of E_3 corresponding to the character $\tau = \tau_2 + \tau_6$. By the character table of E_3 we know that τ_2 is the only non-trivial linear character of E. Its corresponding representation $\mathfrak{W}_1 : E_3 \to \mathrm{GL}_1(K)$ is given by $\mathfrak{W}_1(x_1) = \mathfrak{W}_1(y_1) = 12$ and $\mathfrak{W}_1(e_1) = 1$. In order to find the irreducible representation $\mathfrak{W}_2$ of τ_6 we determine the irreducible constituents of the faithful permutation representation $(1_{T_1})^{E_3}$ of E_3 of degree 1024 with stabilizer A_{22} constructed in Lemma 2.5. An application of MAGMA shows that it has four irreducible constituents of degrees 1, 22, 231, and 770. Using the Meataxe Algorithm implemented in MAGMA the matrices $\mathfrak{J}x_1, \mathfrak{J}y_1, \mathfrak{J}e_1 \in \mathrm{GL}_{22}(13)$ of the generators of E_3 in the 22-dimensional irreducible constituent $\mathfrak{X}$ of $(1_{T_1})^E$ over $K = \mathrm{GF}(13)$ and their traces have been determined. Now it follows from the character table of E_3 that τ_6 is the character of the tensor product $\mathfrak{X} \otimes \mathfrak{W}_1$. Another application of the Meataxe Algorithm and MAGMA implies that the restriction of $\mathfrak{W}_2$ to $D_E = \langle x_1, y_1 \rangle$ splits into two irreducible KD_E-modules $\mathfrak{W}_{21}$ and $\mathfrak{W}_{22}$ having respective bases B_1 and B_2 such that the generators x_1 and y_1 of D have the following matrices with respect to B_1 and B_2, respectively:

$$\mathfrak{W}_{21}(x_1) = \begin{pmatrix} 12 & 1 & 1 & 0 & 0 & 0 \\ 0 & 1 & 0 & 0 & 12 & 1 \\ 12 & 0 & 0 & 0 & 0 & 0 \\ 0 & 0 & 12 & 0 & 0 & 0 \\ 1 & 0 & 0 & 1 & 0 & 12 \\ 0 & 0 & 0 & 1 & 1 & 12 \end{pmatrix} \quad \text{and} \quad \mathfrak{W}_{21}(y_1) = \begin{pmatrix} 12 & 0 & 1 & 12 & 0 & 0 \\ 12 & 0 & 0 & 12 & 0 & 1 \\ 0 & 0 & 1 & 0 & 1 & 12 \\ 0 & 0 & 1 & 1 & 1 & 12 \\ 12 & 1 & 0 & 12 & 12 & 1 \\ 0 & 0 & 0 & 0 & 0 & 12 \end{pmatrix} ;$$

$$\mathfrak{W}_{22}(x_1) = \begin{pmatrix} 1 & 12 & 0 & 0 & 0 & 1 & 12 & 0 & 12 & 12 & 1 & 1 & 1 & 12 & 12 \\ 2 & 0 & 1 & 0 & 1 & 12 & 1 & 11 & 12 & 12 & 12 & 0 & 1 & 12 & 1 & 12 \\ 3 & 12 & 2 & 1 & 1 & 12 & 0 & 11 & 11 & 12 & 0 & 12 & 1 & 12 & 2 & 11 \\ 3 & 11 & 2 & 2 & 2 & 12 & 0 & 10 & 12 & 12 & 1 & 11 & 0 & 0 & 2 & 10 \\ 0 & 0 & 0 & 0 & 0 & 12 & 0 & 1 & 12 & 0 & 1 & 0 & 12 & 12 & 1 & 0 \\ 12 & 1 & 0 & 12 & 12 & 0 & 1 & 0 & 1 & 0 & 11 & 1 & 1 & 1 & 12 & 1 \\ 0 & 12 & 0 & 1 & 0 & 0 & 0 & 12 & 0 & 0 & 1 & 12 & 0 & 1 & 0 & 12 \\ 2 & 0 & 1 & 1 & 1 & 12 & 12 & 12 & 11 & 0 & 1 & 11 & 0 & 12 & 2 & 12 \\ 1 & 1 & 10 & 1 & 1 & 1 & 1 & 12 & 0 & 12 & 0 & 1 & 1 & 1 & 12 & 11 \\ 2 & 10 & 1 & 2 & 1 & 0 & 0 & 11 & 12 & 12 & 2 & 12 & 1 & 1 & 0 & 10 \\ 2 & 11 & 1 & 1 & 1 & 0 & 1 & 11 & 12 & 12 & 0 & 0 & 1 & 0 & 0 & 11 \\ 3 & 12 & 2 & 0 & 1 & 12 & 1 & 11 & 12 & 12 & 12 & 0 & 1 & 12 & 1 & 11 \\ 2 & 10 & 1 & 3 & 2 & 0 & 12 & 11 & 11 & 0 & 3 & 11 & 0 & 1 & 1 & 10 \\ 1 & 0 & 0 & 0 & 12 & 0 & 12 & 0 & 0 & 0 & 0 & 12 & 1 & 0 \\ 0 & 0 & 0 & 0 & 0 & 0 & 0 & 1 & 0 & 0 & 0 & 0 & 0 \\ 0 & 11 & 0 & 1 & 0 & 0 & 0 & 0 & 0 & 1 & 0 & 12 & 1 & 0 & 12 \end{pmatrix}$$

$$\text{and} \quad \mathfrak{W}_{22}(y_1) = \begin{pmatrix} 3 & 10 & 2 & 3 & 2 & 12 & 0 & 10 & 11 & 12 & 2 & 11 & 0 & 0 & 2 & 10 \\ 2 & 11 & 2 & 2 & 1 & 12 & 0 & 11 & 12 & 12 & 2 & 11 & 12 & 0 & 2 & 11 \\ 11 & 1 & 12 & 0 & 12 & 0 & 12 & 1 & 0 & 1 & 1 & 0 & 12 & 0 & 0 & 1 \\ 2 & 11 & 1 & 2 & 1 & 0 & 12 & 11 & 12 & 0 & 2 & 11 & 0 & 0 & 1 & 11 \\ 12 & 0 & 0 & 0 & 0 & 0 & 0 & 0 & 0 & 0 & 1 & 0 & 0 & 12 & 0 \\ 0 & 2 & 0 & 12 & 0 & 0 & 0 & 0 & 0 & 12 & 0 & 12 & 1 & 1 \\ 2 & 0 & 2 & 0 & 1 & 12 & 0 & 11 & 12 & 0 & 0 & 11 & 0 & 12 & 2 & 12 \\ 12 & 0 & 0 & 1 & 0 & 0 & 12 & 0 & 12 & 1 & 2 & 12 & 12 & 0 & 0 & 0 \\ 12 & 11 & 12 & 2 & 0 & 1 & 0 & 0 & 1 & 0 & 1 & 0 & 0 & 2 & 12 & 12 \\ 1 & 0 & 1 & 1 & 1 & 12 & 12 & 12 & 12 & 1 & 1 & 11 & 12 & 12 & 2 & 12 \\ 2 & 11 & 2 & 2 & 1 & 12 & 0 & 11 & 12 & 12 & 1 & 11 & 0 & 0 & 2 & 11 \\ 1 & 12 & 1 & 2 & 1 & 12 & 12 & 12 & 12 & 0 & 2 & 11 & 12 & 12 & 2 & 12 \\ 1 & 11 & 1 & 2 & 1 & 0 & 12 & 12 & 12 & 0 & 2 & 11 & 0 & 0 & 1 & 11 \\ 0 & 0 & 1 & 0 & 0 & 12 & 0 & 0 & 12 & 0 & 1 & 0 & 12 & 12 & 1 & 0 \\ 12 & 1 & 0 & 0 & 0 & 0 & 12 & 1 & 0 & 1 & 1 & 12 & 12 & 12 & 1 & 1 \\ 2 & 11 & 1 & 1 & 1 & 0 & 0 & 11 & 12 & 12 & 0 & 0 & 2 & 0 & 0 & 11 \end{pmatrix} .$$

Let $B = B_1 \cup B_2$. Then B is a basis of the irreducible KE_3-module $\mathfrak{W}_2$, and with respect to B the third generator e_1 of E_3 has the matrix:

$$\mathfrak{W}_2(e_1) = \begin{pmatrix} 7 & 0 & 6 & 0 & 0 & 0 & 7 & 6 & 7 & 7 & 7 & 0 & 6 & 6 & 6 & 0 & 7 & 6 & 0 & 0 & 7 & 6 \\ 6 & 7 & 0 & 6 & 6 & 1 & 7 & 0 & 7 & 6 & 0 & 0 & 6 & 0 & 6 & 0 & 7 & 0 & 0 & 0 \\ 7 & 6 & 0 & 7 & 7 & 12 & 6 & 6 & 0 & 7 & 0 & 0 & 0 & 0 & 0 & 7 & 0 & 6 & 7 & 0 & 0 \\ 6 & 0 & 7 & 0 & 0 & 12 & 6 & 6 & 1 & 0 & 0 & 6 & 7 & 0 & 7 & 8 & 6 & 12 & 7 & 0 & 0 \\ 7 & 0 & 6 & 0 & 0 & 0 & 0 & 0 & 0 & 0 & 6 & 0 & 7 & 7 & 6 & 0 & 0 & 0 & 0 \\ 7 & 0 & 6 & 0 & 0 & 0 & 12 & 6 & 6 & 1 & 0 & 0 & 6 & 7 & 0 & 7 & 8 & 6 & 12 & 7 & 0 & 0 \\ 6 & 0 & 7 & 0 & 1 & 0 & 11 & 9 & 6 & 11 & 12 & 0 & 7 & 8 & 1 & 7 & 5 & 8 & 0 & 6 & 12 & 2 \\ 7 & 6 & 0 & 7 & 7 & 0 & 7 & 12 & 7 & 7 & 1 & 0 & 1 & 6 & 0 & 12 & 7 & 0 & 7 & 0 & 0 & 12 \\ 7 & 0 & 7 & 0 & 0 & 0 & 7 & 11 & 0 & 8 & 7 & 1 & 7 & 12 & 7 & 12 & 7 & 1 & 8 & 5 & 5 \\ 7 & 0 & 7 & 0 & 0 & 0 & 8 & 0 & 2 & 7 & 7 & 12 & 6 & 12 & 5 & 0 & 7 & 5 & 0 & 5 & 8 & 6 \\ 7 & 6 & 0 & 7 & 7 & 0 & 5 & 0 & 5 & 6 & 12 & 1 & 1 & 7 & 2 & 0 & 5 & 2 & 7 & 2 & 11 & 0 \\ 7 & 12 & 6 & 0 & 0 & 0 & 12 & 0 & 12 & 0 & 0 & 1 & 6 & 1 & 0 & 7 & 7 & 7 & 0 & 0 & 12 & 1 \\ 6 & 0 & 7 & 0 & 1 & 0 & 1 & 12 & 1 & 1 & 1 & 0 & 6 & 12 & 12 & 6 & 8 & 5 & 0 & 12 & 1 & 12 \\ 7 & 0 & 7 & 0 & 0 & 0 & 5 & 7 & 6 & 6 & 6 & 1 & 7 & 7 & 8 & 0 & 6 & 7 & 0 & 1 & 5 & 7 \\ 6 & 7 & 1 & 6 & 7 & 0 & 12 & 1 & 6 & 12 & 5 & 0 & 0 & 8 & 7 & 7 & 6 & 1 & 6 & 7 & 6 & 8 \\ 6 & 7 & 1 & 6 & 7 & 0 & 5 & 7 & 12 & 6 & 12 & 1 & 0 & 1 & 1 & 0 & 6 & 1 & 7 & 7 & 12 & 1 \\ 6 & 0 & 7 & 0 & 1 & 0 & 0 & 0 & 0 & 0 & 0 & 0 & 6 & 7 & 7 & 0 & 0 & 0 & 0 \\ 7 & 0 & 7 & 0 & 0 & 0 & 6 & 7 & 6 & 6 & 6 & 0 & 7 & 7 & 7 & 0 & 6 & 8 & 0 & 0 & 6 & 7 \\ 6 & 7 & 1 & 6 & 7 & 0 & 7 & 0 & 7 & 7 & 0 & 0 & 0 & 6 & 0 & 0 & 7 & 12 & 6 & 0 & 1 & 0 \\ 7 & 6 & 0 & 7 & 7 & 0 & 7 & 6 & 0 & 7 & 1 & 0 & 1 & 12 & 1 & 12 & 6 & 0 & 7 & 7 & 0 & 12 \\ 7 & 0 & 7 & 0 & 0 & 0 & 7 & 6 & 0 & 0 & 6 & 7 & 0 & 7 & 7 & 6 & 12 & 6 & 1 & 1 \\ 6 & 0 & 7 & 0 & 1 & 0 & 1 & 1 & 0 & 0 & 12 & 6 & 12 & 12 & 7 & 6 & 6 & 0 & 12 & 1 & 0 \end{pmatrix} .$$

By construction, the KD-modules $\mathfrak{V}_{|D}$ and $\mathfrak{W}_{|D}$ described by the two pairs $(\mathfrak{V}(x), \mathfrak{V}(y))$ and $(\mathfrak{W}(x_1), \mathfrak{W}(y_1))$ of matrices in $\mathrm{GL}_{23}(13)$ are isomorphic. Let $Y = \mathrm{GL}_{23}(13)$. Applying then Parker's isomorphism test of Proposition 6.1.6 of [11] by means of the MAGMA command

```
IsIsomorphic(GModule(sub<Y|V(x),V(y)>),GModule(sub<Y|W(x1),W(y1)>))
```

we obtain the transformation matrix:

$$\mathcal{T} = \begin{pmatrix}
1 & 0 & 0 & 0 & 0 & 0 & 0 & 0 & 0 & 0 & 0 & 0 & 0 & 0 & 0 & 0 & 0 & 0 & 0 & 0 & 0 & 0 & 0 & 0 \\
0 & 1 & 1 & 11 & 1 & 9 & 0 & 0 & 0 & 0 & 0 & 0 & 0 & 0 & 0 & 0 & 0 & 0 & 0 & 0 & 0 & 0 & 0 & 0 \\
0 & 1 & 1 & 11 & 0 & 8 & 0 & 0 & 0 & 0 & 0 & 0 & 0 & 0 & 0 & 0 & 0 & 0 & 0 & 0 & 0 & 0 & 0 & 0 \\
0 & 4 & 2 & 5 & 3 & 12 & 2 & 0 & 0 & 0 & 0 & 0 & 0 & 0 & 0 & 0 & 0 & 0 & 0 & 0 & 0 & 0 & 0 & 0 \\
0 & 3 & 1 & 7 & 2 & 3 & 11 & 0 & 0 & 0 & 0 & 0 & 0 & 0 & 0 & 0 & 0 & 0 & 0 & 0 & 0 & 0 & 0 & 0 \\
0 & 0 & 0 & 0 & 1 & 12 & 0 & 0 & 0 & 0 & 0 & 0 & 0 & 0 & 0 & 0 & 0 & 0 & 0 & 0 & 0 & 0 & 0 & 0 \\
0 & 1 & 1 & 9 & 10 & 2 & 11 & 0 & 0 & 0 & 0 & 0 & 0 & 0 & 0 & 0 & 0 & 0 & 0 & 0 & 0 & 0 & 0 & 0 \\
0 & 0 & 0 & 0 & 0 & 0 & 0 & 11 & 0 & 0 & 1 & 0 & 0 & 1 & 1 & 0 & 2 & 12 & 0 & 12 & 0 & 12 \\
0 & 0 & 0 & 0 & 0 & 0 & 0 & 11 & 0 & 3 & 2 & 0 & 0 & 11 & 2 & 2 & 11 & 12 & 11 & 3 & 12 \\
0 & 0 & 0 & 0 & 0 & 0 & 12 & 12 & 3 & 11 & 11 & 11 & 1 & 1 & 3 & 12 & 0 & 3 & 1 & 2 & 9 & 1 \\
0 & 0 & 0 & 0 & 0 & 1 & 4 & 7 & 4 & 4 & 3 & 12 & 12 & 7 & 3 & 1 & 7 & 10 & 10 & 10 & 11 \\
0 & 0 & 0 & 0 & 0 & 0 & 2 & 11 & 12 & 0 & 0 & 2 & 11 & 0 & 2 & 1 & 2 & 10 & 1 \\
0 & 0 & 0 & 0 & 0 & 0 & 11 & 2 & 12 & 12 & 11 & 1 & 1 & 2 & 0 & 1 & 2 & 1 & 0 & 9 & 0 \\
0 & 0 & 0 & 0 & 0 & 11 & 1 & 1 & 12 & 10 & 10 & 1 & 12 & 3 & 11 & 11 & 3 & 1 & 3 & 9 & 1 \\
0 & 0 & 0 & 0 & 0 & 0 & 1 & 12 & 0 & 12 & 1 & 1 & 1 & 12 & 0 & 1 & 0 & 12 & 1 \\
0 & 0 & 0 & 0 & 0 & 1 & 11 & 0 & 1 & 2 & 1 & 0 & 2 & 0 & 1 & 2 & 11 & 12 & 10 & 3 & 12 \\
0 & 0 & 0 & 0 & 0 & 12 & 12 & 2 & 12 & 10 & 10 & 0 & 0 & 4 & 11 & 0 & 3 & 1 & 2 & 9 & 1 \\
0 & 0 & 0 & 0 & 0 & 12 & 11 & 1 & 12 & 1 & 0 & 0 & 1 & 2 & 0 & 1 & 1 & 0 & 0 & 12 & 0 \\
0 & 0 & 0 & 0 & 0 & 0 & 12 & 1 & 2 & 1 & 12 & 1 & 12 & 1 & 2 & 11 & 12 & 11 & 3 & 12 \\
0 & 0 & 0 & 0 & 0 & 0 & 0 & 0 & 1 & 0 & 0 & 0 & 0 & 0 & 0 & 12 & 0 & 1 & 1 \\
0 & 0 & 0 & 0 & 0 & 12 & 12 & 2 & 10 & 12 & 12 & 0 & 0 & 2 & 11 & 0 & 3 & 1 & 2 & 9 & 1 \\
0 & 0 & 0 & 0 & 0 & 0 & 12 & 1 & 0 & 1 & 12 & 0 & 1 & 0 & 0 & 1 & 0 & 0 & 11 & 0 & 0 \\
0 & 0 & 0 & 0 & 0 & 1 & 12 & 1 & 1 & 1 & 12 & 12 & 12 & 0 & 0 & 12 & 12 & 1 & 2 & 12 \\
\end{pmatrix}$$

satisfying $\mathfrak{V}(x) = (\mathfrak{W}(x_1))^{\mathcal{T}}$ and $\mathfrak{V}(y) = (\mathfrak{W}(y_1))^{\mathcal{T}}$.

Let $\mathfrak{e} = (\mathfrak{W}(e_1))^{\mathcal{T}}$. Then $\mathfrak{E}_3 = \langle \mathfrak{x}, \mathfrak{y}, \mathfrak{e} \rangle \cong E$. Let $\mathfrak{G}_3 = \langle \mathfrak{h}, \mathfrak{x}, \mathfrak{y}, \mathfrak{e} \rangle$. Using the MAGMA command `CosetAction(G, E)` we obtain a faithful permutation representation of $\mathfrak{G}_3$ of degree 46575 with stabilizer $\mathfrak{E}_3$. In particular $|\mathfrak{G}_3| = 2^{18} \cdot 3^6 \cdot 5^3 \cdot 7 \cdot 11 \cdot 23$. Using the faithful permutation representation of $\mathfrak{G}_3$ of degree 46575 and Kratzer's Algorithm 5.3.18 of [11] we calculated the representatives of all the conjugacy classes of $\mathfrak{G}_3$, see Table A.4. Furthermore, the character table of $\mathfrak{G}_3$ has been calculated by means of the above permutation representation and MAGMA. It coincides with the one of Co_2 in [6, pp. 154–155].

(d) Let $\mathfrak{z} = (\mathfrak{x}\mathfrak{y}\mathfrak{h})^{15}$. Then $C_{\mathfrak{G}_2}(\mathfrak{z})$ contains $\mathfrak{H} = \langle \mathfrak{h}, \mathfrak{x}, \mathfrak{y} \rangle$ which is isomorphic to H. By the table of (c)(4) $|C_{\mathfrak{G}}(\mathfrak{z})| = |H|$. Hence $C_{\mathfrak{G}_3} = \langle \mathfrak{z} \rangle \cong \mathfrak{H}$. The character table of $\mathfrak{G}_3$ implies that $\mathfrak{G}_3$ is a simple group. This completes the proof. □

Praeger and Soicher give in [12] a nice presentation for Conway's simple group Co_2 which is given in the statement of the following corollary.

COROLLARY 4.3. *Keep the notation of Theorem 4.2. The finite simple group $\mathfrak{G}_3$ is isomorphic to the finitely presented group $G = \langle a, b, c, d, e, f, g \rangle$ with set $\mathcal{R}(G)$ of defining relations:*

$$a^2 = b^2 = c^2 = d^2 = e^2 = f^2 = g^2 = 1,$$
$$(ab)^3 = (bc)^5 = (cd)^3 = (df)^3 = (fe)^6 = (ae)^4 = 1,$$
$$(ec)^3 = (cf)^4 = (fg)^4 = (gb)^4 = 1,$$
$$(ac)^2 = (ad)^2 = (af)^2 = (ag)^2 = (bd)^2 = (be)^2 = 1,$$
$$(bf)^2 = (cg)^2 = (dg)^2 = (eg)^2 = 1,$$
$$a = (cf)^2, \quad e = (bg)^2, \quad b = (ef)^3,$$
$$(aecd)^4 = (baefg)^3 = (cef)^7 = 1.$$

PROOF. Using the MAGMA command `FPGroupStrong(PG)` and the faithful permutation representation of degree 46575 of the simple group $\mathfrak{G}_3$ given in Theorem 4.2(c), we obtained a finite presentation of $\mathfrak{G}$ with sixteen generators and too many relations to be stated in this article. Therefore we gave in the statement the presentation of [12, p. 106]. Using its subgroup $Q = \langle a, b, c, d, e, g, (gfdc)^4 \rangle$ as a stabilizer we obtained a permutation representation for their group $G = \langle a, b, c, d, e, f, g \rangle$ of degree 46575. An isomorphism test using MAGMA then showed that $\mathfrak{G}_3 \cong G$. □

5. Construction of Fischer's simple group Fi_{22}

By Lemma 3.4 the amalgam $H \leftarrow D \rightarrow E_2$ constructed in Sections 2 and 3 satisfies the main conditions of Step 5 of Algorithm 2.5 of [10]. Therefore we can apply Algorithm 7.4.8 of [11] to give here a new existence proof for Fischer's sporadic group Fi_{22}, see [7].

THEOREM 5.1. *Keep the notation of Lemma 3.4 and Proposition 3.3. Using the notation of the three character tables B.4, B.5, and B.6 of the groups H, D, and E_2, respectively, the following statements hold:*

(a) *The smallest degree of a nontrivial pair of compatible characters (χ, τ) in $mf\mathrm{char}_{\mathbb{C}}(H) \times mf\mathrm{char}_{\mathbb{C}}(E_2)$ is 78.*

(b) *There is exactly one compatible pair $(\chi, \tau) \in mf\mathrm{char}_{\mathbb{C}}(H) \times mf\mathrm{char}_{\mathbb{C}}(E_2)$ of degree 78 of the groups $H = \langle D, h \rangle$ and $E_2 = \langle D, e \rangle$, namely:*

$$(\chi_4 + \chi_{20} + \chi_{\mathbf{17}}, \tau_1 + \tau_{\mathbf{6}})$$

with common restriction $\tau_{|D} = \chi_{|D} = \psi_1 + \psi_8 + \psi_{41} + \psi_{\mathbf{33}}$, where irreducible characters with boldface indices denote faithful irreducible characters.

(c) *Let $\mathfrak{V}$ and $\mathfrak{W}$ be the uniquely determined (up to isomorphism) faithful semisimple multiplicity-free 78-dimensional modules of H and E_2 over $F = \mathrm{GF}(13)$ corresponding to χ and τ, respectively, where (χ, τ) is the compatible pair. Let $\kappa_{\mathfrak{V}} : H \rightarrow \mathrm{GL}_{78}(13)$ and $\kappa_{\mathfrak{W}} : E_2 \rightarrow \mathrm{GL}_{78}(13)$ be the representations of H and E_2 afforded by the modules $\mathfrak{V}$ and $\mathfrak{W}$, respectively. Let $\mathfrak{h} = \kappa_{\mathfrak{V}}(h)$, $\mathfrak{x} = \kappa_{\mathfrak{V}}(x)$, $\mathfrak{y} = \kappa_{\mathfrak{V}}(y)$ in $\kappa_{\mathfrak{V}}(H) \leq \mathrm{GL}_{78}(13)$. Then the following assertions hold:*

(1) $\mathfrak{V}_{|D} \cong \mathfrak{W}_{|D}$, *and there is a transformation matrix $T \in \mathrm{GL}_{23}(13)$ such that*

$$\mathfrak{x} = T^{-1}\kappa_{\mathfrak{W}}(x_1)T \text{ and } \mathfrak{y} = T^{-1}\kappa_{\mathfrak{W}}(y_1)T.$$

Let $\mathfrak{e} = T^{-1}\kappa_{\mathfrak{W}}(e)T \in \mathrm{GL}_{78}(13)$.

(2) *In $\mathfrak{G}_2 = \langle \mathfrak{h}, \mathfrak{x}, \mathfrak{y}, \mathfrak{e} \rangle$ the subgroup $\mathfrak{E} = \langle \mathfrak{x}, \mathfrak{y}, \mathfrak{e} \rangle$ is the stabilizer of a 1-dimensional subspace $\mathfrak{U}$ of $\mathfrak{V}$ such that the $\mathfrak{G}_2$-orbit $\mathfrak{U}^{\mathfrak{G}_2}$ has degree 142155.*

(3) *The four generating matrices of $\mathfrak{G}_2$ are stated in Appendix C.*

(4) $\mathfrak{G}_2 = \langle \mathfrak{h}, \mathfrak{x}, \mathfrak{y}, \mathfrak{e} \rangle$, *and $\mathfrak{G}_2$ has 65 conjugacy classes $\mathfrak{g}_i^{\mathfrak{G}_2}$ with representatives $\mathfrak{g}_i$ and centralizer orders $|C_{\mathfrak{G}_2}(\mathfrak{g}_i)|$ as given in Table A.8.*

(5) *The character table of $\mathfrak{G}_2$ coincides with that of Fi_{22} in the Atlas [6, pp. 156–157].*

(d) $\mathfrak{G}_2$ *is a finite simple group with 2-central involution $\kappa_{\mathfrak{V}}(z) = (\mathfrak{x}\mathfrak{y}\mathfrak{x}\mathfrak{h})^{10}$ such that $C_{\mathfrak{G}_2}(\kappa_{\mathfrak{V}}(z)) = \kappa_{\mathfrak{V}}(H)$, and $|\mathfrak{G}_2| = 2^{17} \cdot 3^9 \cdot 5^2 \cdot 7 \cdot 11 \cdot 13$.*

PROOF. (a) The character tables of the groups H, D, and E_2 are stated in the Appendix; we will follow the notation used there. Using the character tables of H, D, and E_2 and the fusion of the classes of $D(H)$ in H and $D(E_2)$ in E_2 an application of Kratzer's Algorithm 7.3.10 of [11] yields the compatible pair stated in assertion (a).

(b) Kratzer's Algorithm also shows that the pair (χ, τ) of (b) is the unique compatible pair of degree 78 with respect to the fusion of the D-classes into the H- and E_2-classes.

(c) In order to construct the semisimple faithful representation $\mathfrak{V}$ corresponding to the character $\chi = \chi_4 + \chi_{20} + \chi_{\mathbf{17}}$ we determine the irreducible constituents of the faithful permutation representation $(1_T)^H$ of H with stabilizer $T = \langle h_1, h_2, h_3, h_4 \rangle$ given in Lemma 3.4. Calculating inner products with the irreducible characters of H it follows $(1_T)^H$ has seven irreducible constituents of degrees 1, 32, 40, 120, 135, 216, and 480. By the character table of H we know that $\chi_{\mathbf{17}}$ is the irreducible character of H of degree 32 occurring in the permutation character $(1_T)^H$. Constructing the permutation matrices of the three generators h, x, and y of H over the field $K = \mathrm{GF}(13)$ and applying the Meataxe Algorithm implemented in MAGMA we obtain the matrices $\mathcal{X}_2, \mathcal{Y}_2, \mathcal{H}_2 \in \mathrm{GL}_{32}(13)$ of the generators of H in the 32-dimensional representation $\mathfrak{V}_1$. These three matrices are given in Appendix C. Furthermore, it has been checked that $\mathfrak{V}_1$ restricted to $D(H) = \langle x, y \rangle$ is irreducible.

In order to find the irreducible modules $\mathfrak{V}_2$ and $\mathfrak{V}_3$ corresponding to the characters χ_4 of degree 6 and χ_{20} of degree 40 we applied Theorem 2.5.4 of [11] to the faithful character $\chi_{\mathbf{17}}$. It follows that both missing representations are constituents of the 6th tensor power of $\mathfrak{V}_1$. A character calculation shows that the irreducible representations $\mathfrak{V}_2$ corresponding to χ_{20} and $\mathfrak{V}_3$ corresponding to χ_4 can be constructed by the Meataxe decompositions of the representations of the tensor products belonging to the following character products: the 135-dimensional irreducible character χ_{42} occurs in $\chi_{\mathbf{17}}^2$. Their tensor product $\chi_{42} \otimes \chi_{\mathbf{17}}$ contains the 640-dimensional character χ_{80}, and χ_{59} of degree 270 is an irreducible constituent of $\chi_{80} \otimes \chi_{\mathbf{17}}$. Their tensor product $\chi_{59} \otimes \chi_{\mathbf{17}}$ has the 192-dimensional constituent χ_{47}. The desired character χ_4 of degree 6 is then an irreducible composition factor of $\chi_{47} \otimes \chi_{\mathbf{17}}$.

The faithful 480-dimensional irreducible character $\chi_{\mathbf{74}}$ occurs in $\chi_{\mathbf{17}}^3$. The tensor product $\chi_{\mathbf{74}} \otimes \chi_{\mathbf{17}}$ contains the 135-dimensional character χ_{41}, and χ_{45} of degree 160 is an irreducible constituent of $\chi_{41} \otimes \chi_{\mathbf{17}}$. The tensor product $\chi_{45} \otimes \chi_{\mathbf{17}}$ has the desired character χ_{20} of degree 40 as an irreducible composition factor.

An application of the Meataxe Algorithm implemented in MAGMA produces then the six matrices of the three generators of H with respect to fixed bases in $\mathfrak{V}_2$ and $\mathfrak{V}_3$. Their 46-dimensional diagonal joins $\mathcal{H}_1$, $\mathcal{X}_1$, and $\mathcal{Y}_1$ are given in Appendix C.

Now we construct the faithful representation $\mathfrak{W}$ of E_2 corresponding to the character $\tau = \tau_1 + \tau_6$. Clearly τ_1 corresponds to the trivial representation $\mathfrak{W}_1$ of E_2. By means of the character table of E_2 we see that the irreducible representation $\mathfrak{W}_2$ of τ_6 occurs as an irreducible constituent of the faithful permutation representation $(1_{T_1})^{E_2}$ of E_2 of degree 1024 with stabilizer $\mathcal{M}_{22}$ constructed in Lemma 2.4. Using the Meataxe Algorithm implemented in MAGMA the 77-dimensional irreducible constituent $\mathfrak{W}_2$ of $(1_{T_1})^E$ over $K = \mathrm{GF}(13)$ has been determined. Its restriction $(\mathfrak{W}_2)_{|D}$ to $D = \langle x_1, y_1 \rangle$ decomposes into three irreducible constituents of dimensions 5, 40, and 32. Taking a basis B_1 of the 1-dimensional restriction $(\mathfrak{W}_1)_{|D}$ to D and bases B_2, B_3, and B_4 in the three irreducible constituents of $(\mathfrak{W}_2)_{|D}$ we obtain the two matrices $\mathfrak{W}(x_1)$ and $\mathfrak{W}(y_1)$ of the generators x_1 and y_1 of $D = D_{E_2}$ in $\mathrm{GL}_{78}(13)$. Let $B = B_1 \cup B_2 \cup B_3 \cup B_4$. Then B is a basis of the semisimple KE_2-module $\mathfrak{W}$. Let $\mathfrak{W}(e_1)$ be the matrix of the third generator e_1 of E_2 with respect to B. The two KD-modules $\mathfrak{V}_{|D}$ and $\mathfrak{W}_{|D}$ described by the pairs $(\mathfrak{V}(x), \mathfrak{V}(y))$ and $(\mathfrak{W}(x_1), \mathfrak{W}(y_1))$ of matrices in $\mathrm{GL}_{78}(13)$ are isomorphic by construction. Let $Y = \mathrm{GL}_{78}(13)$. Applying then Parker's isomorphism test of Proposition 6.1.6 of [11] by means of the MAGMA command

```
IsIsomorphic(GModule(sub<Y|V(x),V(y)>),GModule(sub<Y|W(x1),W(y1)>))
```

we obtain the transformation matrix T satisfying $\mathfrak{V}(x) = (\mathfrak{W}(x_1))^T$ and $\mathfrak{V}(y) = (\mathfrak{W}(y_1))^T$. In view of its size this transformation matrix is decomposed into four block matrices $\mathcal{A} \in Mat_{38,38}$, $\mathcal{B} \in Mat_{38,40}$, $\mathcal{C} \in Mat_{40,38}$ and, $\mathcal{D} \in Mat_{40,40}$ such that:

$$T = \begin{pmatrix} \mathcal{A} & \mathcal{B} \\ \mathcal{C} & \mathcal{D} \end{pmatrix}$$

has the following four blocks:

Matrix $\mathcal{A}$:

Matrix $\mathcal{B}$:

Matrix $\mathcal{C}$:

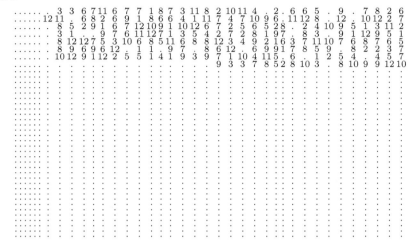

Matrix $\mathcal{D}$:

Let $\mathfrak{e} = (\mathfrak{W}(e_1))^T$. Then $\mathfrak{E}_2 = \langle \mathfrak{r}, \mathfrak{n}, \mathfrak{e} \rangle \cong E_2$. Let $\mathfrak{G}_2 = \langle \mathfrak{h}, \mathfrak{r}, \mathfrak{n}, \mathfrak{e} \rangle$. The four generating matrices of $\mathfrak{G}_2$ are given in Appendix C. Using the MAGMA command CosetAction(G, E) we obtain a faithful permutation representation of $\mathfrak{G}_2$ of degree 142155 with stabilizer $\mathfrak{E}_2$. In particular $|\mathfrak{G}_2| = 2^{17} \cdot 3^9 \cdot 5^2 \cdot 7 \cdot 11 \cdot 13$.

Using the faithful permutation representation of $\mathfrak{G}_2$ of degree 142155 and Kratzer's Algorithm 5.3.18 of [11] we calculated the representatives of the 65 conjugacy classes of $\mathfrak{G}_2$, see Table A.8. Furthermore, the character table of $\mathfrak{G}_2$ has been calculated by means of the above permutation representation and MAGMA. It coincides with the one of Fi$_{22}$ in [6, pp. 156–157].

(d) Let $\mathfrak{z} = (\mathfrak{r}\mathfrak{n}\mathfrak{r}\mathfrak{h})^{10}$. Then $C_{\mathfrak{G}_2}(\mathfrak{z})$ contains $\mathfrak{H} = \langle \mathfrak{h}, \mathfrak{r}, \mathfrak{n} \rangle$ which is isomorphic to H. Now $|C_{\mathfrak{G}_2}(\mathfrak{z})| = |H|$ by Table A.8. Hence $C_{\mathfrak{E}_2}(\mathfrak{z}) \cong \mathfrak{H}$. The character table of $\mathfrak{G}_2$ implies that $\mathfrak{G}_2$ is a simple group. This completes the proof. $\qquad\square$

Praeger and Soicher have given in [12] a nice presentation for Fischer's simple group Fi_{22} which we now also verify for $\mathfrak{G}_2$.

COROLLARY 5.2. *Keep the notation of Theorem 5.1. The finite simple group $\mathfrak{G}_2$ is isomorphic to the finitely presented group $G = \langle a, b, c, d, e, f, g, h, i \rangle$ with the following set $\mathcal{R}(G)$ of defining relations:*

$$a^2 = b^2 = c^2 = d^2 = e^2 = f^2 = g^2 = h^2 = i^2 = 1,$$
$$(ab)^3 = (bc)^3 = (cd)^3 = (de)^3 = (ef)^3 = (fg)^3 = (dh)^3 = (hi)^3 = 1,$$
$$(ac)^2 = (ad)^2 = (ae)^2 = (af)^2 = (ag)^2 = (ah)^2 = (ai)^2 = 1,$$
$$(bd)^2 = (be)^2 = (bf)^2 = (bg)^2 = (bh)^2 = (bi)^2 = 1,$$
$$(ce)^2 = (cf)^2 = (cg)^2 = (ch)^2 = (ci)^2 = (df)^2 = (dg)^2 = (di)^2 = 1,$$
$$(eg)^2 = (eh)^2 = (ei)^2 = (fh)^2 = (fi)^2 = (gh)^2 = (gi)^2 = 1,$$
$$(dcbdefdhi)^{10} = (abcdefh)^9 = (bcdefgh)^9 = 1.$$

PROOF. Using the MAGMA command `FPGroupStrong(PG)` and the faithful permutation representation of degree 142155 of the simple group $\mathfrak{G}_2$ given in Theorem 5.1(c) we obtained a finite presentation of $\mathfrak{G}_2$ with 16 generators and too many relations to be stated in this article. Therefore we stated in the assertion the presentation of [12, p. 110]. Using its subgroup

$$Q = \langle a, c, e, g, h, bacb, dced, fegf, dchd, dehd, (cdehi)^4 \rangle$$

as a stabilizer we obtained a faithful permutation representation of degree 142155 for the group $G = \langle a, b, c, d, e, f, g \rangle$. An isomorphism test in MAGMA established that $\mathfrak{G}_2 \cong G$.   □

6. The remaining cases: E_1, E_4, and E_5

In this section we summarize our results on the remaining cases. Using Theorem 5.1 we prove that the application of Algorithm 2.5 of [10] to the extension group E_4 constructs the automorphism group $\mathrm{Aut}(\mathrm{Fi}_{22})$. In the cases of E_1 and E_5 the algorithm terminates without success.

COROLLARY 6.1. *Keep the notation of Lemma 2.5. Let $E_4 = \langle p_1, q_1, v_1 \rangle$ be the split extension of $A_{22} = \mathrm{Aut}(\mathcal{M}_{22})$ by its simple module V_4 of dimension 10 over $F = \mathrm{GF}(2)$. Let $\mathfrak{A}_2$ be the automorphism group of the simple group $\mathfrak{G}_2$, and let $\mathfrak{H} = C_{\mathfrak{G}_2}(\mathfrak{z})$ be the centralizer of a 2-central involution $\mathfrak{z}$ of $\mathfrak{G}_2$. Then there is a unique maximal elementary abelian normal subgroup $\mathfrak{B}$ of a Sylow 2-subgroup $\mathfrak{S}$ of $\mathfrak{H}$ such that $N_{\mathfrak{G}_2}(\mathfrak{B}) \cong E_4$.*

PROOF. In order to simplify the notation in the proof we replace Gothic letters by Roman letters and $\mathfrak{G}_2$, $\mathfrak{A}_2$ by G, A, respectively. Then $G = \langle x, y, h, e \rangle$ by Theorem 5.1 and $t = (xye)^7$ is an involution of G with a centralizer $C_G(t)$ of index $|G : C_G(t)| = 3510$ by Table A.8. Since G is simple it has a faithful permutation representation PG of degree 3510. Using it, MAGMA has been able to calculate the automorphism group A of G and its order $|A| = 2^{18} \cdot 3^9 \cdot 5^2 \cdot 7 \cdot 11 \cdot 13$. Using the command `PermutationRepresentation(AutG)`, MAGMA produces a faithful permutation representation PA of A of degree 3510. Let PU be the derived subgroup of PA. Then MAGMA verifies that $PU \cong PG$ and that $|PA : PU| = 2$. Let PS be a Sylow 2-subgroup of PU and pz an involution in the center of PS. Let $PH = C_{PU}(pz)$. An isomorphism test shows that $PH \cong C_G(z)$ where z is a 2-central involution of the simple group G. Let $NH = N_{PA}(PH)$. Then $|NH : PH| = 2$. Further-more, PH has a complement $PC = \langle pu \rangle$ of order 2 in NH, as has been checked by means of MAGMA. Now Sylow's Theorem asserts that pu can be chosen so that $PS = PS^{pu}$. Another application of MAGMA shows that PS has a unique elementary abelian normal subgroup PB of order 2^{10}. Since PB is a characteristic subgroup of PS it is normalized by pu. Furthermore, $PA = \langle PU, pu \rangle$. MAGMA also shows that $PD = N_{PA}(PB)$ has order $2^{18} \cdot 3 \cdot 5$. Another isomorphism test verifies that $PD \cong C_{E_4}(y)$ where y is the 2-central involution $y = (p_1^2 q_1 v)^{10}$ of E_4 given in Lemma 2.6. Therefore Algorithm 2.5 of [**10**] and Theorem 5.1 imply that the application of that algorithm to the extension group E_4 returns the automorphism group A of the simple group G. $\square$

REMARK 6.2. Let $E_1 = V_1 \rtimes \mathcal{M}_{22} = \langle a, b, c, d, t, g, h, i, v_1 \rangle$ be the split exten-sion of $\mathcal{M}_{22}$ by its simple module V_1 of dimension 10 over $F = \mathrm{GF}(2)$ defined in Lemma 2.4. By Lemma 2.6 E_1 has a unique class of 2-central involutions rep-resented by $z = (tv_1)^3$. Its centralizer $D = C_{E_1}(z)$ has a uniquely determined nonabelian normal subgroup Q of order 512 such that $V = Q/Z(Q)$ is elementary abelian of order 2^8, where $Z(Q) = \langle z \rangle$ denotes the center of Q. Furthermore, Q has a complement W in D. The Fitting subgroup B of W has order 16 and its complement L in W is isomorphic to the alternating group A_6 and not to S_5 as in Proposition 3.3. Since the center of W has order 2 we applied Algorithm 7.4.8 of [**11**] to construct a subgroup K of $\mathrm{GL}_8(2)$ such that $|K : W|$ is odd. However, this application was not successful.

REMARK 6.3. Let E_5 be the non-split extension of A_{22} by its simple module V_3 of dimension 10 over $F = \mathrm{GF}(2)$ defined in Lemma 2.5. By Lemma 2.6, E_5 has a unique class of 2-central involutions represented by $z = p_2^4$. Its centralizer $D = C_{E_5}(z)$ has a uniquely determined nonabelian normal subgroup Q of order 512 such that $V = Q/Z(Q)$ is elementary abelian of order 2^8, where $Z(Q) = \langle z \rangle$ denotes the center of Q. This time Q does not have a complement in D. The Fitting subgroup B of $W = D/Q$ has order 32 and its factor group $L = W/B$ is isomorphic to the symmetric group S_6. Since the center of W has order 2 we applied Algorithm 7.4.8 of [**11**] to construct a subgroup K of $\mathrm{GL}_8(2)$ such that $|K : W|$ is odd. However, that application was not successful.

Appendix A. Representatives of conjugacy classes

A.1. *Conjugacy classes of* $H(\mathrm{Co}_2) = \langle x, y, h \rangle$

$Class$	$Representative$	$|Class|$	$|Centralizer|$	2P	3P	5P	7P
1	1	1	$2^{18} \cdot 3^4 \cdot 5 \cdot 7$	1	1	1	1
2_1	$(xyh)^{15}$	1	$2^{18} \cdot 3^4 \cdot 5 \cdot 7$	1	2_1	2_1	2_1
2_2	$(yh)^7$	270	$2^{17} \cdot 3 \cdot 7$	1	2_2	2_2	2_2
2_3	$(y)^3$	1008	$2^{14} \cdot 3^2 \cdot 5$	1	2_3	2_3	2_3
2_4	$(xyhy)^5$	1008	$2^{14} \cdot 3^2 \cdot 5$	1	2_4	2_4	2_4
2_5	$(xy)^4$	1260	$2^{16} \cdot 3^2$	1	2_5	2_5	2_5
2_6	$(xyxh^2)^6$	1260	$2^{16} \cdot 3^2$	1	2_6	2_6	2_6
2_7	$(x)^6$	15120	$2^{14} \cdot 3$	1	2_7	2_7	2_7
2_8	$(x^2y^4)^3$	15120	$2^{14} \cdot 3$	1	2_8	2_8	2_8
2_9	$(xy^4)^2$	22680	2^{15}	1	2_9	2_9	2_9
2_{10}	$(xh^2)^3$	120960	$2^{11} \cdot 3$	1	2_{10}	2_{10}	2_{10}
3_1	$(x^2h)^4$	143360	$2^6 \cdot 3^4$	3_1	1	3_1	3_1
3_2	$(x)^4$	172032	$2^5 \cdot 3^3 \cdot 5$	3_2	1	3_2	3_2
3_3	$(xhy)^4$	215040	$2^7 \cdot 3^3$	3_3	1	3_3	3_3
4_1	$(yh^2)^7$	240	$2^{14} \cdot 3^3 \cdot 7$	2_1	4_1	4_1	4_1
4_2	$(x^2yh)^4$	15120	$2^{14} \cdot 3$	2_1	4_2	4_2	4_2
4_3	$(xh^2y)^3$	30240	$2^{13} \cdot 3$	2_2	4_3	4_3	4_3
4_4	$(xyxh^2)^3$	60480	$2^{12} \cdot 3$	2_6	4_4	4_4	4_4
4_5	$(xh^2y^2)^3$	60480	$2^{12} \cdot 3$	2_5	4_5	4_5	4_5
4_6	$(xhyxh^2)^3$	60480	$2^{12} \cdot 3$	2_6	4_6	4_6	4_6
4_7	$(x^2yxhxy)^3$	60480	$2^{12} \cdot 3$	2_2	4_7	4_7	4_7
4_8	$(x^4y^3)^2$	90720	2^{13}	2_2	4_8	4_8	4_8
4_9	$(x^2y)^2$	120960	$2^{11} \cdot 3$	2_5	4_9	4_9	4_9
4_{10}	$(xhy)^3$	120960	$2^{11} \cdot 3$	2_2	4_{10}	4_{10}	4_{10}
4_{11}	$(xy)^2$	181440	2^{12}	2_5	4_{11}	4_{11}	4_{11}
4_{12}	$x^2yxhyxyh$	181440	2^{12}	2_5	4_{12}	4_{12}	4_{12}
4_{13}	$(yh^3)^3$	241920	$2^{10} \cdot 3$	2_5	4_{13}	4_{13}	4_{13}
4_{14}	$(xh^3y)^2$	362880	2^{11}	2_2	4_{14}	4_{14}	4_{14}
4_{15}	$(x^3yhxy)^2$	362880	2^{11}	2_6	4_{15}	4_{15}	4_{15}
4_{16}	$(x)^3$	483840	$2^9 \cdot 3$	2_7	4_{16}	4_{16}	4_{16}
4_{17}	$(xy^2)^3$	483840	$2^9 \cdot 3$	2_7	4_{17}	4_{17}	4_{17}
4_{18}	xhy^2	483840	$2^9 \cdot 3$	2_7	4_{18}	4_{18}	4_{18}
4_{19}	$(yh^4)^3$	483840	$2^9 \cdot 3$	2_7	4_{19}	4_{19}	4_{19}
4_{20}	x^2y^2	725760	2^{10}	2_2	4_{20}	4_{20}	4_{20}
4_{21}	x^2hxy	725760	2^{10}	2_5	4_{21}	4_{21}	4_{21}
4_{22}	xy^4	2903040	2^8	2_9	4_{22}	4_{22}	4_{22}
4_{23}	$xyxh^3y$	2903040	2^8	2_7	4_{23}	4_{23}	4_{23}
4_{24}	xh^5y	2903040	2^8	2_7	4_{24}	4_{24}	4_{24}
5	xh	12386304	$2^2 \cdot 3 \cdot 5$	5	5	1	5
6_1	$(x^2hyh)^3$	143360	$2^6 \cdot 3^4$	3_1	2_1	6_1	6_1
6_2	$(xyh)^5$	172032	$2^5 \cdot 3^3 \cdot 5$	3_2	2_1	6_2	6_2
6_3	$(x^3y)^2$	215040	$2^7 \cdot 3^3$	3_3	2_1	6_3	6_3
6_4	$(x^2h)^2$	1290240	$2^6 \cdot 3^2$	3_1	2_5	6_4	6_4
6_5	$(xyxh^2)^2$	1290240	$2^6 \cdot 3^2$	3_1	2_6	6_5	6_5
6_6	y	2580480	$2^5 \cdot 3^2$	3_2	2_3	6_6	6_6
6_7	x^4y	2580480	$2^5 \cdot 3^2$	3_2	2_4	6_7	6_7
6_8	x^3hxy	2580480	$2^5 \cdot 3^2$	3_3	2_5	6_8	6_8
6_9	x^2yh^3	2580480	$2^5 \cdot 3^2$	3_3	2_4	6_9	6_9
6_{10}	x^2y^4h	2580480	$2^5 \cdot 3^2$	3_3	2_6	6_{10}	6_{10}
6_{11}	x^2hyhxh	2580480	$2^5 \cdot 3^2$	3_2	2_5	6_{11}	6_{11}
6_{12}	x^2h^2yxy	2580480	$2^5 \cdot 3^2$	3_2	2_6	6_{12}	6_{12}
6_{13}	xyh^2yh^2	2580480	$2^5 \cdot 3^2$	3_3	2_3	6_{13}	6_{13}
6_{14}	$(xhy)^2$	3870720	$2^6 \cdot 3$	3_3	2_2	6_{14}	6_{14}
6_{15}	$(x)^2$	7741440	$2^5 \cdot 3$	3_2	2_7	6_{15}	6_{15}
6_{16}	x^2y^4	7741440	$2^5 \cdot 3$	3_2	2_8	6_{16}	6_{16}
6_{17}	xh^2	15482880	$2^4 \cdot 3$	3_3	2_{10}	6_{17}	6_{17}
7	h	13271040	$2^3 \cdot 7$	7	7	7	1

Conjugacy classes of $H(\mathrm{Co_2}) = \langle x, y, h \rangle$ (continued)

Class	Representative	$|Class|$	$|Centralizer|$	2P	3P	5P	7P
8_1	$(xy^3h)^3$	967680	$2^8 \cdot 3$	4_1	8_1	8_1	8_1
8_2	$(x^2yh)^2$	1451520	2^9	4_2	8_2	8_2	8_2
8_3	$(xhyh)^2$	1451520	2^9	4_2	8_3	8_3	8_3
8_4	x^4y^3	2903040	2^8	4_8	8_4	8_4	8_4
8_5	$x^2h^2xy^2$	2903040	2^8	4_8	8_5	8_5	8_5
8_6	x^2y	5806080	2^7	4_9	8_6	8_6	8_6
8_7	x^3yhxy	5806080	2^7	4_{15}	8_7	8_7	8_7
8_8	x^2y^2hyh	5806080	2^7	4_8	8_8	8_8	8_8
8_9	xh^4yh	5806080	2^7	4_9	8_9	8_9	8_9
8_{10}	x^2hxy^3h	5806080	2^7	4_{15}	8_{10}	8_{10}	8_{10}
8_{11}	xy	11612160	2^6	4_{11}	8_{11}	8_{11}	8_{11}
8_{12}	xh^3y	11612160	2^6	4_{14}	8_{12}	8_{12}	8_{12}
8_{13}	xh^2y^2h	11612160	2^6	4_4	8_{13}	8_{13}	8_{13}
9	y^2h	41287680	$2 \cdot 3^2$	9	3_1	9	9
10_1	$(xyh)^3$	12386304	$2^2 \cdot 3 \cdot 5$	5	10_1	2_1	10_1
10_2	$xyhy$	37158912	$2^2 \cdot 5$	5	10_2	2_4	10_2
10_3	y^2h^2	37158912	$2^2 \cdot 5$	5	10_3	2_3	10_3
12_1	$xh^2xh^2y^2$	860160	$2^5 \cdot 3^3$	6_1	4_1	12_1	12_1
12_2	x^3y	2580480	$2^5 \cdot 3^2$	6_3	4_1	12_2	12_2
12_3	$xyxh^2$	7741440	$2^5 \cdot 3$	6_5	4_4	12_3	12_3
12_4	xh^2y^2	7741440	$2^5 \cdot 3$	6_{14}	4_5	12_4	12_4
12_5	xy^4h	7741440	$2^5 \cdot 3$	6_1	4_2	12_5	12_5
12_6	$xhyxh^2$	7741440	$2^5 \cdot 3$	6_5	4_6	12_6	12_6
12_7	x^2yxhxy	7741440	$2^5 \cdot 3$	6_{14}	4_7	12_7	12_7
12_8	x	15482880	$2^4 \cdot 3$	6_{15}	4_{16}	12_8	12_8
12_9	x^2h	15482880	$2^4 \cdot 3$	6_4	4_9	12_9	12_9
12_{10}	xy^2	15482880	$2^4 \cdot 3$	6_{15}	4_{17}	12_{10}	12_{10}
12_{11}	xhy	15482880	$2^4 \cdot 3$	6_{14}	4_{10}	12_{11}	12_{11}
12_{12}	xh^2y	15482880	$2^4 \cdot 3$	6_{14}	4_3	12_{12}	12_{12}
12_{13}	yh^3	15482880	$2^4 \cdot 3$	6_4	4_{13}	12_{14}	12_{13}
12_{14}	$(yh^3)^5$	15482880	$2^4 \cdot 3$	6_4	4_{13}	12_{13}	12_{14}
12_{15}	yh^4	15482880	$2^4 \cdot 3$	6_{15}	4_{19}	12_{15}	12_{15}
12_{16}	x^2y^2hy	15482880	$2^4 \cdot 3$	6_{15}	4_{18}	12_{16}	12_{16}
14_1	$(yh^2)^2$	13271040	$2^3 \cdot 7$	7	14_1	14_1	2_1
14_2	yh	26542080	$2^2 \cdot 7$	7	14_3	14_3	2_2
14_3	$(yh)^3$	26542080	$2^2 \cdot 7$	7	14_2	14_2	2_2
15	$(xyh)^2$	24772608	$2 \cdot 3 \cdot 5$	15	5	3_2	15
16_1	x^2yh	23224320	2^5	8_2	16_1	16_1	16_1
16_2	$xhyh$	23224320	2^5	8_3	16_2	16_2	16_2
18	x^2hyh	41287680	$2 \cdot 3^2$	9	6_1	18	18
24	xy^3h	30965760	$2^3 \cdot 3$	12_2	8_1	24	24
28	yh^2	26542080	$2^2 \cdot 7$	14_1	28	28	4_1
30	xyh	24772608	$2 \cdot 3 \cdot 5$	15	10_1	6_2	30

A.2. *Conjugacy classes of* $D(\mathrm{Co}_2) = \langle x, y \rangle$

Class	Representative	$|Class|$	$|Centralizer|$	2P	3P	5P
1	1	1	$2^{18} \cdot 3^2 \cdot 5$	1	1	1
2_1	$(x^3y)^6$	1	$2^{18} \cdot 3^2 \cdot 5$	1	2_1	2_1
2_2	$(x^5y)^5$	16	$2^{14} \cdot 3^2 \cdot 5$	1	2_2	2_2
2_3	$(x^4yxy^2)^5$	16	$2^{14} \cdot 3^2 \cdot 5$	1	2_3	2_3
2_4	$(x^4y^3)^4$	30	$2^{17} \cdot 3$	1	2_4	2_4
2_5	$(xy)^4$	60	$2^{16} \cdot 3$	1	2_5	2_5
2_6	$(x^4yxy)^4$	60	$2^{16} \cdot 3$	1	2_6	2_6
2_7	$(x)^6$	240	$2^{14} \cdot 3$	1	2_7	2_7
2_8	$(x^2y^2)^2$	240	$2^{14} \cdot 3$	1	2_8	2_8
2_9	$(x^2y^4)^3$	240	$2^{14} \cdot 3$	1	2_9	2_9
2_{10}	$(x^4yxy^3)^3$	240	$2^{14} \cdot 3$	1	2_{10}	2_{10}
2_{11}	$(x^2y^3xy^3xy)^3$	240	$2^{14} \cdot 3$	1	2_{11}	2_{11}
2_{12}	$(xy^4)^2$	360	2^{15}	1	2_{12}	2_{12}
2_{13}	$(y)^3$	480	$2^{13} \cdot 3$	1	2_{13}	2_{13}
2_{14}	$(x^4y)^3$	480	$2^{13} \cdot 3$	1	2_{14}	2_{14}
2_{15}	$(x^2yxyxy)^3$	480	$2^{13} \cdot 3$	1	2_{15}	2_{15}
2_{16}	$(xy^2xy^3)^3$	480	$2^{13} \cdot 3$	1	2_{16}	2_{16}
2_{17}	$(x^2yxyxy^2x^2y^2)^2$	720	2^{14}	1	2_{17}	2_{17}
2_{18}	$x^3yxy^2x^2yxyx^2y^3$	1440	2^{13}	1	2_{18}	2_{18}
2_{19}	$(x^2y^3xy)^3$	1920	$2^{11} \cdot 3$	1	2_{19}	2_{19}
2_{20}	$(x^3y^3xy)^2$	2880	2^{12}	1	2_{20}	2_{20}
2_{21}	$x^2yxyxy^2xy^4$	2880	2^{12}	1	2_{21}	2_{21}
2_{22}	$x^3y^4x^2yx^2yxy^2$	2880	2^{12}	1	2_{22}	2_{22}
2_{23}	$xyxyxyxy^4$	5760	2^{11}	1	2_{23}	2_{23}
2_{24}	$x^3yx^2yxyxyx^2y$	5760	2^{11}	1	2_{24}	2_{24}
3_1	$(x^3y)^4$	10240	$2^7 \cdot 3^2$	3_1	1	3_1
3_2	$(x)^4$	40960	$2^5 \cdot 3^2$	3_2	1	3_2
4_1	$(x^3y)^3$	240	$2^{14} \cdot 3$	2_1	4_1	4_1
4_2	$(x^2yxyxy^2)^3$	480	$2^{13} \cdot 3$	2_4	4_2	4_2
4_3	$(x^2yxy^2)^4$	720	2^{14}	2_1	4_3	4_3
4_4	$(x^4y^3xy)^3$	960	$2^{12} \cdot 3$	2_4	4_4	4_4
4_5	$(x^2yxy^3xy^3)^3$	960	$2^{12} \cdot 3$	2_4	4_5	4_5
4_6	$(x^4yx^2y)^2$	1440	2^{13}	2_4	4_6	4_6
4_7	$(x^2yxy^3)^3$	1920	$2^{11} \cdot 3$	2_4	4_7	4_7
4_8	$(xy)^2$	2880	2^{12}	2_5	4_8	4_8
4_9	$(x^4yxy)^2$	2880	2^{12}	2_6	4_9	4_9
4_{10}	x^7yxy	2880	2^{12}	2_4	4_{10}	4_{10}
4_{11}	$x^3yxy^3x^2y^2$	2880	2^{12}	2_5	4_{11}	4_{11}
4_{12}	x^7yxy^4	2880	2^{12}	2_4	4_{12}	4_{12}
4_{13}	x^6yxyxy^3	2880	2^{12}	2_4	4_{13}	4_{13}
4_{14}	$x^2y^2xyxy^4xy$	2880	2^{12}	2_1	4_{14}	4_{14}
4_{15}	$x^5yx^2yx^2yx^2y$	2880	2^{12}	2_4	4_{15}	4_{15}
4_{16}	$x^3yx^3yx^2y^3xy$	2880	2^{12}	2_4	4_{16}	4_{16}
4_{17}	$x^3yxyx^2yx^2y^2xy^2$	2880	2^{12}	2_6	4_{17}	4_{17}
4_{18}	$x^3y^2x^2yxyx^2yxy^2$	2880	2^{12}	2_4	4_{18}	4_{18}
4_{19}	$(x^2y)^2$	5760	2^{11}	2_5	4_{19}	4_{19}
4_{20}	$(x^4y^3)^2$	5760	2^{11}	2_4	4_{20}	4_{20}
4_{21}	x^2yxy^3xy	5760	2^{11}	2_5	4_{21}	4_{21}
4_{22}	$(x^5yx^3y)^2$	5760	2^{11}	2_6	4_{22}	4_{22}
4_{23}	$x^3yxyx^3y^2$	5760	2^{11}	2_5	4_{23}	4_{23}
4_{24}	$x^6yx^2y^2x^2y$	5760	2^{11}	2_4	4_{24}	4_{24}
4_{25}	$x^2yxyxyxyxy^2x^2y^2$	5760	2^{11}	2_4	4_{25}	4_{25}
4_{26}	$x^5y^5xy^4xy$	5760	2^{11}	2_4	4_{26}	4_{26}
4_{27}	$(x)^3$	7680	$2^9 \cdot 3$	2_7	4_{27}	4_{27}
4_{28}	$(xy^2)^3$	7680	$2^9 \cdot 3$	2_7	4_{28}	4_{28}
4_{29}	$(x^6y)^3$	7680	$2^9 \cdot 3$	2_7	4_{29}	4_{29}
4_{30}	$xyxy^4$	7680	$2^9 \cdot 3$	2_7	4_{30}	4_{30}

Conjugacy classes of $D(Co_2) = \langle x, y \rangle$ (continued)

Class	Representative	$|Class|$	$|Centralizer|$	2P	3P	5P
4_{31}	$x^4 y^4$	11520	2^{10}	2_8	4_{31}	4_{31}
4_{32}	$(x^3 y^2 xy^3)^2$	11520	2^{10}	2_8	4_{32}	4_{32}
4_{33}	$x^2 y^2 xy^4$	11520	2^{10}	2_5	4_{33}	4_{33}
4_{34}	$x^2 y^2 x^2 y^5$	11520	2^{10}	2_8	4_{34}	4_{34}
4_{35}	$x^4 yx^4 yxy$	11520	2^{10}	2_5	4_{35}	4_{35}
4_{36}	$x^4 yxyx^2 yxy$	11520	2^{10}	2_5	4_{36}	4_{36}
4_{37}	$x^2 yxy^2 x^2 y^2 xy$	11520	2^{10}	2_4	4_{37}	4_{37}
4_{38}	$x^2 yxy^2 xyxy^3$	11520	2^{10}	2_4	4_{38}	4_{38}
4_{39}	$x^4 yxyx^3 yxy$	11520	2^{10}	2_4	4_{39}	4_{39}
4_{40}	$x^2 yxyx^2 y^3 xy^2$	11520	2^{10}	2_6	4_{40}	4_{40}
4_{41}	$x^2 yxy^4 xyxy^2$	11520	2^{10}	2_4	4_{41}	4_{41}
4_{42}	$x^2 y^2 x^2 y^3 xy^3$	11520	2^{10}	2_5	4_{42}	4_{42}
4_{43}	$x^4 yx^2 yx^3 y^2 xy$	11520	2^{10}	2_8	4_{43}	4_{43}
4_{44}	$(x^3 yxyxy^2)^3$	15360	$2^8 \cdot 3$	2_8	4_{44}	4_{44}
4_{45}	$x^2 y^2$	23040	2^9	2_8	4_{45}	4_{45}
4_{46}	$x^3 yxyxyx^3 y^2$	23040	2^9	2_8	4_{46}	4_{46}
4_{47}	xy^4	46080	2^8	2_{12}	4_{47}	4_{47}
4_{48}	$x^3 y^3 xy$	46080	2^8	2_{20}	4_{48}	4_{48}
4_{49}	$x^8 y$	46080	2^8	2_{20}	4_{49}	4_{49}
4_{50}	$x^2 y^3 xyxy$	46080	2^8	2_{20}	4_{50}	4_{50}
4_{51}	$x^4 yxy^2 xy$	46080	2^8	2_7	4_{51}	4_{51}
4_{52}	$x^7 yxy^2$	46080	2^8	2_{20}	4_{52}	4_{52}
4_{53}	$x^6 yxyxy$	46080	2^8	2_{20}	4_{53}	4_{53}
4_{54}	$x^8 yx^2 y$	46080	2^8	2_7	4_{54}	4_{54}
4_{55}	$x^7 yxy^3$	46080	2^8	2_{20}	4_{55}	4_{55}
4_{56}	$x^7 y^3 xy$	46080	2^8	2_{20}	4_{56}	4_{56}
4_{57}	$x^6 y^2 xyxy$	46080	2^8	2_{20}	4_{57}	4_{57}
4_{58}	$x^2 yxyxy^2 x^2 y^2$	46080	2^8	2_{17}	4_{58}	4_{58}
4_{59}	$x^3 yx^3 yx^2 y^3$	46080	2^8	2_{17}	4_{59}	4_{59}
4_{60}	$x^5 yx^3 y^3 xy$	46080	2^8	2_8	4_{60}	4_{60}
5	xy^3	589824	$2^2 \cdot 5$	5	5	1
6_1	$(x^3 y)^2$	10240	$2^7 \cdot 3^2$	3_1	2_1	6_1
6_2	$x^5 y^2 xy^2$	40960	$2^5 \cdot 3^2$	3_2	2_3	6_2
6_3	$xyxyxyxy^3$	40960	$2^5 \cdot 3^2$	3_2	2_2	6_3
6_4	$x^6 y^2 x^2 y^2$	40960	$2^5 \cdot 3^2$	3_2	2_1	6_4
6_5	$x^3 y^2 xy^3 xy^2$	40960	$2^5 \cdot 3^2$	3_1	2_2	6_5
6_6	$x^2 yx^2 y^2 xyxy^2$	40960	$2^5 \cdot 3^2$	3_1	2_3	6_6
6_7	$(x^2 yxy^3)^2$	61440	$2^6 \cdot 3$	3_1	2_4	6_7
6_8	$(x)^2$	122880	$2^5 \cdot 3$	3_2	2_7	6_8
6_9	$x^2 y^4$	122880	$2^5 \cdot 3$	3_2	2_9	6_9
6_{10}	$x^4 yxy^3$	122880	$2^5 \cdot 3$	3_1	2_{10}	6_{10}
6_{11}	$(x^3 yxyxy^2)^2$	122880	$2^5 \cdot 3$	3_1	2_8	6_{11}
6_{12}	$x^3 yxyxy^3$	122880	$2^5 \cdot 3$	3_2	2_6	6_{12}
6_{13}	$x^3 y^2 xy^4$	122880	$2^5 \cdot 3$	3_2	2_5	6_{13}
6_{14}	$x^2 y^3 xy^3 xy$	122880	$2^5 \cdot 3$	3_1	2_{11}	6_{14}
6_{15}	y	245760	$2^4 \cdot 3$	3_2	2_{13}	6_{15}
6_{16}	$x^4 y$	245760	$2^4 \cdot 3$	3_2	2_{14}	6_{16}
6_{17}	$x^2 yxyxy$	245760	$2^4 \cdot 3$	3_2	2_{15}	6_{17}
6_{18}	$x^2 y^3 xy$	245760	$2^4 \cdot 3$	3_1	2_{19}	6_{18}
6_{19}	$xy^2 xy^3$	245760	$2^4 \cdot 3$	3_2	2_{16}	6_{19}

Conjugacy classes of $D(\mathrm{Co}_2) = \langle x, y \rangle$ (continued)

Class	Representative	$\lvert Class \rvert$	$\lvert Centralizer \rvert$	2P	3P	5P
8_1	$(x^3y^2xyxy)^3$	15360	$2^8 \cdot 3$	4_1	8_1	8_1
8_2	$(x^2yxy^2)^2$	23040	2^9	4_3	8_2	8_2
8_3	$(x^3yxy^2)^2$	23040	2^9	4_3	8_3	8_3
8_4	$x^3yx^3y^2$	46080	2^8	4_6	8_4	8_4
8_5	$x^3yx^2yxyxy^2$	46080	2^8	4_3	8_5	8_5
8_6	$x^2yx^2y^2x^2y^3$	46080	2^8	4_1	8_6	8_6
8_7	$x^5yx^2y^2xy^2$	46080	2^8	4_1	8_7	8_7
8_8	$x^4yx^2yxyx^2y$	46080	2^8	4_6	8_8	8_8
8_9	$x^3yx^3y^3x^2y$	46080	2^8	4_3	8_9	8_9
8_{10}	x^2y	92160	2^7	4_{19}	8_{10}	8_{10}
8_{11}	x^4y^3	92160	2^7	4_{20}	8_{11}	8_{11}
8_{12}	x^4yx^2y	92160	2^7	4_6	8_{12}	8_{12}
8_{13}	x^3yxy^3	92160	2^7	4_{20}	8_{13}	8_{13}
8_{14}	x^2yxyxy^3	92160	2^7	4_{20}	8_{14}	8_{14}
8_{15}	x^5yx^3y	92160	2^7	4_{22}	8_{15}	8_{15}
8_{16}	$x^3yxy^2xy^2$	92160	2^7	4_{22}	8_{16}	8_{16}
8_{17}	x^2yxyxy^4	92160	2^7	4_{20}	8_{17}	8_{17}
8_{18}	$x^5y^2x^3y$	92160	2^7	4_{19}	8_{18}	8_{18}
8_{19}	xy	184320	2^6	4_8	8_{19}	8_{19}
8_{20}	x^4yxy	184320	2^6	4_9	8_{20}	8_{20}
8_{21}	$x^3y^2xy^3$	184320	2^6	4_{32}	8_{21}	8_{21}
8_{22}	$x^4y^2xy^3$	184320	2^6	4_{32}	8_{22}	8_{22}
8_{23}	$x^4y^3xy^2$	184320	2^6	4_{31}	8_{23}	8_{23}
8_{24}	$x^4yx^2yxy^2$	184320	2^6	4_{31}	8_{24}	8_{24}
8_{25}	$x^5yx^4y^2xy$	184320	2^6	4_{24}	8_{25}	8_{25}
10_1	x^5y	589824	$2^2 \cdot 5$	5	10_1	2_2
10_2	x^4yxy^2	589824	$2^2 \cdot 5$	5	10_2	2_3
10_3	x^3yxyxy	589824	$2^2 \cdot 5$	5	10_3	2_1
12_1	x^3y	122880	$2^5 \cdot 3$	6_1	4_1	12_1
12_2	x^4y^3xy	122880	$2^5 \cdot 3$	6_7	4_4	12_2
12_3	$x^2yxy^3xy^3$	122880	$2^5 \cdot 3$	6_7	4_5	12_3
12_4	x	245760	$2^4 \cdot 3$	6_8	4_{27}	12_4
12_5	xy^2	245760	$2^4 \cdot 3$	6_8	4_{28}	12_5
12_6	x^6y	245760	$2^4 \cdot 3$	6_8	4_{29}	12_6
12_7	x^2yxy^3	245760	$2^4 \cdot 3$	6_7	4_7	12_7
12_8	x^2yxyxy^2	245760	$2^4 \cdot 3$	6_7	4_2	12_8
12_9	$x^3y^2xy^5$	245760	$2^4 \cdot 3$	6_8	4_{30}	12_9
12_{10}	x^3yxyxy^2	491520	$2^3 \cdot 3$	6_{11}	4_{44}	12_{10}
16_1	x^2yxy^2	368640	2^5	8_2	16_1	16_1
16_2	x^3yxy^2	368640	2^5	8_3	16_2	16_2
24	x^3y^2xyxy	491520	$2^3 \cdot 3$	12_1	8_1	24

A.3. Conjugacy classes of $E_3 = E(\text{Co}_2) = \langle x, y, e \rangle$

Class	Representative	|Class|	|Centralizer|	2P	3P	5P	7P	11P
1	1	1	$2^{18} \cdot 3^2 \cdot 5 \cdot 7 \cdot 11$	1	1	1	1	1
2_1	$(xye)^5$	77	$2^{18} \cdot 3^2 \cdot 5$	1	2_1	2_1	2_1	2_1
2_2	$(x)^6$	330	$2^{17} \cdot 3 \cdot 7$	1	2_2	2_2	2_2	2_2
2_3	$(xe)^{10}$	616	$2^{15} \cdot 3^2 \cdot 5$	1	2_3	2_3	2_3	2_3
2_4	$(x^2ey)^7$	2640	$2^{14} \cdot 3 \cdot 7$	1	2_4	2_4	2_4	2_4
2_5	$(xy^2e)^7$	2640	$2^{14} \cdot 3 \cdot 7$	1	2_5	2_5	2_5	2_5
2_6	$(e)^2$	18480	$2^{14} \cdot 3$	1	2_6	2_6	2_6	2_6
2_7	$(y)^3$	36960	$2^{13} \cdot 3$	1	2_7	2_7	2_7	2_7
2_8	$(xe^3)^5$	44352	$2^{12} \cdot 5$	1	2_8	2_8	2_8	2_8
2_9	$(x^2ye^3y)^2$	55440	2^{14}	1	2_9	2_9	2_9	2_9
3	$(x)^4$	788480	$2^7 \cdot 3^2$	3	1	3	3	3
4_1	$(x^3y)^3$	18480	$2^{14} \cdot 3$	2_1	4_1	4_1	4_1	4_1
4_2	$(y^2e^2)^3$	36960	$2^{13} \cdot 3$	2_2	4_2	4_2	4_2	4_2
4_3	$(xe^2)^4$	55440	2^{14}	2_1	4_3	4_3	4_3	4_3
4_4	$(ye)^3$	73920	$2^{12} \cdot 3$	2_2	4_4	4_4	4_4	4_4
4_5	$(xy^2)^3$	73920	$2^{12} \cdot 3$	2_2	4_5	4_5	4_5	4_5
4_6	$(xy)^2$	110880	2^{13}	2_2	4_6	4_6	4_6	4_6
4_7	$(x)^3$	147840	$2^{11} \cdot 3$	2_2	4_7	4_7	4_7	4_7
4_8	x^2yx^2eyxe	221760	2^{12}	2_1	4_8	4_8	4_8	4_8
4_9	$(x^3yey^2)^2$	443520	2^{11}	2_2	4_9	4_9	4_9	4_9
4_{10}	$xyxe^3xe$	443520	2^{11}	2_2	4_{10}	4_{10}	4_{10}	4_{10}
4_{11}	xy^3exe^2	443520	2^{11}	2_2	4_{11}	4_{11}	4_{11}	4_{11}
4_{12}	$(xe)^5$	709632	$2^8 \cdot 5$	2_3	4_{12}	4_{12}	4_{12}	4_{12}
4_{13}	e	887040	2^{10}	2_6	4_{13}	4_{13}	4_{13}	4_{13}
4_{14}	$(xy^2e^2)^2$	887040	2^{10}	2_6	4_{14}	4_{14}	4_{14}	4_{14}
4_{15}	$x^2e^2y^3e$	887040	2^{10}	2_6	4_{15}	4_{15}	4_{15}	4_{15}
4_{16}	x^8y	887040	2^{10}	2_6	4_{16}	4_{16}	4_{16}	4_{16}
4_{17}	$(x^3y^2e)^3$	1182720	$2^8 \cdot 3$	2_6	4_{17}	4_{17}	4_{17}	4_{17}
4_{18}	x^2y^2	1774080	2^9	2_6	4_{18}	4_{18}	4_{18}	4_{18}
4_{19}	$x^2y^2xe^3$	1774080	2^9	2_6	4_{19}	4_{19}	4_{19}	4_{19}
4_{20}	x^2ye^3y	3548160	2^8	2_9	4_{20}	4_{20}	4_{20}	4_{20}
4_{21}	x^2exeye	3548160	2^8	2_6	4_{21}	4_{21}	4_{21}	4_{21}
4_{22}	xye^2y^3	3548160	2^8	2_9	4_{22}	4_{22}	4_{22}	4_{22}
5	$(xe)^4$	22708224	$2^3 \cdot 5$	5	5	1	5	5
6_1	$(x^3y)^2$	788480	$2^7 \cdot 3^2$	3	2_1	6_1	6_1	6_1
6_2	x^2y^4	3153920	$2^5 \cdot 3^2$	3	2_3	6_2	6_2	6_2
6_3	$xyxe^2y$	3153920	$2^5 \cdot 3^2$	3	2_1	6_3	6_3	6_3
6_4	$(x)^2$	4730880	$2^6 \cdot 3$	3	2_2	6_4	6_4	6_4
6_5	x^4y	9461760	$2^5 \cdot 3$	3	2_4	6_5	6_5	6_5
6_6	x^2yxe	9461760	$2^5 \cdot 3$	3	2_6	6_6	6_6	6_6
6_7	x^3e^3	9461760	$2^5 \cdot 3$	3	2_5	6_7	6_7	6_7
6_8	y	18923520	$2^4 \cdot 3$	3	2_7	6_8	6_8	6_8
7_1	x^2e	32440320	$2^2 \cdot 7$	7_1	7_2	7_2	1	7_1
7_2	$(x^2e)^3$	32440320	$2^2 \cdot 7$	7_2	7_1	7_1	1	7_2

Conjugacy classes of $E_3 = E(Co_2) = \langle x, y, e \rangle$ (continued)

Class	Representative	|Class|	|Centralizer|	2P	3P	5P	7P	11P
8_1	$(x^2exe^2)^3$	1182720	$2^8 \cdot 3$	4_1	8_1	8_1	8_1	8_1
8_2	$(xe^2)^2$	1774080	2^9	4_3	8_2	8_2	8_2	8_2
8_3	x^3ye	1774080	2^9	4_3	8_3	8_3	8_3	8_3
8_4	x^2y	3548160	2^8	4_6	8_4	8_4	8_4	8_4
8_5	x^3e^2	3548160	2^8	4_6	8_5	8_5	8_5	8_5
8_6	x^2yey^2	3548160	2^8	4_1	8_6	8_6	8_6	8_6
8_7	x^2eyxe	3548160	2^8	4_3	8_7	8_7	8_7	8_7
8_8	xy^4e^2	3548160	2^8	4_3	8_8	8_8	8_8	8_8
8_9	$x^3y^2e^2y$	3548160	2^8	4_1	8_9	8_9	8_9	8_9
8_{10}	xy	7096320	2^7	4_6	8_{10}	8_{10}	8_{10}	8_{10}
8_{11}	x^2e^2y	14192640	2^6	4_{13}	8_{11}	8_{11}	8_{11}	8_{11}
8_{12}	xy^2e^2	14192640	2^6	4_{14}	8_{12}	8_{12}	8_{12}	8_{12}
8_{13}	$yeye^2$	14192640	2^6	4_{13}	8_{13}	8_{13}	8_{13}	8_{13}
8_{14}	$x^2e^2y^2$	14192640	2^6	4_{14}	8_{14}	8_{14}	8_{14}	8_{14}
8_{15}	x^3yey^2	14192640	2^6	4_9	8_{15}	8_{15}	8_{15}	8_{15}
10_1	$(xe)^2$	22708224	$2^3 \cdot 5$	5	10_1	2_3	10_1	10_1
10_2	xye	45416448	$2^2 \cdot 5$	5	10_2	2_1	10_2	10_2
10_3	xe^3	45416448	$2^2 \cdot 5$	5	10_3	2_8	10_3	10_3
11	xy^2xe	82575360	11	11	11	11	11	1
12_1	ye	9461760	$2^5 \cdot 3$	6_4	4_4	12_1	12_1	12_1
12_2	xy^2	9461760	$2^5 \cdot 3$	6_4	4_5	12_2	12_2	12_2
12_3	x^3y	9461760	$2^5 \cdot 3$	6_1	4_1	12_3	12_3	12_3
12_4	x	18923520	$2^4 \cdot 3$	6_4	4_7	12_4	12_4	12_4
12_5	y^2e^2	18923520	$2^4 \cdot 3$	6_4	4_2	12_5	12_5	12_5
12_6	x^3y^2e	37847040	$2^3 \cdot 3$	6_6	4_{17}	12_6	12_6	12_6
14_1	x^2ey	32440320	$2^2 \cdot 7$	7_1	14_2	14_2	2_4	14_1
14_2	$(x^2ey)^3$	32440320	$2^2 \cdot 7$	7_2	14_1	14_1	2_4	14_2
14_3	xy^2e	32440320	$2^2 \cdot 7$	7_2	14_4	14_4	2_5	14_3
14_4	$(xy^2e)^3$	32440320	$2^2 \cdot 7$	7_1	14_3	14_3	2_5	14_4
14_5	$xeye$	32440320	$2^2 \cdot 7$	7_2	14_6	14_6	2_2	14_5
14_6	$(xeye)^3$	32440320	$2^2 \cdot 7$	7_1	14_5	14_5	2_2	14_6
16_1	xe^2	28385280	2^5	8_2	16_1	16_1	16_1	16_1
16_2	$xyxey$	28385280	2^5	8_3	16_2	16_2	16_2	16_2
20	xe	45416448	$2^2 \cdot 5$	10_1	20	4_{12}	20	20
24	x^2exe^2	37847040	$2^3 \cdot 3$	12_3	8_1	24	24	24

A.4. Conjugacy classes of $\mathfrak{G}_3 = \langle h, x, y, e \rangle \cong \mathrm{Co}_2$

Class	Representative	$\lvert Class\rvert$	$\lvert Centralizer\rvert$	2P	3P	5P	7P	11P	23P
1	1	1	$2^{18} \cdot 3^6 \cdot 5^3 \cdot 7 \cdot 11 \cdot 23$	1	1	1	1	1	1
2_1	$(xyh)^{15}$	56925	$2^{18} \cdot 3^4 \cdot 5 \cdot 7$	1	2_1	2_1	2_1	2_1	2_1
2_2	$(x)^6$	1024650	$2^{17} \cdot 3^2 \cdot 5 \cdot 7$	1	2_2	2_2	2_2	2_2	2_2
2_3	$(y)^3$	28690200	$2^{15} \cdot 3^2 \cdot 5$	1	2_3	2_3	2_3	2_3	2_3
3_1	$(x^2 h)^4$	90675200	$2^7 \cdot 3^6 \cdot 5$	3_1	1	3_1	3_1	3_1	3_1
3_2	$(x)^4$	272025600	$2^7 \cdot 3^5 \cdot 5$	3_2	1	3_2	3_2	3_2	3_2
4_1	$(yh^2)^7$	13662000	$2^{14} \cdot 3^3 \cdot 7$	2_1	4_1	4_1	4_1	4_1	4_1
4_2	$(xy^2)^3$	344282400	$2^{13} \cdot 3 \cdot 5$	2_2	4_2	4_2	4_2	4_2	4_2
4_3	e	573804000	$2^{13} \cdot 3^2$	2_2	4_3	4_3	4_3	4_3	4_3
4_4	$(xe^2)^4$	860706000	$2^{14} \cdot 3$	2_1	4_4	4_4	4_4	4_4	4_4
4_5	$(x)^3$	6885648000	$2^{11} \cdot 3$	2_2	4_5	4_5	4_5	4_5	4_5
4_6	$(xheh)^2$	6885648000	$2^{11} \cdot 3$	2_2	4_6	4_6	4_6	4_6	4_6
4_7	$(xe)^5$	33051110400	$2^8 \cdot 5$	2_3	4_7	4_7	4_7	4_7	4_7
5_1	$(xhye)^2$	14101807104	$2^3 \cdot 3 \cdot 5^3$	5_1	5_1	1	5_1	5_1	5_1
5_2	xh	70509035520	$2^3 \cdot 3 \cdot 5^2$	5_2	5_2	1	5_2	5_2	5_2
6_1	$(x^2 h)^2$	7344691200	$2^7 \cdot 3^2 \cdot 5$	3_1	2_2	6_1	6_1	6_1	6_1
6_2	$(x^2 eh)^2$	8160768000	$2^6 \cdot 3^4$	3_1	2_1	6_2	6_2	6_2	6_2
6_3	$(xyh)^5$	9792921600	$2^5 \cdot 3^3 \cdot 5$	3_2	2_1	6_3	6_3	6_3	6_3
6_4	$(yhe)^4$	12241152000	$2^7 \cdot 3^3$	3_2	2_1	6_4	6_4	6_4	6_4
6_5	$(x)^2$	73446912000	$2^6 \cdot 3^2$	3_2	2_2	6_5	6_5	6_5	6_5
6_6	y	146893824000	$2^5 \cdot 3^2$	3_2	2_3	6_6	6_6	6_6	6_6
7	h	755453952000	$2^3 \cdot 7$	7	7	7	1	7	7
8_1	$(yhe)^3$	55085184000	$2^8 \cdot 3$	4_1	8_1	8_1	8_1	8_1	8_1
8_2	$(h^2 e)^3$	55085184000	$2^8 \cdot 3$	4_3	8_2	8_2	8_2	8_2	8_2
8_3	$(xe^2)^2$	82627776000	2^9	4_4	8_3	8_3	8_3	8_3	8_3
8_4	$(x^2 yh)^2$	82627776000	2^9	4_4	8_4	8_4	8_4	8_4	8_4
8_5	xy	165255552000	2^8	4_3	8_5	8_5	8_5	8_5	8_5
8_6	$xheh$	661022208000	2^6	4_6	8_6	8_6	8_6	8_6	8_6
9	$y^2 h$	783433728000	$2 \cdot 3^3$	9	3_1	9	9	9	9
10_1	$xhye$	352545177600	$2^3 \cdot 3 \cdot 5$	5_1	10_1	2_2	10_1	10_1	10_1
10_2	$(xyh)^3$	705090355200	$2^2 \cdot 3 \cdot 5$	5_2	10_2	2_1	10_2	10_2	10_2
10_3	$(xe)^2$	1057635532800	$2^3 \cdot 5$	5_2	10_3	2_3	10_3	10_3	10_3
11	xhe	3845947392000	11	11	11	11	11	1	11
12_1	$xyxh^2$	48964608000	$2^5 \cdot 3^3$	6_2	4_1	12_1	12_1	12_1	12_1
12_2	ye	146893824000	$2^5 \cdot 3^2$	6_5	4_3	12_2	12_2	12_2	12_2
12_3	$x^2 h$	146893824000	$2^5 \cdot 3^2$	6_1	4_3	12_3	12_3	12_3	12_3
12_4	$(yhe)^2$	146893824000	$2^5 \cdot 3^2$	6_4	4_1	12_4	12_4	12_4	12_4
12_5	xy^2	440681472000	$2^5 \cdot 3$	6_5	4_2	12_5	12_5	12_5	12_5
12_6	$x^2 eh$	440681472000	$2^5 \cdot 3$	6_2	4_4	12_6	12_6	12_6	12_6
12_7	x	881362944000	$2^4 \cdot 3$	6_5	4_5	12_7	12_7	12_7	12_7
12_8	yh^3	881362944000	$2^4 \cdot 3$	6_1	4_6	12_8	12_8	12_8	12_8
14_1	$(yh^2)^2$	755453952000	$2^3 \cdot 7$	7	14_1	14_1	2_1	14_1	14_1
14_2	yh	1510907904000	$2^2 \cdot 7$	7	14_3	14_3	2_2	14_2	14_2
14_3	$(yh)^3$	1510907904000	$2^2 \cdot 7$	7	14_2	14_2	2_2	14_3	14_3
15_1	he	1410180710400	$2 \cdot 3 \cdot 5$	15_1	5_2	3_2	15_1	15_1	15_1
15_2	$(ye^2 h)^2$	1410180710400	$2 \cdot 3 \cdot 5$	15_2	5_1	3_1	15_3	15_3	15_2
15_3	$(ye^2 h)^{14}$	1410180710400	$2 \cdot 3 \cdot 5$	15_3	5_1	3_1	15_2	15_2	15_3
16_1	xe^2	1322044416000	2^5	8_3	16_1	16_1	16_1	16_1	16_1
16_2	$x^2 yh$	1322044416000	2^5	8_4	16_2	16_2	16_2	16_2	16_2
18	yeh^2	2350301184000	$2 \cdot 3^2$	9	6_2	18	18	18	18
20_1	xe	2115271065600	$2^2 \cdot 5$	10_3	20_1	4_7	20_1	20_1	20_1
20_2	$xehe$	2115271065600	$2^2 \cdot 5$	10_1	20_2	4_2	20_2	20_2	20_2
23_1	$x^2 yhe$	1839366144000	23	23_1	23_1	23_2	23_2	23_2	1
23_2	$(x^2 yhe)^5$	1839366144000	23	23_2	23_2	23_1	23_1	23_1	1
24_1	yhe	1762725888000	$2^3 \cdot 3$	12_4	8_1	24_1	24_1	24_1	24_1
24_2	$h^2 e$	1762725888000	$2^3 \cdot 3$	12_3	8_2	24_2	24_2	24_2	24_2
28	yh^2	1510907904000	$2^2 \cdot 7$	14_1	28	28	4_1	28	28
30_1	xyh	1410180710400	$2 \cdot 3 \cdot 5$	15_1	10_2	6_3	30_1	30_1	30_1
30_2	$ye^2 h$	1410180710400	$2 \cdot 3 \cdot 5$	15_2	10_1	6_1	30_3	30_3	30_2
30_3	$(ye^2 h)^7$	1410180710400	$2 \cdot 3 \cdot 5$	15_3	10_1	6_1	30_2	30_2	30_3

A.5. *Conjugacy classes of* $H(\mathrm{Fi}_{22}) = \langle x, y, h \rangle$

Class	Representative	$|Class|$	$|Centralizer|$	2P	3P	5P
1	1	1	$2^{17} \cdot 3^4 \cdot 5$	1	1	1
2_1	$(xyxh)^{10}$	1	$2^{17} \cdot 3^4 \cdot 5$	1	2_1	2_1
2_2	$(xhy^2)^9$	2	$2^{16} \cdot 3^4 \cdot 5$	1	2_2	2_2
2_3	$(xh)^6$	180	$2^{15} \cdot 3^2$	1	2_3	2_3
2_4	$(xyhyxh^2)^3$	180	$2^{15} \cdot 3^2$	1	2_4	2_4
2_5	$(yh)^6$	270	$2^{16} \cdot 3$	1	2_5	2_5
2_6	$(xy^2xy^2xh)^4$	270	$2^{16} \cdot 3$	1	2_6	2_6
2_7	$(xyxyxh)^3$	360	$2^{14} \cdot 3^2$	1	2_7	2_7
2_8	$(y)^3$	1152	$2^{10} \cdot 3^2 \cdot 5$	1	2_8	2_8
2_9	x	4320	$2^{12} \cdot 3$	1	2_9	2_9
2_{10}	$(xhyh)^6$	4320	$2^{12} \cdot 3$	1	2_{10}	2_{10}
2_{11}	$xyhxyhxyhyh$	6480	2^{13}	1	2_{11}	2_{11}
2_{12}	$(xyxyxy^2h^2)^3$	8640	$2^{11} \cdot 3$	1	2_{12}	2_{12}
2_{13}	$(xhy^2hy)^3$	17280	$2^{10} \cdot 3$	1	2_{13}	2_{13}
3_1	h	5120	$2^7 \cdot 3^4$	3_1	1	3_1
3_2	$(yh)^4$	7680	$2^8 \cdot 3^3$	3_2	1	3_2
3_3	$(y)^2$	61440	$2^5 \cdot 3^3$	3_3	1	3_3
4_1	$(xy^3h)^6$	480	$2^{12} \cdot 3^3$	2_1	4_1	4_1
4_2	$(xyxh)^5$	1152	$2^{10} \cdot 3^2 \cdot 5$	2_1	4_2	4_2
4_3	$(xyh^2xh)^4$	4320	$2^{12} \cdot 3$	2_1	4_3	4_3
4_4	$(yh)^3$	17280	$2^{10} \cdot 3$	2_5	4_4	4_4
4_5	$(xyxhy^2)^3$	17280	$2^{10} \cdot 3$	2_1	4_5	4_5
4_6	$(xy^2xh^2)^3$	17280	$2^{10} \cdot 3$	2_5	4_6	4_6
4_7	$(xy^2h^2y)^3$	17280	$2^{10} \cdot 3$	2_5	4_7	4_7
4_8	$(xhyhy^2)^3$	17280	$2^{10} \cdot 3$	2_5	4_8	4_8
4_9	y^2hyh^2	17280	$2^{10} \cdot 3$	2_5	4_9	4_9
4_{10}	$(xy^2xh^2y^2)^3$	17280	$2^{10} \cdot 3$	2_5	4_{10}	4_{10}
4_{11}	$(xhy)^2$	25920	2^{11}	2_5	4_{11}	4_{11}
4_{12}	$(xy^2hyh^2)^2$	25920	2^{11}	2_5	4_{12}	4_{12}
4_{13}	$(xh)^3$	34560	$2^9 \cdot 3$	2_3	4_{13}	4_{13}
4_{14}	$(xyxy^2)^2$	34560	$2^9 \cdot 3$	2_3	4_{14}	4_{14}
4_{15}	$(xy^3xhy)^3$	34560	$2^9 \cdot 3$	2_7	4_{15}	4_{15}
4_{16}	$(xyhxh^2y)^3$	34560	$2^9 \cdot 3$	2_7	4_{16}	4_{16}
4_{17}	$(xy)^3$	69120	$2^8 \cdot 3$	2_9	4_{17}	4_{17}
4_{18}	$(xhyh)^3$	69120	$2^8 \cdot 3$	2_{10}	4_{18}	4_{18}
4_{19}	xhy^2hyh	103680	2^9	2_5	4_{19}	4_{19}
4_{20}	$xhyhyhy$	103680	2^9	2_5	4_{20}	4_{20}
4_{21}	$(xy^2xy^2xh)^2$	103680	2^9	2_6	4_{21}	4_{21}
4_{22}	$xy^2hxhyhyh$	103680	2^9	2_5	4_{22}	4_{22}
4_{23}	xy^3	207360	2^8	2_9	4_{23}	4_{23}
4_{24}	$xyhyh$	207360	2^8	2_9	4_{24}	4_{24}
4_{25}	$xyxhxh^2$	207360	2^8	2_{10}	4_{25}	4_{25}
4_{26}	$xyhxyhxh$	207360	2^8	2_9	4_{26}	4_{26}
4_{27}	xy^2xyhyh	414720	2^7	2_9	4_{27}	4_{27}
5	xy^2	1327104	$2^3 \cdot 5$	5	5	1
6_1	$(xy^2h)^3$	5120	$2^7 \cdot 3^4$	3_1	2_1	6_1
6_2	$(xhy^2)^3$	5120	$2^7 \cdot 3^4$	3_1	2_2	6_3
6_3	$(xhy^2)^{15}$	5120	$2^7 \cdot 3^4$	3_1	2_2	6_2
6_4	$(xy^3h)^4$	7680	$2^8 \cdot 3^3$	3_2	2_1	6_4
6_5	$xyxhyhyhy$	15360	$2^7 \cdot 3^3$	3_2	2_2	6_5
6_6	$(xh)^2$	46080	$2^7 \cdot 3^2$	3_1	2_3	6_6
6_7	$(xy^3xhy)^2$	46080	$2^7 \cdot 3^2$	3_1	2_7	6_8
6_8	$(xy^3xhy)^{10}$	46080	$2^7 \cdot 3^2$	3_1	2_7	6_7
6_9	xy^2h^2yxyh	46080	$2^7 \cdot 3^2$	3_1	2_4	6_9
6_{10}	$(xy^2xhyh)^2$	61440	$2^5 \cdot 3^3$	3_3	2_1	6_{10}
6_{11}	$xyxh^2xhy$	92160	$2^6 \cdot 3^2$	3_2	2_3	6_{11}
6_{12}	$xyhxyhyhyh$	92160	$2^6 \cdot 3^2$	3_2	2_4	6_{12}
6_{13}	xyh^2xh^2yh	122880	$2^4 \cdot 3^3$	3_3	2_2	6_{13}

Conjugacy classes of $H(\mathrm{Fi}_{22}) = \langle x, y, h \rangle$ (continued)

Class	Representative	$|Class|$	$|Centralizer|$	2P	3P	5P
6_{14}	$(yh)^2$	138240	$2^7 \cdot 3$	3_2	2_5	6_{14}
6_{15}	$xyxyhxy^2h$	138240	$2^7 \cdot 3$	3_2	2_6	6_{15}
6_{16}	y^2hyh	184320	$2^5 \cdot 3^2$	3_2	2_8	6_{16}
6_{17}	$xyxyxh$	184320	$2^5 \cdot 3^2$	3_2	2_7	6_{17}
6_{18}	$xyhxhy$	368640	$2^4 \cdot 3^2$	3_3	2_7	6_{19}
6_{19}	$(xyhxhy)^5$	368640	$2^4 \cdot 3^2$	3_3	2_7	6_{18}
6_{20}	$xyhyxh^2$	368640	$2^4 \cdot 3^2$	3_3	2_4	6_{20}
6_{21}	$xyxy^2h^2y$	368640	$2^4 \cdot 3^2$	3_3	2_3	6_{21}
6_{22}	$(xy)^2$	552960	$2^5 \cdot 3$	3_3	2_9	6_{22}
6_{23}	$(xhyh)^2$	552960	$2^5 \cdot 3$	3_3	2_{10}	6_{23}
6_{24}	xhy^2hy	552960	$2^5 \cdot 3$	3_2	2_{13}	6_{24}
6_{25}	y	737280	$2^3 \cdot 3^2$	3_3	2_8	6_{25}
6_{26}	$xyxyxy^2h^2$	1105920	$2^4 \cdot 3$	3_3	2_{12}	6_{26}
8_1	$(xy^3h)^3$	138240	$2^7 \cdot 3$	4_1	8_1	8_1
8_2	$(xyxy^2h)^3$	138240	$2^7 \cdot 3$	4_1	8_2	8_2
8_3	xhy	414720	2^7	4_{11}	8_3	8_3
8_4	$xhyhy$	414720	2^7	4_{11}	8_4	8_4
8_5	$xyxy^3$	414720	2^7	4_{11}	8_5	8_5
8_6	$(xyh^2xh)^2$	414720	2^7	4_3	8_6	8_6
8_7	xy^2xy^3	414720	2^7	4_{11}	8_7	8_7
8_8	xy^2hyh^2	829440	2^6	4_{12}	8_8	8_8
8_9	xyh	1658880	2^5	4_{13}	8_9	8_9
8_{10}	$xyxy^2$	1658880	2^5	4_{14}	8_{10}	8_{10}
8_{11}	xy^2xy^2xh	1658880	2^5	4_{21}	8_{11}	8_{11}
9	$(xy^2h)^2$	1474560	$2^2 \cdot 3^2$	9	3_1	9
10_1	$(xyxh)^2$	1327104	$2^3 \cdot 5$	5	10_1	2_1
10_2	xyh^2	2654208	$2^2 \cdot 5$	5	10_2	2_8
10_3	xhy^2h^2	2654208	$2^2 \cdot 5$	5	10_3	2_2
12_1	y^2hyhyh^2	61440	$2^5 \cdot 3^3$	6_1	4_1	12_1
12_2	$(xy^3h)^2$	92160	$2^6 \cdot 3^2$	6_4	4_1	12_2
12_3	$xyxhy^2hyh$	92160	$2^6 \cdot 3^2$	6_4	4_1	12_3
12_4	$xyxhyhxhy$	184320	$2^5 \cdot 3^2$	6_4	4_2	12_4
12_5	yh	552960	$2^5 \cdot 3$	6_{14}	4_4	12_5
12_6	$xyxhy^2$	552960	$2^5 \cdot 3$	6_4	4_5	12_6
12_7	xy^2xh^2	552960	$2^5 \cdot 3$	6_{14}	4_6	12_7
12_8	xy^2h^2y	552960	$2^5 \cdot 3$	6_{14}	4_7	12_8
12_9	$xhyhy^2$	552960	$2^5 \cdot 3$	6_{14}	4_8	12_9
12_{10}	xy^3xhy	552960	$2^5 \cdot 3$	6_7	4_{15}	12_{11}
12_{11}	$(xy^3xhy)^5$	552960	$2^5 \cdot 3$	6_8	4_{15}	12_{10}
12_{12}	xy^2h^2xh	552960	$2^5 \cdot 3$	6_1	4_3	12_{12}
12_{13}	$xyhxh^2y$	552960	$2^5 \cdot 3$	6_7	4_{16}	12_{14}
12_{14}	$(xyhxh^2y)^5$	552960	$2^5 \cdot 3$	6_8	4_{16}	12_{13}
12_{15}	$xy^2xh^2y^2$	552960	$2^5 \cdot 3$	6_{14}	4_{10}	12_{15}
12_{16}	$xyhxhxh^2$	552960	$2^5 \cdot 3$	6_{14}	4_9	12_{16}
12_{17}	xy^2xhyh	737280	$2^3 \cdot 3^2$	6_{10}	4_2	12_{17}
12_{18}	xh	1105920	$2^4 \cdot 3$	6_6	4_{13}	12_{18}
12_{19}	xy^2xh	1105920	$2^4 \cdot 3$	6_6	4_{14}	12_{19}
12_{20}	xy	2211840	$2^3 \cdot 3$	6_{22}	4_{17}	12_{20}
12_{21}	$xhyh$	2211840	$2^3 \cdot 3$	6_{23}	4_{18}	12_{21}
16_1	xyh^2xh	1658880	2^5	8_6	16_1	16_2
16_2	$(xyh^2xh)^5$	1658880	2^5	8_6	16_2	16_1
18_1	xy^2h	1474560	$2^2 \cdot 3^2$	9	6_1	18_1
18_2	xhy^2	1474560	$2^2 \cdot 3^2$	9	6_2	18_3
18_3	$(xhy^2)^5$	1474560	$2^2 \cdot 3^2$	9	6_3	18_2
20	$xyxh$	2654208	$2^2 \cdot 5$	10_1	20	4_2
24_1	xy^3h	1105920	$2^4 \cdot 3$	12_2	8_1	24_1
24_2	$xyxy^2h$	1105920	$2^4 \cdot 3$	12_2	8_2	24_2

A.6. Conjugacy classes of $D(\mathrm{Fi}_{22}) = \langle x, y \rangle$

Class	Representative	$|Class|$	$|Centralizer|$	2P	3P	5P
1	1	1	$2^{17} \cdot 3 \cdot 5$	1	1	1
2_1	$(xyxyxy^2xy^3)^8$	1	$2^{17} \cdot 3 \cdot 5$	1	2_1	2_1
2_2	$(xyxyxyxy^2xy^5)^3$	2	$2^{16} \cdot 3 \cdot 5$	1	2_2	2_2
2_3	$(xyxy^2)^4$	20	$2^{15} \cdot 3$	1	2_3	2_3
2_4	$(xyxyxyxy^3)^3$	20	$2^{15} \cdot 3$	1	2_4	2_4
2_5	$(xyxy^3)^4$	30	2^{16}	1	2_5	2_5
2_6	$(xyxy^3xy^4)^4$	30	2^{16}	1	2_6	2_6
2_7	$(xyxyxyxyxy^2)^2$	40	$2^{14} \cdot 3$	1	2_7	2_7
2_8	$(xy)^6$	160	$2^{12} \cdot 3$	1	2_8	2_8
2_9	$(xyxyxyxy^2xy^4)^6$	160	$2^{12} \cdot 3$	1	2_9	2_9
2_{10}	$(xyxy^2xy^4)^4$	240	2^{13}	1	2_{10}	2_{10}
2_{11}	$(xyxyxy^3xy^2xy^4)^2$	240	2^{13}	1	2_{11}	2_{11}
2_{12}	$xyxyxyxy^2xy^5xy^2xyxyxyxy^3$	240	2^{13}	1	2_{12}	2_{12}
2_{13}	$(xyxyxyxy^2xyxy^4xy^2)^3$	320	$2^{11} \cdot 3$	1	2_{13}	2_{13}
2_{14}	$xyxyxy^2xy^4xy^2$	480	2^{12}	1	2_{14}	2_{14}
2_{15}	$(y)^3$	640	$2^{10} \cdot 3$	1	2_{15}	2_{15}
2_{16}	x	960	2^{11}	1	2_{16}	2_{16}
2_{17}	$(xyxyxyxyxy^5)^2$	960	2^{11}	1	2_{17}	2_{17}
2_{18}	$xyxy^2xy^4xy^2xy^4$	1920	2^{10}	1	2_{18}	2_{18}
2_{19}	$xyxyxyxyxy^2xyxy^2xyxy^2$	1920	2^{10}	1	2_{19}	2_{19}
3	$(y)^2$	20480	$2^5 \cdot 3$	3	1	3
4_1	$(xyxyxy^2xy^3)^4$	480	2^{12}	2_1	4_1	4_1
4_2	$(xyxyxy^3xy^4)^2$	480	2^{12}	2_1	4_2	4_2
4_3	$(xyxy^2xy^3xy^2xy^3)^3$	640	$2^{10} \cdot 3$	2_1	4_3	4_3
4_4	$(xyxy^3)^2$	960	2^{11}	2_5	4_4	4_4
4_5	$(xyxyxy^2xyxy^4)^2$	960	2^{11}	2_5	4_5	4_5
4_6	$xyxy^3xy^4xy^3$	1920	2^{10}	2_5	4_6	4_6
4_7	$(xyxyxyxyxy^2xyxy^4)^2$	1920	2^{10}	2_5	4_7	4_7
4_8	$xyxy^3xyxy^3xy^2xy^3$	1920	2^{10}	2_5	4_8	4_8
4_9	$xyxy^2xy^2xy^3xy^2xy^3$	1920	2^{10}	2_5	4_9	4_9
4_{10}	$xyxyxy^3xy^2xy^2xyxy^4$	1920	2^{10}	2_5	4_{10}	4_{10}
4_{11}	$xyxyxyxy^4xyxy^2xy^5$	1920	2^{10}	2_5	4_{11}	4_{11}
4_{12}	$xyxyxyxyxyxy^3xyxyxy^5$	1920	2^{10}	2_1	4_{12}	4_{12}
4_{13}	$xyxyxyxy^2xy^2xy^3xyxy^2xy^3$	1920	2^{10}	2_5	4_{13}	4_{13}
4_{14}	$xyxyxyxyxy^2xyxyxyxy^2xyxy^3$	1920	2^{10}	2_5	4_{14}	4_{14}
4_{15}	$(xy)^3$	2560	$2^8 \cdot 3$	2_8	4_{15}	4_{15}
4_{16}	$(xyxyxyxy^2xy^4)^3$	2560	$2^8 \cdot 3$	2_9	4_{16}	4_{16}
4_{17}	$(xyxy^2)^2$	3840	2^9	2_3	4_{17}	4_{17}
4_{18}	$(xyxy^4)^2$	3840	2^9	2_3	4_{18}	4_{18}
4_{19}	$(xyxy^2xy^4)^2$	3840	2^9	2_{10}	4_{19}	4_{19}
4_{20}	$xyxyxyxyxy^2$	3840	2^9	2_7	4_{20}	4_{20}
4_{21}	$(xyxy^3xy^4)^2$	3840	2^9	2_6	4_{21}	4_{21}
4_{22}	$(xyxyxyxy^3xy^5)^2$	3840	2^9	2_{10}	4_{22}	4_{22}
4_{23}	$xyxyxy^2xy^3xy^3xyxy^3$	3840	2^9	2_5	4_{23}	4_{23}
4_{24}	$xyxyxyxy^2xyxy^5xy^3$	3840	2^9	2_5	4_{24}	4_{24}
4_{25}	$xyxyxyxy^2xy^2xyxy^2xy^3xy^2$	3840	2^9	2_5	4_{25}	4_{25}
4_{26}	$xyxyxyxy^2xyxy^2xy^3xy^2xy^3$	3840	2^9	2_7	4_{26}	4_{26}
4_{27}	$xyxyxy^2xyxy^2$	7680	2^8	2_{10}	4_{27}	4_{27}
4_{28}	$xyxyxyxyxyxy^4$	7680	2^8	2_8	4_{28}	4_{28}
4_{29}	$xyxyxyxy^5xy^5$	7680	2^8	2_{10}	4_{29}	4_{29}
4_{30}	$xyxyxy^2xyxy^4xy^3$	7680	2^8	2_6	4_{30}	4_{30}

Conjugacy classes of $D(Fi_{22}) = \langle x, y \rangle$ (continued)

Class	Representative	$|Class|$	$|Centralizer|$	2P	3P	5P
4_{31}	$xyxy^2xy^3xyxy^3xy^2$	7680	2^8	2_8	4_{31}	4_{31}
4_{32}	$xyxy^2xyxy^2xy^4xy^3$	7680	2^8	2_{10}	4_{32}	4_{32}
4_{33}	$xyxyxy^2xy^2xyxy^2xyxy^3$	7680	2^8	2_{10}	4_{33}	4_{33}
4_{34}	$xyxyxyxyxy^2xyxy^5xy^2$	7680	2^8	2_{10}	4_{34}	4_{34}
4_{35}	$xyxyxyxy^2xy^5xy^2xy^3$	7680	2^8	2_9	4_{35}	4_{35}
4_{36}	$xyxyxyxy^5xy^2xyxy^2xy^5$	7680	2^8	2_8	4_{36}	4_{36}
4_{37}	$xyxyxyxy^3xy^2xy^4$	15360	2^7	2_{11}	4_{37}	4_{37}
4_{38}	$xyxyxyxyxy^2xy^2xyxy$	15360	2^7	2_8	4_{38}	4_{38}
4_{39}	$xyxyxyxy^2xyxy^3xy^2xy^3$	15360	2^7	2_{10}	4_{39}	4_{39}
4_{40}	xy^3	30720	2^6	2_{16}	4_{40}	4_{40}
4_{41}	$xyxyxyxyxy^5$	30720	2^6	2_{17}	4_{41}	4_{41}
5	xy^2	98304	$2^2 \cdot 5$	5	5	1
6_1	$(xy)^2$	20480	$2^5 \cdot 3$	3	2_8	6_1
6_2	$(xyxyxyxy^2xy^4)^2$	20480	$2^5 \cdot 3$	3	2_9	6_2
6_3	$xyxyxyxy^2xy^3xy^2$	20480	$2^5 \cdot 3$	3	2_1	6_3
6_4	$xyxyxyxy^3$	40960	$2^4 \cdot 3$	3	2_4	6_4
6_5	$xyxy^2xy^5$	40960	$2^4 \cdot 3$	3	2_7	6_6
6_6	$(xyxy^2xy^5)^5$	40960	$2^4 \cdot 3$	3	2_7	6_5
6_7	$xyxyxyxy^2xy^5$	40960	$2^4 \cdot 3$	3	2_2	6_7
6_8	$xyxy^2xyxy^2xy^4$	40960	$2^4 \cdot 3$	3	2_3	6_8
6_9	$xyxyxyxy^2xyxy^4xy^2$	40960	$2^4 \cdot 3$	3	2_{13}	6_9
6_{10}	y	81920	$2^3 \cdot 3$	3	2_{15}	6_{10}
8_1	$xyxy^3$	15360	2^7	4_4	8_1	8_1
8_2	xy^2xy^3	15360	2^7	4_4	8_2	8_2
8_3	$(xyxyxy^2xy^3)^2$	15360	2^7	4_1	8_3	8_3
8_4	$xyxyxy^3xy^4$	15360	2^7	4_2	8_4	8_4
8_5	$xyxyxyxyxy^2xy^4$	15360	2^7	4_4	8_5	8_5
8_6	$xyxyxyxy^2xy^2xy^4$	15360	2^7	4_4	8_6	8_6
8_7	$xyxyxyxy^2xy^3xy^2xy^3$	15360	2^7	4_2	8_7	8_7
8_8	$xyxyxy^2xyxy^3$	30720	2^6	4_1	8_8	8_8
8_9	$xyxyxy^2xyxy^4$	30720	2^6	4_5	8_9	8_9
8_{10}	$xyxyxyxyxy^2xyxy^4$	30720	2^6	4_7	8_{11}	8_{10}
8_{11}	$(xyxyxyxyxy^2xyxy^4)^3$	30720	2^6	4_7	8_{10}	8_{11}
8_{12}	$xyxy^2$	61440	2^5	4_{17}	8_{12}	8_{12}
8_{13}	$xyxy^4$	61440	2^5	4_{18}	8_{13}	8_{13}
8_{14}	$xyxy^2xy^4$	61440	2^5	4_{19}	8_{14}	8_{14}
8_{15}	$xyxy^3xy^4$	61440	2^5	4_{21}	8_{15}	8_{15}
8_{16}	$xyxyxyxy^3xy^5$	61440	2^5	4_{22}	8_{16}	8_{16}
10_1	$xyxyxy^3xy^5$	98304	$2^2 \cdot 5$	5	10_1	2_1
10_2	$xyxy^2xy^3xyxy^3$	98304	$2^2 \cdot 5$	5	10_3	2_2
10_3	$(xyxy^2xy^3xyxy^3)^3$	98304	$2^2 \cdot 5$	5	10_2	2_2
12_1	xy	81920	$2^3 \cdot 3$	6_1	4_{15}	12_1
12_2	$xyxyxyxy^2xy^4$	81920	$2^3 \cdot 3$	6_2	4_{16}	12_2
12_3	$xyxy^2xy^3xy^2xy^3$	81920	$2^3 \cdot 3$	6_3	4_3	12_3
16_1	$xyxyxy^2xy^3$	61440	2^5	8_3	16_1	16_2
16_2	$(xyxyxy^2xy^3)^5$	61440	2^5	8_3	16_2	16_1

A.7. *Conjugacy classes of* $E_2 = E(\mathrm{Fi}_{22}) = \langle x, y, e \rangle$

Class	Representative	$|Class|$	$|Centralizer|$	2P	3P	5P	7P	11P
1	1	1	$2^{17} \cdot 3^2 \cdot 5 \cdot 7 \cdot 11$	1	1	1	1	1
2_1	$(xye)^7$	22	$2^{16} \cdot 3^2 \cdot 5 \cdot 7$	1	2_1	2_1	2_1	2_1
2_2	$(xy)^6$	231	$2^{17} \cdot 3 \cdot 5$	1	2_2	2_2	2_2	2_2
2_3	$(ye^2)^6$	770	$2^{16} \cdot 3^2$	1	2_3	2_3	2_3	2_3
2_4	x	18480	$2^{13} \cdot 3$	1	2_4	2_4	2_4	2_4
2_5	$(xeyxe^2)^2$	55440	2^{13}	1	2_5	2_5	2_5	2_5
3	$(y)^2$	788480	$2^6 \cdot 3^2$	3	1	3	3	3
4_1	$(xyey)^4$	110880	2^{12}	2_2	4_1	4_1	4_1	4_1
4_2	$(xyxy^3)^2$	110880	2^{12}	2_2	4_2	4_2	4_2	4_2
4_3	$(xy)^3$	147840	$2^{10} \cdot 3$	2_2	4_3	4_3	4_3	4_3
4_4	$(ye^2)^3$	295680	$2^9 \cdot 3$	2_3	4_4	4_4	4_4	4_4
4_5	$xeyey^2e^2$	443520	2^{10}	2_2	4_5	4_5	4_5	4_5
4_6	$(xy^2e^2)^2$	887040	2^9	2_4	4_6	4_6	4_6	4_6
4_7	$(xy^2xe^2)^2$	887040	2^9	2_4	4_7	4_7	4_7	4_7
4_8	$xeye^2$	1774080	2^8	2_4	4_8	4_8	4_8	4_8
4_9	xy^3	3548160	2^7	2_4	4_9	4_9	4_9	4_9
4_{10}	$xeyxe^2$	3548160	2^7	2_5	4_{10}	4_{10}	4_{10}	4_{10}
5	xy^2	22708224	$2^2 \cdot 5$	5	5	1	5	5
6_1	$(ye^2)^2$	788480	$2^6 \cdot 3^2$	3	2_3	6_1	6_1	6_1
6_2	$xyxye$	3153920	$2^4 \cdot 3^2$	3	2_1	6_2	6_2	6_2
6_3	$xyeyey$	3153920	$2^4 \cdot 3^2$	3	2_3	6_3	6_3	6_3
6_4	$(xy)^2$	4730880	$2^5 \cdot 3$	3	2_2	6_4	6_4	6_4
6_5	y	9461760	$2^4 \cdot 3$	3	2_4	6_5	6_5	6_5
7_1	$(xye)^2$	32440320	$2 \cdot 7$	7_1	7_2	7_2	1	7_1
7_2	$(xye)^6$	32440320	$2 \cdot 7$	7_2	7_1	7_1	1	7_2
8_1	$(xyey)^2$	3548160	2^7	4_1	8_1	8_1	8_1	8_1
8_2	$xyxy^3$	3548160	2^7	4_2	8_2	8_2	8_2	8_2
8_3	$xyxy^4e$	3548160	2^7	4_2	8_3	8_3	8_3	8_3
8_4	xe^2ye^2	7096320	2^6	4_1	8_4	8_4	8_4	8_4
8_5	$xyxy^2$	14192640	2^5	4_4	8_5	8_5	8_5	8_5
8_6	xy^2e^2	14192640	2^5	4_6	8_6	8_6	8_6	8_6
8_7	xy^2xe^2	14192640	2^5	4_7	8_7	8_7	8_7	8_7
10_1	y^2e	22708224	$2^2 \cdot 5$	5	10_2	2_1	10_2	10_1
10_2	$(y^2e)^3$	22708224	$2^2 \cdot 5$	5	10_1	2_1	10_1	10_2
10_3	xy^2e	22708224	$2^2 \cdot 5$	5	10_3	2_2	10_3	10_3
11_1	xe	41287680	11	11_2	11_1	11_1	11_2	1
11_2	$(xe)^2$	41287680	11	11_1	11_2	11_2	11_1	1
12_1	ye^2	9461760	$2^4 \cdot 3$	6_1	4_4	12_1	12_1	12_1
12_2	xy	18923520	$2^3 \cdot 3$	6_4	4_3	12_2	12_2	12_2
14_1	xye	32440320	$2 \cdot 7$	7_1	14_2	14_2	2_1	14_1
14_2	$(xye)^3$	32440320	$2 \cdot 7$	7_2	14_1	14_1	2_1	14_2
16_1	$xyey$	14192640	2^5	8_1	16_1	16_2	16_2	16_1
16_2	$(xyey)^5$	14192640	2^5	8_1	16_2	16_1	16_1	16_2

A.8. *Conjugacy classes of* $\mathfrak{G}_2 = \langle h, x, y, e \rangle \cong \mathrm{Fi}_{22}$

Class	Representative	$|Class|$	$|Centralizer|$	2P	3P	5P	7P	11P	13P
1	1	1	$2^{17} \cdot 3^9 \cdot 5^2 \cdot 7 \cdot 11 \cdot 13$	1	1	1	1	1	1
2_1	$(xye)^7$	3510	$2^{16} \cdot 3^6 \cdot 5 \cdot 7 \cdot 11$	1	2_1	2_1	2_1	2_1	2_1
2_2	x	1216215	$2^{17} \cdot 3^4 \cdot 5$	1	2_2	2_2	2_2	2_2	2_2
2_3	$(xh)^6$	36486450	$2^{16} \cdot 3^3$	1	2_3	2_3	2_3	2_3	2_3
3_1	$(yh)^4$	3294720	$2^8 \cdot 3^7 \cdot 5 \cdot 7$	3_1	1	3_1	3_1	3_1	3_1
3_2	h	25625600	$2^7 \cdot 3^9$	3_2	1	3_2	3_2	3_2	3_2
3_3	$(y)^2$	461260800	$2^6 \cdot 3^7$	3_3	1	3_3	3_3	3_3	3_3
3_4	$(xy^2eh)^3$	3690086400	$2^3 \cdot 3^7$	3_4	1	3_4	3_4	3_4	3_4
4_1	$(he)^3$	583783200	$2^{12} \cdot 3^3$	2_2	4_1	4_1	4_1	4_1	4_1
4_2	$(xy)^3$	1401079680	$2^{10} \cdot 3^2 \cdot 5$	2_2	4_2	4_2	4_2	4_2	4_2
4_3	$(xyey)^4$	5254048800	$2^{12} \cdot 3$	2_2	4_3	4_3	4_3	4_3	4_3
4_4	$(xh)^3$	14010796800	$2^9 \cdot 3^2$	2_3	4_4	4_4	4_4	4_4	4_4
4_5	$(yh)^3$	21016195200	$2^{10} \cdot 3$	2_2	4_5	4_5	4_5	4_5	4_5
5	xy^2	107602919424	$2^3 \cdot 3 \cdot 5^2$	5	5	1	5	5	5
6_1	$(xyxeh)^5$	415134720	$2^7 \cdot 3^5 \cdot 5$	3_1	2_1	6_1	6_1	6_1	6_1
6_2	$(xhy^2)^3$	691891200	$2^7 \cdot 3^6$	3_2	2_1	6_2	6_2	6_2	6_2
6_3	$(he)^2$	6227020800	$2^7 \cdot 3^4$	3_2	2_2	6_3	6_3	6_3	6_3
6_4	$(yh)^2$	9340531200	$2^8 \cdot 3^3$	3_1	2_2	6_4	6_4	6_4	6_4
6_5	$xyxye$	16605388800	$2^4 \cdot 3^5$	3_3	2_1	6_5	6_5	6_5	6_5
6_6	$(xh)^2$	18681062400	$2^7 \cdot 3^3$	3_2	2_3	6_6	6_6	6_6	6_6
6_7	$xyeyxh$	18681062400	$2^7 \cdot 3^3$	3_1	2_3	6_7	6_7	6_7	6_7
6_8	$(ye^2)^2$	37362124800	$2^6 \cdot 3^3$	3_3	2_3	6_8	6_8	6_8	6_8
6_9	y	74724249600	$2^5 \cdot 3^3$	3_3	2_2	6_9	6_9	6_9	6_9
6_{10}	$xhey$	149448499200	$2^4 \cdot 3^3$	3_3	2_3	6_{10}	6_{10}	6_{10}	6_{10}
6_{11}	$xheh^2$	298896998400	$2^3 \cdot 3^3$	3_4	2_3	6_{11}	6_{11}	6_{11}	6_{11}
7	$(xye)^2$	1537184563200	$2 \cdot 3 \cdot 7$	7	7	7	1	7	7
8_1	xhy	168129561600	$2^7 \cdot 3$	4_1	8_1	8_1	8_1	8_1	8_1
8_2	$(xyhye)^3$	168129561600	$2^7 \cdot 3$	4_1	8_2	8_2	8_2	8_2	8_2
8_3	$(xyey)^2$	504388684800	2^7	4_3	8_3	8_3	8_3	8_3	8_3
8_4	xyh	2017554739200	2^5	4_4	8_4	8_4	8_4	8_4	8_4

Conjugacy classes of $\mathfrak{G}_2 = \langle h, x, y, e \rangle \cong \mathrm{Fi}_{22}$ *(continued)*

Class	Representative	$\|Class\|$	$\|Centralizer\|$	2P	3P	5P	7P	11P	13P
9_1	$(xy^2h)^2$	199264665600	$2^2 \cdot 3^4$	9_1	3_2	9_1	9_1	9_1	9_1
9_2	$xheye$	398529331200	$2 \cdot 3^4$	9_2	3_2	9_2	9_2	9_2	9_2
9_3	xy^2eh	2391175987200	3^3	9_3	3_4	9_3	9_3	9_3	9_3
10_1	y^2e	1076029194240	$2^2 \cdot 3 \cdot 5$	5	10_1	2_1	10_1	10_1	10_1
10_2	xhe	1614043791360	$2^3 \cdot 5$	5	10_2	2_2	10_2	10_2	10_2
11_1	xe	2934625075200	$2 \cdot 11$	11_2	11_1	11_1	11_2	1	11_2
11_2	$(xe)^2$	2934625075200	$2 \cdot 11$	11_1	11_2	11_2	11_1	1	11_1
12_1	he	74724249600	$2^5 \cdot 3^3$	6_3	4_1	12_1	12_1	12_1	12_1
12_2	$(xy^3h)^2$	112086374400	$2^6 \cdot 3^2$	6_4	4_1	12_2	12_2	12_2	12_2
12_3	$xyhxhxh^2$	112086374400	$2^6 \cdot 3^2$	6_4	4_1	12_3	12_3	12_3	12_3
12_4	$xexeh$	224172748800	$2^5 \cdot 3^2$	6_4	4_2	12_4	12_4	12_4	12_4
12_5	xh	448345497600	$2^4 \cdot 3^2$	6_6	4_4	12_5	12_5	12_5	12_5
12_6	ye^2	448345497600	$2^4 \cdot 3^2$	6_8	4_4	12_6	12_6	12_6	12_6
12_7	xy^2xh	448345497600	$2^4 \cdot 3^2$	6_6	4_4	12_7	12_7	12_7	12_7
12_8	yh	672518246400	$2^5 \cdot 3$	6_4	4_5	12_8	12_8	12_8	12_8
12_9	xhe^2	672518246400	$2^5 \cdot 3$	6_3	4_3	12_9	12_9	12_9	12_9
12_{10}	xy	896690995200	$2^3 \cdot 3^2$	6_9	4_2	12_{10}	12_{10}	12_{10}	12_{10}
12_{11}	$xehye$	1793381990400	$2^2 \cdot 3^2$	6_{11}	4_4	12_{11}	12_{11}	12_{11}	12_{11}
13_1	yhe	4966288588800	13	13_2	13_1	13_2	13_2	13_2	1
13_2	$(yhe)^2$	4966288588800	13	13_1	13_2	13_1	13_1	13_1	1
14	xye	4611553689600	$2 \cdot 7$	7	14	14	2_1	14	14
15	$xeyh$	2152058388480	$2 \cdot 3 \cdot 5$	15	5	3_1	15	15	15
16_1	$xyey$	2017554739200	2^5	8_3	16_1	16_2	16_2	16_1	16_2
16_2	$(xyey)^5$	2017554739200	2^5	8_3	16_2	16_1	16_1	16_2	16_1
18_1	xhy^2	597793996800	$2^2 \cdot 3^3$	9_1	6_2	18_2	18_1	18_2	18_1
18_2	$(xhy^2)^5$	597793996800	$2^2 \cdot 3^3$	9_1	6_2	18_1	18_2	18_1	18_2
18_3	$xyheyh$	1195587993600	$2 \cdot 3^3$	9_2	6_2	18_3	18_3	18_3	18_3
18_4	xy^2h	1793381990400	$2^2 \cdot 3^2$	9_1	6_3	18_4	18_4	18_4	18_4
20	$xyxh$	3228087582720	$2^2 \cdot 5$	10_2	20	4_2	20	20	20
21	$xyhe$	3074369126400	$3 \cdot 7$	21	7	21	3_1	21	21
22_1	yeh	2934625075200	$2 \cdot 11$	11_2	22_1	22_1	22_2	2_1	22_2
22_2	$(yeh)^7$	2934625075200	$2 \cdot 11$	11_1	22_2	22_2	22_1	2_1	22_1
24_1	xy^3h	1345036492800	$2^4 \cdot 3$	12_2	8_1	24_1	24_1	24_1	24_1
24_2	$xyhye$	1345036492800	$2^4 \cdot 3$	12_2	8_2	24_2	24_2	24_2	24_2
30	$xyxeh$	2152058388480	$2 \cdot 3 \cdot 5$	15	10_1	6_1	30	30	30

Appendix B. Character Tables of Local Subgroups of Co$_2$ and Fi$_{22}$

B.1. *Character table of* $H(\mathrm{Co}_2) = \langle x, y, h \rangle$

	1a	2a	2b	2c	2d	2e	2f	2g	2h	2i	2j	3a	3b	3c	4a	4b	4c	4d	4e	4f	4g
2	18	18	17	14	14	16	16	14	14	15	11	6	5	7	14	14	13	12	12	12	12
3	4	4	1	2	2	2	2	1	1	.	1	4	3	3	3	1	1	1	1	1	1
5	1	1	1	1	1	.	.	.	.	.	.	.	.	.	.	.	.	.	.	.	.
7	1	1	1	.	.	.	.	.	.	.	.	1	.	.	.	.	.	.	.	.	.
2P	1a	1a	1a	1a	1a	1a	1a	1a	1a	1a	1a	3a	3b	3c	2a	2a	2b	2f	2b	2f	2b
3P	1a	2a	2b	2c	2d	2e	2f	2g	2h	2i	2j	1a	1a	1a	4a	4b	4c	4d	4e	4f	4g
5P	1a	2a	2b	2c	2d	2e	2f	2g	2h	2i	2j	3a	3b	3c	4a	4b	4c	4d	4e	4f	4g
7P	1a	2a	2b	2c	2d	2e	2f	2g	2h	2i	2j	3a	3b	3c	4a	4b	4c	4d	4e	4f	4g
X.1	1	1	1	1	1	1	1	1	1	1	1	1	1	1	1	1	1	1	1	1	1
X.2	7	7	7	-5	-5	-1	-1	3	3	-1	-1	-2	4	1	7	-1	-5	3	-1	3	-1
X.3	15	15	15	-5	-5	7	7	3	3	7	-1	-3	.	3	15	7	-5	-1	-1	-1	-1
X.4	16	-16	.	4	-4	8	-8	-4	-4	.	.	-2	1	4	.	.	.	4	-4	-4	4
X.5	21	21	21	-11	-11	5	5	5	5	5	-3	3	6	.	21	5	-11	1	-3	1	-3
X.6	21	21	21	9	9	-3	-3	1	1	-3	-3	3	6	.	21	-3	9	5	-3	5	-3
X.7	27	27	27	15	15	3	3	7	7	3	3	.	9	.	27	3	15	3	3	3	3
X.8	35	35	35	-5	-5	3	3	3	-5	-5	3	-1	5	2	35	3	-5	7	3	7	3
X.9	35	35	35	15	15	11	11	7	7	11	3	-1	5	2	35	11	15	-1	3	-1	3
X.10	56	56	56	-24	-24	-8	-8	8	8	-8	.	2	11	2	56	-8	-24	.	.	.	.
X.11	70	70	70	-10	-10	-10	-10	6	6	-10	-2	7	-5	1	70	-10	-10	2	-2	2	-2
X.12	84	84	84	4	4	20	20	4	4	20	4	3	-6	3	84	20	4	4	4	4	4
X.13	105	105	105	25	25	-7	-7	9	9	-7	1	6	.	3	105	-7	25	-3	1	-3	1
X.14	105	105	105	5	5	17	17	-3	-3	17	-7	6	.	3	105	17	5	-3	-7	-3	-7
X.15	105	105	105	-35	-35	1	1	5	5	1	1	-3	15	-3	105	1	-35	5	1	5	1
X.16	112	-112	.	.	-20	20	-8	8	12	-12	.	.	4	4	.	.	.	12	4	-12	-4
X.17	120	120	-8	.	.	24	24	.	.	-8	-8	3	.	6	8	8	.	12	8	12	8
X.18	120	120	120	40	40	-8	-8	8	8	-8	.	-6	15	.	120	-8	40	.	.	.	.
X.19	120	120	-8	.	.	24	24	.	.	-8	8	3	.	6	8	8	.	-4	-8	-4	-8
X.20	135	135	7	15	15	39	39	15	15	7	7	.	.	9	-9	-9	-1	3	7	3	7
X.21	168	168	168	40	40	8	8	8	8	8	8	6	6	-3	168	8	40	.	8	.	8
X.22	189	189	189	21	21	-3	-3	-11	-11	-3	-3	.	9	.	189	-3	21	9	-3	9	-3
X.23	189	189	189	-39	-39	21	21	1	1	21	-3	.	9	.	189	21	-39	-3	-3	-3	-3
X.24	189	189	189	-51	-51	-3	-3	13	13	-3	-3	.	9	.	189	-3	-51	-3	-3	-3	-3
X.25	210	210	210	10	10	-14	-14	10	10	-14	2	-6	-15	3	210	-14	10	6	2	6	2
X.26	210	210	210	50	50	2	2	2	2	2	-6	3	15	.	210	2	50	-2	-6	-2	-6
X.27	216	216	216	-24	-24	24	24	8	8	24	.	.	-9	.	216	24	-24	.	.	.	.
X.28	240	-240	.	.	-20	20	56	-56	12	-12	.	.	6	.	12	.	.	-4	4	4	-4
X.29	280	280	280	40	40	24	24	8	8	24	.	-8	-5	-2	280	24	40	.	.	.	.
X.30	280	280	280	-40	-40	-8	-8	-8	-8	-8	8	10	10	1	280	-8	-40	.	8	.	8
X.31	315	315	315	-45	-45	-21	-21	3	3	-21	3	-9	.	.	315	-21	-45	-5	3	-5	3
X.32	336	-336	.	.	36	-36	-24	24	-4	-4	.	.	-6	6	.	.	.	20	-12	-20	-12
X.33	336	-336	.	.	-44	44	40	-40	20	-20	.	.	-6	6	.	.	.	4	12	-4	-12
X.34	336	336	336	-16	-16	16	16	-16	-16	16	16	.	-6	6	336	16	-16	.	.	.	.
X.35	378	378	378	-30	-30	-6	-6	2	2	-6	-6	.	.	-9	378	-6	-30	6	-6	6	-6
X.36	405	405	405	45	45	-27	-27	-3	-3	-27	-3	.	.	.	405	-27	45	-3	-3	-3	-3
X.37	405	405	21	45	45	-27	-27	-3	-3	5	-3	.	.	.	-27	21	-3	-3	-3	-3	-3
X.38	405	405	21	45	45	-27	-27	-3	-3	5	-3	.	.	.	-27	21	-3	-3	-3	-3	-3
X.39	420	420	420	20	20	4	4	-12	-12	4	4	-3	.	3	420	4	20	-4	4	-4	4
X.40	432	-432	.	.	60	-60	24	-24	28	-28	.	.	.	9	.	.	.	12	-12	-12	12
X.41	512	512	512	.	.	.	.	.	.	.	.	8	-16	-4	512	.	.	.	.	.	.
X.42	560	-560	.	.	60	-60	88	-88	28	-28	.	.	2	5	8	.	.	-4	-12	4	12
X.43	560	-560	.	.	-20	20	24	-24	-20	20	.	.	2	5	8	.	.	28	-12	-28	12
X.44	720	720	-48	.	.	-48	-48	.	.	16	.	-9	.	.	48	-16	.	-8	.	-8	.
X.45	720	720	-48	.	.	-48	-48	.	.	16	.	-9	.	.	48	-16	.	-8	.	-8	.
X.46	810	810	42	90	90	90	90	42	42	26	18	.	.	.	-54	-6	-6	6	18	6	18
X.47	840	840	-56	.	.	-24	-24	.	.	8	8	-6	.	6	56	-8	.	36	-8	36	-8
X.48	840	840	-56	.	.	-24	-24	.	.	8	8	-6	.	6	56	-8	.	-12	8	-12	8
X.49	896	-896	.	.	-96	96	-64	64	32	-32	.	.	-4	11	8	.	.	-12	28	-12	28
X.50	945	945	49	-75	-75	105	105	-3	-3	9	-7	.	.	9	-63	-39	5	-3	-7	-3	-7
X.51	945	945	49	-75	-75	-39	-39	45	45	-7	-7	.	.	9	-63	9	5	9	-7	9	-7
X.52	945	945	49	-15	-15	129	129	33	33	-7	.	.	.	9	-63	-15	1	-3	-7	-3	-7
X.53	945	945	49	-15	-15	-15	-15	-15	-15	17	17	.	.	9	-63	33	1	9	17	9	17
X.54	945	945	49	105	105	-15	-15	9	9	17	1	.	.	9	-63	33	-7	-3	1	-3	1
X.55	1080	1080	56	120	120	24	24	24	24	24	8	.	.	-9	-72	24	-8	.	8	.	8
X.56	1120	-1120	.	.	-40	40	-80	80	24	-24	.	.	-14	-5	4	.	.	8	8	-8	-8
X.57	1344	-1344	.	.	16	-16	160	-160	16	-16	.	.	-6	-6	12	.	.	16	-16	-16	16
X.58	1680	1680	-112	.	.	.	.	144	144	.	-48	.	.	-12	12	112	48	.	-8	.	-8
X.59	1680	-1680	.	.	100	-100	-56	56	36	-36	.	.	-12	.	12	.	.	-12	-4	12	4
X.60	1680	1680	-112	.	.	-48	-48	.	.	16	16	15	.	-6	112	-16	.	24	-16	24	-16
X.61	1680	1680	-112	.	.	-48	-48	.	.	16	-16	15	.	-6	112	-16	.	-8	16	-8	16
X.62	1680	-1680	.	.	-140	140	8	-8	20	-20	.	.	6	15	-12	.	.	20	-4	-20	4
X.63	1680	-1680	.	.	20	-20	136	-136	-12	12	.	.	-12	.	12	.	.	-12	28	12	-28
X.64	1890	1890	98	-30	-30	114	114	18	18	50	10	.	.	-9	-126	18	2	6	10	6	10
X.65	1890	1890	98	-150	-150	66	66	42	42	2	-14	.	.	-9	-126	-30	10	6	-14	6	-14
X.66	1920	-1920	.	.	160	-160	-64	64	32	-32	.	.	12	15	.	.	.	.	.	.	.
X.67	2520	2520	-168	.	.	120	120	.	.	-40	-24	9	.	.	168	40	.	-4	24	-4	24
X.68	2520	2520	-168	.	.	120	120	.	.	-40	24	9	.	.	168	40	.	12	-24	12	-24
X.69	2688	-2688	.	.	160	-160	64	-64	32	-32	.	.	-12	6	-12	.	.	-32	.	.	32
X.70	2835	2835	147	135	135	171	171	15	15	11	3	.	.	.	-189	-69	-9	-9	3	-9	3
X.71	2835	2835	147	135	135	27	27	63	63	-5	3	.	.	.	-189	-21	-9	3	3	3	3
X.72	2835	2835	147	-45	-45	-45	-45	-45	-45	51	3	.	.	.	-189	99	3	3	3	3	3
X.73	2835	2835	147	-225	-225	27	27	-9	-9	-5	3	.	.	.	-189	-21	15	-9	3	-9	3
X.74	2835	2835	147	135	135	27	27	-33	-33	-5	-21	.	.	.	-189	-21	-9	3	-21	3	-21
X.75	2835	2835	147	-225	-225	-117	-117	39	39	-21	3	.	.	.	-189	27	15	3	3	3	3
X.76	2835	2835	147	135	135	-117	-117	15	15	-21	-21	.	.	.	-189	27	-9	15	-21	15	-21
X.77	2835	2835	147	-45	-45	99	99	3	3	67	-21	.	.	.	-189	51	3	-9	-21	-9	-21
X.78	3024	-3024	.	.	-204	204	-24	24	52	-52	.	.	.	9	.	.	.	-12	12	12	-12
X.79	3024	-3024	.	.	-156	156	168	-168	4	-4	.	.	.	9	.	.	.	-12	12	12	-12
X.80	3024	-3024	.	.	84	-84	-24	24	-44	44	.	.	.	9	.	.	.	36	12	-36	-12
X.81	3240	3240	-216	.	.	72	72	.	.	-24	-24	.	.	.	216	24	.	36	24	36	24
X.82	3240	3240	-216	.	.	72	72	.	.	-24	24	.	.	.	216	24	.	-12	-24	-12	-24

HYUN KYU KIM AND GERHARD O. MICHLER

Character table of $H(Co_2)$ (continued)

	4h	4i	4j	4k	4l	4m	4n	4o	4p	4q	4r	4s	4t	4u	4v	4w	4x	5a	6a	6b	6c	6d	6e	6f	6g
2	13	11	11	12	12	10	11	11	9	9	9	9	10	10	8	8	8	2	6	5	7	6	6	5	5
3	.	1	1	.	.	1	.	.	1	1	1	1	.	.	.	.	.	1	4	3	3	2	2	2	2
5	.	.	.	.	.	.	.	.	.	.	.	.	.	.	.	.	.	1	1	1	.	.	.	.	.
7	.	.	.	.	.	.	.	.	.	.	.	.	.	.	.	.	.	.	.	.	.	.	.	.	.
2P	2b	2e	2b	2e	2e	2e	2b	2f	2g	2g	2g	2g	2b	2e	2i	2g	2g	5a	3a	3b	3c	3a	3a	3b	3b
3P	4h	4i	4j	4k	4l	4m	4n	4o	4p	4q	4r	4s	4t	4u	4v	4w	4x	5a	2a	2a	2a	2e	2f	2c	2d
5P	4h	4i	4j	4k	4l	4m	4n	4o	4p	4q	4r	4s	4t	4u	4v	4w	4x	1a	6a	6b	6c	6d	6e	6f	6g
7P	4h	4i	4j	4k	4l	4m	4n	4o	4p	4q	4r	4s	4t	4u	4v	4w	4x	5a	6a	6b	6c	6d	6e	6f	6g
X.1	1	1	1	1	1	1	1	1	1	1	1	1	1	1	1	1	1	1	1	1	1	1	1	1	1
X.2	3	3	-1	-1	-1	3	3	-1	1	-3	1	-3	-1	-1	-1	1	1	2	-2	4	1	2	2	-2	-2
X.3	3	-1	7	3	3	-1	3	3	-3	1	-3	1	-1	3	3	1	1	.	-3	.	3	1	1	-2	-2
X.4	.	.	.	4	-4	.	.	.	-2	-2	2	2	.	.	.	2	-2	1	2	-1	-4	2	-2	1	-1
X.5	5	1	5	1	1	1	1	1	-3	-3	-3	-3	-3	1	1	1	1	1	3	6	.	-1	-1	-2	-2
X.6	1	5	-3	1	1	5	1	1	-1	3	-1	3	-3	1	1	-1	-1	1	3	6	.	3	3	.	.
X.7	7	3	-1	-1	-1	3	7	-1	1	1	1	5	3	-1	-1	1	1	2	.	9	.	.	.	3	3
X.8	-5	7	3	-1	-1	7	-5	-1	-1	-1	-1	-1	3	-1	-1	-1	-1	.	-1	5	2	3	3	1	1
X.9	7	-1	11	3	3	-1	7	3	5	1	5	1	3	3	3	1	1	.	-1	5	2	-1	-1	3	3
X.10	8	.	-8	.	.	.	8	.	4	-4	4	-4	.	.	.	.	.	1	2	11	2	-2	-2	-3	-3
X.11	6	2	-10	2	2	2	6	2	2	2	2	2	-2	2	2	-2	-2	.	7	-5	1	-1	-1	-1	-1
X.12	4	4	20	4	4	4	4	4	.	.	.	.	4	4	4	.	.	-1	3	-6	3	-1	-1	-2	-2
X.13	9	-3	-7	-3	-3	-3	9	-3	-3	-3	-3	-3	1	-3	-3	1	1	.	6	.	3	2	2	4	4
X.14	-3	-3	17	1	1	-3	-3	1	3	1	3	-1	-7	1	1	-1	-1	.	6	.	3	2	2	2	2
X.15	5	5	1	1	1	5	5	1	-1	-5	-1	-5	1	1	1	-1	-1	.	-3	15	-3	1	1	1	1
X.16	.	.	.	-4	4	.	.	.	-2	6	2	-6	.	.	.	2	-2	2	-4	-4	-4	-4	-2	2	
X.17	.	-4	.	4	4	-4	.	4	.	.	.	.	.	.	.	-4	.	.	3	.	6	3	3	.	.
X.18	8	.	-8	.	.	.	8	.	-4	4	-4	4	.	.	.	.	.	.	-6	15	.	-2	-2	1	1
X.19	.	.	12	4	4	-4	.	4	.	.	.	.	.	.	-4	.	.	.	3	.	6	3	3	.	.
X.20	-1	3	-1	11	11	3	-1	-5	3	3	3	3	-1	3	-1	3	3	.	9	.	.	.	.	.	.
X.21	8	.	8	.	.	.	8	.	.	.	.	8	.	.	.	.	.	-2	6	6	-3	2	2	-2	-2
X.22	-11	9	-3	1	1	9	-11	1	1	1	1	-3	1	1	1	1	-1	.	9	.	.	.	.	-3	-3
X.23	1	-3	21	1	1	-3	1	1	-5	-1	-5	-1	3	1	1	-1	-1	-1	.	9	.	.	.	3	3
X.24	13	-3	-3	-3	-3	-3	13	-3	1	1	1	1	-3	-3	-3	1	1	-1	.	9	.	.	.	-3	-3
X.25	10	6	-14	-2	-2	6	10	-2	-2	-2	-2	-2	2	-2	-2	-2	-2	.	-6	-15	3	-2	-2	1	1
X.26	2	-2	2	-2	-2	-2	2	-2	2	2	2	2	-6	-2	-2	-2	-2	.	3	15	-1	-1	-1	-1	-1
X.27	8	.	24	.	.	.	8	.	-4	-4	-4	4	.	.	.	.	.	1	-9	.	.	.	.	-3	-3
X.28	.	.	.	12	-12	.	.	.	6	-2	-6	2	.	.	.	.	.	2	-2	-6	.	-12	2	-2	-2	2
X.29	8	.	24	.	.	.	8	.	4	-4	4	-4	.	.	.	.	.	.	-8	-5	-2	.	.	1	1
X.30	-8	.	-8	.	.	.	-8	.	.	.	.	.	8	.	.	.	.	.	10	10	1	-2	-2	2	2
X.31	3	-5	-21	3	3	-5	3	3	3	3	3	3	3	3	3	-1	-1	.	-9	.	.	3	3	.	.
X.32	.	.	.	4	-4	.	.	.	2	-6	-2	6	.	.	.	-2	2	1	6	-6	.	6	-6	.	.
X.33	.	.	.	4	-4	.	.	.	6	6	-6	-6	.	.	.	2	-2	1	6	-6	.	-2	2	-2	2
X.34	-16	.	16	.	.	.	-16	.	.	.	.	.	.	.	.	.	.	1	-6	6	.	-2	-2	2	2
X.35	2	6	-6	-2	-2	6	2	-2	2	2	2	2	-6	-2	-2	2	2	-2	.	-9	.	.	.	3	3
X.36	-3	-3	-27	5	5	-3	-3	5	-3	-3	-3	-3	-3	5	5	1	1	.	.	.	.	.	.	.	.
X.37	13	-3	-3	5	5	-3	-3	-11	-3	-3	-3	-3	5	-3	1	1	1	.	.	.	.	.	.	.	.
X.38	-11	-3	-3	5	5	-3	-3	-11	-3	-3	-3	-3	5	-3	1	1	1	.	.	.	.	.	.	.	.
X.39	-12	-4	4	-4	-4	-4	-12	-4	.	.	.	.	4	-4	-4	.	.	-3	.	-9	.	3	1	1	-4	-4
X.40	.	.	.	-4	4	.	.	.	-2	-10	2	10	.	.	.	2	-2	2	2	8	-16	-4	.	.	3	-3
X.41	.	.	.	.	.	.	.	.	.	.	.	.	.	.	.	.	.	2	8	-16	-4	.	.	.	.
X.42	.	.	.	12	-12	.	.	.	-10	-2	10	2	.	.	.	2	2	.	-2	-5	-8	-2	2	3	-1
X.43	.	.	.	-4	4	.	.	.	2	2	-2	-2	.	.	.	-2	2	.	-2	-5	-8	6	-6	1	-1
X.44	.	-8	.	8	8	8	.	8	.	.	.	.	.	.	.	-8	.	.	-9	.	.	3	3	.	.
X.45	.	-8	.	8	8	8	.	8	.	.	.	.	.	.	.	-8	.	.	-9	.	.	3	3	.	.
X.46	10	6	-6	6	6	-6	-6	6	6	6	6	6	2	6	-2	2	2	.	.	.	.	.	.	.	.
X.47	.	-12	.	-4	-4	-12	.	-4	.	.	.	.	.	.	.	4	.	.	-6	.	.	6	6	6	
X.48	.	36	.	-4	-4	-12	.	-4	.	.	.	.	.	.	.	4	.	.	-6	.	.	6	6	6	
X.49	.	.	.	.	.	.	.	.	-8	8	8	-8	.	.	.	.	.	1	4	-11	-8	-4	4	-3	3
X.50	13	-3	1	9	9	-3	-3	-7	-9	3	-9	3	1	1	-3	-1	-1	.	9	.	.	.	.	.	.
X.51	-3	9	1	-11	-11	9	-3	5	3	-9	3	-9	1	3	3	.	.	.	9	.	.	.	.	.	.
X.52	-15	-3	-7	13	13	-3	1	-3	-3	-3	-3	-3	1	5	1	1	1	.	9	.	.	.	.	.	.
X.53	1	9	-7	-7	-7	9	1	9	-3	-3	-3	-3	-7	1	-3	-3	.	.	9	.	.	.	.	.	.
X.54	25	-3	-7	-11	-11	-3	-7	5	-3	-3	-3	-3	9	-3	1	-3	-3	.	9	.	.	.	.	.	.
X.55	24	.	-8	.	.	.	.	.	.	.	.	.	8	.	.	.	.	.	-9	.	.	.	.	.	.
X.56	.	.	.	8	-8	.	.	.	-4	-4	4	4	.	.	.	-4	4	.	14	5	-4	-2	2	-1	1
X.57	.	.	.	16	-16	.	.	.	.	.	.	.	.	.	.	.	.	-1	6	-12	-2	2	-2	2	
X.58	.	-8	.	8	8	8	.	8	.	.	.	.	.	.	.	-8	.	.	-12	.	12	.	.	.	.
X.59	.	.	.	-12	12	.	.	.	6	6	-6	-6	.	.	.	2	-2	.	12	.	-12	4	-4	4	-4
X.60	.	-8	.	8	8	-8	.	8	.	.	.	.	.	.	.	-8	.	.	15	.	-6	3	3	.	.
X.61	.	24	.	8	8	-8	.	8	.	.	.	.	.	.	.	-8	.	.	15	.	-6	3	3	.	.
X.62	.	.	.	4	-4	.	.	.	2	10	-2	-10	.	.	.	-2	2	.	-6	-15	12	2	-2	1	-1
X.63	.	.	.	4	-4	.	.	.	-6	2	6	-2	.	.	.	-2	2	.	12	.	-12	4	-4	2	-2
X.64	-14	6	-14	6	6	6	2	6	-6	-6	-6	-6	6	-2	-2	-2	-2	.	-9	.	.	.	.	.	.
X.65	10	6	2	-2	-2	6	-6	-2	-6	-6	-6	-6	2	-2	-2	2	2	.	-9	.	.	.	.	.	.
X.66	.	.	.	.	.	.	.	.	8	-8	-8	8	.	.	.	.	.	.	-12	-15	.	-4	4	1	-1
X.67	.	12	.	4	4	-4	.	4	.	.	.	.	.	.	.	-4	.	.	9	.	.	-3	-3	.	.
X.68	.	-4	.	4	4	-4	.	4	.	.	.	.	.	.	.	-4	.	.	9	.	.	-3	-3	.	.
X.69	.	.	.	.	.	.	.	.	.	.	.	.	.	.	.	.	.	-2	12	-6	12	4	-4	-2	2
X.70	-1	-9	3	3	3	-9	-1	-13	9	-3	9	-3	-5	-5	-1	-3	-3	.	.	.	.	.	.	.	.
X.71	-17	3	3	-17	-17	3	-1	-1	-3	9	-3	9	-5	-9	3	1	1	.	.	.	.	.	.	.	.
X.72	3	3	-21	3	3	3	-13	-13	3	3	3	3	-5	-5	-1	3	3	.	.	.	.	.	.	.	.
X.73	39	-9	3	-5	-5	-9	-9	11	-3	9	-3	9	-5	3	-1	1	1	.	.	.	.	.	.	.	.
X.74	15	3	3	-1	-1	3	-1	15	9	-3	9	-3	3	7	-5	1	1	.	.	.	.	.	.	.	.
X.75	23	3	3	7	7	3	-9	-9	9	-3	9	-3	-5	-1	3	-3	-3	.	.	.	.	.	.	.	.
X.76	-1	15	3	11	11	15	-1	-5	-3	9	-3	9	3	3	-1	-3	-3	.	.	.	.	.	.	.	.
X.77	-13	-9	-21	-9	-9	9	3	7	3	3	3	3	3	-1	3	-1	-1	.	.	.	.	.	.	.	.
X.78	.	.	.	-12	12	.	.	.	-2	-2	2	2	.	.	.	2	-2	-1	.	-9	.	.	.	-3	3
X.79	.	.	.	4	-4	.	.	.	10	2	-10	-2	.	.	.	-2	2	-1	.	-9	.	.	.	3	-3
X.80	.	.	.	4	-4	.	.	.	-2	-2	2	2	.	.	.	2	-2	-1	.	-9	.	.	.	-3	3
X.81	.	-12	.	-4	-4	-12	.	-4	.	.	.	.	.	.	.	4	.	.	.	.	.	.	.	.	.
X.82	.	36	.	-4	-4	-12	.	-4	.	.	.	.	.	.	.	4	.	.	.	.	.	.	.	.	.

Character table of $H(Co_2)$ (continued)

2	5	5	5	5	5	5	6	5	5	4	3	8	9	9	8	8	7	7	7	7	7	6	6	6	1	2	2	2	5
3	2	2	2	2	2	2	1	1	1	1	.	1	.	.	.	.	.	.	.	.	.	.	.	.	2	1	.	.	3
5	.	.	.	.	.	.	.	.	.	.	.	.	.	.	.	.	.	.	.	.	.	.	.	.	.	1	1	1	.
7	.	.	.	.	.	.	.	.	.	.	1	.	.	.	.	.	.	.	.	.	.	.	.	.	.	.	.	.	.
	6h	6i	6j	6k	6l	6m	6n	6o	6p	6q	7a	8a	8b	8c	8d	8e	8f	8g	8h	8i	8j	8k	8l	8m	9a	10a	10b	10c	12a
2P	3c	3c	3c	3b	3b	3c	3c	3b	3b	3c	7a	4a	4b	4b	4h	4h	4i	4o	4h	4i	4o	4k	4n	4d	9a	5a	5a	5a	6a
3P	2e	2d	2f	2e	2f	2c	2b	2g	2h	2j	7a	8a	8b	8c	8d	8e	8f	8g	8h	8i	8j	8k	8l	8m	3a	10a	10b	10c	4a
5P	6h	6i	6j	6k	6l	6m	6n	6o	6p	6q	7a	8a	8b	8c	8d	8e	8f	8g	8h	8i	8j	8k	8l	8m	9a	2a	2d	2c	12a
7P	6h	6i	6j	6k	6l	6m	6n	6o	6p	6q	1a	8a	8b	8c	8d	8e	8f	8g	8h	8i	8j	8k	8l	8m	9a	10a	10b	10c	12a
X.1	1	1	1	1	1	1	1	1	1	1	1	1	1	1	1	1	1	1	1	1	1	1	1	1	1	1	1	1	1
X.2	−1	1	−1	2	2	1	1	.	.	−1	.	−1	−1	3	−3	1	−1	1	1	−1	1	1	1	−1	1	2	.	.	−2
X.3	1	1	1	−2	−2	1	3	.	.	−1	1	−1	3	−1	1	−3	1	−1	1	1	−1	−1	1	1	.	.	.	.	−3
X.4	2	2	−2	−1	1	−2	.	1	−1	.	2	.	.	.	.	.	2	−2	.	−2	2	.	.	.	1	−1	1	−1	.
X.5	2	−2	2	2	2	−2	.	2	2	.	.	−3	1	1	−3	−3	−1	−1	1	−1	−1	−1	1	−1	.	1	−1	−1	3
X.6	.	.	.	.	.	.	.	−2	−2	.	.	−3	1	5	3	−1	−1	1	−1	−1	1	1	−1	−1	.	1	−1	−1	3
X.7	.	.	.	3	3	.	.	1	1	.	−1	3	−1	3	5	1	1	−1	1	1	−1	−1	1	1	.	2	.	.	.
X.8	.	−2	.	−3	−3	−2	2	1	1	.	.	3	−1	7	−1	−1	1	1	−1	1	1	1	−1	1	−1	.	.	.	−1
X.9	2	.	2	−1	−1	.	2	1	1	.	.	3	3	−1	1	5	−1	1	1	−1	1	1	1	−1	−1	.	.	.	−1
X.10	−2	.	−2	1	1	.	2	−1	−1	.	.	.	.	.	−4	4	.	.	.	.	.	.	.	.	−1	1	1	1	2
X.11	−1	−1	−1	−1	−1	−1	1	3	3	1	.	−2	2	2	2	2	.	.	−2	.	.	.	−2	.	1	.	.	.	7
X.12	−1	1	−1	2	2	1	3	−2	−2	1	.	4	4	4	.	.	.	.	.	.	.	.	.	.	.	−1	−1	−1	3
X.13	−1	1	−1	−4	−4	1	3	.	.	1	.	1	−3	−3	−3	−3	−1	−1	1	−1	−1	−1	1	−1	.	.	.	.	6
X.14	−1	−1	−1	2	2	−1	3	.	.	−1	.	−7	1	−3	−1	3	1	−1	−1	1	−1	−1	−1	1	.	.	.	.	6
X.15	1	1	1	1	1	1	−3	−1	−1	1	.	1	1	5	−5	−1	1	−1	−1	1	−1	−1	−1	1	.	.	.	.	−3
X.16	−2	2	2	−2	2	−2	.	.	.	.	.	.	.	.	.	.	−2	−2	.	2	2	.	.	.	1	−2	.	.	.
X.17	.	.	.	.	.	.	−2	.	.	−2	1	.	.	.	.	.	2	2	.	2	2	−2	.	−2	.	.	.	.	−1
X.18	−2	−2	−2	1	1	−2	.	−1	−1	.	1	.	.	.	4	−4	.	.	.	.	.	.	.	.	.	.	.	.	−6
X.19	.	.	.	.	.	.	−2	.	.	2	1	.	.	.	.	.	2	−2	.	2	−2	2	.	−2	.	.	.	.	−1
X.20	3	3	3	.	.	3	1	.	.	1	2	−1	−1	−1	−1	−1	1	1	−1	1	1	1	−1	1	.	.	.	.	.
X.21	−1	1	−1	2	2	1	−3	2	2	−1	.	8	.	.	.	.	.	.	.	.	.	.	.	.	.	−2	.	.	6
X.22	.	.	.	−3	−3	.	.	1	1	.	.	−3	1	9	1	1	−1	−1	1	−1	−1	−1	1	−1	.	−1	1	1	.
X.23	.	.	.	3	3	.	.	1	1	.	.	−3	1	−3	−1	−5	−1	1	−1	−1	1	1	−1	−1	.	−1	1	1	.
X.24	.	.	.	−3	−3	.	.	1	1	.	.	−3	−3	−3	1	1	1	1	1	1	1	1	1	1	.	−1	−1	−1	.
X.25	1	1	1	1	1	1	3	1	1	−1	.	2	−2	6	−2	−2	.	.	−2	.	.	.	−2	.	.	.	.	.	−6
X.26	2	2	2	−1	−1	2	.	−1	−1	.	.	−6	−2	−2	2	2	.	.	−2	.	.	.	−2	.	.	.	.	.	3
X.27	.	.	.	−3	−3	.	.	−1	−1	.	−1	.	.	.	4	−4	.	.	.	.	.	.	.	.	.	1	1	1	.
X.28	2	2	−2	2	−2	−2	.	.	.	.	2	.	.	.	.	.	2	2	.	−2	−2	.	.	.	.	.	.	.	.
X.29	.	−2	.	−3	−3	−2	−2	−1	−1	.	.	.	.	.	−4	4	.	.	.	.	.	.	.	.	1	.	.	.	−8
X.30	1	−1	1	−2	−2	−1	1	−2	−2	−1	.	8	.	.	.	.	.	.	.	.	.	.	.	.	1	.	.	.	10
X.31	.	.	.	.	.	.	.	.	.	.	.	3	3	−5	3	3	−1	−1	−1	−1	−1	−1	−1	−1	.	.	.	.	−9
X.32	.	.	.	.	.	.	.	−2	2	.	.	.	.	.	.	.	−2	−2	.	2	2	.	.	.	.	−1	−1	1	.
X.33	4	−4	−4	−2	2	4	.	2	−2	.	.	.	.	.	.	.	−2	2	.	2	−2	.	.	.	.	−1	−1	1	.
X.34	−2	2	−2	−2	−2	2	.	2	2	.	.	.	.	.	.	.	.	.	.	.	.	.	.	.	.	1	−1	−1	−6
X.35	.	.	.	3	3	.	.	−1	−1	.	.	−6	−2	6	2	2	.	.	2	.	.	.	2	.	.	−2	.	.	.
X.36	.	.	.	.	.	.	.	.	.	.	−1	−3	5	−3	−3	−3	1	1	1	1	1	1	1	1	.	.	.	.	.
X.37	.	.	.	.	.	.	.	.	.	.	−1	−3	1	1	1	1	1	1	−3	1	1	1	1	1	.	.	.	.	.
X.38	.	.	.	.	.	.	.	.	.	.	−1	−3	1	1	1	1	1	1	−3	1	1	1	1	1	.	.	.	.	.
X.39	1	−1	1	4	4	−1	3	.	.	1	.	4	−4	−4	.	.	.	.	.	.	.	.	.	.	.	.	.	.	−3
X.40	.	.	.	−3	3	.	.	1	−1	.	−2	.	.	.	.	.	2	2	.	−2	−2	.	.	.	.	−2	.	.	.
X.41	.	.	.	.	.	.	−4	.	.	.	1	.	.	.	.	.	.	.	.	.	.	.	.	.	−1	2	.	.	8
X.42	4	.	−4	1	−1	.	.	1	−1	.	.	.	.	.	.	.	−2	−2	.	2	2	.	.	.	−1	.	.	.	.
X.43	.	−4	.	3	−3	4	.	1	−1	.	.	.	.	.	.	.	2	−2	.	−2	2	.	.	.	−1	.	.	.	.
X.44	.	.	.	.	.	.	.	.	.	.	−1	.	.	.	.	.	.	.	.	.	.	.	.	.	.	.	.	.	3
X.45	.	.	.	.	.	.	.	.	.	.	−1	.	.	.	.	.	.	.	.	.	.	.	.	.	.	.	.	.	3
X.46	.	.	.	.	.	.	.	.	.	.	−2	−6	−2	−2	−2	−2	.	.	2	.	.	.	−2	.	.	.	.	.	.
X.47	.	.	.	.	.	.	−2	.	.	2	.	.	.	.	.	.	−2	2	.	−2	2	−2	.	2	.	.	.	.	2
X.48	.	.	.	.	.	.	−2	.	.	−2	.	.	.	.	.	.	−2	−2	.	−2	−2	2	.	2	.	.	.	.	2
X.49	−4	.	4	−1	1	.	.	−1	1	.	.	.	.	.	.	.	.	.	.	.	.	.	.	.	−1	−1	1	−1	.
X.50	−3	3	−3	.	.	3	1	.	.	−1	.	1	5	1	−1	3	1	−1	3	1	−1	−1	−1	1	.	.	.	.	.
X.51	−3	3	−3	.	.	3	1	.	.	−1	.	1	1	−3	3	−1	−1	1	−1	−1	1	1	−1	−1	.	.	.	.	.
X.52	3	−3	3	.	.	−3	1	.	.	−1	.	1	−7	1	1	1	−1	−1	−3	−1	−1	−1	1	−1	.	.	.	.	.
X.53	3	−3	3	.	.	−3	1	.	.	−1	.	1	5	−3	1	1	1	1	1	1	1	1	1	1	.	.	.	.	.
X.54	3	3	3	.	.	3	1	.	.	1	.	−7	1	1	1	1	−1	−1	1	−1	−1	−1	1	−1	.	.	.	.	.
X.55	−3	−3	−3	.	.	−3	−1	.	.	−1	2	−8	.	.	.	.	.	.	.	.	.	.	.	.	.	.	.	.	.
X.56	−2	−2	2	1	−1	2	.	3	−3	.	.	.	.	.	.	.	.	.	.	.	.	.	.	.	1	.	.	.	.
X.57	−2	2	2	−2	2	−2	.	−2	2	.	.	.	.	.	.	.	.	.	.	.	.	.	.	.	.	1	−1	1	.
X.58	.	.	.	.	.	.	−4	.	.	.	.	.	.	.	.	.	.	.	.	.	.	.	.	.	.	.	.	.	4
X.59	−2	2	2	4	−4	−2	.	.	.	.	.	.	.	.	.	.	−2	2	.	2	−2	.	.	.	.	.	.	.	.
X.60	.	.	.	.	.	.	2	.	.	−2	.	.	.	.	.	.	.	.	.	.	.	.	.	.	.	.	.	.	−5
X.61	.	.	.	.	.	.	2	.	.	2	.	.	.	.	.	.	.	.	.	.	.	.	.	.	.	.	.	.	−5
X.62	2	2	−2	−1	1	−2	.	−1	1	.	.	.	.	.	.	.	2	2	.	−2	−2	.	.	.	.	.	.	.	.
X.63	−2	−2	2	−2	2	2	.	.	.	.	.	.	.	.	.	.	2	2	.	−2	−2	.	.	.	.	.	.	.	.
X.64	−3	3	−3	.	.	3	−1	.	.	1	.	2	−2	−2	2	2	.	.	−2	.	.	.	2	.	.	.	.	.	.
X.65	3	−3	3	.	.	−3	−1	.	.	1	.	2	6	−2	2	2	.	.	2	.	.	.	−2	.	.	.	.	.	.
X.66	−4	−4	4	−1	1	4	.	−1	1	.	2	.	.	.	.	.	.	.	.	.	.	.	.	.	.	.	.	.	.
X.67	.	.	.	.	.	.	.	.	.	.	.	.	.	.	.	.	−2	2	.	−2	2	−2	.	2	.	.	.	.	−3
X.68	.	.	.	.	.	.	.	.	.	.	.	.	.	.	.	.	−2	−2	.	−2	−2	2	.	2	.	.	.	.	−3
X.69	−2	2	2	−2	2	−2	.	2	−2	.	.	.	.	.	.	.	.	.	.	.	.	.	.	.	.	2	.	.	.
X.70	.	.	.	.	.	.	.	.	.	.	.	3	7	3	1	−3	−1	1	1	−1	1	1	1	−1	.	.	.	.	.
X.71	.	.	.	.	.	.	.	.	.	.	.	3	3	−1	−3	1	1	−1	−3	1	−1	−1	1	1	.	.	.	.	.
X.72	.	.	.	.	.	.	.	.	.	.	.	3	7	−1	−1	−1	−1	−1	−1	−1	−1	−1	−1	−1	.	.	.	.	.
X.73	.	.	.	.	.	.	.	.	.	.	.	3	−1	3	−3	1	−1	1	−3	−1	1	1	1	−1	.	.	.	.	.
X.74	.	.	.	.	.	.	.	.	.	.	.	3	3	−1	1	−3	1	−1	−3	1	−1	−1	1	1	.	.	.	.	.
X.75	.	.	.	.	.	.	.	.	.	.	.	3	−5	−1	1	−3	1	−1	1	1	−1	−1	1	1	.	.	.	.	.
X.76	.	.	.	.	.	.	.	.	.	.	.	3	−1	−5	−3	1	−1	1	1	−1	1	1	1	−1	.	.	.	.	.
X.77	.	.	.	.	.	.	.	.	.	.	.	3	−5	3	−1	−1	1	1	3	1	1	1	−1	1	.	.	.	.	.
X.78	.	.	.	3	−3	.	.	1	−1	.	.	.	.	.	.	.	2	−2	.	−2	2	.	.	.	.	1	−1	1	.
X.79	.	.	.	−3	3	.	.	1	−1	.	.	.	.	.	.	.	−2	−2	.	2	2	.	.	.	.	1	1	−1	.
X.80	.	.	.	3	−3	.	.	1	−1	.	.	.	.	.	.	.	−2	2	.	2	−2	.	.	.	.	1	1	−1	.
X.81	.	.	.	.	.	.	.	.	.	.	−1	.	.	.	.	.	2	−2	.	2	−2	2	.	−2	.	.	.	.	.
X.82	.	.	.	.	.	.	.	.	.	.	−1	.	.	.	.	.	2	2	.	2	2	−2	.	−2	.	.	.	.	.

Character table of $H(\mathrm{Co}_2)$ (continued)

	12b	12c	12d	12e	12f	12g	12h	12i	12j	12k	12l	12m	12n	12o	12p	14a	14b	14c	15a	16a	16b	18a	24a	28a	30a
2	5	5	5	5	5	4	4	4	4	4	4	4	4	4	3	2	2	1	5	5	1	3	2	1	
3	2	1	1	1	1	1	1	1	1	1	1	1	1	1	1	1	.	.	1	.	2	1	.	1	
5	.	.	.	.	.	.	.	.	.	.	.	.	.	.	.	.	.	1	.	.	1	.	.	1	
7	.	.	.	.	.	.	.	.	.	.	.	.	.	1	1	1	.	.	.	1					
2P	6c	6e	6n	6a	6e	6n	6o	6d	6o	6n	6n	6d	6d	6o	6o	7a	7a	7a	15a	8b	8c	9a	12b	14a	15a
3P	4a	4d	4e	4b	4f	4g	4p	4i	4q	4j	4c	4m	4m	4s	4r	14a	14c	14b	5a	16a	16b	6a	8a	28a	10a
5P	12b	12c	12d	12e	12f	12g	12h	12i	12j	12k	12l	12n	12m	12o	12p	14a	14c	14b	3b	16a	16b	18a	24a	28a	6b
7P	12b	12c	12d	12e	12f	12g	12h	12i	12j	12k	12l	12m	12n	12o	12p	2a	2b	2b	15a	16a	16b	18a	24a	4a	30a
X.1	1	1	1	1	1	1	1	1	1	1	1	1	1	1	1	1	1	1	1	1	1	1	1	1	1
X.2	1	.	-1	2	.	-1	-2	.	.	-1	1	.	.	-2	.	.	.	-1	1	-1	1	-1	1	-1	-1
X.3	3	-1	-1	1	-1	-1	.	-1	-2	1	1	-1	-1	-2	.	1	1	1	.	-1	1	.	-1	1	.
X.4	.	-2	2	.	2	-2	1	.	1	.	.	.	-1	-1	-2	.	.	1	.	.	-1	.	.	.	-1
X.5	.	1	.	-1	1	.	.	1	.	2	-2	1	1	.	.	.	1	-1	-1	.	.	.	.	.	1
X.6	.	-1	.	3	-1	.	2	-1	.	.	-1	-1	.	2	.	.	1	1	-1	.	.	.	.	.	1
X.7	.	.	.	.	.	.	1	.	-1	.	.	.	.	-1	1	-1	-1	-1	-1	-1	1	.	.	-1	-1
X.8	2	1	.	3	1	.	-1	1	-1	.	-2	.	1	-1	-1	.	.	.	1	1	-1	.	.	.	.
X.9	2	-1	.	-1	-1	.	-1	-1	1	2	.	-1	-1	1	-1	.	.	.	1	-1	-1	.	.	.	.
X.10	2	.	.	-2	.	.	-1	-2	.	.	.	.	-1	1	.	.	.	.	1	.	-1	.	.	.	1
X.11	1	-1	1	-1	-1	1	-1	-1	-1	-1	-1	-1	-1	-1	.	.	.	.	.	.	1	1	.	.	.
X.12	3	1	-1	1	1	1	.	1	.	-1	1	1	1	.	.	.	.	-1	.	.	1	.	-1	.	-1
X.13	3	.	1	2	.	1	.	.	.	.	.	-1	1	.	.	.	.	.	-1	-1	1	.	.	.	.
X.14	3	.	-1	2	.	-1	.	.	2	-1	-1	.	2	.	.	.	.	.	-1	1	1	-1	.	.	.
X.15	-3	-1	1	1	-1	1	-1	-1	1	1	1	-1	1	-1	.	.	.	.	-1	1	1	1	.	.	1
X.16	.	-2	.	.	.	2	-2	.	-1	.	.	.	.	2	.	.	.	.	-1	.	.	-1	.	.	1
X.17	2	3	2	-1	3	2	.	-1	.	.	.	-1	-1	.	.	1	-1	-1	.	.	.	.	.	1	.
X.18	.	.	-2	.	.	-2	.	-1	.	1	-2	-2	.	.	1	-1	1	1	1	.	.	.	.	1	.
X.19	2	-1	-2	-1	-1	-2	.	3	.	.	.	.	-1	.	.	1	-1	-1	.	.	.	.	.	1	.
X.20	-3	.	1	.	.	1	.	.	.	.	-1	-1	.	.	.	.	2	.	.	-1	-1	.	.	-1	-2
X.21	-3	.	-1	2	.	-1	.	.	.	.	-1	1	.	.	.	.	.	.	.	.	.	-1	.	.	1
X.22	.	.	.	.	.	.	1	.	1	.	.	.	.	1	1	.	.	.	.	-1	-1	-1	.	.	-1
X.23	.	.	.	.	.	.	1	.	-1	.	.	.	.	-1	1	.	.	.	.	-1	1	-1	.	.	-1
X.24	.	.	.	.	.	.	1	.	1	.	.	.	.	1	1	.	.	.	.	-1	1	1	.	.	-1
X.25	3	.	-1	-2	.	-1	.	-1	1	1	1	1	1	1	.	.	.	.	.	.	.	-1	.	.	.
X.26	.	1	.	-1	1	.	-1	1	-1	2	2	1	1	-1	-1	.	.	.	.	.	.	.	.	.	.
X.27	.	.	.	.	.	.	-1	.	1	.	.	.	1	-1	-1	-1	-1	1	.	.	.	.	.	-1	1
X.28	.	2	-2	.	-2	2	.	.	-2	.	.	.	2	-2	.	.	.	.	.	.	.	.	.	.	.
X.29	-2	.	.	.	.	.	1	.	-1	.	-2	.	-1	1	.	.	.	.	.	.	1	.	.	.	.
X.30	1	.	-1	-2	.	-1	.	.	.	1	-1	.	.	.	.	.	.	.	.	.	1	-1	.	.	.
X.31	.	1	.	3	1	.	.	1	.	.	.	1	1	.	.	.	.	.	-1	-1	.	.	.	.	.
X.32	.	2	.	.	-2	.	2	.	.	.	.	.	.	.	-2	.	.	.	1	.	.	.	.	.	-1
X.33	.	-2	.	.	2	.	.	.	.	.	.	.	.	.	.	.	.	.	1	.	.	.	.	.	-1
X.34	.	.	-2	.	.	.	-2	2	.	.	.	.	.	.	.	.	.	.	1	.	.	.	.	.	1
X.35	.	.	.	.	.	-1	.	-1	.	.	.	.	-1	-1	.	.	.	.	1	.	.	.	.	.	1
X.36	.	.	.	.	.	.	.	.	.	.	.	.	.	.	.	-1	-1	-1	.	1	1	.	.	-1	.
X.37	.	.	.	.	.	.	.	.	.	.	.	.	.	.	.	-1	A	$\bar{A}$	.	-1	-1	.	.	1	.
X.38	.	.	.	.	.	.	.	.	.	.	.	.	.	.	.	-1	$\bar{A}$	A	.	-1	-1	.	.	1	.
X.39	3	-1	1	1	-1	1	.	-1	.	1	-1	-1	-1	-1	.	.	.	.	.	1	.	.	.	.	.
X.40	.	.	.	.	.	.	1	.	-1	.	.	.	1	-1	2	.	-1	.	.	.	.	.	.	.	1
X.41	-4	.	.	.	.	.	.	.	.	.	.	.	.	.	1	1	1	-1	.	.	-1	.	.	1	-1
X.42	.	2	.	.	-2	.	-1	.	1	.	.	.	-1	1	.	.	.	.	.	1	.	.	.	.	.
X.43	.	-2	.	.	2	.	-1	.	-1	.	.	.	1	1	.	.	.	.	.	1	.	.	.	.	.
X.44	.	1	.	-1	1	.	.	1	.	.	B	$\bar{B}$	.	-1	1	1	.	.	.	.	.	-1	.	.	.
X.45	.	1	.	-1	1	.	.	1	.	.	$\bar{B}$	B	.	-1	1	1	.	.	.	.	.	-1	.	.	.
X.46	.	.	.	.	.	.	.	.	.	.	.	.	.	-2	.	.	.	.	.	.	.	.	.	2	.
X.47	2	.	-2	-2	.	-2	.	.	.	.	.	.	.	.	.	.	.	.	.	.	.	.	.	.	.
X.48	2	.	2	-2	.	2	.	.	.	.	.	.	.	.	.	.	.	.	.	.	.	.	.	.	.
X.49	.	.	.	.	.	.	1	.	-1	.	.	.	1	-1	.	.	.	.	1	.	.	1	.	.	-1
X.50	-3	.	-1	.	.	-1	.	.	1	-1	.	.	.	.	.	.	.	.	1	-1	1	.	.	.	.
X.51	-3	.	-1	.	.	-1	.	.	1	-1	.	.	.	.	.	.	.	.	-1	1	1	.	.	.	.
X.52	-3	.	-1	.	.	-1	.	.	-1	1	.	.	.	.	.	.	.	.	1	1	1	.	.	.	.
X.53	-3	.	-1	.	.	-1	.	.	-1	1	.	.	.	.	.	.	.	.	-1	-1	1	.	.	.	.
X.54	-3	.	1	.	.	1	.	.	-1	-1	.	.	.	.	.	.	.	.	1	1	.	-1	.	.	.
X.55	3	.	-1	.	.	-1	.	.	1	1	.	.	.	.	.	.	2	.	.	.	.	1	-2	.	.
X.56	.	2	2	.	-2	-2	-1	.	-1	.	.	.	1	1	.	.	.	.	.	.	.	-1	.	.	.
X.57	.	-2	2	.	2	-2	.	.	.	.	.	.	.	.	-1	.	.	.	.	.	.	.	.	.	1
X.58	4	-2	.	.	.	-2	.	.	-2	.	.	.	2	2	.	.	.	.	.	.	.	.	.	.	.
X.59	.	.	2	.	.	-2	.	.	.	.	.	.	.	.	.	.	.	.	.	.	.	.	.	.	.
X.60	-2	-3	2	-1	-3	2	.	1	.	.	.	1	1	.	.	.	.	.	.	.	.	.	.	.	.
X.61	-2	1	-2	-1	1	-2	.	-3	.	.	.	1	1	.	.	.	.	.	.	.	.	.	.	.	.
X.62	.	2	2	.	-2	-2	-1	.	1	.	.	.	-1	1	.	.	.	.	.	.	.	.	.	.	.
X.63	.	-2	.	.	.	2	.	.	2	.	.	.	-2	.	.	.	.	.	.	.	.	.	.	.	.
X.64	3	.	1	.	.	1	.	.	1	-1	.	.	.	.	.	.	.	.	.	.	.	.	.	-1	.
X.65	3	.	1	.	.	1	.	.	-1	1	.	.	.	.	.	.	.	.	.	.	.	.	.	-1	.
X.66	.	.	.	.	.	.	-1	.	1	.	.	.	-1	1	-2	.	.	.	.	.	.	.	.	.	.
X.67	.	-1	.	1	-1	.	.	3	.	.	.	-1	-1	.	.	.	.	.	.	.	.	.	.	.	.
X.68	.	3	.	1	3	.	.	-1	.	.	.	-1	-1	.	.	.	.	.	.	.	.	.	.	.	.
X.69	.	.	-2	.	.	.	2	.	.	.	.	.	.	.	.	.	.	.	1	.	.	.	.	.	-1
X.70	.	.	.	.	.	.	.	.	.	.	.	.	.	.	.	.	.	.	.	-1	1	.	.	.	.
X.71	.	.	.	.	.	.	.	.	.	.	.	.	.	.	.	.	.	.	.	1	-1	.	.	.	.
X.72	.	.	.	.	.	.	.	.	.	.	.	.	.	.	.	.	.	.	.	1	1	.	.	.	.
X.73	.	.	.	.	.	.	.	.	.	.	.	.	.	.	.	.	.	.	.	-1	1	.	.	.	.
X.74	.	.	.	.	.	.	.	.	.	.	.	.	.	.	.	.	.	.	.	1	-1	.	.	.	.
X.75	.	.	.	.	.	.	.	.	.	.	.	.	.	.	.	.	.	.	.	1	1	.	.	.	.
X.76	.	.	.	.	.	.	.	.	.	.	.	.	.	.	.	.	.	.	.	-1	1	.	.	.	.
X.77	.	.	.	.	.	.	.	.	.	.	.	.	.	.	.	.	.	.	.	-1	-1	.	.	.	.
X.78	.	.	.	.	.	.	1	.	1	.	.	.	.	-1	-1	.	.	.	.	-1	.	.	.	.	1
X.79	.	.	.	.	.	.	1	.	-1	.	.	.	.	1	-1	.	.	.	.	-1	.	.	.	.	1
X.80	.	.	.	.	.	.	1	.	1	.	.	.	.	-1	-1	.	.	.	.	-1	.	.	.	.	1
X.81	.	.	.	.	.	.	.	.	.	.	.	.	.	.	.	-1	1	1	.	.	.	.	.	-1	.
X.82	.	.	.	.	.	.	.	.	.	.	.	.	.	.	.	-1	1	1	.	.	.	.	.	-1	.

Character table of $H(\mathrm{Co}_2)$ (continued)

2|	18	18	17	14	14	16	16	14	14	15	11	6	5	7	14	14	13	12	12	12	12
3|	4	4	1	2	2	2	2	1	1	.	1	4	3	3	3	1	1	1	1	1	1
5|	1	1	.	.	1	.	.	.	.	.	.	.	.	1	.	.	.	.	.	.	.
7|	.	.	1	.	.	.	.	.	.	.	.	.	.	.	.	.	.	.	.	.	.
	1a	2a	2b	2c	2d	2e	2f	2g	2h	2i	2j	3a	3b	3c	4a	4b	4c	4d	4e	4f	4g
2P	1a	1a	1a	1a	1a	1a	1a	1a	1a	1a	1a	3a	3b	3c	2a	2a	2b	2f	2b	2f	2b
3P	1a	2a	2b	2c	2d	2e	2f	2g	2h	2i	2j	1a	1a	1a	4a	4b	4c	4d	4e	4f	4g
5P	1a	2a	2b	2c	2d	2e	2f	2g	2h	2i	2j	3a	3b	3c	4a	4b	4c	4d	4e	4f	4g
7P	1a	2a	2b	2c	2d	2e	2f	2g	2h	2i	2j	3a	3b	3c	4a	4b	4c	4d	4e	4f	4g
X.83	3360	−3360	.	40	−40	−112	112	40	−40	.	.	12	−15	12	.	.	.	24	−8	−24	8
X.84	3360	−3360	.	200	−200	16	−16	8	−8	.	.	−6	15	.	.	.	.	−8	24	8	−24
X.85	3456	−3456	.	−96	96	192	−192	32	−32	.	.	.	−9	.	.	.	.	.	.	.	.
X.86	3780	3780	196	−60	−60	132	132	−60	−60	4	4	.	.	9	−252	−60	4	12	4	12	4
X.87	3780	3780	196	−60	−60	−156	−156	36	36	−28	4	.	.	9	−252	36	4	−12	4	−12	4
X.88	4480	−4480	.	−160	160	−64	64	−32	32	.	.	−20	10	4	.	.	.	−32	.	.	32
X.89	4480	−4480	.	160	−160	192	−192	32	−32	.	.	16	−5	−8	.	.	.	−24	.	−24	.
X.90	5040	5040	−336	.	.	48	48	.	.	−16	.	18	.	.	336	16	.	−24	.	−24	.
X.91	5040	−5040	.	−180	180	−168	168	12	−12	.	.	18	.	.	.	.	.	−20	−12	20	12
X.92	5376	−5376	.	−64	64	128	−128	−64	64	.	.	12	6	.	.	.	.	.	.	.	.
X.93	5670	5670	294	270	270	−90	−90	−18	−18	−26	6	.	.	.	−378	6	−18	−6	6	−6	6
X.94	6048	−6048	.	−120	120	−48	48	8	−8	.	.	.	−9	.	.	.	.	24	24	−24	−24
X.95	6480	−6480	.	180	−180	−216	216	−12	12	.	.	.	.	.	.	.	.	−12	12	12	−12
X.96	6720	−6720	−448	.	.	−192	−192	.	.	.	.	64	.	.	6	.	12	448	−64	.	.
X.97	6720	−6720	−448	.	.	80	−80	32	−32	−48	48	.	6	.	.	12	.	.	.	.	.
X.98	7560	7560	392	−120	−120	−24	−24	−24	−24	−24	8	.	.	.	−9	−504	−24	8	.	8	.
X.99	7680	−7680	−512	.	.	.	.	.	.	.	.	−24	.	−12	512	.	.	.	.	.	.
X.100	8192	−8192	.	.	.	.	.	.	.	.	.	−16	−16	−16	.	.	.	.	.	.	.

Character table of $H(\mathrm{Co}_2)$ (continued)

2|	13	11	11	12	12	10	11	11	9	9	9	9	10	10	8	8	8	2	6	5	7	6	6	5	5
3|	.	1	1	.	.	1	.	.	1	1	1	1	.	.	.	.	.	1	4	3	3	2	2	2	2
5|	.	.	.	.	.	.	.	.	.	.	.	.	.	.	.	.	.	1	.	1	.	.	.	.	.
7|	.	.	.	.	.	.	.	.	.	.	.	.	.	.	.	.	.	.	1	.	.	.	.	.	.
	4h	4i	4j	4k	4l	4m	4n	4o	4p	4q	4r	4s	4t	4u	4v	4w	4x	5a	6a	6b	6c	6d	6e	6f	6g
2P	2b	2e	2b	2e	2e	2e	2b	2f	2g	2g	2g	2g	2b	2e	2i	2g	2g	5a	3a	3b	3c	3a	3a	3b	3b
3P	4h	4i	4j	4k	4l	4m	4n	4o	4p	4q	4r	4s	4t	4u	4v	4w	4x	5a	2a	2a	2a	2e	2f	2c	2d
5P	4h	4i	4j	4k	4l	4m	4n	4o	4p	4q	4r	4s	4t	4u	4v	4w	4x	1a	6a	6b	6c	6d	6e	6f	6g
7P	4h	4i	4j	4k	4l	4m	4n	4o	4p	4q	4r	4s	4t	4u	4v	4w	4x	5a	6a	6b	6c	6d	6e	6f	6g
X.83	.	.	.	−8	8	.	.	.	4	4	−4	−4	.	.	.	−4	4	.	−12	15	−12	−4	4	1	−1
X.84	.	.	.	−8	8	.	.	.	−4	−4	4	4	.	.	.	−4	4	.	6	−15	9	.	.	−3	3
X.85	.	.	.	.	.	.	.	.	8	−8	−8	8	.	.	.	.	.	1	.	9	.	.	.	−3	3
X.86	4	12	4	−4	−4	12	4	−4	.	.	.	.	4	−4	4	.	.	.	.	9	.	.	.	.	.
X.87	−28	−12	4	4	4	−12	4	4	.	.	.	.	4	4	−4	.	.	.	.	9	.	.	.	.	.
X.88	.	.	.	.	.	.	.	.	−8	8	8	−8	.	.	.	.	.	.	20	−10	−4	−4	4	2	−2
X.89	.	.	.	.	.	.	.	.	−8	8	8	−8	.	.	.	.	.	.	−16	5	8	.	.	1	−1
X.90	.	−24	.	−8	−8	24	.	−8	.	.	.	.	8	.	.	.	.	.	18	.	.	.	.	6	6
X.91	.	.	.	12	−12	.	.	.	−6	−6	6	6	.	.	.	.	.	.	−18	.	.	.	.	6	−6
X.92	.	.	.	.	.	.	.	.	.	.	.	.	.	.	.	.	.	1	−12	−6	.	−4	4	2	−2
X.93	14	−6	6	2	2	−6	−2	2	−6	−6	−6	−6	−10	2	2	2	2	.	.	.	.	.	.	3	−3
X.94	.	.	.	−8	8	.	.	.	−4	−4	4	4	.	.	.	.	.	.	4	−4	−2	.	9	.	.
X.95	.	.	.	20	−20	.	.	.	6	6	−6	−6	.	.	.	.	.	.	2	−2	.	.	.	.	.
X.96	.	.	.	.	.	.	.	.	.	.	.	.	.	.	.	.	.	.	6	.	.	12	−6	−6	.
X.97	.	.	.	−16	16	.	.	.	.	.	.	.	.	.	.	.	.	.	−6	.	.	−12	2	−2	−4	4
X.98	−24	.	8	.	.	.	.	.	8	.	.	.	.	.	.	.	.	8	.	.	.	.	−9	.	.
X.99	.	.	.	.	.	.	.	.	.	.	.	.	.	.	.	.	.	.	−24	.	−12	.	.	.	.
X.100	.	.	.	.	.	.	.	.	.	.	.	.	.	.	.	.	.	.	2	16	16	16	.	.	.

Character table of $H(\mathrm{Co}_2)$ (continued)

2|	5	5	5	5	5	5	6	5	5	4	3	8	9	9	8	8	7	7	7	7	7	6	6	6	1	2	2	2	5
3|	2	2	2	2	2	2	1	1	1	1	.	1	.	.	.	.	.	.	.	.	.	.	.	.	2	1	1	1	3
5|	.	.	.	.	.	.	.	.	.	.	.	.	.	.	1	.	.	.	.	.	.	.	.	.	.	1	1	1	.
7|	.	.	.	.	.	.	.	.	.	.	1	.	.	.	.	.	.	.	.	.	.	.	.	.	.	.	.	.	.
	6h	6i	6j	6k	6l	6m	6n	6o	6p	6q	7a	8a	8b	8c	8d	8e	8f	8g	8h	8i	8j	8k	8l	8m	9a	10a	10b	10c	12a
2P	3c	3c	3c	3b	3b	3c	3c	3b	3b	3c	7a	4a	4b	4h	4h	4i	4o	4h	4i	4o	4k	4n	4d	4d	9a	5a	5a	5a	6a
3P	2e	2d	2f	2e	2f	2c	2b	2g	2h	2j	7a	8a	8b	8c	8d	8e	8f	8g	8h	8i	8j	8k	8l	8m	3a	10a	10b	10c	4a
5P	6h	6i	6j	6k	6l	6m	6n	6o	6p	6q	7a	8a	8b	8c	8d	8e	8f	8g	8h	8i	8j	8k	8l	8m	9a	2a	2d	2c	12a
7P	6h	6i	6j	6k	6l	6m	6n	6o	6p	6q	1a	8a	8b	8c	8d	8e	8f	8g	8h	8i	8j	8k	8l	8m	9a	10a	10b	10c	12a
X.83	2	2	−2	−1	1	−2	.	1	−1	.	.	.	.	.	.	.	.	.	.	.	.	.	.	.	.	−1	1	−1	.
X.84	4	4	−4	1	−1	−4	.	−1	1	.	.	.	.	.	.	.	.	.	.	.	.	.	.	.	.	.	.	.	.
X.85	.	.	.	3	−3	.	.	−1	1	.	−2	.	.	.	.	.	.	.	.	.	.	.	.	.	.	−1	1	−1	.
X.86	−3	−3	−3	.	.	−3	1	.	.	−4	−4	−4	.	.	.	.	.	.	.	.	.	.	.	.	1	.	.	.	.
X.87	−3	−3	−3	.	.	−3	1	.	.	1	−4	4	4	.	.	.	.	.	.	.	.	.	.	.	1	.	.	.	.
X.88	2	−2	−2	2	−2	2	.	−2	2	.	.	.	.	.	.	.	.	.	.	.	.	.	.	1	.	.	.	.	−6
X.89	.	−4	.	3	−3	4	.	−1	1	.	.	.	.	.	.	.	.	.	.	.	.	.	.	1	.	.	.	.	.
X.90	.	.	.	.	.	.	.	.	.	.	.	.	.	.	.	.	.	.	.	.	.	.	.	.	.	.	.	.	−6
X.91	.	.	.	.	.	.	.	.	.	.	.	.	.	.	.	.	.	.	.	−2	2	.	2	−2	.	−1	−1	1	.
X.92	−4	4	4	2	−2	−4	.	2	−2	.	.	.	.	.	.	.	6	−6	2	2	2	.	2	.	.	−1	−1	1	.
X.93	.	.	.	−3	3	.	.	−1	1	.	.	.	.	.	.	.	.	.	.	.	.	−2	.	.	2	.	.	.	.
X.94	.	.	.	−3	3	.	.	−1	1	.	.	.	.	.	.	.	2	−2	.	−2	2	.	.	.	.	.	.	.	−2
X.95	.	.	.	.	.	.	.	−4	.	.	.	.	.	.	.	.	.	.	.	.	.	.	.	.	.	.	.	.	−2
X.96	2	−2	−2	−4	4	2	.	.	.	.	.	−4	.	.	.	.	.	.	.	.	.	.	.	.	.	.	.	.	.
X.97	2	−2	−2	−4	4	2	.	3	−1	.	.	−1	.	−8	.	.	.	.	.	.	.	.	.	.	.	.	.	.	8
X.98	3	3	3	.	.	3	−1	.	−8	.	.	.	.	.	.	.	.	.	.	.	.	.	.	.	.	.	.	.	8
X.99	.	.	.	4	.	.	.	1	.	.	.	.	.	.	.	.	.	.	.	.	.	.	.	.	.	−1	−2	.	.
X.100	.	.	.	.	.	.	.	.	2	.	.	.	.	.	.	.	.	.	.	.	.	.	.	.	.	.	.	.	.

Character table of $H(\mathrm{Co}_2)$ *(continued)*

	12b	12c	12d	12e	12f	12g	12h	12i	12j	12k	12l	12m	12n	12o	12p	14a	14b	14c	15a	16a	16b	18a	24a	28a	30a
2	5	5	5	5	5	5	4	4	4	4	4	4	4	4	4	3	2	2	1	5	5	1	3	2	1
3	2	1	1	1	1	1	1	1	1	1	1	1	1	1	1	.	.	.	1	.	.	2	1	.	1
5	.	.	.	.	.	.	.	.	.	.	.	.	.	.	.	.	.	.	1	.	.	.	.	.	1
7	.	.	.	.	.	.	.	.	.	.	.	.	.	.	.	1	1	1	.	.	.	.	.	1	.
2P	6c	6e	6n	6a	6e	6n	6o	6d	6o	6n	6n	6d	6d	6o	6o	7a	7a	7a	15a	8b	8c	9a	12b	14a	15a
3P	4a	4d	4e	4b	4f	4g	4p	4i	4q	4j	4c	4m	4m	4s	4r	14a	14c	14b	5a	16a	16b	6a	8a	28a	10a
5P	12b	12c	12d	12e	12f	12g	12h	12i	12j	12k	12l	12n	12m	12o	12p	14a	14c	14b	3b	16a	16b	18a	24a	28a	6b
7P	12b	12c	12d	12e	12f	12g	12h	12i	12j	12k	12l	12m	12n	12o	12p	2a	2b	2b	15a	16a	16b	18a	24a	4a	30a
X.83	.	.	-2	.	.	2	1	.	1	.	.	.	-1	-1	.	.	.	.	.	.	.	.	.	.	.
X.84	.	-2	.	.	2	.	-1	.	-1	.	.	.	1	1	.	.	.	.	.	.	.	.	.	.	.
X.85	.	.	.	.	.	.	-1	.	1	.	.	.	-1	1	2	.	.	.	.	1	.	.	.	.	-1
X.86	-3	.	1	.	.	1	.	.	.	1	1	.	.	.	.	.	.	.	.	.	-1	.	.	.	.
X.87	-3	.	1	.	.	1	.	.	.	1	1	.	.	.	.	.	.	.	.	.	-1	.	.	.	.
X.88	.	.	-2	.	.	2	.	.	.	.	.	.	.	.	.	.	.	.	.	-1	.	.	.	.	.
X.89	.	.	.	.	.	1	.	-1	.	.	.	.	.	.	1	-1	.	.	.	.	-1	.	.	.	.
X.90	.	.	.	-2	.	.	.	.	.	.	.	.	.	.	.	.	.	.	.	.	.	.	.	.	.
X.91	.	-2	.	.	2	.	.	.	.	.	.	.	.	.	.	.	.	.	.	.	.	.	.	.	.
X.92	.	.	.	.	.	.	.	.	.	.	.	.	.	.	.	.	.	1	.	.	.	.	.	.	-1
X.93	.	.	.	.	.	.	.	.	.	.	.	.	.	.	.	.	.	.	.	.	.	.	.	.	-1
X.94	.	.	.	.	.	-1	.	-1	.	.	.	.	.	.	1	1	.	.	1	.	.	.	.	.	-1
X.95	.	.	.	.	.	.	.	.	.	.	.	.	.	.	.	.	.	2	.	.	.	.	.	.	.
X.96	4	.	.	2	.	.	.	.	.	.	.	.	.	.	.	.	.	.	.	.	.	.	.	.	.
X.97	.	2	2	.	-2	-2	.	.	.	.	.	.	.	.	.	.	.	.	.	.	.	.	.	.	.
X.98	3	.	-1	.	.	-1	.	.	.	.	.	.	-1	-1	.	.	.	.	.	.	.	.	.	1	.
X.99	-4	.	.	.	.	.	.	.	.	.	.	.	.	.	.	1	-1	-1	.	.	.	.	.	1	.
X.100	.	.	.	.	.	.	.	.	.	.	.	.	.	.	.	-2	.	.	.	-1	.	.	1	.	1

where $A = i\sqrt{7}, \quad B = -1 - i\sqrt{3}$.

B.2. Character table of $D(\mathrm{Co_2}) = \langle x, y \rangle$

	1a	2a	2b	2c	2d	2e	2f	2g	2h	2i	2j	2k	2l	2m	2n	2o	2p	2q	2r	2s	2t	2u	2v
2	18	18	14	14	17	16	16	14	14	14	14	14	15	13	13	13	13	14	13	11	12	12	12
3	2	2	2	2	1	1	1	1	1	1	1	1	.	1	1	1	1	.	.	1	.	.	.
5	1	1	1	.	1	.	.	.	.	.	.	.	.	.	.	.	.	.	.	.	.	.	.
2P	1a	1a	1a	1a	1a	1a	1a	1a	1a	1a	1a	1a	1a	1a	1a	1a	1a	1a	1a	1a	1a	1a	1a
3P	1a	2a	2b	2c	2d	2e	2f	2g	2h	2i	2j	2k	2l	2m	2n	2o	2p	2q	2r	2s	2t	2u	2v
5P	1a	2a	2b	2c	2d	2e	2f	2g	2h	2i	2j	2k	2l	2m	2n	2o	2p	2q	2r	2s	2t	2u	2v
X.1	1	1	1	1	1	1	1	1	1	1	1	1	1	1	1	1	1	1	1	1	1	1	1
X.2	1	1	1	1	1	1	1	1	1	1	-1	-1	1	-1	-1	-1	-1	1	-1	-1	1	-1	1
X.3	5	5	5	5	5	5	5	5	5	5	3	3	5	-1	-1	-1	-1	5	3	3	1	3	1
X.4	5	5	5	5	5	5	5	5	5	5	1	1	5	-3	-3	-3	-3	5	1	1	1	1	1
X.5	5	5	5	5	5	5	5	5	5	5	-1	-1	5	3	3	3	3	5	-1	-1	1	-1	1
X.6	5	5	5	5	5	5	5	5	5	5	-3	-3	5	1	1	1	1	5	-3	-3	1	-3	1
X.7	6	6	-6	-6	6	-2	-2	2	6	2	.	.	-2	4	4	-4	-4	-2	.	.	2	.	2
X.8	6	6	-6	-6	6	-2	-2	2	6	2	.	.	-2	-4	-4	4	4	-2	.	.	2	.	2
X.9	9	9	9	9	9	9	9	9	9	9	-3	-3	9	-3	-3	-3	-3	9	-3	-3	1	-3	1
X.10	9	9	9	9	9	9	9	9	9	9	3	3	9	3	3	3	3	9	3	3	1	3	1
X.11	10	10	10	10	10	10	10	10	10	10	-2	-2	10	2	2	2	2	10	-2	-2	-2	-2	-2
X.12	10	10	-10	-10	10	2	2	-2	10	-2	4	4	2	4	4	-4	-4	2	4	-4	2	4	2
X.13	10	10	10	10	10	10	10	10	10	10	2	2	10	-2	-2	-2	-2	10	2	2	-2	2	-2
X.14	10	10	-10	-10	10	2	2	-2	10	-2	4	4	2	-4	-4	4	4	2	-4	4	2	-4	2
X.15	10	10	-10	-10	10	2	2	-2	10	-2	-4	-4	2	4	4	-4	-4	2	-4	4	2	-4	2
X.16	10	10	-10	-10	10	2	2	-2	10	-2	-4	-4	2	-4	-4	4	4	2	4	-4	2	4	2
X.17	15	15	15	15	15	15	15	15	-1	15	7	7	15	3	3	3	3	-1	7	7	3	-1	3
X.18	15	15	15	15	15	15	15	15	-1	15	5	5	15	-3	-3	-3	-3	-1	5	5	-1	3	-1
X.19	15	15	15	15	15	15	15	15	-1	15	-5	-5	15	3	3	3	3	-1	-5	-5	-1	3	-1
X.20	15	15	15	15	15	15	15	15	-1	15	-7	-7	15	-3	-3	-3	-3	-1	-7	-7	3	1	3
X.21	15	15	15	15	15	-1	-1	-1	15	-1	-3	-3	-1	-7	-7	-7	-7	-1	-3	-3	3	3	3
X.22	15	15	15	15	15	-1	-1	-1	15	-1	-3	-3	-1	5	5	5	5	-1	-3	-3	-1	-3	-1
X.23	15	15	15	15	15	-1	-1	-1	15	-1	3	3	-1	7	7	7	7	-1	3	3	3	3	3
X.24	15	15	15	15	15	-1	-1	-1	15	-1	3	3	-1	-5	-5	-5	-5	-1	3	3	-1	3	-1
X.25	16	-16	4	4	.	8	-8	4	.	-4	8	-8	.	-4	4	-4	-4	.	.	.	4	.	-4
X.26	16	-16	-4	4	.	8	-8	4	.	-4	-8	8	.	4	-4	-4	4	.	.	.	4	.	-4
X.27	16	16	16	16	16	16	16	16	16	16	.	.	16	.	.	.	.	16	.	.	.	.	.
X.28	20	20	-20	-20	20	4	4	-4	20	-4	.	.	4	.	.	.	.	4	.	.	-4	.	-4
X.29	24	24	-24	-24	24	-8	-8	8	24	8	.	.	-8	8	8	-8	-8	-8	.	.	.	.	.
X.30	24	24	-24	-24	24	-8	-8	8	24	8	.	.	-8	-8	-8	8	8	-8	.	.	.	.	.
X.31	30	30	-30	-30	30	-10	-10	10	30	10	.	.	-10	-4	-4	4	4	-10	.	.	2	.	2
X.32	30	30	30	30	30	30	30	30	-2	30	-2	-2	30	-6	-6	-6	-6	-2	-2	-2	2	-2	2
X.33	30	30	30	30	30	30	30	30	-2	30	2	2	30	6	6	6	6	-2	2	2	2	2	2
X.34	30	30	-30	-30	30	-10	-10	10	30	10	.	.	-10	4	4	-4	-4	-10	.	.	2	.	2
X.35	30	30	30	30	30	-2	-2	-2	30	-2	6	6	-2	2	2	2	2	-2	6	6	2	6	2
X.36	30	30	30	30	30	-2	-2	-2	30	-2	-6	-6	-2	-2	-2	-2	-2	-2	-6	-6	2	-6	2
X.37	36	36	-36	-36	36	-12	-12	12	36	12	.	.	-12	.	.	.	.	-12	.	.	-4	.	-4
X.38	40	40	-40	-40	40	8	8	-8	40	-8	.	.	8	8	8	-8	-8	8	.	.	.	.	.
X.39	40	40	-40	-40	40	8	8	-8	40	-8	-8	-8	8	.	.	.	.	8	-8	8	.	-8	.
X.40	40	40	-40	-40	40	8	8	-8	40	-8	8	8	8	.	.	.	.	8	8	-8	.	8	.
X.41	40	40	-40	-40	40	8	8	-8	40	-8	.	.	8	-8	-8	8	8	8	.	.	.	.	.
X.42	45	45	45	45	45	-3	-3	-3	45	-3	-3	-3	45	9	9	9	9	-3	-3	-3	1	-3	1
X.43	45	45	45	45	45	-3	-3	-3	-3	-3	9	9	-3	-3	-3	-3	-3	9	13	13	1	1	1
X.44	45	45	45	45	45	-3	-3	-3	-3	-3	9	9	-3	9	9	9	9	9	13	-3	1	5	1
X.45	45	45	45	45	45	45	45	45	-3	45	9	9	45	-3	-3	-3	-3	-3	-3	-3	1	1	1
X.46	45	45	45	45	45	-3	-3	-3	45	-3	-3	-3	-3	-3	-3	-3	-3	13	-3	-3	-3	-3	-3
X.47	45	45	45	45	45	-3	-3	-3	45	-3	3	3	-3	-9	-9	-9	-9	13	3	3	1	-5	1
X.48	45	45	45	45	45	-3	-3	-3	45	-3	3	3	-3	-9	-9	-9	-9	-3	3	3	1	1	1
X.49	45	45	45	45	45	-3	-3	-3	-3	-3	-9	-9	-3	3	3	3	3	13	-9	-9	1	-1	1
X.50	45	45	45	45	45	-3	-3	-3	45	-3	3	3	45	3	3	3	3	-3	3	3	3	3	3
X.51	45	45	45	45	45	45	45	45	-3	45	-9	-9	45	3	3	3	3	-3	-9	-9	1	-1	1
X.52	45	45	45	45	45	-3	-3	-3	45	-3	3	3	45	3	3	3	3	13	3	3	-3	3	-3
X.53	45	45	45	45	45	-3	-3	-3	-3	-3	3	3	-3	9	9	9	9	13	9	9	5	1	5
X.54	45	45	45	45	45	45	45	45	-3	45	-3	-3	45	9	9	9	9	13	-3	-3	5	1	5
X.55	45	45	45	45	45	45	45	45	-3	45	-3	-3	45	-3	-3	-3	-3	13	-3	-3	-3	-3	-3
X.56	45	45	45	45	45	-3	-3	-3	-3	-3	-3	-3	-3	-3	-3	-3	-3	13	-3	-3	-3	-3	-3
X.57	45	45	45	45	45	-3	-3	-3	-3	-3	-9	-9	-3	-9	-9	-9	-9	13	-9	-9	5	1	5
X.58	60	60	-60	-60	60	12	12	-12	-4	-12	-16	-16	12	.	.	.	.	12	-16	16	4	.	4
X.59	60	60	-60	-60	60	12	12	-12	-4	-12	-8	-8	12	.	.	.	.	12	-8	8	4	-8	4
X.60	60	60	-60	-60	60	12	12	-12	-4	-12	16	16	12	.	.	.	.	12	16	-16	4	.	4
X.61	60	60	-60	-60	60	12	12	-12	-4	-12	8	8	12	.	.	.	.	12	8	-8	4	-8	4
X.62	80	-80	-20	20	.	40	-40	20	.	-20	-24	24	.	-4	4	4	-4	.	.	.	4	.	-4
X.63	80	-80	-20	20	.	40	-40	20	.	-20	24	-24	.	4	-4	-4	4	.	.	.	4	.	-4
X.64	80	-80	-20	20	.	40	-40	20	.	-20	8	-8	.	-12	12	12	-12	.	.	.	4	.	-4
X.65	80	-80	-20	20	.	40	-40	20	.	-20	-8	8	.	12	-12	-12	12	.	.	.	4	.	-4
X.66	90	90	-90	-90	90	-30	-30	30	-6	30	.	.	-30	12	12	-12	-12	-12	2	.	6	.	6
X.67	90	90	-90	-90	90	-30	-30	30	-6	30	.	.	-30	-12	-12	12	12	-12	2	.	6	.	6
X.68	90	90	-90	-90	90	-30	-30	30	-6	30	.	.	-30	12	12	-12	-12	-12	2	.	2	.	2
X.69	90	90	90	90	90	-6	-6	-6	-6	-6	6	6	-6	-6	-6	-6	-6	26	6	6	-2	6	-2
X.70	90	90	-90	-90	90	-30	-30	30	-6	30	.	.	-30	12	12	-12	-12	-12	2	.	6	.	6
X.71	90	90	-90	-90	90	18	18	-18	-6	-18	12	12	18	12	12	-12	-12	-14	12	-12	2	-4	2
X.72	90	90	-90	-90	90	18	18	-18	-6	-18	12	12	18	-12	-12	12	12	-14	12	-12	2	-4	2
X.73	90	90	90	90	90	-6	-6	-6	-6	-6	-6	-6	-6	6	6	6	6	26	-6	-6	-6	-6	-6
X.74	90	90	-90	-90	90	18	18	-18	-6	-18	-12	-12	18	12	12	-12	-12	-14	-12	12	2	4	2
X.75	90	90	-90	-90	90	18	18	-18	-6	-18	-12	-12	18	-12	-12	12	12	-14	-12	12	2	4	2
X.76	96	-96	24	-24	.	-16	16	8	.	-8	.	.	.	16	-16	16	-16	.	.	.	8	.	-8
X.77	96	-96	24	-24	.	-16	16	8	.	-8	.	.	.	-16	16	-16	16	.	.	.	8	.	-8
X.78	120	120	.	.	-8	24	24	.	-8	.	-24	-24	-8	.	.	.	.	-8	8	-8	.	8	.
X.79	120	120	.	.	-8	24	24	.	-8	.	16	16	-8	-12	-12	-12	-12	.	.	.	-4	8	-4
X.80	120	120	-120	-120	120	24	24	-24	-8	-24	8	8	24	.	.	.	.	24	8	-8	.	8	.
X.81	120	120	.	.	-8	24	24	.	-8	.	24	24	-8	.	.	.	.	-8	8	-8	.	8	.
X.82	120	120	.	.	-8	24	24	.	8	.	-16	-16	-8	12	12	12	12	8	16	.	-4	-8	-4
X.83	120	120	.	.	-8	24	24	.	8	.	32	32	-8	12	12	12	12	8	.	.	12	8	12
X.84	120	120	.	.	-8	24	24	.	-8	.	-24	-24	-8	.	.	.	.	-8	8	-8	.	8	.

Character table of $D(\mathrm{Co}_2)$ (continued)

	$2w$	$2x$	$3a$	$3b$	4_1	4_2	4_3	4_4	4_5	4_6	4_7	4_8	4_9	4_{10}	4_{11}	4_{12}	4_{13}	4_{14}	4_{15}	4_{16}	4_{17}	4_{18}	4_{19}	4_{20}
2	11	11	7	5	14	13	14	12	12	13	11	12	12	12	12	12	12	12	12	12	12	12	11	11
3	.	.	2	2	1	1	.	1	1	.	1	.	.	.	.	.	.	.	.	.	.	.	.	.
5	.	.	.	.	.	.	.	.	.	.	.	.	.	.	.	.	.	.	.	.	.	.	.	.
2P	$1a$	$1a$	$3a$	$3b$	$2a$	$2d$	$2a$	$2d$	$2d$	$2d$	$2d$	$2e$	$2f$	$2d$	$2e$	$2d$	$2d$	$2a$	$2d$	$2d$	$2f$	$2d$	$2e$	$2d$
3P	$2w$	$2x$	$1a$	$1a$	4_1	4_2	4_3	4_4	4_5	4_6	4_7	4_8	4_9	4_{10}	4_{11}	4_{12}	4_{13}	4_{14}	4_{15}	4_{16}	4_{17}	4_{18}	4_{19}	4_{20}
5P	$2w$	$2x$	$3a$	$3b$	4_1	4_2	4_3	4_4	4_5	4_6	4_7	4_8	4_9	4_{10}	4_{11}	4_{12}	4_{13}	4_{14}	4_{15}	4_{16}	4_{17}	4_{18}	4_{19}	4_{20}
X.1	1	1	1	1	1	1	1	1	1	1	1	1	1	1	1	1	1	1	1	1	1	1	1	1
X.2	−1	1	1	1	1	1	1	−1	−1	1	−1	1	1	1	1	−1	−1	−1	−1	−1	1	1	1	1
X.3	−1	1	2	−1	5	5	5	3	3	5	3	1	1	1	1	−1	−1	3	−1	−1	1	1	1	1
X.4	−3	1	−1	2	5	5	5	1	1	5	1	1	1	1	1	−3	−3	1	−3	−3	1	1	1	1
X.5	3	1	−1	2	5	5	5	−1	−1	5	−1	1	1	1	1	3	3	−1	3	3	1	1	1	1
X.6	1	1	2	−1	5	5	5	−3	−3	5	−3	1	1	1	1	1	1	−3	1	1	1	1	1	1
X.7	.	−2	.	3	6	−6	−2	.	.	2	.	−2	2	−2	−2	.	−4	.	.	4	2	−2	2	2
X.8	.	−2	.	3	6	−6	−2	.	.	2	.	−2	2	−2	−2	.	4	.	.	−4	2	−2	2	2
X.9	−3	1	.	.	9	9	9	−3	−3	9	−3	1	1	1	1	−3	−3	−3	−3	−3	1	1	1	1
X.10	3	1	.	.	9	9	9	3	3	9	3	1	1	1	1	3	3	3	3	3	1	1	1	1
X.11	2	−2	1	1	10	10	10	−2	−2	10	−2	−2	−2	−2	−2	2	2	−2	2	2	−2	−2	−2	−2
X.12	.	−2	1	1	10	−10	2	−4	−4	−2	4	2	−2	2	2	.	−4	4	.	4	−2	−2	−2	2
X.13	−2	−2	1	1	10	10	10	2	2	10	2	−2	−2	−2	−2	−2	−2	2	−2	−2	−2	−2	−2	−2
X.14	.	−2	1	1	10	−10	2	−4	−4	−2	4	2	−2	2	2	.	4	4	.	−4	−2	−2	−2	2
X.15	.	−2	1	1	10	−10	2	4	4	−2	−4	2	−2	2	2	.	−4	−4	.	4	−2	−2	−2	2
X.16	.	−2	1	1	10	−10	2	4	4	−2	−4	2	−2	2	2	.	4	−4	.	−4	−2	−2	−2	2
X.17	3	3	3	.	−1	−1	−1	7	7	−1	7	3	3	3	3	3	3	−1	3	3	3	3	3	3
X.18	−3	−1	3	.	−1	−1	−1	5	5	−1	5	−1	−1	−1	−1	−3	−3	−3	−3	−3	−1	−1	−1	−1
X.19	3	−1	3	.	−1	−1	−1	−5	−5	−1	−5	−1	−1	−1	−1	3	3	3	3	3	−1	−1	−1	−1
X.20	−3	3	3	.	−1	−1	−1	−7	−7	−1	−7	3	3	3	3	−3	−3	−1	−3	−3	3	3	3	3
X.21	1	3	.	3	15	15	−1	−3	−3	−1	−3	−1	−1	3	−1	1	−7	−3	1	−7	−1	3	−1	3
X.22	−3	−1	.	3	15	15	−1	−3	−3	−1	−3	3	3	−1	3	−3	5	−3	−3	5	3	−1	3	1
X.23	−1	3	.	3	15	15	−1	3	3	−1	3	−1	−1	3	−1	−1	7	3	−1	7	−1	3	−1	3
X.24	3	−1	.	3	15	15	−1	3	3	−1	3	3	3	−1	3	3	−5	3	3	−5	3	−1	3	−1
X.25	.	4	1	.	.	.	.	.	.	.	4	4	4	−4	−4	−4	.	.	.	4	.	−4	−4	.
X.26	.	4	1	.	.	.	.	.	.	.	4	4	4	−4	.	.	−4	.	.	4	.	−4	−4	.
X.27	.	.	−2	−2	16	16	16	.	.	16	.	.	.	.	.	.	.	.	.	.	.	.	.	.
X.28	.	4	2	2	20	−20	4	.	.	−4	.	−4	4	4	−4	4	.	.	.	.	4	4	4	−4
X.29	.	.	.	3	24	−24	−8	.	.	8	.	.	.	.	.	.	−8	.	.	8	.	.	.	.
X.30	.	.	.	3	24	−24	−8	.	.	8	.	.	.	.	.	.	8	.	.	−8	.	.	.	.
X.31	.	−2	.	−3	30	−30	−10	.	.	10	.	−2	2	−2	−2	.	4	.	.	−4	2	−2	2	2
X.32	−6	2	−3	.	−2	−2	−2	−2	−2	−2	−2	2	2	2	2	−6	−6	−2	−6	−6	2	2	2	2
X.33	6	2	−3	.	−2	−2	−2	2	2	−2	2	2	2	2	2	6	6	2	6	6	2	2	2	2
X.34	.	−2	.	−3	30	−30	−10	.	.	10	.	−2	2	−2	−2	.	−4	.	.	4	2	−2	2	2
X.35	2	2	.	−3	30	30	−2	6	6	−2	6	2	2	2	2	2	2	6	2	2	2	2	2	2
X.36	−2	−2	.	−3	30	30	−2	−6	−6	−2	−6	2	2	2	2	−2	−2	−6	−2	−2	2	2	2	2
X.37	.	4	.	−3	36	−36	−12	.	.	12	.	4	−4	4	4	.	.	.	.	.	−4	4	−4	−4
X.38	.	−2	1	.	40	−40	8	.	.	−8	.	.	.	.	.	.	−8	.	.	8	.	.	.	.
X.39	.	.	1	−2	40	−40	8	8	8	−8	−8	.	.	.	.	.	.	.	.	−8	.	.	.	.
X.40	.	.	1	−2	40	−40	8	−8	−8	8	8	.	.	.	.	.	.	.	.	8	.	.	.	.
X.41	.	−2	1	.	40	−40	8	.	.	−8	.	.	.	.	.	.	8	.	.	−8	.	.	.	.
X.42	1	1	.	.	45	45	−3	−3	−3	13	−3	−3	−3	−3	1	−3	1	9	−3	1	9	−3	1	−3
X.43	5	1	.	.	−3	−3	13	9	9	13	9	5	5	1	5	5	−3	1	5	−3	5	1	5	1
X.44	1	1	.	.	−3	−3	13	−3	−3	13	−3	−3	−3	1	−3	1	9	5	1	9	−3	5	1	−3
X.45	−3	1	.	.	−3	−3	−3	9	9	−3	9	1	1	1	1	−3	−3	1	−3	−3	1	1	1	1
X.46	5	−3	.	.	45	45	−3	−3	−3	−3	−3	1	1	−3	1	5	−3	−3	5	−3	1	−3	1	−3
X.47	−1	1	.	.	−3	−3	13	3	3	13	3	−3	−3	1	−3	−1	−9	−5	−1	−9	−3	1	−3	1
X.48	−1	1	.	.	45	45	−3	3	3	−3	3	−3	−3	1	−3	−1	−9	3	−1	−9	−3	1	−3	1
X.49	−5	1	.	.	−3	−3	13	−9	−9	13	−9	5	5	1	5	−5	3	−1	−5	3	5	1	−3	1
X.50	−5	−3	.	.	45	45	−3	3	3	−3	3	1	1	−3	1	−5	3	3	−5	3	1	−3	1	−3
X.51	3	1	.	.	−3	−3	−3	−9	−9	−3	−9	1	1	1	1	3	3	−3	3	3	1	1	1	1
X.52	3	−3	.	.	−3	−3	−3	3	3	−3	3	−3	−3	−3	−3	3	3	−5	3	3	−3	−3	−3	−3
X.53	−5	−3	.	.	−3	−3	13	3	3	13	3	1	1	1	−3	−5	3	−5	3	1	−3	1	−3	−3
X.54	1	5	.	.	−3	−3	13	9	9	13	9	1	1	1	5	1	9	1	1	9	1	5	1	5
X.55	−3	−3	.	.	−3	−3	13	−3	−3	13	−3	−3	−3	−3	−3	−3	−3	5	−3	−3	−3	−3	−3	−3
X.56	5	−3	.	.	−3	−3	13	−3	−3	13	−3	1	1	1	−3	1	5	−3	5	5	1	−3	1	−3
X.57	−1	5	.	.	−3	−3	13	−9	−9	13	−9	1	1	1	5	1	−9	−1	−1	−9	1	5	1	5
X.58	.	−4	3	.	−4	4	12	16	16	−12	−16	4	−4	−4	4	.	.	.	.	.	−4	−4	−4	4
X.59	.	4	3	.	−4	4	12	8	8	−12	−8	−4	4	4	−4	.	.	.	.	8	4	4	4	−4
X.60	.	−4	3	.	−4	4	12	−16	−16	−12	16	4	−4	−4	4	.	.	.	.	.	−4	−4	−4	4
X.61	.	4	3	.	−4	4	12	−8	−8	−12	8	−4	4	4	−4	.	.	.	.	−8	4	4	4	−4
X.62	.	.	8	−1	.	.	−12	12	.	.	.	4	4	4	−4	−4	.	.	.	4	.	−4	−4	.
X.63	.	.	8	−1	.	.	12	−12	.	.	.	4	4	4	−4	4	.	−4	.	−4	.	−4	−4	.
X.64	.	.	−4	2	.	.	−4	4	.	.	.	4	4	4	−4	12	.	12	.	−12	.	−4	−4	.
X.65	.	.	−4	2	.	.	4	−4	.	.	.	4	4	4	−4	−12	.	−12	.	12	.	−4	−4	.
X.66	.	2	.	.	−6	6	2	.	.	−2	.	2	−2	2	2	.	−12	.	.	12	2	−2	−2	−2
X.67	.	−6	.	.	−6	6	2	.	.	−2	.	−6	6	−6	−6	.	12	.	.	−12	6	−6	6	6
X.68	.	2	.	.	−6	6	2	.	.	−2	.	2	−2	2	2	.	12	.	.	−12	6	−6	6	6
X.69	−6	−2	.	.	−6	−6	26	6	6	26	6	−2	−2	−2	−2	−6	−6	6	−6	−6	−2	−2	−2	−2
X.70	.	−6	.	.	−6	6	2	.	.	−2	.	−6	6	−6	−6	.	−12	.	.	12	6	−6	6	6
X.71	.	−2	.	.	−6	6	−14	−12	−12	14	12	2	−2	−2	2	.	−12	−4	.	12	−2	−2	−2	2
X.72	.	−2	.	.	−6	6	−14	−12	−12	14	12	2	−2	−2	2	.	12	−4	.	−12	−2	−2	−2	2
X.73	6	−2	.	.	−6	−6	26	−6	−6	26	−6	−2	−2	−2	−2	6	6	−6	6	6	−2	−2	−2	−2
X.74	.	−2	.	.	−6	6	−14	12	12	14	−12	2	−2	−2	2	.	−12	4	.	12	−2	−2	−2	2
X.75	.	−2	.	.	−6	6	−14	12	12	14	−12	2	−2	−2	2	.	12	4	.	−12	−2	−2	−2	2
X.76	.	.	3	.	.	.	.	.	.	.	.	−8	8	−8	8	.	.	.	.	.	−8	8	.	.
X.77	.	.	3	.	.	.	.	.	.	.	.	−8	8	−8	8	.	.	.	.	.	−8	8	.	.
X.78	−8	−8	6	.	8	.	8	8	8	.	4	−4	−8	4	8	.	−8	8	.	−4	−8	12	.	.
X.79	−4	4	6	.	.	−8	.	−8	.	.	8	−8	.	4	−8	−4	4	−8	−4	4	.	4	.	12
X.80	.	.	−3	.	−8	8	24	−8	−8	−24	8	.	.	.	.	.	8	.	.	.	.	.	.	.
X.81	−8	−8	6	.	8	.	8	8	8	.	4	12	8	4	−8	.	−8	8	.	−4	8	12	8	−4
X.82	4	4	6	.	−8	.	.	−8	.	.	−8	−8	.	4	−8	4	−4	8	4	−4	.	4	.	12
X.83	4	4	6	.	−8	.	.	−8	.	.	−8	8	.	4	8	4	−4	−8	4	−4	.	4	.	−4
X.84	8	−8	6	.	8	.	8	−8	−8	.	4	12	8	4	−8	.	−8	−8	.	12	8	−4	.	.

Character table of $D(\mathrm{Co_2})$ (continued)

	4_{21}	4_{22}	4_{23}	4_{24}	4_{25}	4_{26}	4_{27}	4_{28}	4_{29}	4_{30}	4_{31}	4_{32}	4_{33}	4_{34}	4_{35}	4_{36}	4_{37}	4_{38}	4_{39}	4_{40}	4_{41}	4_{42}	4_{43}	4_{44}	4_{45}
2	11	11	11	11	11	11	9	9	9	9	10	10	10	10	10	10	10	10	10	10	10	10	10	8	9
3	.	.	.	.	.	.	1	1	1	1	.	.	.	.	.	.	.	.	.	.	.	.	.	.	.
5	.	.	.	.	.	.	.	.	.	.	.	.	.	.	.	.	.	.	.	.	.	.	1	.	.
2P	2e	2f	2e	2d	2d	2d	2g	2g	2g	2g	2h	2h	2e	2h	2e	2e	2d	2d	2d	2f	2d	2e	2h	2h	2h
3P	4_{21}	4_{22}	4_{23}	4_{24}	4_{25}	4_{26}	4_{27}	4_{28}	4_{29}	4_{30}	4_{31}	4_{32}	4_{33}	4_{34}	4_{35}	4_{36}	4_{37}	4_{38}	4_{39}	4_{40}	4_{41}	4_{42}	4_{43}	4_{44}	4_{45}
5P	4_{21}	4_{22}	4_{23}	4_{24}	4_{25}	4_{26}	4_{27}	4_{28}	4_{29}	4_{30}	4_{31}	4_{32}	4_{33}	4_{34}	4_{35}	4_{36}	4_{37}	4_{38}	4_{39}	4_{40}	4_{41}	4_{42}	4_{43}	4_{44}	4_{45}
X.1	1	1	1	1	1	1	1	1	1	1	1	1	1	1	1	1	1	1	1	1	1	1	1	1	1
X.2	-1	1	-1	1	1	-1	-1	-1	-1	-1	1	1	-1	-1	1	1	1	1	-1	-1	-1	-1	-1	-1	1
X.3	3	1	3	5	5	3	-1	-1	-1	-1	1	1	3	-1	1	1	1	1	-1	3	3	3	-1	3	1
X.4	1	1	1	5	5	1	-3	-3	-3	-3	1	1	1	-3	1	1	1	1	-3	1	1	1	-3	1	1
X.5	-1	1	-1	5	5	-1	3	3	3	3	1	1	-1	3	1	1	1	1	3	-1	-1	-1	3	-1	1
X.6	-3	1	-3	5	5	-3	1	1	1	1	1	1	-3	1	1	1	1	1	1	-3	-3	-3	1	-3	1
X.7	.	-2	.	2	-2	.	-2	2	2	-2	2	2	.	4	-2	2	-2	2	.	.	.	.	-4	.	-2
X.8	.	-2	.	2	-2	.	2	-2	-2	2	2	2	.	-4	-2	2	-2	2	.	.	.	.	4	.	-2
X.9	-3	1	-3	9	9	-3	-3	-3	-3	-3	1	1	-3	-3	1	1	1	1	-3	-3	-3	-3	-3	-3	1
X.10	3	1	3	9	9	3	3	3	3	3	1	1	3	3	1	1	1	1	3	3	3	3	3	3	1
X.11	-2	-2	-2	10	10	-2	2	2	2	2	-2	-2	-2	2	-2	-2	-2	-2	2	-2	-2	-2	-2	2	-2
X.12	.	2	.	-2	2	.	4	-2	-2	2	2	2	.	4	2	-2	-2	2	.	.	.	.	-4	.	-2
X.13	2	-2	2	10	10	2	-2	-2	-2	-2	-2	-2	2	-2	-2	-2	-2	-2	-2	2	2	2	-2	2	-2
X.14	.	2	.	-2	2	.	4	-2	-2	2	2	2	.	4	2	-2	-2	2	.	.	.	.	4	.	-2
X.15	.	2	.	-2	2	.	-2	-2	-2	-2	2	2	.	4	2	-2	-2	2	.	.	4	.	-4	-4	-2
X.16	.	2	.	-2	2	.	-4	-2	-2	2	2	2	.	4	2	-2	-2	2	.	.	4	.	-4	-4	-2
X.17	7	3	7	-1	-1	-1	3	3	3	3	-1	-1	-1	-1	3	3	3	3	3	-1	-1	7	-1	-1	-1
X.18	5	-1	5	-1	-3	-3	-3	-3	-3	-3	3	3	-1	-1	-1	-1	-1	-1	-3	-3	-3	5	1	1	3
X.19	-5	-1	-5	-1	-1	3	3	3	3	3	3	3	-1	-1	-1	-1	-1	-3	3	3	-5	-1	-1	3	3
X.20	-7	3	-7	-1	-1	3	-3	-3	-3	-3	1	1	3	3	-1	-1	-1	-1	3	3	1	1	-7	1	-1
X.21	1	-1	1	-1	-1	-3	1	1	1	1	3	3	1	-7	-1	-1	3	3	1	1	3	1	-7	-3	3
X.22	1	3	1	-1	-1	-3	1	1	1	1	-1	-1	1	5	3	3	-1	-1	3	1	-3	1	5	-3	-1
X.23	-1	-1	-1	-1	-1	3	-1	-1	-1	-1	-1	-1	3	1	-1	-1	-1	-1	3	1	3	-1	1	7	3
X.24	-1	3	-1	-1	-1	3	-1	-1	-1	-1	-1	-1	-1	-5	3	3	-1	-1	3	-1	3	-1	-5	3	-1
X.25	-4	.	.	4	.	.	.	2	2	-2	-2	.	.	.	.	.	.	.	.	.	.	.	.	.	.
X.26	4	.	.	-4	.	.	.	-2	-2	2	2	.	.	.	.	.	.	.	.	.	.	.	.	.	.
X.27	.	.	.	16	16	.	.	.	.	.	.	.	.	.	.	.	.	.	.	.	.	.	.	.	.
X.28	.	-4	.	-4	4	.	.	.	.	.	-4	-4	.	.	8	.	-4	4	4	-4	.	.	.	.	4
X.29	.	.	.	8	-8	.	-4	4	4	-4	.	.	.	8	.	.	-4	4	4	-4	.	.	.	-8	.
X.30	.	.	.	8	-8	.	4	-4	-4	4	.	.	.	-8	.	.	.	.	.	.	.	.	.	8	.
X.31	.	-2	.	10	-10	.	2	-2	-2	2	2	2	.	-4	-2	2	-2	2	.	.	.	.	4	.	-2
X.32	-2	2	-2	-2	-2	-2	-6	-6	-6	-6	2	2	-2	2	2	2	2	2	-6	-2	-2	-2	2	2	-2
X.33	2	2	2	-2	-2	2	6	6	6	6	2	2	-2	2	2	2	2	2	6	2	2	-2	-2	-2	2
X.34	.	-2	.	10	-10	.	-2	2	2	-2	2	2	.	4	-2	2	-2	2	.	.	.	.	-4	.	-2
X.35	-2	2	-2	-2	-2	6	-2	-2	-2	-2	2	2	-2	2	2	2	2	2	2	-2	6	-2	2	6	2
X.36	2	2	2	-2	-2	-6	2	2	2	2	2	2	-2	2	2	2	2	2	-2	2	-6	2	-2	-6	2
X.37	.	4	.	12	-12	.	.	.	.	.	-4	-4	.	.	8	.	4	-4	4	-4	.	.	.	.	4
X.38	.	.	.	-8	8	.	4	-4	-4	4	.	.	.	8	.	.	.	.	.	.	.	.	.	-8	.
X.39	.	.	.	-8	8	-8	.	.	.	.	.	.	.	.	.	.	.	.	.	.	.	.	8	.	-8
X.40	.	.	.	-8	8	8	.	.	.	.	.	.	.	.	.	.	.	.	.	.	.	.	-8	.	8
X.41	.	.	.	-8	8	.	-4	4	4	-4	.	.	.	-8	.	.	.	.	.	.	.	.	8	.	.
X.42	1	-3	1	-3	-3	-3	-3	-3	-3	-3	1	1	1	9	-3	-3	1	1	1	1	-3	1	9	-3	1
X.43	-3	5	-3	-3	-3	1	-3	-3	-3	-3	-3	-3	5	1	5	5	1	1	5	5	1	-3	1	-3	-3
X.44	1	-3	1	-3	-3	5	-3	-3	-3	-3	5	5	-7	-3	-3	-3	1	1	1	1	-7	1	5	-3	-3
X.45	9	1	9	-3	-3	1	-3	-3	-3	-3	1	1	1	1	1	1	-3	1	1	9	1	1	-3	-3	-3
X.46	1	1	1	-3	-3	-3	-3	-3	-3	-3	1	-3	1	1	-3	-3	1	1	5	1	-3	-3	-3	-3	-3
X.47	-1	-3	-1	-3	-3	-5	3	3	3	3	5	5	7	3	-3	-3	1	1	-1	7	-5	-1	3	3	5
X.48	-1	-3	-1	-3	-3	3	3	3	3	3	1	1	-1	-9	-3	-3	1	1	-1	-1	1	1	-9	3	1
X.49	3	5	3	-3	-3	-1	3	3	3	3	-3	-3	-5	-1	5	5	1	1	-5	-5	-1	3	-1	3	-3
X.50	-1	1	-1	-3	-3	3	3	3	3	3	-3	-3	-1	3	1	1	-3	-3	-5	-1	3	1	3	3	-3
X.51	-9	1	-9	-3	-3	1	3	3	3	3	-3	-3	-1	-1	1	1	1	1	5	1	-1	-1	-9	-1	1
X.52	3	-3	3	-3	-3	-5	1	1	1	1	-5	-1	-3	-3	-3	-3	3	-5	-5	3	-1	3	1	.	.
X.53	-1	1	-1	-3	-3	-5	3	3	3	3	1	1	7	-1	1	1	5	-5	-1	-1	1	.	3	.	.
X.54	-3	1	-3	-3	-3	1	-3	-3	-3	-3	1	1	5	-3	1	1	5	5	1	-3	-3	-3	3	1	1
X.55	-3	-3	-3	-3	-3	-3	-3	-3	-3	-3	1	1	5	1	-3	-3	-3	-3	5	5	1	-3	1	1	1
X.56	1	1	1	-3	-3	5	-3	-3	-3	-3	1	1	-7	1	1	1	-3	-3	5	-7	5	1	-3	1	1
X.57	3	1	3	-3	-3	-1	3	3	3	3	1	1	-5	3	1	1	1	5	5	-1	-5	-1	3	3	1
X.58	.	4	.	4	-4	.	.	.	.	.	-4	-4	.	.	4	.	-4	-4	4	4	.	.	.	.	4
X.59	.	-4	.	4	-4	8	.	.	.	.	.	.	.	.	4	4	.	.	.	.	.	-8	.	-4	-4
X.60	.	4	.	4	-4	.	.	.	.	.	-4	-4	.	.	4	.	-4	-4	4	4	.	.	.	.	4
X.61	.	-4	.	4	-4	-8	.	.	.	.	4	4	.	.	-4	4	4	-4	.	.	.	8	.	4	-4
X.62	12	.	-12	.	.	.	2	2	-2	-2	.	.	.	.	.	.	.	.	.	.	.	.	.	.	.
X.63	-12	.	12	.	.	.	-2	-2	2	2	.	.	.	.	.	.	.	.	.	.	.	.	.	.	.
X.64	4	.	-4	.	.	.	6	6	-6	-6	.	.	.	.	.	.	.	.	.	.	.	.	.	.	.
X.65	-4	.	4	.	.	.	-6	-6	6	6	.	.	.	.	.	.	.	.	.	.	.	.	.	.	.
X.66	.	2	.	-2	2	.	-6	6	6	-6	6	6	.	-4	2	-2	2	-2	.	.	.	.	4	.	-6
X.67	.	-6	.	-2	2	.	6	-6	-6	6	-2	-2	.	4	-6	6	-6	6	.	.	.	.	-4	.	2
X.68	.	2	.	-2	2	.	6	-6	-6	6	6	6	.	4	2	-2	2	-2	.	.	.	.	-4	.	-6
X.69	-2	-2	-2	-6	-6	6	6	6	6	6	-2	-2	-2	2	-2	-2	-2	-2	-6	-2	6	-2	2	-6	-2
X.70	.	-6	.	-2	2	.	6	-6	-6	6	-2	-2	.	-4	-6	6	-6	6	.	.	.	.	4	.	2
X.71	.	2	.	-2	2	-4	6	-6	-6	6	2	2	.	-4	2	-2	2	-2	.	.	.	4	.	4	-2
X.72	.	-2	.	-2	2	-4	6	-6	-6	6	2	2	.	4	-2	2	-2	2	.	.	.	4	.	4	-2
X.73	2	-2	2	-6	-6	-6	-6	-6	-6	-6	-2	-2	2	-2	-2	-2	-2	-2	-6	2	-6	2	-2	6	-2
X.74	.	2	.	-2	2	4	6	-6	-6	6	2	2	.	-4	2	-2	2	-2	.	.	.	.	-4	4	-2
X.75	.	2	.	-2	2	4	6	-6	-6	6	2	2	.	4	2	-2	2	-2	.	.	.	.	-4	4	-2
X.76	.	.	.	.	.	.	4	-4	4	-4	.	.	.	.	.	.	.	.	.	.	.	.	.	.	.
X.77	.	.	.	.	.	.	-4	4	-4	4	.	.	.	.	.	.	.	.	.	.	.	.	.	.	.
X.78	-4	4	-4	.	.	.	.	.	.	.	.	.	.	4	.	-4	-4	.	.	.	.	-4	.	4	.
X.79	4	8	4	.	.	.	.	.	.	.	.	.	.	4	.	.	-4	-4	4	-4	.	.	-4	-4	.
X.80	.	.	.	8	-8	8	.	.	.	.	.	.	.	.	.	.	.	.	.	.	.	.	-8	.	-8
X.81	4	4	4	.	.	.	.	.	.	.	.	.	.	-4	.	-4	-4	.	.	.	.	4	.	-4	.
X.82	-4	8	-4	.	.	.	.	.	.	.	.	.	.	-4	.	.	-4	-4	-4	4	.	.	4	.	.
X.83	4	-8	4	.	.	.	.	.	.	.	.	.	.	4	.	-4	-4	-4	-4	.	.	-4	.	.	.
X.84	-4	4	-4	.	.	.	.	.	.	.	.	.	.	4	.	-4	-4	.	.	.	.	-4	.	4	.

Character table of $D(\mathrm{Co}_2)$ (continued)

2	9	8	8	8	8	8	8	8	8	8	8	8	8	8	8	2	7	5	5	5	5	5	6	5	5	5	5	5
3	.	.	.	.	.	.	.	.	.	.	.	.	.	.	.	.	2	2	2	2	2	2	1	1	1	1	1	1
5	.	.	.	.	.	.	.	.	.	.	.	.	.	.	.	1	.	.	.	.	.	.	.	.	.	.	.	.
	4_{46}	4_{47}	4_{48}	4_{49}	4_{50}	4_{51}	4_{52}	4_{53}	4_{54}	4_{55}	4_{56}	4_{57}	4_{58}	4_{59}	4_{60}	$5a$	$6a$	$6b$	$6c$	$6d$	$6e$	$6f$	$6g$	$6h$	$6i$	$6j$	$6k$	$6l$
$2P$	$2h$	$2l$	$2t$	$2t$	$2t$	$2g$	$2t$	$2t$	$2g$	$2t$	$2t$	$2t$	$2q$	$2q$	$2h$	$5a$	$3a$	$3b$	$3b$	$3b$	$3a$	$3a$	$3a$	$3b$	$3b$	$3a$	$3a$	$3b$
$3P$	4_{46}	4_{47}	4_{48}	4_{49}	4_{50}	4_{51}	4_{52}	4_{53}	4_{54}	4_{55}	4_{56}	4_{57}	4_{58}	4_{59}	4_{60}	$5a$	$2a$	$2c$	$2b$	$2a$	$2b$	$2c$	$2d$	$2g$	$2i$	$2j$	$2h$	$2f$
$5P$	4_{46}	4_{47}	4_{48}	4_{49}	4_{50}	4_{51}	4_{52}	4_{53}	4_{54}	4_{55}	4_{56}	4_{57}	4_{58}	4_{59}	4_{60}	$1a$	$6a$	$6b$	$6c$	$6d$	$6e$	$6f$	$6g$	$6h$	$6i$	$6j$	$6k$	$6l$
$X.1$	1	1	1	1	1	1	1	1	1	1	1	1	1	1	1	1	1	1	1	1	1	1	1	1	1	1	1	1
$X.2$	-1	-1	1	-1	-1	1	-1	-1	1	1	1	1	1	-1	1	1	1	1	1	1	1	1	1	1	1	-1	1	1
$X.3$	-1	3	-1	1	1	1	1	1	1	-1	-1	-1	1	3	1	.	2	-1	-1	-1	2	2	2	-1	-1	.	2	-1
$X.4$	-3	1	-1	-1	-1	1	-1	-1	1	-1	-1	-1	1	1	1	.	-1	2	2	2	-1	-1	-1	2	2	1	-1	2
$X.5$	3	-1	-1	1	1	1	1	1	1	-1	-1	-1	1	-1	1	.	-1	2	2	2	-1	-1	-1	2	2	-1	-1	2
$X.6$	1	-3	-1	-1	-1	1	-1	-1	1	-1	-1	-1	1	-3	1	.	2	-1	-1	-1	2	2	2	-1	-1	.	2	-1
$X.7$	.	.	.	2	-2	.	2	-2	.	.	.	.	-2	.	2	1	.	-3	-3	3	.	.	.	-1	-1	.	.	1
$X.8$	.	.	.	-2	2	.	-2	2	.	.	.	.	-2	.	2	1	.	-3	-3	3	.	.	.	-1	-1	.	.	1
$X.9$	-3	-3	1	1	1	1	1	1	1	1	1	1	1	-3	1	-1	.	.	.	.	.	.	.	.	.	.	.	.
$X.10$	3	3	1	-1	-1	1	-1	-1	1	1	1	1	1	3	1	-1	.	.	.	.	.	.	.	.	.	.	.	.
$X.11$	2	-2	.	.	.	-2	.	.	-2	.	.	.	-2	-2	-2	.	1	1	1	1	1	1	1	1	1	1	1	1
$X.12$	.	.	2	.	.	.	.	.	.	-2	2	-2	2	.	2	.	1	-1	-1	1	-1	-1	1	1	1	1	1	-1
$X.13$	-2	2	.	.	.	-2	.	.	-2	.	.	.	-2	2	-2	.	1	1	1	1	1	1	1	1	1	-1	1	1
$X.14$	.	.	-2	.	.	.	.	.	.	2	-2	2	2	.	2	.	1	-1	-1	1	-1	-1	1	1	1	1	1	-1
$X.15$	.	.	-2	.	.	.	.	.	.	2	-2	2	2	.	2	.	1	-1	-1	1	-1	-1	1	1	1	-1	1	-1
$X.16$	.	.	2	.	.	.	.	.	.	-2	2	-2	2	.	2	.	1	-1	-1	1	-1	-1	1	1	1	-1	1	-1
$X.17$	-1	-1	1	1	1	3	1	1	3	1	1	1	-1	-1	-1	.	3	.	.	.	3	3	3	.	.	1	-1	.
$X.18$	1	-3	-1	1	1	-1	1	1	-1	-1	-1	-1	-1	1	-1	.	3	.	.	.	3	3	3	.	.	-1	-1	.
$X.19$	-1	3	-1	-1	-1	-1	-1	-1	-1	-1	-1	-1	-1	-1	-1	.	3	.	.	.	3	3	3	.	.	1	-1	.
$X.20$	1	1	1	-1	-1	3	-1	-1	3	1	1	1	-1	1	-1	.	3	.	.	.	3	3	3	.	.	-1	-1	.
$X.21$	1	1	1	-1	-1	-1	-1	-1	-1	1	1	1	-1	1	3	.	.	3	3	3	.	.	.	-1	-1	.	.	-1
$X.22$	-3	1	-1	1	1	-1	1	1	-1	-1	-1	-1	3	1	-1	.	.	3	3	3	.	.	.	-1	-1	.	.	-1
$X.23$	-1	-1	1	1	1	-1	1	1	-1	1	1	1	-1	-1	3	.	.	3	3	3	.	.	.	-1	-1	.	.	-1
$X.24$	3	-1	-1	-1	-1	-1	-1	-1	-1	-1	-1	-1	3	-1	-1	.	.	3	3	3	.	.	.	-1	-1	.	.	-1
$X.25$	.	.	2	2	-2	2	-2	2	-2	2	-2	-2	.	.	.	1	-4	1	-1	-1	2	-2	.	1	-1	2	.	1
$X.26$	.	.	2	-2	2	2	2	-2	-2	2	-2	-2	.	.	.	1	-4	1	-1	-1	2	-2	.	1	-1	-2	.	1
$X.27$	.	.	.	.	.	.	.	.	.	.	.	.	.	.	.	1	-2	-2	-2	-2	-2	-2	-2	-2	-2	.	-2	-2
$X.28$	.	.	.	.	.	.	.	.	.	.	.	.	-4	.	-4	.	2	-2	-2	2	-2	-2	2	2	2	.	2	-2
$X.29$	.	.	.	.	.	.	.	.	.	.	.	.	.	.	.	-1	.	-3	-3	3	.	.	.	-1	-1	.	.	1
$X.30$	.	.	.	.	.	.	.	.	.	.	.	.	.	.	.	-1	.	-3	-3	3	.	.	.	-1	-1	.	.	1
$X.31$	.	.	.	2	-2	.	2	-2	.	.	.	.	-2	.	2	.	.	3	3	-3	.	.	.	1	1	.	.	-1
$X.32$	2	-2	.	.	.	2	.	.	2	.	.	.	-2	2	-2	.	-3	.	.	.	-3	-3	-3	.	.	1	1	.
$X.33$	-2	2	.	.	.	2	.	.	2	.	.	.	-2	-2	-2	.	-3	.	.	.	-3	-3	-3	.	.	-1	1	.
$X.34$	.	.	.	-2	2	.	-2	2	.	.	.	.	-2	.	2	.	.	3	3	-3	.	.	.	1	1	.	.	-1
$X.35$	2	-2	.	.	.	-2	.	.	-2	.	.	.	2	-2	2	.	.	-3	-3	-3	.	.	.	1	1	.	.	1
$X.36$	-2	2	.	.	.	-2	.	.	-2	.	.	.	2	2	2	.	.	-3	-3	-3	.	.	.	1	1	.	.	1
$X.37$	.	.	.	.	.	.	.	.	.	.	.	.	4	.	-4	1	.	.	.	.	.	.	.	.	.	.	.	.
$X.38$	.	.	.	.	.	.	.	.	.	.	.	.	.	.	.	.	-2	-1	-1	1	2	2	-2	1	1	.	-2	-1
$X.39$	.	.	.	.	.	.	.	.	.	.	.	.	.	.	.	.	1	2	2	-2	-1	-1	1	-2	-2	1	1	2
$X.40$	.	.	.	.	.	.	.	.	.	.	.	.	.	.	.	.	1	2	2	-2	-1	-1	1	-2	-2	-1	1	2
$X.41$	.	.	.	.	.	.	.	.	.	.	.	.	.	.	.	.	-2	-1	-1	1	2	2	-2	1	1	.	-2	-1
$X.42$	1	1	-1	-1	-1	1	-1	-1	1	-1	-1	-1	-3	1	1	.	.	.	.	.	.	.	.	.	.	.	.	.
$X.43$	-7	-3	-1	-1	-1	-3	-1	-1	-3	-1	-1	-1	-3	1	1	.	.	.	.	.	.	.	.	.	.	.	.	.
$X.44$	5	1	-1	-1	-1	1	-1	-1	1	-1	-1	-1	1	1	-3	.	.	.	.	.	.	.	.	.	.	.	.	.
$X.45$	1	1	-1	-1	-1	1	-1	-1	1	-1	-1	-1	1	-3	1	.	.	.	.	.	.	.	.	.	.	.	.	.
$X.46$	5	1	1	1	1	1	1	1	1	1	1	1	1	1	-3	.	.	.	.	.	.	.	.	.	.	.	.	.
$X.47$	-5	-1	-1	1	1	1	1	1	1	-1	-1	-1	1	-1	-3	.	.	.	.	.	.	.	.	.	.	.	.	.
$X.48$	-1	-1	-1	1	1	1	1	1	1	-1	-1	-1	-3	-1	1	.	.	.	.	.	.	.	.	.	.	.	.	.
$X.49$	7	3	-1	1	1	-3	1	1	-3	-1	-1	-1	-3	-1	1	.	.	.	.	.	.	.	.	.	.	.	.	.
$X.50$	-5	-1	1	-1	-1	1	-1	-1	1	1	1	1	1	-1	-3	.	.	.	.	.	.	.	.	.	.	.	.	.
$X.51$	-1	-1	-1	1	1	1	1	1	1	-1	-1	-1	1	3	1	.	.	.	.	.	.	.	.	.	.	.	.	.
$X.52$	-1	-5	1	-1	-1	-3	-1	-1	-3	1	1	1	1	3	1	.	.	.	.	.	.	.	.	.	.	.	.	.
$X.53$	7	-1	1	-1	-1	1	-1	-1	1	1	1	1	-3	-1	1	.	.	.	.	.	.	.	.	.	.	.	.	.
$X.54$	5	-3	1	1	1	-3	1	1	-3	1	1	1	1	1	-3	.	.	.	.	.	.	.	.	.	.	.	.	.
$X.55$	1	5	1	1	1	-3	1	1	-3	1	1	1	1	-3	1	.	.	.	.	.	.	.	.	.	.	.	.	.
$X.56$	-7	1	1	1	1	1	1	1	1	1	1	1	-3	1	1	.	.	.	.	.	.	.	.	.	.	.	.	.
$X.57$	-5	3	1	-1	-1	-3	-1	-1	-3	1	1	1	1	-1	-3	.	.	.	.	.	.	.	.	.	.	.	.	.
$X.58$	.	.	.	.	.	.	.	.	.	.	.	.	.	-4	.	.	3	.	.	.	-3	-3	3	.	.	-1	-1	.
$X.59$	.	.	.	.	.	.	.	.	.	.	.	.	.	-4	.	.	3	.	.	.	-3	-3	3	.	.	1	-1	.
$X.60$	.	.	.	.	.	.	.	.	.	.	.	.	.	4	.	.	3	.	.	.	-3	-3	3	.	.	1	-1	.
$X.61$	.	.	.	.	.	.	.	.	.	.	.	.	.	4	.	.	3	.	.	.	-3	-3	3	.	.	-1	-1	.
$X.62$	.	.	-2	-2	2	2	2	-2	-2	-2	2	2	.	.	.	.	-8	-1	1	1	4	-4	.	-1	1	.	.	-1
$X.63$	.	.	-2	2	-2	2	-2	2	-2	-2	2	2	.	.	.	.	-8	-1	1	1	4	-4	.	-1	1	.	.	-1
$X.64$	.	.	-2	2	-2	2	-2	2	-2	-2	2	2	.	.	.	.	4	2	-2	-2	-2	2	.	2	-2	-2	.	2
$X.65$	.	.	-2	-2	2	2	2	-2	-2	-2	2	2	.	.	.	.	4	2	-2	-2	-2	2	.	2	-2	2	.	2
$X.66$	.	.	.	-2	2	.	-2	2	.	.	.	.	2	.	-2	.	.	.	.	.	.	.	.	.	.	.	.	.
$X.67$	.	.	.	-2	2	.	-2	2	.	.	.	.	2	.	-2	.	.	.	.	.	.	.	.	.	.	.	.	.
$X.68$	.	.	.	2	-2	.	2	-2	.	.	.	.	2	.	-2	.	.	.	.	.	.	.	.	.	.	.	.	.
$X.69$	2	-2	.	.	.	2	.	.	2	.	.	.	2	2	2	.	.	.	.	.	.	.	.	.	.	.	.	.
$X.70$	.	.	.	2	-2	.	2	-2	.	.	.	.	2	.	-2	.	.	.	.	.	.	.	.	.	.	.	.	.
$X.71$	.	.	2	.	.	.	.	.	.	-2	2	-2	-2	-4	-2	.	.	.	.	.	.	.	.	.	.	.	.	.
$X.72$	.	.	-2	.	.	.	.	.	.	2	-2	2	-2	-4	-2	.	.	.	.	.	.	.	.	.	.	.	.	.
$X.73$	-2	2	.	.	.	2	.	.	2	.	.	.	2	-2	2	.	.	.	.	.	.	.	.	.	.	.	.	.
$X.74$	.	.	-2	.	.	.	.	.	.	2	-2	2	-2	4	-2	.	.	.	.	.	.	.	.	.	.	.	.	.
$X.75$	.	.	2	.	.	.	.	.	.	-2	2	-2	-2	4	-2	.	.	.	.	.	.	.	.	.	.	.	.	.
$X.76$	.	.	.	-4	-4	.	4	4	.	.	.	.	.	.	.	1	.	-3	3	-3	.	.	.	-1	1	.	.	1
$X.77$	.	.	.	4	4	.	-4	-4	.	.	.	.	.	.	.	1	.	-3	3	-3	.	.	.	-1	1	.	.	1
$X.78$	.	.	.	.	.	.	.	.	.	.	.	.	.	.	.	.	6	.	.	.	.	.	-2	.	.	.	-2	.
$X.79$	.	.	-2	2	2	.	2	2	.	-2	-2	-2	.	.	.	.	6	.	.	.	.	.	-2	.	.	-2	2	.
$X.80$	.	.	.	.	.	.	.	.	.	.	.	.	.	.	.	.	-3	.	.	.	3	3	-3	.	.	-1	1	.
$X.81$	.	.	.	.	.	.	.	.	.	.	.	.	.	.	.	.	6	.	.	.	.	.	-2	.	.	.	-2	.
$X.82$	.	.	-2	-2	-2	.	-2	-2	.	-2	-2	-2	.	.	.	.	6	.	.	.	.	.	-2	.	.	2	2	.
$X.83$	.	.	2	2	2	.	2	2	.	2	2	2	.	.	.	.	6	.	.	.	.	.	-2	.	.	2	2	.
$X.84$	.	.	.	.	.	.	.	.	.	.	.	.	.	.	.	.	6	.	.	.	.	.	-2	.	.	.	-2	.

Character table of $D(Co_2)$ (continued)

2	5	5	4	4	4	4	4	8	9	9	8	8	8	8	8	8	7	7	7	7	7	7	7	7	7	6	6
3	1	1	1	1	1	1	1	1	.	.	.	.	.	.	.	.	.	.	.	.	.	.	.	.	.	.	.
5	.	.	.	.	.	.	.	.	.	.	.	.	.	.	.	.	.	.	.	.	.	.	.	.	.	.	.
	6m	6n	6o	6p	6q	6r	6s	8a	8b	8c	8d	8e	8f	8g	8h	8i	8j	8k	8l	8m	8n	8o	8p	8q	8r	8s	8t
2P	3b	3a	3b	3b	3b	3a	3b	4_1	4_3	4_3	4_6	4_3	4_1	4_1	4_6	4_3	4_{19}	4_{20}	4_6	4_{20}	4_{20}	4_{22}	4_{22}	4_{20}	4_{19}	4_8	4_9
3P	2e	2k	2m	2n	2o	2s	2p	8a	8b	8c	8d	8e	8f	8g	8h	8i	8j	8k	8l	8m	8n	8o	8p	8q	8r	8s	8t
5P	6m	6n	6o	6p	6q	6r	6s	8a	8b	8c	8d	8e	8f	8g	8h	8i	8j	8k	8l	8m	8n	8o	8p	8q	8r	8s	8t
X.1	1	1	1	1	1	1	1	1	1	1	1	1	1	1	1	1	1	1	1	1	1	1	1	1	1	1	1
X.2	1	−1	−1	−1	−1	−1	−1	−1	1	1	−1	1	1	−1	−1	−1	−1	−1	1	1	−1	1	1	1	−1	1	−1
X.3	−1	.	−1	−1	−1	.	−1	3	1	1	−1	1	1	−1	−1	3	1	1	1	−1	1	−1	−1	−1	1	−1	1
X.4	2	1	.	.	.	1	.	1	1	1	−3	1	1	−3	−3	1	−1	−1	1	−1	−1	−1	−1	−1	−1	−1	−1
X.5	2	−1	.	.	.	−1	.	−1	1	1	3	1	1	3	3	−1	1	1	1	−1	1	−1	−1	−1	1	−1	1
X.6	−1	.	1	1	1	.	1	−3	1	1	1	1	1	1	1	−3	−1	−1	1	−1	−1	−1	−1	−1	−1	−1	−1
X.7	1	.	1	1	−1	.	−1	.	−2	2	−2	2	−2	.	2	.	.	2	.	.	−2	.	.	.	.	.	.
X.8	1	.	−1	−1	1	.	1	.	−2	2	2	2	−2	.	−2	.	.	−2	.	.	2	.	.	.	.	.	.
X.9	.	.	.	.	.	.	.	−3	1	1	−3	1	1	−3	−3	−3	1	1	1	1	1	1	1	1	1	1	1
X.10	.	.	.	.	.	.	.	3	1	1	3	1	1	3	3	3	−1	−1	1	1	−1	1	1	1	−1	1	−1
X.11	1	1	−1	−1	−1	1	−1	−2	−2	−2	2	−2	−2	2	2	−2	.	.	−2	.	.	.	.	.	.	.	.
X.12	−1	1	1	1	−1	−1	−1	−4	2	−2	2	−2	−2	.	−2	.	.	.	.	−2	.	.	.	2	.	.	.
X.13	1	−1	1	1	1	−1	1	2	−2	−2	−2	−2	−2	−2	−2	2	.	.	−2	.	.	.	.	.	.	.	.
X.14	−1	1	−1	−1	1	−1	1	−4	2	−2	−2	−2	−2	.	2	.	.	.	.	2	.	.	.	−2	.	.	.
X.15	−1	−1	1	1	−1	1	−1	4	2	−2	2	−2	−2	.	−2	.	.	.	.	2	.	.	.	−2	.	.	.
X.16	−1	−1	−1	−1	1	1	1	4	2	−2	−2	−2	−2	.	2	.	.	.	.	−2	.	.	.	2	.	.	.
X.17	.	1	.	.	.	1	.	−1	−1	−1	−1	−1	−1	−1	−1	−1	1	1	−1	1	1	1	1	1	1	1	1
X.18	.	−1	.	.	.	−1	.	1	3	3	1	−1	−1	1	1	1	1	1	3	−1	1	−1	−1	−1	1	−1	1
X.19	.	1	.	.	.	1	.	−1	3	3	−1	−1	−1	−1	−1	−1	−1	−1	3	−1	−1	−1	−1	−1	−1	−1	−1
X.20	.	−1	.	.	.	−1	.	1	−1	−1	1	−1	−1	1	1	1	−1	−1	−1	1	−1	1	1	1	−1	1	−1
X.21	−1	.	−1	−1	−1	.	−1	−3	−1	−1	1	−1	3	1	1	1	1	−1	−1	1	−1	−1	−1	1	1	−1	1
X.22	−1	.	−1	−1	−1	.	−1	−3	3	3	1	3	−1	−3	1	1	−1	1	−1	−1	1	1	1	−1	−1	1	−1
X.23	−1	.	1	1	1	.	1	3	−1	−1	−1	−1	3	−1	−1	−1	−1	1	−1	1	1	−1	−1	1	−1	−1	−1
X.24	−1	.	1	1	1	.	1	3	3	3	−1	3	−1	3	−1	−1	1	−1	−1	−1	−1	1	1	−1	1	1	1
X.25	−1	−2	−1	1	1	.	−1	.	.	.	.	.	.	.	.	.	−2	.	.	.	.	−2	2	.	2	.	.
X.26	−1	2	1	−1	−1	.	1	.	.	.	.	.	.	.	.	.	2	.	.	.	.	−2	2	.	−2	.	.
X.27	−2	.	.	.	.	.	.	.	.	.	.	.	.	.	.	.	.	.	.	.	.	.	.	.	.	.	.
X.28	−2	.	.	.	.	.	.	.	−4	4	.	4	4	.	.	.	.	.	.	.	.	.	.	.	.	.	.
X.29	1	.	−1	−1	1	.	1	.	.	.	−4	.	.	.	4	.	.	.	.	.	.	.	.	.	.	.	.
X.30	1	.	1	1	−1	.	−1	.	.	.	4	.	.	.	−4	.	.	.	.	.	.	.	.	.	.	.	.
X.31	−1	.	−1	−1	1	.	1	.	−2	2	2	2	−2	.	−2	.	.	2	.	.	−2	.	.	.	.	.	.
X.32	.	1	.	.	.	1	.	2	2	2	2	−2	−2	2	2	2	.	.	2	.	.	.	.	.	.	.	.
X.33	.	−1	.	.	.	−1	.	−2	2	2	−2	−2	−2	−2	−2	−2	.	.	2	.	.	.	.	.	.	.	.
X.34	−1	.	1	1	−1	.	−1	.	−2	2	−2	2	−2	.	2	.	.	−2	.	.	2	.	.	.	.	.	.
X.35	1	.	−1	−1	−1	.	−1	6	2	2	−2	2	2	2	−2	−2	.	.	−2	.	.	.	.	.	.	.	.
X.36	1	.	1	1	1	.	1	−6	2	2	2	2	2	−2	2	2	.	.	−2	.	.	.	.	.	.	.	.
X.37	.	.	.	.	.	.	.	.	4	−4	.	−4	4	.	.	.	.	.	.	.	.	.	.	.	.	.	.
X.38	−1	.	−1	−1	1	.	1	.	.	.	4	.	.	.	−4	.	.	.	.	.	.	.	.	.	.	.	.
X.39	2	1	.	.	.	−1	.	8	.	.	.	.	.	.	.	.	.	.	.	.	.	.	.	.	.	.	.
X.40	2	−1	.	.	.	1	.	−8	.	.	.	.	.	.	.	.	.	.	.	.	.	.	.	.	.	.	.
X.41	−1	.	1	1	−1	.	−1	.	.	.	−4	.	.	.	4	.	.	.	.	.	.	.	.	.	.	.	.
X.42	.	.	.	.	.	.	.	−3	−3	−3	−3	−3	1	1	−3	1	1	−1	1	−1	−1	1	1	−1	1	1	1
X.43	.	.	.	.	.	.	.	−3	1	1	1	−3	1	1	1	1	1	−1	1	−1	−1	1	1	−1	1	1	1
X.44	.	.	.	.	.	.	.	−3	1	1	1	1	−3	−3	1	1	1	−1	−3	−1	−1	1	1	−1	1	1	1
X.45	.	.	.	.	.	.	.	−3	−3	−3	1	1	1	1	1	−3	−1	−1	−3	−1	−1	−1	−1	−1	−1	−1	−1
X.46	.	.	.	.	.	.	.	−3	1	1	−3	1	−3	5	−3	1	−1	1	1	1	1	−1	−1	1	−1	−1	−1
X.47	.	.	.	.	.	.	.	3	1	1	−1	1	−3	3	−1	−1	−1	1	−3	−1	1	1	1	−1	−1	1	−1
X.48	.	.	.	.	.	.	.	3	−3	−3	3	−3	1	−1	3	−1	−1	1	1	−1	1	1	1	−1	−1	1	−1
X.49	.	.	.	.	.	.	.	3	1	1	−1	−3	1	−1	−1	−1	−1	1	1	−1	1	1	1	−1	−1	1	−1
X.50	.	.	.	.	.	.	.	3	1	1	3	1	−3	−5	3	−1	1	−1	1	1	−1	−1	−1	1	1	−1	1
X.51	.	.	.	.	.	.	.	3	−3	−3	−1	1	1	−1	−1	3	1	1	−3	−1	1	−1	−1	−1	1	−1	1
X.52	.	.	.	.	.	.	.	3	1	1	−1	1	1	−1	−1	3	−1	−1	1	1	−1	1	1	1	−1	1	−1
X.53	.	.	.	.	.	.	.	3	5	5	−1	−3	1	−1	−1	−1	1	−1	−3	1	−1	−1	−1	1	1	−1	1
X.54	.	.	.	.	.	.	.	−3	−3	−3	1	1	−3	−3	1	1	−1	1	1	1	1	−1	−1	1	−1	−1	−1
X.55	.	.	.	.	.	.	.	−3	1	1	1	1	1	1	1	−3	1	1	1	1	1	1	1	1	1	1	1
X.56	.	.	.	.	.	.	.	−3	5	5	1	−3	1	1	1	1	−1	1	−3	1	1	−1	−1	1	−1	−1	−1
X.57	.	.	.	.	.	.	.	3	−3	−3	−1	1	−3	3	−1	−1	1	−1	1	1	−1	−1	−1	1	1	−1	1
X.58	.	−1	.	.	.	1	.	−4	−4	4	.	.	.	.	.	4	.	.	.	.	.	.	.	.	.	.	.
X.59	.	1	.	.	.	−1	.	4	4	−4	.	.	.	.	.	4	.	.	.	.	.	.	.	.	.	.	.
X.60	.	1	.	.	.	−1	.	4	−4	4	.	.	.	.	.	−4	.	.	.	.	.	.	.	.	.	.	.
X.61	.	−1	.	.	.	1	.	−4	4	−4	.	.	.	.	.	−4	.	.	.	.	.	.	.	.	.	.	.
X.62	1	.	−1	1	1	.	−1	.	.	.	.	.	.	.	.	.	2	.	.	.	.	2	−2	.	−2	.	.
X.63	1	.	1	−1	−1	.	1	.	.	.	.	.	.	.	.	.	−2	.	.	.	.	2	−2	.	2	.	.
X.64	−2	2	.	.	.	.	.	.	.	.	.	.	.	.	.	.	−2	.	.	.	.	2	−2	.	2	.	.
X.65	−2	−2	.	.	.	.	.	.	.	.	.	.	.	.	.	.	2	.	.	.	.	2	−2	.	−2	.	.
X.66	.	.	.	.	.	.	.	.	−6	6	2	−2	2	.	−2	.	.	−2	.	.	2	.	.	.	.	.	.
X.67	.	.	.	.	.	.	.	.	2	−2	−2	−2	2	.	2	.	.	−2	.	.	2	.	.	.	.	.	.
X.68	.	.	.	.	.	.	.	.	−6	6	−2	−2	2	.	2	.	.	2	.	.	−2	.	.	.	.	.	.
X.69	.	.	.	.	.	.	.	−6	−2	−2	−2	2	2	2	−2	2	.	.	2	.	.	.	.	.	.	.	.
X.70	.	.	.	.	.	.	.	.	2	−2	2	−2	2	.	−2	.	.	2	.	.	−2	.	.	.	.	.	.
X.71	.	.	.	.	.	.	.	.	2	−2	−2	2	2	.	2	4	.	.	.	−2	.	.	.	2	.	.	.
X.72	.	.	.	.	.	.	.	.	2	−2	2	2	2	.	−2	4	.	.	.	2	.	.	.	−2	.	.	.
X.73	.	.	.	.	.	.	.	6	−2	−2	2	2	2	−2	2	−2	.	.	2	.	.	.	.	.	.	.	.
X.74	.	.	.	.	.	.	.	.	2	−2	−2	2	2	.	2	−4	.	.	.	2	.	.	.	−2	.	.	.
X.75	.	.	.	.	.	.	.	.	2	−2	2	2	2	.	−2	−4	.	.	.	−2	.	.	.	2	.	.	.
X.76	−1	.	1	−1	1	.	−1	.	.	.	.	.	.	.	.	.	.	.	.	.	.	.	.	.	.	.	.
X.77	−1	.	−1	1	−1	.	1	.	.	.	.	.	.	.	.	.	.	.	.	.	.	.	.	.	.	.	.
X.78	.	.	.	.	.	−2	.	.	.	.	.	.	.	.	.	.	−2	.	.	.	.	−2	−2	.	−2	2	2
X.79	.	−2	.	.	.	.	.	.	.	.	.	.	.	.	.	.	.	−2	.	2	−2	.	.	2	.	.	.
X.80	.	−1	.	.	.	1	.	8	.	.	.	.	.	.	.	.	.	.	.	.	.	.	.	.	.	.	.
X.81	.	.	.	.	.	−2	.	.	.	.	.	.	.	.	.	.	2	.	.	.	.	2	2	.	2	−2	−2
X.82	.	2	.	.	.	.	.	.	.	.	.	.	.	.	.	.	.	2	.	2	2	.	.	2	.	.	.
X.83	.	2	.	.	.	.	.	.	.	.	.	.	.	.	.	.	.	−2	.	−2	−2	.	.	−2	.	.	.
X.84	.	.	.	.	.	2	.	.	.	.	.	.	.	.	.	.	−2	.	.	.	.	2	2	.	−2	−2	2

Character table of $D(Co_2)$ (continued)

	8u	8v	8w	8x	8y	10a	10b	10c	12a	12b	12c	12d	12e	12f	12g	12h	12i	12j	16a	16b	24a
2	6	6	6	6	6	2	2	2	5	5	5	4	4	4	4	4	4	3	5	5	3
3	.	.	.	.	.	.	.	.	1	1	1	1	1	1	1	1	1	1	.	.	1
5	.	.	.	.	.	1	1	1	.	.	.	.	.	.	.	.	.	.	.	.	.
2P	4_{32}	4_{32}	4_{31}	4_{31}	4_{24}	5a	5a	5a	6a	6g	6g	6h	6h	6h	6g	6g	6h	6k	8b	8c	12a
3P	8u	8v	8w	8x	8y	10a	10b	10c	4_1	4_4	4_5	4_{27}	4_{28}	4_{29}	4_7	4_2	4_{30}	4_{44}	16a	16b	8a
5P	8u	8v	8w	8x	8y	2b	2c	2a	12a	12b	12c	12d	12e	12f	12g	12h	12i	12j	16a	16b	24a
X.1	1	1	1	1	1	1	1	1	1	1	1	1	1	1	1	1	1	1	1	1	1
X.2	−1	1	1	−1	1	1	1	1	1	−1	−1	−1	−1	−1	−1	1	−1	−1	1	−1	−1
X.3	1	−1	−1	1	1	.	.	.	2	.	.	−1	−1	−1	.	2	−1	.	−1	1	.
X.4	−1	−1	−1	−1	1	.	.	.	.	−1	1	1	.	.	1	−1	.	1	−1	−1	1
X.5	1	−1	−1	1	1	.	.	.	.	−1	−1	−1	.	.	−1	−1	.	−1	−1	1	−1
X.6	−1	−1	−1	−1	1	.	.	.	2	.	.	1	1	1	.	2	1	.	−1	−1	.
X.7	2	.	.	−2	.	.	−1	−1	.	.	.	1	−1	−1	.	.	1	.	.	.	.
X.8	−2	.	.	2	.	−1	−1	1	.	.	.	−1	1	1	.	.	−1	.	.	.	.
X.9	1	1	1	1	1	−1	−1	−1	.	.	.	.	.	.	.	.	.	.	1	1	.
X.10	−1	1	1	−1	1	−1	−1	−1	.	.	.	.	.	.	.	.	.	.	1	−1	.
X.11	.	.	.	−2	.	.	.	.	1	1	1	−1	−1	−1	1	1	−1	1	.	.	1
X.12	.	2	−2	.	.	.	.	.	1	−1	−1	1	1	1	−1	1	−1	1	.	.	−1
X.13	.	.	.	−2	.	.	.	.	1	−1	−1	1	1	1	−1	1	1	−1	.	.	−1
X.14	.	−2	2	.	.	.	.	.	1	−1	−1	1	1	1	−1	1	1	1	.	.	−1
X.15	.	−2	2	.	.	.	.	.	1	1	1	−1	1	1	−1	−1	−1	−1	.	.	1
X.16	.	2	−2	.	.	.	.	.	1	1	1	1	−1	−1	1	1	−1	1	.	.	1
X.17	−1	−1	−1	−1	−1	.	.	.	−1	1	1	.	.	.	1	−1	.	.	−1	−1	−1
X.18	−1	1	1	−1	−1	.	.	.	−1	1	1	.	.	.	1	−1	.	.	1	1	−1
X.19	1	1	1	1	−1	.	.	.	−1	1	1	.	.	.	1	−1	.	.	1	1	−1
X.20	1	−1	−1	1	−1	.	.	.	−1	−1	−1	.	.	.	1	−1	.	.	1	1	1
X.21	−1	1	1	−1	−1	.	.	.	.	.	.	1	1	1	.	.	1	.	−1	1	1
X.22	1	1	1	1	−1	.	.	.	.	.	.	1	1	1	.	.	1	.	1	−1	.
X.23	1	1	1	1	−1	.	.	.	.	.	.	−1	−1	−1	.	.	−1	.	−1	−1	.
X.24	−1	−1	−1	−1	−1	.	.	.	.	.	.	−1	−1	−1	.	.	−1	.	1	1	.
X.25	.	.	.	.	.	1	−1	−1	.	−2	2	−1	−1	1	.	.	1	.	.	.	.
X.26	.	.	.	.	.	1	−1	−1	.	2	−2	1	1	−1	.	.	−1	.	.	.	.
X.27	.	.	.	.	.	1	1	1	−2	.	.	.	.	.	.	.	−2	.	.	.	.
X.28	.	.	.	.	.	2	.	.	2	.	.	.	.	.	.	.	−2	.	.	.	.
X.29	.	.	.	.	.	1	1	−1	.	.	.	−1	1	1	.	.	−1	.	.	.	.
X.30	.	.	.	.	.	1	1	−1	.	.	.	1	−1	−1	.	.	1	.	.	.	.
X.31	2	.	.	−2	.	.	.	.	.	.	.	−1	1	1	.	.	−1	.	.	.	.
X.32	.	.	.	−2	.	.	.	.	1	1	1	.	.	.	1	1	.	−1	.	.	−1
X.33	.	.	.	−2	.	.	.	.	1	−1	−1	.	.	.	−1	1	.	1	.	.	1
X.34	−2	.	.	2	.	.	.	.	.	.	.	1	−1	−1	.	.	1	.	.	.	.
X.35	.	.	.	−2	.	.	.	.	.	.	.	1	1	1	.	.	1	.	.	.	.
X.36	.	.	.	−2	.	.	.	.	.	.	.	−1	−1	−1	.	.	−1	.	.	.	.
X.37	.	.	.	.	.	−1	−1	1	.	.	.	.	.	.	.	.	.	.	.	.	.
X.38	.	.	.	.	.	.	.	.	−2	.	.	1	−1	−1	.	.	2	1	.	.	.
X.39	.	.	.	.	.	.	.	.	1	−1	−1	.	.	.	1	−1	.	1	.	.	−1
X.40	.	.	.	.	.	.	.	.	1	1	1	.	.	.	−1	−1	.	−1	.	.	1
X.41	.	.	.	.	.	.	.	.	−2	.	.	−1	1	1	.	.	2	−1	.	.	.
X.42	−1	−1	−1	−1	1	.	.	.	.	.	.	.	.	.	.	.	.	.	1	1	.
X.43	1	1	1	1	1	.	.	.	.	.	.	.	.	.	.	.	.	.	−1	−1	.
X.44	1	1	1	1	1	.	.	.	.	.	.	.	.	.	.	.	.	.	−1	−1	.
X.45	1	1	1	1	1	.	.	.	.	.	.	.	.	.	.	.	.	.	1	1	.
X.46	1	1	1	1	1	.	.	.	.	.	.	.	.	.	.	.	.	.	−1	−1	.
X.47	−1	1	1	−1	1	.	.	.	.	.	.	.	.	.	.	.	.	.	−1	1	.
X.48	1	−1	−1	1	1	.	.	.	.	.	.	.	.	.	.	.	.	.	1	−1	.
X.49	−1	1	1	−1	1	.	.	.	.	.	.	.	.	.	.	.	.	.	−1	1	.
X.50	−1	1	1	−1	1	.	.	.	.	.	.	.	.	.	.	.	.	.	−1	1	.
X.51	−1	1	1	−1	1	.	.	.	.	.	.	.	.	.	.	.	.	.	1	−1	.
X.52	1	−1	−1	1	1	.	.	.	.	.	.	.	.	.	.	.	.	.	−1	1	.
X.53	1	−1	−1	1	1	.	.	.	.	.	.	.	.	.	.	.	.	.	1	−1	.
X.54	−1	−1	−1	−1	1	.	.	.	.	.	.	.	.	.	.	.	.	.	1	1	.
X.55	−1	−1	−1	−1	1	.	.	.	.	.	.	.	.	.	.	.	.	.	−1	−1	.
X.56	−1	−1	−1	−1	1	.	.	.	.	.	.	.	.	.	.	.	.	.	1	1	.
X.57	1	−1	−1	1	1	.	.	.	.	.	.	.	.	.	.	.	.	.	1	−1	.
X.58	.	.	.	.	.	.	.	.	−1	1	1	.	.	.	−1	1	.	1	.	.	−1
X.59	.	.	.	.	.	.	.	.	−1	−1	−1	.	.	.	1	1	.	−1	.	.	1
X.60	.	.	.	.	.	.	.	.	−1	−1	−1	.	.	.	1	1	.	−1	.	.	1
X.61	.	.	.	.	.	.	.	.	−1	1	1	.	.	.	−1	1	.	1	.	.	−1
X.62	.	.	.	.	.	.	.	.	.	.	.	−1	−1	1	.	.	1	.	.	.	.
X.63	.	.	.	.	.	.	.	.	.	.	.	1	1	−1	.	.	−1	.	.	.	.
X.64	.	.	.	.	.	.	.	.	.	2	−2	.	.	.	.	.	.	.	.	.	.
X.65	.	.	.	.	.	.	.	.	.	−2	2	.	.	.	.	.	.	.	.	.	.
X.66	2	.	.	−2	.	.	.	.	.	.	.	.	.	.	.	.	.	.	.	.	.
X.67	2	.	.	−2	.	.	.	.	.	.	.	.	.	.	.	.	.	.	.	.	.
X.68	−2	.	.	2	.	.	.	.	.	.	.	.	.	.	.	.	.	.	.	.	.
X.69	.	.	.	−2	.	.	.	.	.	.	.	.	.	.	.	.	.	.	.	.	.
X.70	−2	.	.	2	.	.	.	.	.	.	.	.	.	.	.	.	.	.	.	.	.
X.71	.	−2	2	.	.	.	.	.	.	.	.	.	.	.	.	.	.	.	.	.	.
X.72	.	2	−2	.	.	.	.	.	.	.	.	.	.	.	.	.	.	.	.	.	.
X.73	.	.	.	−2	.	.	.	.	.	.	.	.	.	.	.	.	.	.	.	.	.
X.74	.	2	−2	.	.	.	.	.	.	.	.	.	.	.	.	.	.	.	.	.	.
X.75	.	−2	2	.	.	.	.	.	.	.	.	.	.	.	.	.	.	.	.	.	.
X.76	.	.	.	.	.	−1	1	−1	.	.	.	1	−1	1	.	.	−1	.	.	.	.
X.77	.	.	.	.	.	−1	1	−1	.	.	.	−1	1	−1	.	.	1	.	.	.	.
X.78	.	.	.	.	.	.	.	.	2	2	2	.	.	.	.	.	.	.	.	.	.
X.79	.	.	.	.	.	.	.	.	−2	.	.	.	.	.	.	2	.	.	.	.	.
X.80	.	.	.	.	.	.	.	.	1	1	1	.	.	.	−1	−1	.	1	.	.	−1
X.81	.	.	.	.	.	.	.	.	2	2	2	.	.	.	.	.	.	.	.	.	.
X.82	.	.	.	.	.	.	.	.	−2	.	.	.	.	.	.	−2	.	.	.	.	.
X.83	.	.	.	.	.	.	.	.	−2	.	.	.	.	.	.	−2	.	.	.	.	.
X.84	.	.	.	.	.	.	.	.	2	−2	−2	.	.	.	.	.	.	.	.	.	.

Character table of $D(\mathrm{Co_2})$ (continued)

	1a	2a	2b	2c	2d	2e	2f	2g	2h	2i	2j	2k	2l	2m	2n	2o	2p	2q	2r	2s	2t
2	18	18	14	14	17	16	16	14	14	14	14	14	15	13	13	13	13	14	13	11	12
3	2	2	2	2	1	1	1	1	1	1	1	1	.	1	1	1	1	.	.	1	.
5	1	1	1	1	.	.	.	.	.	.	.	.	.	.	.	.	.	.	.	.	.
2P	1a	1a	1a	1a	1a	1a	1a	1a	1a	1a	1a	1a	1a	1a	1a	1a	1a	1a	1a	1a	1a
3P	1a	2a	2b	2c	2d	2e	2f	2g	2h	2i	2j	2k	2l	2m	2n	2o	2p	2q	2r	2s	2t
5P	1a	2a	2b	2c	2d	2e	2f	2g	2h	2i	2j	2k	2l	2m	2n	2o	2p	2q	2r	2s	2t
X.85	120	120	.	.	-8	24	24	.	-8	.	24	24	-8	.	.	.	.	-8	-8	8	.
X.86	120	120	-120	-120	120	24	24	-24	-8	-24	-8	-8	24	.	.	.	.	24	-8	8	.
X.87	120	120	.	.	-8	24	24	.	8	.	-32	-32	-8	-12	-12	-12	-12	8	.	.	12
X.88	144	-144	-36	36	.	72	-72	36	.	-36	24	-24	.	-12	12	12	-12	.	.	.	4
X.89	144	-144	-36	36	.	72	-72	36	.	-36	-24	24	.	12	-12	-12	12	.	.	.	4
X.90	160	-160	-40	40	.	80	-80	40	.	-40	-16	16	.	-8	8	8	-8	.	.	.	-8
X.91	160	-160	40	-40	.	16	-16	-8	.	8	32	-32	.	16	-16	16	-16	.	.	.	8
X.92	160	-160	40	-40	.	16	-16	-8	.	8	32	-32	.	-16	16	-16	16	.	.	.	8
X.93	160	-160	-40	40	.	80	-80	40	.	-40	16	-16	.	8	-8	-8	8	.	.	.	-8
X.94	160	-160	40	-40	.	16	-16	-8	.	8	-32	32	.	16	-16	16	-16	.	.	.	8
X.95	160	-160	40	-40	.	16	-16	-8	.	8	-32	32	.	-16	16	-16	16	.	.	.	8
X.96	180	180	180	180	180	-12	-12	-12	-12	-12	12	12	-12	12	12	12	12	-12	12	12	-4
X.97	180	180	180	180	180	-12	-12	-12	-12	-12	-12	-12	-12	12	12	12	12	-12	-12	-12	-4
X.98	180	180	180	180	180	-12	-12	-12	-12	-12	-12	-12	-12	-12	-12	-12	-12	-12	-12	-12	4
X.99	180	180	180	180	180	-12	-12	-12	-12	-12	12	12	-12	-12	-12	-12	-12	-12	12	12	-4
X.100	180	180	-180	-180	180	36	36	-12	-36	.	36	.	.	.	.	.	.	-28	.	.	-4
X.101	180	180	-180	-180	180	-60	-60	60	-12	60	.	-60	.	.	.	.	.	4	.	.	-4
X.102	240	240	.	.	-16	48	48	.	16	.	-16	-16	-16	-24	-24	-24	-24	16	-16	8	.
X.103	240	240	.	.	-16	48	48	.	16	.	16	16	-16	24	24	24	24	16	16	8	.
X.104	240	240	.	.	-16	48	48	.	-16	.	.	.	-16	.	.	.	.	.	-16	.	-16
X.105	240	-240	-60	60	.	-8	8	-4	.	4	-24	24	.	28	-28	-28	28	.	.	.	12
X.106	240	-240	-60	60	.	-8	8	-4	.	4	24	-24	.	20	-20	-20	20	.	.	.	-4
X.107	240	-240	-60	60	.	-8	8	-4	.	4	24	-24	.	-28	28	28	-28	.	.	.	12
X.108	240	-240	-60	60	.	-8	8	-4	.	4	-24	24	.	-20	20	20	-20	.	.	.	-4
X.109	240	240	.	.	-16	48	48	.	-16	.	.	.	-16	.	.	.	.	.	-16	.	16
X.110	256	-256	-64	64	.	128	-128	64	.	-64	.	.	.	.	.	.	.	.	.	.	-16
X.111	320	-320	80	-80	.	32	-32	-16	.	16	.	.	.	.	.	.	.	.	.	.	-16
X.112	360	360	.	.	-24	-24	-24	.	24	.	-24	-24	8	36	36	-12	-12	-8	8	.	-4
X.113	360	360	.	.	-24	72	72	.	-24	.	-24	-24	-24	.	.	.	.	-24	8	24	.
X.114	360	360	.	.	-24	-24	-24	.	24	.	-24	-24	8	-12	-12	36	36	-8	8	.	-4
X.115	360	360	.	.	-24	72	72	.	-24	.	24	24	-24	.	.	.	.	-24	-8	-24	.
X.116	360	360	.	.	-24	72	72	.	24	.	48	48	-24	-12	-12	-12	-12	24	16	.	4
X.117	360	360	.	.	-24	-24	-24	.	24	.	24	24	8	-36	-36	12	12	-8	-8	.	-4
X.118	360	360	.	.	-24	72	72	.	-24	.	24	24	-24	.	.	.	.	-24	-8	24	.
X.119	360	360	.	.	-24	-24	-24	.	24	.	24	24	8	12	12	-36	-36	-8	8	.	-4
X.120	360	360	.	.	-24	72	72	.	-24	.	-24	-24	-24	.	.	.	.	-24	8	-24	.
X.121	360	360	.	.	-24	-24	-24	.	24	.	-24	-24	8	12	12	-36	-36	-8	8	.	12
X.122	360	360	.	.	-24	-24	-24	.	24	.	-24	-24	8	-36	-36	12	12	-8	8	.	12
X.123	360	360	.	.	-24	-24	-24	.	24	.	24	24	8	-12	-12	36	36	-8	-8	.	12
X.124	360	360	.	.	-24	-24	-24	.	24	.	24	24	8	36	36	-12	-12	-8	-8	.	12
X.125	360	360	.	.	-24	72	72	.	24	.	.	.	-24	-12	-12	-12	-12	24	32	.	-12
X.126	360	360	.	.	-24	72	72	.	24	.	.	.	-24	12	12	12	12	24	-32	.	-12
X.127	360	360	.	.	-24	72	72	.	24	.	-48	-48	-24	12	12	12	12	24	-16	.	4
X.128	384	-384	96	-96	.	-64	64	32	.	-32	.	.	.	-32	-32	-32	32	.	.	.	.
X.129	384	-384	96	-96	.	-64	64	32	.	-32	.	.	.	32	-32	32	-32	.	.	.	.
X.130	480	-480	120	-120	.	-80	80	40	.	-40	.	.	.	-16	-16	-16	16	.	.	.	8
X.131	480	-480	-120	120	.	-16	16	-8	.	8	-48	48	.	8	-8	-8	8	.	.	.	8
X.132	480	-480	-120	120	.	-16	16	-8	.	8	48	-48	.	-8	8	8	-8	.	.	.	8
X.133	480	-480	120	-120	.	-80	80	40	.	-40	.	.	.	16	-16	16	-16	.	.	.	8
X.134	576	-576	144	-144	.	-96	96	48	.	-48	.	.	.	.	.	.	.	.	.	.	-16
X.135	640	-640	160	-160	.	64	-64	-32	.	32	64	-64	.	.	.	.	.	.	.	.	.
X.136	640	-640	160	-160	.	64	-64	-32	.	32	.	.	.	32	-32	32	-32	.	.	.	.
X.137	640	-640	160	-160	.	64	-64	-32	.	32	.	.	.	-32	32	-32	32	.	.	.	.
X.138	640	-640	160	-160	.	64	-64	-32	.	32	-64	64	.	.	.	.	.	.	.	.	.
X.139	720	720	.	.	-48	-48	-48	.	-48	.	48	48	16	.	.	.	.	.	16	-16	.
X.140	720	720	.	.	-48	-48	-48	.	-48	.	-48	-48	16	.	.	.	.	.	16	16	.
X.141	720	720	.	.	-48	-48	-48	.	48	.	.	.	16	24	24	24	24	-16	.	.	-8
X.142	720	720	.	.	-48	-48	-48	.	48	.	.	.	16	-24	-24	-24	-24	-16	.	.	-8
X.143	720	720	.	.	-48	-48	-48	.	-48	.	.	.	16	.	.	.	.	.	16	.	.
X.144	720	-720	-180	180	.	-24	24	-12	.	12	24	-24	.	36	-36	-36	36	.	.	.	4
X.145	720	-720	-180	180	.	-24	24	-12	.	12	24	-24	.	-12	12	12	-12	.	.	.	-12
X.146	720	720	.	.	-48	-48	-48	.	-48	.	.	.	16	.	.	.	.	.	16	.	.
X.147	720	-720	-180	180	.	-24	24	-12	.	12	-24	24	.	12	-12	-12	12	.	.	.	-12
X.148	720	-720	-180	180	.	-24	24	-12	.	12	-24	24	.	-36	36	36	-36	.	.	.	4

Character table of $D(Co_2)$ (continued)

	$2u$	$2v$	$2w$	$2x$	$3a$	$3b$	4_1	4_2	4_3	4_4	4_5	4_6	4_7	4_8	4_9	4_{10}	4_{11}	4_{12}	4_{13}	4_{14}	4_{15}	4_{16}
2	12	12	11	11	7	5	14	13	14	12	12	13	11	12	12	12	12	12	12	12	12	12
3	.	.	.	.	2	2	1	1	.	1	1	.	1	.	.	.	.	.	.	.	.	.
5	.	.	.	.	.	.	.	.	.	.	.	.	.	.	.	.	.	.	.	.	.	.
2P	$1a$	$1a$	$1a$	$1a$	$3a$	$3b$	$2a$	$2d$	$2a$	$2d$	$2d$	$2d$	$2d$	$2e$	$2f$	$2d$	$2e$	$2d$	$2d$	$2a$	$2d$	$2d$
3P	$2u$	$2v$	$2w$	$2x$	$1a$	$1a$	4_1	4_2	4_3	4_4	4_5	4_6	4_7	4_8	4_9	4_{10}	4_{11}	4_{12}	4_{13}	4_{14}	4_{15}	4_{16}
5P	$2u$	$2v$	$2w$	$2x$	$3a$	$3b$	4_1	4_2	4_3	4_4	4_5	4_6	4_7	4_8	4_9	4_{10}	4_{11}	4_{12}	4_{13}	4_{14}	4_{15}	4_{16}
X.85	-8	.	8	8	6	.	8	.	8	-8	-8	.	.	4	-4	-8	4	-8	.	.	8	-8
X.86	-8	.	.	.	-3	.	-8	8	24	8	8	-24	-8	.	.	.	.	.	.	.	-8	.
X.87	-8	12	-4	4	6	.	-8	.	-8	.	.	.	8	8	.	4	8	-4	4	8	-4	4
X.88	.	-4	.	.	.	.	.	.	.	12	-12	.	.	4	4	4	-4	-12	.	.	12	.
X.89	.	-4	.	.	.	.	.	.	.	-12	12	.	.	4	4	4	-4	12	.	.	-12	.
X.90	.	8	.	.	4	1	.	.	.	-8	8	.	.	-8	-8	-8	8	-8	.	.	8	.
X.91	.	-8	.	.	4	1	.	.	.	-16	16	.	.	8	-8	-8	-8	.	.	.	.	.
X.92	.	-8	.	.	4	1	.	.	.	-16	16	.	.	8	-8	-8	-8	.	.	.	.	.
X.93	.	8	.	.	4	1	.	.	.	8	-8	.	.	-8	-8	-8	8	8	.	.	-8	.
X.94	.	-8	.	.	4	1	.	.	.	16	-16	.	.	8	-8	-8	-8	.	.	.	.	.
X.95	.	-8	.	.	4	1	.	.	.	16	-16	.	.	8	-8	-8	-8	.	.	.	.	.
X.96	-4	4	-4	4	.	.	-12	-12	-12	12	12	-12	12	-4	-4	4	-4	-4	12	-4	-4	12
X.97	4	-4	-4	-4	.	.	-12	-12	-12	-12	-12	-12	-12	4	4	-4	4	-4	12	4	-4	12
X.98	4	4	4	4	.	.	-12	-12	-12	-12	-12	-12	-12	-4	-4	4	-4	4	-12	4	4	-12
X.99	-4	-4	4	-4	.	.	-12	-12	-12	12	12	-12	12	4	4	-4	4	4	-12	-4	4	-12
X.100	.	-4	.	4	.	.	-12	12	-28	.	.	28	.	-4	-4	4	-4	.	.	.	.	.
X.101	.	-4	.	4	.	.	-12	12	4	.	.	-4	.	4	-4	4	4	.	.	.	.	.
X.102	.	8	-8	-6	.	.	-16	.	-16	.	.	.	16	.	8	.	-8	8	.	-8	8	.
X.103	.	8	8	8	-6	.	-16	.	-16	.	.	.	-16	.	8	.	8	-8	.	8	-8	.
X.104	.	.	-16	.	-6	.	16	.	16	16	16	.	.	8	8	.	8	16	.	.	16	.
X.105	.	-12	.	.	.	3	.	.	.	-12	12	.	.	-4	-4	12	4	-4	.	.	4	.
X.106	.	4	.	.	.	3	.	.	.	12	-12	.	.	12	12	-4	-12	-12	.	.	12	.
X.107	.	-12	.	.	.	3	.	.	.	12	-12	.	.	-4	-4	12	4	4	.	.	-4	.
X.108	.	4	.	.	.	3	.	.	.	-12	12	.	.	12	12	-4	-12	12	.	.	-12	.
X.109	.	.	16	.	-6	.	16	.	16	-16	-16	.	.	8	8	.	8	-16	.	.	-16	.
X.110	.	.	.	.	-8	-2	.	.	.	.	.	.	.	.	.	.	.	.	.	.	.	.
X.111	.	16	.	.	8	2	.	.	.	.	.	.	.	-16	16	16	16	.	.	.	.	.
X.112	.	-4	-4	-4	.	.	-24	.	8	.	.	.	.	8	.	-4	8	-4	4	16	-4	-12
X.113	8	.	-8	-8	.	.	24	.	24	-24	-24	.	.	-4	4	8	-4	8	.	-8	8	.
X.114	.	-4	-4	-4	.	.	-24	.	8	.	.	.	.	8	.	-4	8	-4	-12	16	-4	4
X.115	-8	.	-8	-8	.	.	24	.	24	24	24	.	.	-4	4	8	-4	-8	.	-8	-8	.
X.116	8	4	-4	-4	.	.	-24	.	-24	.	.	.	-24	8	.	-4	8	-4	4	-8	-4	4
X.117	.	-4	4	-4	.	.	-24	.	8	.	.	.	.	8	.	8	8	-4	-16	4	12	.
X.118	-8	.	-8	8	.	.	24	.	24	-24	-24	.	.	-4	-12	-8	-4	8	.	8	8	.
X.119	.	-4	4	-4	.	.	-24	.	8	.	.	.	.	8	.	-4	8	4	12	-16	4	-4
X.120	8	.	8	8	.	.	24	.	24	24	24	.	.	-4	-12	-8	-4	-8	.	-8	-8	.
X.121	-16	12	4	-4	.	.	-24	.	8	.	.	.	.	-8	.	-4	-8	4	12	.	4	-4
X.122	-16	12	4	-4	.	.	-24	.	8	.	.	.	.	-8	.	-4	-8	4	-4	.	4	12
X.123	16	12	-4	-4	.	.	-24	.	8	.	.	.	.	-8	.	-4	-8	-4	-12	.	-4	4
X.124	16	12	-4	-4	.	.	-24	.	8	.	.	.	.	-8	.	-4	-8	-4	4	.	-4	-12
X.125	-8	-12	-4	-4	.	.	-24	.	-24	.	.	.	-24	-8	.	-4	-8	-4	4	8	-4	4
X.126	8	-12	4	-4	.	.	-24	.	-24	.	.	.	24	-8	.	-4	-8	4	-4	-8	4	-4
X.127	-8	4	4	-4	.	.	-24	.	-24	.	.	.	24	8	.	-4	8	4	-4	8	4	-4
X.128	.	.	.	.	3	.	.	.	.	.	.	.	.	.	.	.	.	.	.	.	.	.
X.129	.	.	.	.	3	.	.	.	.	.	.	.	.	.	.	.	.	.	.	.	.	.
X.130	.	-8	.	.	-3	.	.	.	.	.	.	.	.	-8	8	-8	8	.	.	.	.	.
X.131	.	-8	.	.	-3	.	.	.	.	-24	24	.	.	8	8	8	-8	8	.	.	-8	.
X.132	.	-8	.	.	-3	.	.	.	.	24	-24	.	.	8	8	8	-8	-8	.	.	8	.
X.133	.	-8	.	.	-3	.	.	.	.	.	.	.	.	-8	8	-8	8	.	.	.	.	.
X.134	.	16	.	.	-3	.	.	.	.	.	.	.	.	16	-16	16	-16	.	.	.	.	.
X.135	.	.	.	4	-2	.	.	.	.	-32	32	.	.	.	.	.	.	.	.	.	.	.
X.136	.	.	.	-8	1	.	.	.	.	.	.	.	.	.	.	.	.	.	.	.	.	.
X.137	.	.	.	-8	1	.	.	.	.	.	.	.	.	.	.	.	.	.	.	.	.	.
X.138	.	.	.	4	-2	.	.	.	.	32	-32	.	.	.	.	.	.	.	.	.	.	.
X.139	-16	.	.	.	48	.	-16	.	.	.	.	.	.	8	-8	.	8	.	.	16	.	.
X.140	16	.	.	.	48	.	-16	.	.	.	.	.	.	8	-8	.	8	.	.	-16	.	.
X.141	-16	-8	-8	8	.	.	-48	.	16	.	.	.	.	.	8	.	-8	-8	-16	-8	-8	.
X.142	16	-8	8	8	.	.	-48	.	16	.	.	.	.	.	8	.	8	8	16	8	8	.
X.143	.	.	16	.	.	.	48	.	-16	.	.	.	.	-8	24	-16	-8	.	.	.	.	.
X.144	.	-4	.	.	.	.	.	.	.	12	-12	.	.	-12	-12	4	12	4	.	.	-4	.
X.145	.	12	.	.	.	.	.	.	.	12	-12	.	.	4	4	-12	-4	20	.	.	-20	.
X.146	.	.	-16	.	48	.	-16	.	.	.	.	.	.	-8	-8	16	-8	.	.	.	.	.
X.147	.	12	.	.	.	.	.	.	.	-12	12	.	.	4	4	-12	-4	-20	.	.	20	.
X.148	.	-4	.	.	.	.	.	.	.	-12	12	.	.	-12	-12	4	12	-4	.	.	4	.

Character table of $D(\mathrm{Co}_2)$ (continued)

	4_{17}	4_{18}	4_{19}	4_{20}	4_{21}	4_{22}	4_{23}	4_{24}	4_{25}	4_{26}	4_{27}	4_{28}	4_{29}	4_{30}	4_{31}	4_{32}	4_{33}	4_{34}	4_{35}	4_{36}	4_{37}	4_{38}	4_{39}	4_{40}	4_{41}
2	12	12	11	11	11	11	11	11	11	11	9	9	9	9	10	10	10	10	10	10	10	10	10	10	10
3	.	.	.	.	.	.	.	.	.	.	1	1	1	1	.	.	.	.	.	.	.	.	.	.	.
5	.	.	.	.	.	.	.	.	.	.	.	.	.	.	.	.	.	.	.	.	.	.	.	.	.
2P	$2f$	$2d$	$2e$	$2d$	$2e$	$2f$	$2e$	$2d$	$2d$	$2d$	$2g$	$2g$	$2g$	$2g$	$2h$	$2h$	$2e$	$2h$	$2e$	$2e$	$2d$	$2d$	$2d$	$2f$	$2d$
3P	4_{17}	4_{18}	4_{19}	4_{20}	4_{21}	4_{22}	4_{23}	4_{24}	4_{25}	4_{26}	4_{27}	4_{28}	4_{29}	4_{30}	4_{31}	4_{32}	4_{33}	4_{34}	4_{35}	4_{36}	4_{37}	4_{38}	4_{39}	4_{40}	4_{41}
5P	4_{17}	4_{18}	4_{19}	4_{20}	4_{21}	4_{22}	4_{23}	4_{24}	4_{25}	4_{26}	4_{27}	4_{28}	4_{29}	4_{30}	4_{31}	4_{32}	4_{33}	4_{34}	4_{35}	4_{36}	4_{37}	4_{38}	4_{39}	4_{40}	4_{41}
X.85	−4	−8	12	.	4	4	4	.	.	.	.	.	.	.	.	.	.	−4	.	−4	−4	.	.	.	4
X.86	.	.	.	.	.	.	.	8	−8	−8	.	.	.	.	.	.	.	.	.	.	.	.	.	.	8
X.87	.	4	.	−4	−4	−8	−4	.	.	.	.	.	.	.	.	.	−4	.	.	−4	−4	4	4	.	.
X.88	−4	−4	.	.	−12	.	12	.	.	.	.	6	6	−6	−6	.	.	.	.	.	.	.	.	.	.
X.89	−4	−4	.	.	12	.	−12	.	.	.	−6	−6	6	6	.	.	.	.	.	.	.	.	.	.	.
X.90	8	8	.	.	8	.	−8	.	.	.	4	4	−4	−4	.	.	.	.	.	.	.	.	.	.	.
X.91	8	8	.	.	.	.	.	.	.	.	−4	4	−4	4	.	.	.	.	.	.	.	.	.	.	.
X.92	8	8	.	.	.	.	.	.	.	.	4	4	4	−4	.	.	.	.	.	.	.	.	.	.	.
X.93	8	8	.	.	−8	.	8	.	.	.	−4	−4	4	4	.	.	.	.	.	.	.	.	.	.	.
X.94	8	8	.	.	.	.	.	.	.	.	−4	4	−4	4	.	.	.	.	.	.	.	.	.	.	.
X.95	8	8	.	.	.	.	.	.	.	.	4	−4	4	−4	.	.	.	.	.	.	.	.	.	.	.
X.96	−4	4	−4	4	−4	−4	−4	4	4	−4	.	.	.	.	−4	−4	−4	−4	−4	−4	4	4	−4	−4	−4
X.97	4	−4	4	−4	4	4	4	4	4	4	.	.	.	.	4	4	4	−4	4	4	−4	−4	−4	4	4
X.98	−4	−4	4	4	4	4	4	4	4	4	.	.	.	.	−4	−4	4	4	−4	−4	4	4	4	4	4
X.99	4	−4	4	−4	−4	4	−4	4	4	−4	.	.	.	.	4	4	−4	4	4	4	−4	−4	4	−4	−4
X.100	4	4	4	−4	.	−4	.	−4	4	4	.	.	.	.	−4	−4	.	.	−4	4	4	4	.	.	.
X.101	−4	4	−4	−4	.	4	.	−4	4	.	.	.	.	.	−4	−4	.	4	−4	4	−4	.	.	.	.
X.102	.	8	.	8	.	.	.	.	.	.	.	.	.	.	.	.	.	.	.	.	.	−8	−8	8	.
X.103	.	8	.	8	.	.	.	.	.	.	.	.	.	.	.	.	.	.	.	.	.	−8	−8	−8	.
X.104	8	.	8	.	8	.	.	.	.	.	.	.	.	.	.	.	.	.	.	.	−8	−8	.	.	.
X.105	4	−12	.	.	−4	.	4	.	.	.	2	2	−2	−2	.	.	.	.	.	.	.	.	.	.	.
X.106	−12	4	.	.	4	.	−4	.	.	.	−2	−2	2	2	.	.	.	.	.	.	.	.	.	.	.
X.107	4	−12	.	.	4	.	−4	.	.	.	−2	−2	2	2	.	.	.	.	.	.	.	.	.	.	.
X.108	−12	4	.	.	−4	.	4	.	.	.	2	2	−2	−2	.	.	.	.	.	.	.	.	.	.	.
X.109	8	.	8	.	8	.	.	.	.	.	.	.	.	.	.	.	.	.	.	.	−8	−8	.	.	.
X.110	.	.	.	.	.	.	.	.	.	.	.	.	.	.	.	.	.	.	.	.	.	.	.	.	.
X.111	−16	−16	.	.	.	.	.	.	.	.	.	.	.	.	.	.	.	.	.	.	.	.	.	.	.
X.112	.	−4	.	12	4	−8	4	.	.	.	−8	.	.	.	−8	8	4	.	.	.	4	−4	4	−4	.
X.113	4	8	−12	.	−4	−4	−4	.	.	.	.	.	.	.	.	4	.	4	4	.	.	.	.	−4	.
X.114	.	−4	.	12	4	−8	4	.	.	.	−8	.	.	.	8	−8	4	.	.	.	4	−4	4	−4	.
X.115	4	8	−12	.	4	−4	4	.	.	.	.	.	.	.	.	−4	.	4	4	.	.	.	.	4	.
X.116	.	−4	.	−12	4	−8	4	.	.	.	.	.	.	.	.	4	.	.	.	.	4	4	4	−4	.
X.117	.	−4	.	12	−4	−8	−4	.	.	.	8	.	.	.	−8	8	−4	.	.	.	4	−4	−4	4	.
X.118	−12	−8	4	.	−4	−4	−4	.	.	.	.	.	.	.	.	.	−4	4	4	.	.	.	.	4	.
X.119	.	−4	.	12	−4	−8	−4	.	.	.	8	.	.	.	8	−8	−4	.	.	.	4	−4	−4	4	.
X.120	−12	−8	4	.	−4	−4	−4	.	.	.	.	.	.	.	.	4	.	4	4	.	.	.	.	−4	.
X.121	.	−4	.	−4	4	8	4	.	.	.	8	.	.	.	8	−8	4	.	.	.	4	−4	−4	−4	.
X.122	.	−4	.	−4	4	8	4	.	.	.	8	.	.	.	−8	8	4	.	.	.	4	−4	−4	−4	.
X.123	.	−4	.	−4	−4	8	−4	.	.	.	−8	.	.	.	8	−8	−4	.	.	.	4	−4	4	4	.
X.124	.	−4	.	−4	−4	8	−4	.	.	.	−8	.	.	.	−8	8	−4	.	.	.	4	−4	4	4	.
X.125	.	−4	.	4	−4	8	−4	.	.	.	.	.	.	.	.	−4	.	.	.	.	4	4	4	4	.
X.126	.	−4	.	4	4	8	4	.	.	.	.	.	.	.	.	4	.	.	.	.	4	4	−4	−4	.
X.127	.	−4	.	−12	−4	−8	−4	.	.	.	.	.	.	.	.	−4	.	.	.	.	4	4	−4	4	.
X.128	.	.	.	.	.	.	.	.	.	.	−8	8	−8	8	.	.	.	.	.	.	.	.	.	.	.
X.129	.	.	.	.	.	.	.	.	.	.	8	−8	8	−8	.	.	.	.	.	.	.	.	.	.	.
X.130	−8	8	.	.	.	.	.	.	.	.	−4	4	−4	4	.	.	.	.	.	.	.	.	.	.	.
X.131	−8	−8	.	.	−8	.	8	.	.	.	4	4	−4	−4	.	.	.	.	.	.	.	.	.	.	.
X.132	−8	−8	.	.	8	.	−8	.	.	.	−4	−4	4	4	.	.	.	.	.	.	.	.	.	.	.
X.133	−8	8	.	.	.	.	.	.	.	.	4	−4	4	−4	.	.	.	.	.	.	.	.	.	.	.
X.134	16	−16	.	.	.	.	.	.	.	.	.	.	.	.	.	.	.	.	.	.	.	.	.	.	.
X.135	.	.	.	.	.	.	.	.	.	.	.	.	.	.	.	.	.	.	.	.	.	.	.	.	.
X.136	.	.	.	.	.	.	.	.	.	.	−8	8	−8	8	.	.	.	.	.	.	.	.	.	.	.
X.137	.	.	.	.	.	.	.	.	.	.	8	−8	8	−8	.	.	.	.	.	.	.	.	.	.	.
X.138	.	.	.	.	.	.	.	.	.	.	.	.	.	.	.	.	.	.	.	.	.	.	.	.	.
X.139	−8	.	.	−8	.	−8	8	−8	.	.	.	.	.	.	.	.	.	.	8	.	−8	8	.	.	−8
X.140	−8	.	.	−8	.	−8	8	8	.	.	.	.	.	.	.	.	.	.	−8	.	−8	8	.	.	8
X.141	.	8	.	.	−8	.	.	.	.	.	16	.	.	.	.	.	.	.	.	.	−8	8	8	.	.
X.142	.	8	.	.	−8	.	.	.	.	.	−16	.	.	.	.	.	.	.	.	.	−8	8	−8	.	.
X.143	24	−16	−8	.	.	−8	.	.	.	.	.	.	.	.	.	.	.	.	8	−8	.	.	.	.	.
X.144	12	−4	.	.	4	.	−4	.	.	.	6	6	−6	−6	.	.	.	.	.	.	.	.	.	.	.
X.145	−4	12	.	.	4	.	−4	.	.	.	6	6	−6	−6	.	.	.	.	.	.	.	.	.	.	.
X.146	−8	16	24	.	−8	.	.	.	.	.	.	.	.	.	.	.	.	.	8	−8	.	.	.	.	.
X.147	−4	12	.	.	−4	.	4	.	.	.	−6	−6	6	6	.	.	.	.	.	.	.	.	.	.	.
X.148	12	−4	.	.	−4	.	4	.	.	.	−6	−6	6	6	.	.	.	.	.	.	.	.	.	.	.

Character table of $D(\mathrm{Co}_2)$ *(continued)*

	4_{42}	4_{43}	4_{44}	4_{45}	4_{46}	4_{47}	4_{48}	4_{49}	4_{50}	4_{51}	4_{52}	4_{53}	4_{54}	4_{55}	4_{56}	4_{57}	4_{58}	4_{59}	4_{60}	$5a$	$6a$	$6b$	$6c$	$6d$	$6e$	$6f$	$6g$	$6h$
2	10	10	8	9	9	8	8	8	8	8	8	8	8	8	8	8	8	8	8	2	7	5	5	5	5	5	6	5
3	.	.	1	.	.	.	.	.	.	.	.	.	.	.	.	.	.	.	.	1	2	2	2	2	2	2	1	1
5	.	.	.	.	.	.	.	.	.	.	.	.	.	.	.	.	.	.	.	.	.	.	.	.	.	.	.	.
2P	$2e$	$2h$	$2h$	$2h$	$2h$	$2l$	$2t$	$2t$	$2t$	$2g$	$2t$	$2t$	$2g$	$2t$	$2t$	$2t$	$2q$	$2q$	$2h$	$5a$	$3a$	$3b$	$3b$	$3b$	$3a$	$3a$	$3b$	
3P	4_{42}	4_{43}	4_{44}	4_{45}	4_{46}	4_{47}	4_{48}	4_{49}	4_{50}	4_{51}	4_{52}	4_{53}	4_{54}	4_{55}	4_{56}	4_{57}	4_{58}	4_{59}	4_{60}	$5a$	$2a$	$2c$	$2b$	$2a$	$2b$	$2c$	$2d$	$2g$
5P	4_{42}	4_{43}	4_{44}	4_{45}	4_{46}	4_{47}	4_{48}	4_{49}	4_{50}	4_{51}	4_{52}	4_{53}	4_{54}	4_{55}	4_{56}	4_{57}	4_{58}	4_{59}	4_{60}	$1a$	$6a$	$6b$	$6c$	$6d$	$6e$	$6f$	$6g$	$6h$
X.85	−4	.	.	.	.	.	.	.	.	.	.	.	.	.	.	.	.	.	.	.	6	.	.	.	.	.	−2	.
X.86	.	.	8	.	.	.	.	.	.	.	.	.	.	.	.	.	.	.	.	.	−3	.	.	.	3	3	−3	.
X.87	4	.	.	.	.	.	2	−2	−2	.	−2	−2	.	2	2	2	.	.	.	.	6	.	.	.	.	.	−2	.
X.88	.	.	.	.	.	.	2	−2	2	2	2	−2	−2	2	−2	−2	.	.	.	.	−1	.	.	.	.	.	.	.
X.89	.	.	.	.	.	.	2	2	−2	2	−2	2	−2	2	−2	−2	.	.	.	.	−1	.	.	.	.	.	.	.
X.90	.	.	.	.	.	.	.	.	.	.	−4	.	.	4	.	.	.	.	.	.	−4	1	−1	−1	2	−2	.	1
X.91	.	.	.	.	.	.	−4	.	.	.	.	.	.	4	4	−4	.	.	.	.	−4	−1	1	−1	−2	2	.	1
X.92	.	.	.	.	.	.	4	.	.	.	.	.	.	−4	−4	4	.	.	.	.	−4	−1	1	−1	−2	2	.	1
X.93	.	.	.	.	.	.	.	.	.	.	−4	.	.	4	.	.	.	.	.	.	−4	1	−1	−1	2	−2	.	1
X.94	.	.	.	.	.	.	4	.	.	.	.	.	.	−4	−4	4	.	.	.	.	−4	1	−1	−1	2	−2	.	1
X.95	.	.	.	.	.	.	−4	.	.	.	.	.	.	4	4	−4	.	.	.	.	−4	−1	1	−1	−2	2	.	1
X.96	−4	−4	.	−4	−4	4	.	.	.	.	.	.	.	.	.	.	.	.	.	.	.	.	.	.	.	.	.	.
X.97	4	−4	.	4	−4	−4	.	.	.	.	.	.	.	.	.	.	.	.	.	.	.	.	.	.	.	.	.	.
X.98	4	4	.	4	−4	−4	.	.	.	.	.	.	.	.	.	.	.	.	.	.	.	.	.	.	.	.	.	.
X.99	−4	4	.	4	4	4	.	.	.	.	.	.	.	.	.	.	.	.	.	.	.	.	.	.	.	.	.	.
X.100	.	.	4	.	.	.	.	.	.	.	.	.	.	.	.	.	4	.	4	.	.	.	.	.	.	.	.	.
X.101	.	.	4	.	.	.	.	.	.	.	.	.	.	.	.	.	−4	.	4	.	.	.	.	.	.	.	.	.
X.102	.	.	.	.	.	.	.	.	.	.	.	.	.	.	.	.	.	.	.	.	−6	.	.	.	.	.	2	.
X.103	.	.	.	.	.	.	.	.	.	.	.	.	.	.	.	.	.	.	.	.	−6	.	.	.	.	.	2	.
X.104	.	.	.	.	.	.	.	.	.	.	.	.	.	.	.	.	.	.	.	.	−6	.	.	.	.	.	2	.
X.105	.	.	.	.	.	.	2	−2	2	−2	2	−2	2	2	−2	−2	.	.	.	.	.	3	−3	−3	.	.	.	−1
X.106	.	.	.	.	.	.	−2	−2	2	−2	2	−2	2	−2	2	2	.	.	.	.	.	3	−3	−3	.	.	.	−1
X.107	.	.	.	.	.	.	2	2	−2	−2	−2	2	2	2	−2	2	.	.	.	.	.	3	−3	−3	.	.	.	−1
X.108	.	.	.	.	.	.	−2	2	−2	−2	−2	2	2	−2	2	2	.	.	.	.	.	3	−3	−3	.	.	.	−1
X.109	.	.	.	.	.	.	.	.	.	.	.	.	.	.	.	.	.	.	.	.	−6	.	.	.	.	.	.	2
X.110	.	.	.	.	.	.	.	.	.	.	.	.	.	.	.	.	.	.	.	1	8	−2	2	2	−4	4	.	−2
X.111	.	.	.	.	.	.	.	.	.	.	.	.	.	.	.	.	.	.	.	.	−8	−2	2	−2	−4	4	.	2
X.112	−4	.	.	.	.	.	−2	−2	2	.	−2	2	.	2	−2	2	.	.	.	.	.	.	.	.	.	.	.	.
X.113	4	.	.	.	.	.	.	.	.	.	.	.	.	.	.	.	.	.	.	.	.	.	.	.	.	.	.	.
X.114	−4	.	.	.	.	.	2	2	−2	.	2	−2	.	−2	2	−2	.	.	.	.	.	.	.	.	.	.	.	.
X.115	−4	.	.	.	.	.	.	.	.	.	.	.	.	.	.	.	.	.	.	.	.	.	.	.	.	.	.	.
X.116	−4	.	.	.	.	.	−2	−2	−2	.	−2	−2	.	−2	−2	−2	.	.	.	.	.	.	.	.	.	.	.	.
X.117	4	.	.	.	.	.	−2	2	−2	.	2	−2	.	2	−2	2	.	.	.	.	.	.	.	.	.	.	.	.
X.118	−4	.	.	.	.	.	.	.	.	.	.	.	.	.	.	.	.	.	.	.	.	.	.	.	.	.	.	.
X.119	4	.	.	.	.	.	2	−2	2	.	−2	2	.	−2	2	−2	.	.	.	.	.	.	.	.	.	.	.	.
X.120	4	.	.	.	.	.	.	.	.	.	.	.	.	.	.	.	.	.	.	.	.	.	.	.	.	.	.	.
X.121	−4	.	.	.	.	.	−2	2	−2	.	2	−2	.	2	−2	2	.	.	.	.	.	.	.	.	.	.	.	.
X.122	−4	.	.	.	.	.	2	2	−2	.	−2	2	.	−2	2	2	.	.	.	.	.	.	.	.	.	.	.	.
X.123	4	.	.	.	.	.	−2	−2	2	.	−2	2	.	2	−2	2	.	.	.	.	.	.	.	.	.	.	.	.
X.124	4	.	.	.	.	.	2	2	−2	.	2	−2	.	−2	2	2	.	.	.	.	.	.	.	.	.	.	.	.
X.125	4	.	.	.	.	.	2	2	2	.	2	2	.	2	2	2	.	.	.	.	.	.	.	.	.	.	.	.
X.126	−4	.	.	.	.	.	2	2	2	.	−2	−2	.	2	2	2	.	.	.	.	.	.	.	.	.	.	.	.
X.127	4	.	.	.	.	.	−2	2	2	.	2	2	.	−2	−2	−2	.	.	.	.	.	.	.	.	.	.	.	.
X.128	.	.	.	.	.	.	.	.	.	.	.	.	.	.	.	.	.	.	.	−1	.	−3	3	−3	.	.	.	−1
X.129	.	.	.	.	.	.	.	.	.	.	.	.	.	.	.	.	.	.	.	−1	.	−3	3	−3	.	.	.	−1
X.130	.	.	.	.	.	.	−4	−4	.	.	4	4	.	.	.	.	.	.	.	.	.	3	−3	3	.	.	.	1
X.131	.	.	.	.	.	.	.	.	.	−4	.	.	4	.	.	.	.	.	.	.	.	3	−3	3	.	.	.	1
X.132	.	.	.	.	.	.	.	.	.	−4	.	.	4	.	.	.	.	.	.	.	.	3	−3	3	.	.	.	1
X.133	.	.	.	.	.	.	4	4	.	−4	−4	.	.	.	.	.	.	.	.	.	.	3	−3	3	.	.	.	1
X.134	.	.	.	.	.	.	.	.	.	.	.	.	.	.	.	.	.	.	.	1	.	.	.	.	.	.	.	.
X.135	.	.	.	.	.	.	.	.	.	.	.	.	.	.	.	.	.	.	.	.	−4	2	−2	2	−2	2	.	−2
X.136	.	.	.	.	.	.	.	.	.	.	.	.	.	.	.	.	.	.	.	.	8	−1	1	−1	4	−4	.	1
X.137	.	.	.	.	.	.	.	.	.	.	.	.	.	.	.	.	.	.	.	.	8	−1	1	−1	4	−4	.	1
X.138	.	.	.	.	.	.	.	.	.	.	.	.	.	.	.	.	.	.	.	.	−4	2	−2	2	−2	2	.	−2
X.139	8	.	.	.	.	.	.	.	.	.	.	.	.	.	.	.	.	.	.	.	.	.	.	.	.	.	.	.
X.140	−8	.	.	.	.	.	.	.	.	.	.	.	.	.	.	.	.	.	.	.	.	.	.	.	.	.	.	.
X.141	.	.	.	.	.	.	.	.	.	.	.	.	.	.	.	.	.	.	.	.	.	.	.	.	.	.	.	.
X.142	.	.	.	.	.	.	.	.	.	.	.	.	.	.	.	.	.	.	.	.	.	.	.	.	.	.	.	.
X.143	.	.	.	.	.	.	.	.	.	.	.	.	.	.	.	.	.	.	.	.	.	.	.	.	.	.	.	.
X.144	.	.	.	.	.	.	−2	2	−2	2	−2	2	−2	−2	2	2	.	.	.	.	.	.	.	.	.	.	.	.
X.145	.	.	.	.	.	.	2	−2	2	2	2	−2	−2	2	−2	−2	.	.	.	.	.	.	.	.	.	.	.	.
X.146	.	.	.	.	.	.	.	.	.	.	.	.	.	.	.	.	.	.	.	.	.	.	.	.	.	.	.	.
X.147	.	.	.	.	.	.	2	2	−2	2	−2	2	−2	2	−2	−2	.	.	.	.	.	.	.	.	.	.	.	.
X.148	.	.	.	.	.	.	−2	−2	2	2	2	−2	−2	−2	2	2	.	.	.	.	.	.	.	.	.	.	.	.

Character table of $D(\mathrm{Co}_2)$ (continued)

	6i	6j	6k	6l	6m	6n	6o	6p	6q	6r	6s	8a	8b	8c	8d	8e	8f	8g	8h	8i	8j	8k	8l	8m	8n	8o	8p
2	5	5	5	5	5	5	4	4	4	4	4	8	9	9	8	8	8	8	8	8	7	7	7	7	7	7	7
3	1	1	1	1	1	1	1	1	1	1	1	1	.	.	.	.	.	.	.	.	.	.	.	.	.	.	.
5																											
2P	3b	3a	3a	3b	3b	3a	3b	3b	3b	3a	3b	4_1	4_3	4_3	4_6	4_3	4_1	4_1	4_6	4_3	4_{19}	4_{20}	4_6	4_{20}	4_{20}	4_{22}	4_{22}
3P	2i	2j	2h	2f	2e	2k	2m	2n	2o	2s	2p	8a	8b	8c	8d	8e	8f	8g	8h	8i	8j	8k	8l	8m	8n	8o	8p
5P	6i	6j	6k	6l	6m	6n	6o	6p	6q	6r	6s	8a	8b	8c	8d	8e	8f	8g	8h	8i	8j	8k	8l	8m	8n	8o	8p
X.85	.	.	−2	.	.	.	.	.	.	2	.	.	.	.	.	.	.	.	.	.	2	.	.	.	.	−2	−2
X.86	.	1	1	.	.	1	.	.	.	−1	.	−8	.	.	.	.	.	.	.	.	.	.	.	.	.	.	.
X.87	.	−2	2	.	.	−2	.	.	.	.	.	.	.	.	.	.	.	.	.	.	.	2	.	−2	2	−2	2
X.88	.	.	.	.	.	.	.	.	.	.	.	.	.	.	.	.	.	.	.	.	2	.	.	.	.	−2	2
X.89	.	.	.	.	.	.	.	.	.	.	.	.	.	.	.	.	.	.	.	.	−2	.	.	.	.	−2	2
X.90	−1	2	.	1	−1	−2	1	−1	−1	.	1	.	.	.	.	.	.	.	.	.	.	.	.	.	.	.	.
X.91	−1	2	.	−1	1	−2	1	−1	1	.	−1	.	.	.	.	.	.	.	.	.	.	.	.	.	.	.	.
X.92	−1	2	.	−1	1	−2	−1	1	−1	.	1	.	.	.	.	.	.	.	.	.	.	.	.	.	.	.	.
X.93	−1	−2	.	1	−1	2	−1	1	1	.	−1	.	.	.	.	.	.	.	.	.	.	.	.	.	.	.	.
X.94	−1	−2	.	−1	1	2	1	−1	1	.	−1	.	.	.	.	.	.	.	.	.	.	.	.	.	.	.	.
X.95	−1	−2	.	−1	1	2	−1	1	−1	.	1	.	.	.	.	.	.	.	.	.	.	.	.	.	.	.	.
X.96	.	.	.	.	.	.	.	.	.	.	.	.	4	4	.	.	.	4	.	.	.	.	.	.	.	.	.
X.97	.	.	.	.	.	.	.	.	.	.	.	.	−4	−4	.	.	.	4	.	.	.	.	.	.	.	.	.
X.98	.	.	.	.	.	.	.	.	.	.	.	.	4	4	.	.	.	.	−4	.	.	.	.	.	.	.	.
X.99	.	.	.	.	.	.	.	.	.	.	.	.	−4	−4	.	.	.	.	−4	.	.	.	.	.	.	.	.
X.100	.	.	.	.	.	.	.	.	.	.	.	.	−4	4	.	−4	−4	.	.	.	.	.	.	.	.	.	.
X.101	.	.	.	.	.	.	.	.	.	.	.	.	4	−4	.	4	−4	.	.	.	.	.	.	.	.	.	.
X.102	.	2	−2	.	.	2	.	.	.	.	.	.	.	.	.	.	.	.	.	.	.	.	.	.	.	.	.
X.103	.	−2	−2	.	.	−2	.	.	.	.	.	.	.	.	.	.	.	.	.	.	.	.	.	.	.	.	.
X.104	.	.	2	.	.	.	.	.	2	.	.	.	.	.	.	2	.	.	.	.	.	.	.	.	.	.	.
X.105	1	.	.	−1	1	.	1	−1	−1	.	1	.	.	.	.	.	.	.	.	.	.	−2	.	.	.	.	2	−2
X.106	1	.	.	−1	1	.	−1	1	1	.	−1	.	.	.	.	.	.	.	.	.	.	−2	.	.	.	.	−2	2
X.107	1	.	.	−1	1	.	−1	1	1	.	−1	.	.	.	.	.	.	.	.	.	.	2	.	.	.	.	2	−2
X.108	1	.	.	−1	1	.	1	−1	−1	.	1	.	.	.	.	.	.	.	.	.	.	2	.	.	.	.	−2	2
X.109	.	.	2	.	.	.	.	.	−2	.	.	.	.	.	.	−2	.	.	.	.	.	.	.	.	.	.	.
X.110	2	.	−2	2	.	.	.	.	.	.	.	.	.	.	.	.	.	.	.	.	.	.	.	.	.	.	.
X.111	−2	.	−2	2	.	.	.	.	.	.	.	.	.	.	.	.	.	.	.	.	.	.	.	.	.	.	.
X.112	.	.	.	.	.	.	.	.	.	.	.	.	.	.	.	.	.	.	.	.	.	2	.	−2	−2	.	.
X.113	.	.	.	.	.	.	.	.	.	.	.	.	.	.	.	.	.	.	.	.	2	.	.	.	.	−2	−2
X.114	.	.	.	.	.	.	.	.	.	.	.	.	.	.	.	.	.	.	.	.	.	−2	.	2	2	.	.
X.115	.	.	.	.	.	.	.	.	.	.	.	.	.	.	.	.	.	.	.	.	−2	.	.	.	.	−2	−2
X.116	.	.	.	.	.	.	.	.	.	.	.	.	.	.	.	.	.	.	.	.	.	.	−2	.	−2	2	.
X.117	.	.	.	.	.	.	.	.	.	.	.	.	.	.	.	.	.	.	.	.	.	−2	.	−2	2	.	.
X.118	.	.	.	.	.	.	.	.	.	.	.	.	.	.	.	.	.	.	.	.	.	−2	.	.	.	2	2
X.119	.	.	.	.	.	.	.	.	.	.	.	.	.	.	.	.	.	.	.	.	.	.	2	2	−2	.	.
X.120	.	.	.	.	.	.	.	.	.	.	.	.	.	.	.	.	.	.	.	.	2	.	.	.	.	2	2
X.121	.	.	.	.	.	.	.	.	.	.	.	.	.	.	.	.	.	.	.	.	.	.	−2	−2	2	.	.
X.122	.	.	.	.	.	.	.	.	.	.	.	.	.	.	.	.	.	.	.	.	.	.	2	2	−2	.	.
X.123	.	.	.	.	.	.	.	.	.	.	.	.	.	.	.	.	.	.	.	.	.	.	2	−2	−2	.	.
X.124	.	.	.	.	.	.	.	.	.	.	.	.	.	.	.	.	.	.	.	.	.	.	−2	2	2	.	.
X.125	.	.	.	.	.	.	.	.	.	.	.	.	.	.	.	.	.	.	.	.	.	.	−2	−2	2	.	.
X.126	.	.	.	.	.	.	.	.	.	.	.	.	.	.	.	.	.	.	.	.	.	.	2	2	−2	.	.
X.127	.	.	.	.	.	.	.	.	.	.	.	.	.	.	.	.	.	.	.	.	.	.	−2	−2	−2	.	.
X.128	1	.	.	1	−1	.	1	−1	1	.	−1	.	.	.	.	.	.	.	.	.	.	.	.	.	.	.	.
X.129	1	.	.	1	−1	.	−1	1	−1	.	1	.	.	.	.	.	.	.	.	.	.	.	.	.	.	.	.
X.130	−1	.	.	−1	1	.	−1	1	−1	.	1	.	.	.	.	.	.	.	.	.	.	.	.	.	.	.	.
X.131	−1	.	.	−1	1	.	1	−1	1	.	−1	.	.	.	.	.	.	.	.	.	.	.	.	.	.	.	.
X.132	−1	.	.	1	−1	.	1	−1	−1	.	1	.	.	.	.	.	.	.	.	.	.	.	.	.	.	.	.
X.133	−1	.	.	−1	1	.	1	−1	1	.	−1	.	.	.	.	.	.	.	.	.	.	.	.	.	.	.	.
X.134	.	.	.	.	.	.	.	.	.	.	.	.	.	.	.	.	.	.	.	.	.	.	.	.	.	.	.
X.135	2	−2	.	2	−2	2	.	.	.	.	.	.	.	.	.	.	.	.	.	.	.	.	.	.	.	.	.
X.136	−1	.	.	−1	1	.	−1	1	−1	.	1	.	.	.	.	.	.	.	.	.	.	.	.	.	.	.	.
X.137	−1	.	.	−1	1	.	1	−1	1	.	−1	.	.	.	.	.	.	.	.	.	.	.	.	.	.	.	.
X.138	2	2	.	2	−2	−2	.	.	.	.	.	.	.	.	.	.	.	.	.	.	.	.	.	.	.	.	.
X.139	.	.	.	.	.	.	.	.	.	.	.	.	.	.	.	.	.	.	.	.	.	.	.	.	.	.	.
X.140	.	.	.	.	.	.	.	.	.	.	.	.	.	.	.	.	.	.	.	.	.	.	.	.	.	.	.
X.141	.	.	.	.	.	.	.	.	.	.	.	.	.	.	.	.	.	.	.	.	.	.	.	.	.	.	.
X.142	.	.	.	.	.	.	.	.	.	.	.	.	.	.	.	.	.	.	.	.	.	.	.	.	.	.	.
X.143	.	.	.	.	.	.	.	.	.	.	.	.	.	.	.	.	.	.	.	.	.	.	.	.	.	.	.
X.144	.	.	.	.	.	.	.	.	.	.	.	.	.	.	.	.	.	.	.	.	.	2	.	.	.	−2	2
X.145	.	.	.	.	.	.	.	.	.	.	.	.	.	.	.	.	.	.	.	.	.	−2	.	.	.	2	−2
X.146	.	.	.	.	.	.	.	.	.	.	.	.	.	.	.	.	.	.	.	.	.	.	.	.	.	.	.
X.147	.	.	.	.	.	.	.	.	.	.	.	.	.	.	.	.	.	.	.	.	.	2	.	.	.	2	−2
X.148	.	.	.	.	.	.	.	.	.	.	.	.	.	.	.	.	.	.	.	.	.	−2	.	.	.	−2	2

Character table of $D(\mathrm{Co}_2)$ (continued)

	8q	8r	8s	8t	8u	8v	8w	8x	8y	10a	10b	10c	12a	12b	12c	12d	12e	12f	12g	12h	12i	12j	16a	16b	24a
2	7	7	6	6	6	6	6	6	6	2	2	2	5	5	5	4	4	4	4	4	4	3	5	5	3
3	.	.	.	.	.	.	.	.	.	.	.	.	1	1	1	1	1	1	1	1	1	1	.	.	1
5	.	.	.	.	.	.	.	.	.	1	1	1	.	.	.	.	.	.	.	.	.	.	.	.	.
	8q	8r	8s	8t	8u	8v	8w	8x	8y	10a	10b	10c	12a	12b	12c	12d	12e	12f	12g	12h	12i	12j	16a	16b	24a
2P	4_{20}	4_{19}	4_8	4_9	4_{32}	4_{32}	4_{31}	4_{31}	4_{24}	5a	5a	5a	6a	6g	6g	6h	6h	6h	6g	6g	6h	6k	8b	8c	12a
3P	8q	8r	8s	8t	8u	8v	8w	8x	8y	10a	10b	10c	4_1	4_4	4_5	4_{27}	4_{28}	4_{29}	4_7	4_2	4_{30}	4_{44}	16a	16b	8a
5P	8q	8r	8s	8t	8u	8v	8w	8x	8y	2b	2c	2a	12a	12b	12c	12d	12e	12f	12g	12h	12i	12j	16a	16b	24a
X.85	.	2	2	-2	.	.	.	.	.	.	.	.	2	-2	-2	.	.	.	.	.	.	.	.	.	.
X.86	.	.	.	.	.	.	.	.	.	.	.	.	1	-1	-1	.	.	.	1	-1	.	-1	.	.	1
X.87	-2	.	.	.	.	.	.	.	.	.	.	.	-2	.	.	.	.	.	2	.	.	.	.	.	.
X.88	.	-2	.	.	.	.	.	.	.	-1	1	1	.	.	.	.	.	.	.	.	.	.	.	.	.
X.89	.	2	.	.	.	.	.	.	.	-1	1	1	.	.	.	.	.	.	.	.	.	.	.	.	.
X.90	.	.	.	.	.	.	.	.	.	.	.	.	.	-2	2	1	1	-1	.	-1	.	.	.	.	.
X.91	.	.	.	.	.	.	.	.	.	.	.	.	.	2	-2	-1	1	-1	.	1	.	.	.	.	.
X.92	.	.	.	.	.	.	.	.	.	.	.	.	.	2	-2	1	-1	1	.	1	.	.	.	.	.
X.93	.	.	.	.	.	.	.	.	.	.	.	.	.	2	-2	-1	-1	1	.	1	.	.	.	.	.
X.94	.	.	.	.	.	.	.	.	.	.	.	.	.	-2	2	-1	1	-1	.	1	.	.	.	.	.
X.95	.	.	.	.	.	.	.	.	.	.	.	.	.	-2	2	1	-1	1	.	-1	.	.	.	.	.
X.96	.	.	.	.	.	.	.	.	.	.	.	.	.	.	.	.	.	.	.	.	.	.	.	.	.
X.97	.	.	.	.	.	.	.	.	.	.	.	.	.	.	.	.	.	.	.	.	.	.	.	.	.
X.98	.	.	.	.	.	.	.	.	.	.	.	.	.	.	.	.	.	.	.	.	.	.	.	.	.
X.99	.	.	.	.	.	.	.	.	.	.	.	.	.	.	.	.	.	.	.	.	.	.	.	.	.
X.100	.	.	.	.	.	.	.	.	.	.	.	.	.	.	.	.	.	.	.	.	.	.	.	.	.
X.101	.	.	.	.	.	.	.	.	.	.	.	.	.	.	.	.	.	.	.	.	.	.	.	.	.
X.102	.	.	.	.	.	.	.	.	.	.	.	.	2	.	.	.	.	.	-2	.	.	.	.	.	.
X.103	.	.	.	.	.	.	.	.	.	.	.	.	2	.	.	.	.	.	2	.	.	.	.	.	.
X.104	.	.	.	.	.	.	.	.	.	.	.	.	-2	-2	-2	.	.	.	.	.	.	.	.	.	.
X.105	.	2	.	.	.	.	.	.	.	.	.	.	.	.	.	-1	-1	1	.	1	.	.	.	.	.
X.106	.	2	.	.	.	.	.	.	.	.	.	.	.	.	.	1	1	-1	.	-1	.	.	.	.	.
X.107	.	-2	.	.	.	.	.	.	.	.	.	.	.	.	.	1	1	-1	.	-1	.	.	.	.	.
X.108	.	-2	.	.	.	.	.	.	.	.	.	.	.	.	.	-1	-1	1	.	1	.	.	.	.	.
X.109	.	.	.	.	.	.	.	.	.	.	.	.	-2	2	2	.	.	.	.	.	.	.	.	.	.
X.110	.	.	.	.	.	.	.	.	.	1	-1	-1	.	.	.	.	.	.	.	.	.	.	.	.	.
X.111	.	.	.	.	.	.	.	.	.	.	.	.	.	.	.	.	.	.	.	.	.	.	.	.	.
X.112	2	.	.	.	.	.	.	.	.	.	.	.	.	.	.	.	.	.	.	.	.	.	.	.	.
X.113	.	2	2	-2	.	.	.	.	.	.	.	.	.	.	.	.	.	.	.	.	.	.	.	.	.
X.114	-2	.	.	.	.	.	.	.	.	.	.	.	.	.	.	.	.	.	.	.	.	.	.	.	.
X.115	.	-2	2	2	.	.	.	.	.	.	.	.	.	.	.	.	.	.	.	.	.	.	.	.	.
X.116	2	.	.	.	.	.	.	.	.	.	.	.	.	.	.	.	.	.	.	.	.	.	.	.	.
X.117	2	.	.	.	.	.	.	.	.	.	.	.	.	.	.	.	.	.	.	.	.	.	.	.	.
X.118	.	-2	-2	2	.	.	.	.	.	.	.	.	.	.	.	.	.	.	.	.	.	.	.	.	.
X.119	-2	.	.	.	.	.	.	.	.	.	.	.	.	.	.	.	.	.	.	.	.	.	.	.	.
X.120	.	2	-2	-2	.	.	.	.	.	.	.	.	.	.	.	.	.	.	.	.	.	.	.	.	.
X.121	2	.	.	.	.	.	.	.	.	.	.	.	.	.	.	.	.	.	.	.	.	.	.	.	.
X.122	-2	.	.	.	.	.	.	.	.	.	.	.	.	.	.	.	.	.	.	.	.	.	.	.	.
X.123	2	.	.	.	.	.	.	.	.	.	.	.	.	.	.	.	.	.	.	.	.	.	.	.	.
X.124	-2	.	.	.	.	.	.	.	.	.	.	.	.	.	.	.	.	.	.	.	.	.	.	.	.
X.125	-2	.	.	.	.	.	.	.	.	.	.	.	.	.	.	.	.	.	.	.	.	.	.	.	.
X.126	-2	.	.	.	.	.	.	.	.	.	.	.	.	.	.	.	.	.	.	.	.	.	.	.	.
X.127	2	.	.	.	.	.	.	.	.	.	.	.	.	.	.	.	.	.	.	.	.	.	.	.	.
X.128	.	.	.	.	.	.	.	.	.	1	-1	1	.	.	.	1	-1	1	.	.	-1	.	.	.	.
X.129	.	.	.	.	.	.	.	.	.	1	-1	1	.	.	.	-1	1	-1	.	.	1	.	.	.	.
X.130	.	.	.	.	.	.	.	.	.	.	.	.	.	.	.	-1	1	-1	.	.	1	.	.	.	.
X.131	.	.	.	.	.	.	.	.	.	.	.	.	.	.	.	1	1	-1	.	.	-1	.	.	.	.
X.132	.	.	.	.	.	.	.	.	.	.	.	.	.	.	.	-1	-1	1	.	.	1	.	.	.	.
X.133	.	.	.	.	.	.	.	.	.	.	.	.	.	.	.	1	-1	1	.	.	-1	.	.	.	.
X.134	.	.	.	.	.	.	.	.	.	-1	1	-1	.	.	.	.	.	.	.	.	.	.	.	.	.
X.135	.	.	.	.	.	.	.	.	.	.	.	.	.	-2	2	.	.	.	.	.	.	.	.	.	.
X.136	.	.	.	.	.	.	.	.	.	.	.	.	.	.	.	1	-1	1	.	.	-1	.	.	.	.
X.137	.	.	.	.	.	.	.	.	.	.	.	.	.	.	.	-1	1	-1	.	.	1	.	.	.	.
X.138	.	.	.	.	.	.	.	.	.	.	.	.	.	2	-2	.	.	.	.	.	.	.	.	.	.
X.139	.	.	.	.	.	.	.	.	.	.	.	.	.	.	.	.	.	.	.	.	.	.	.	.	.
X.140	.	.	.	.	.	.	.	.	.	.	.	.	.	.	.	.	.	.	.	.	.	.	.	.	.
X.141	.	.	.	.	.	.	.	.	.	.	.	.	.	.	.	.	.	.	.	.	.	.	.	.	.
X.142	.	.	.	.	.	.	.	.	.	.	.	.	.	.	.	.	.	.	.	.	.	.	.	.	.
X.143	.	.	.	.	.	.	.	.	.	.	.	.	.	.	.	.	.	.	.	.	.	.	.	.	.
X.144	.	-2	.	.	.	.	.	.	.	.	.	.	.	.	.	.	.	.	.	.	.	.	.	.	.
X.145	.	2	.	.	.	.	.	.	.	.	.	.	.	.	.	.	.	.	.	.	.	.	.	.	.
X.146	.	.	.	.	.	.	.	.	.	.	.	.	.	.	.	.	.	.	.	.	.	.	.	.	.
X.147	.	-2	.	.	.	.	.	.	.	.	.	.	.	.	.	.	.	.	.	.	.	.	.	.	.
X.148	.	2	.	.	.	.	.	.	.	.	.	.	.	.	.	.	.	.	.	.	.	.	.	.	.

B.3. Character table of $E_3 = E(\mathrm{Co}_2) = \langle x, y, e \rangle$

	1a	2a	2b	2c	2d	2e	2f	2g	2h	2i	3a	4a	4b	4c	4d	4e	4f	4g	4h	4i	4j	4k
2	18	18	17	15	14	14	14	13	12	14	7	14	13	14	12	12	13	11	12	11	11	11
3	2	2	1	2	1	1	1	1	1	.	2	1	1	.	1	1	.	1	.	.	.	.
5	1	1	.	1	.	.	.	.	1	.	.	.	.	.	.	.	.	.	.	.	.	.
7	1	.	1	.	1	1	.	.	.	.	.	.	.	.	.	.	.	.	.	.	.	.
11	1	.	.	.	.	.	.	.	.	.	.	.	.	.	.	.	.	.	.	.	.	.
2P	1a	1a	1a	1a	1a	1a	1a	1a	1a	1a	3a	2a	2b	2a	2b	2b	2b	2b	2a	2b	2b	2b
3P	1a	2a	2b	2c	2d	2e	2f	2g	2h	2i	1a	4a	4b	4c	4d	4e	4f	4g	4h	4i	4j	4k
5P	1a	2a	2b	2c	2d	2e	2f	2g	2h	2i	3a	4a	4b	4c	4d	4e	4f	4g	4h	4i	4j	4k
7P	1a	2a	2b	2c	2d	2e	2f	2g	2h	2i	3a	4a	4b	4c	4d	4e	4f	4g	4h	4i	4j	4k
11P	1a	2a	2b	2c	2d	2e	2f	2g	2h	2i	3a	4a	4b	4c	4d	4e	4f	4g	4h	4i	4j	4k
X.1	1	1	1	1	1	1	1	1	1	1	1	1	1	1	1	1	1	1	1	1	1	1
X.2	1	1	1	1	-1	-1	1	-1	-1	1	1	1	1	1	-1	-1	1	-1	-1	1	1	-1
X.3	21	21	21	21	-7	-7	5	-7	1	5	3	5	5	5	-7	-7	5	-7	1	5	5	1
X.4	21	21	21	21	7	7	5	7	-1	5	3	5	5	5	7	7	5	7	-1	5	5	-1
X.5	22	-10	6	-2	8	-8	6	.	.	-2	4	6	-6	-2	-4	4	2	.	.	2	-2	.
X.6	22	-10	6	-2	-8	8	6	.	.	-2	4	6	-6	-2	4	-4	2	.	.	2	-2	.
X.7	45	45	45	45	-3	-3	-3	-3	5	-3	.	-3	-3	-3	-3	-3	-3	-3	5	-3	-3	5
X.8	45	45	45	45	-3	-3	-3	-3	5	-3	.	-3	-3	-3	-3	-3	-3	-3	5	-3	-3	5
X.9	45	45	45	45	3	3	-3	3	-5	-3	.	-3	-3	-3	3	3	-3	3	-5	-3	-3	-5
X.10	45	45	45	45	3	3	-3	3	-5	-3	.	-3	-3	-3	3	3	-3	3	-5	-3	-3	-5
X.11	55	55	55	55	13	13	7	13	5	7	1	7	7	7	13	13	7	13	5	7	7	5
X.12	55	55	55	55	-13	-13	7	-13	-5	7	1	7	7	7	-13	-13	7	-13	-5	7	7	-5
X.13	99	99	99	99	15	15	3	15	1	3	.	3	3	3	15	15	3	15	-1	3	3	-1
X.14	99	99	99	99	-15	-15	3	-15	1	3	.	3	3	3	-15	-15	3	-15	1	3	3	1
X.15	154	154	154	154	14	14	10	14	6	10	1	10	10	10	14	14	10	14	6	10	10	6
X.16	154	154	154	154	-14	-14	10	-14	-6	10	1	10	10	10	-14	-14	10	-14	-6	10	10	-6
X.17	210	210	210	210	-14	-14	2	-14	10	2	3	2	2	2	-14	-14	2	-14	10	2	2	10
X.18	210	210	210	210	14	14	2	14	-10	2	3	2	2	2	14	14	2	14	-10	2	2	-10
X.19	231	39	7	-9	-21	-21	7	11	11	-9	6	23	15	7	-5	-5	-1	3	-5	-1	-1	3
X.20	231	39	7	-9	-35	-35	23	-3	-11	7	6	7	15	-9	-3	-3	-1	5	5	-1	-1	-3
X.21	231	231	231	231	-7	-7	7	-7	9	7	-3	7	7	7	-7	-7	7	-7	9	7	7	9
X.22	231	39	7	-9	35	35	23	3	11	7	6	7	15	-9	3	3	-1	-5	-5	-1	-1	3
X.23	231	39	7	-9	21	21	7	-11	-11	-9	6	23	15	7	5	5	-1	-3	5	-1	-1	-3
X.24	231	231	231	231	7	7	7	7	-9	7	-3	7	7	7	7	7	7	7	-9	7	7	-9
X.25	385	385	385	385	21	21	1	21	5	1	-2	1	1	1	21	21	1	21	5	1	1	5
X.26	385	385	385	385	-21	-21	1	-21	-5	1	-2	1	1	1	-21	-21	1	-21	-5	1	1	-5
X.27	440	-200	120	-40	-48	48	24	.	.	-8	8	24	-24	-8	24	-24	8	.	.	8	-8	.
X.28	440	-200	120	-40	48	-48	24	.	.	-8	8	24	-24	-8	-24	24	8	.	.	8	-8	.
X.29	560	560	560	560	.	.	-16	.	.	-16	2	-16	-16	-16	.	.	-16	.	.	-16	-16	.
X.30	770	-30	-14	10	84	-28	34	-4	20	-6	5	-14	-10	10	-8	-2	4	4	-2	2	2	-4
X.31	770	-350	210	-70	56	-56	18	.	.	-6	-4	18	-18	-6	-28	28	6	.	.	6	-6	.
X.32	770	-30	-14	10	28	-84	34	4	-20	-6	5	-14	-10	10	8	.	8	-2	-4	-4	2	4
X.33	770	-30	-14	10	-84	28	34	4	-20	-6	5	-14	-10	10	.	8	8	-2	-4	-4	2	4
X.34	770	-30	-14	10	-28	84	34	-4	20	-6	5	-14	-10	10	.	8	-8	-2	4	4	2	-4
X.35	770	-350	210	-70	-56	56	18	.	.	-6	-4	18	-18	-6	28	-28	6	.	.	6	-6	.
X.36	924	156	28	-36	-84	-84	28	-20	-4	28	6	-4	12	-4	-4	-4	12	12	-4	-4	-4	-4
X.37	924	156	28	-36	-28	-28	-4	36	4	-4	6	28	12	-12	12	4	4	-4	4	-4	-4	4
X.38	924	156	28	-36	84	84	28	20	4	28	6	-4	12	-4	4	4	12	-12	4	-4	-4	4
X.39	924	156	28	-36	28	28	-4	-36	-4	-4	6	28	12	-12	12	4	4	-4	-4	-4	-4	-4
X.40	990	-450	270	-90	-24	24	-18	.	.	6	.	-18	18	6	12	-12	-6	.	.	-6	6	.
X.41	990	-450	270	-90	-24	24	-18	.	.	6	.	-18	18	6	-12	12	-6	.	.	-6	6	.
X.42	990	-450	270	-90	24	-24	-18	.	.	6	.	-18	18	6	-12	12	-6	.	.	-6	6	.
X.43	990	-450	270	-90	24	-24	-18	.	.	6	.	-18	18	6	12	-12	-6	.	.	-6	6	.
X.44	1155	195	35	-45	63	63	35	31	15	19	-6	19	27	3	-1	-1	11	-9	-1	-5	-5	7
X.45	1155	195	35	-45	7	7	19	39	15	3	-6	35	27	19	-9	-9	11	-1	-1	-5	-5	7
X.46	1155	195	35	-45	-7	-7	19	-39	-15	3	-6	35	27	19	9	9	11	1	1	-5	-5	-7
X.47	1155	195	35	-45	-63	-63	35	-31	-15	19	-6	19	27	3	1	1	11	9	1	-5	-5	-7
X.48	1386	234	42	-54	42	42	-6	42	-14	26	.	-6	-6	26	-6	-6	26	-6	18	-6	-6	2
X.49	1386	234	42	-54	-42	-42	-6	-42	14	26	.	-6	-6	26	6	6	26	6	-18	-6	-6	-2
X.50	1408	-640	384	-128	-64	64	.	.	.	.	4	.	.	.	32	-32	.	.	.	.	.	.
X.51	1408	-640	384	-128	64	-64	.	.	.	.	4	.	.	.	-32	32	.	.	.	.	.	.
X.52	1540	-60	-28	20	.	.	-28	.	.	20	10	68	-20	-12	.	.	-4	.	.	-4	4	.
X.53	1540	-700	420	-140	.	.	36	.	.	-12	-8	36	-36	-12	.	.	12	.	.	12	-12	.
X.54	2772	-1260	756	-252	.	.	-12	.	.	4	-12	12	4	.	.	.	-4	.	.	-4	4	.
X.55	3080	-120	-56	40	56	56	40	-8	40	8	-7	40	-40	8	-8	-8	-8	8	8	-8	8	-8
X.56	3080	-120	-56	40	-112	112	40	.	.	8	2	40	-40	8	-8	-8	-8	.	.	-8	8	.
X.57	3080	-120	-56	40	-56	-56	40	8	-40	8	-7	40	-40	8	8	8	-8	-8	-8	-8	8	8
X.58	3080	-120	-56	40	112	-112	40	.	.	8	2	40	-40	8	8	8	-8	.	.	-8	8	.
X.59	3465	585	105	-135	105	105	9	9	25	25	.	-39	-15	-23	9	9	1	-15	-23	1	1	1
X.60	3465	585	105	-135	-105	-105	9	-9	-25	25	.	-39	-15	-23	-9	-9	1	15	23	1	1	-1
X.61	3465	585	105	-135	63	63	-39	-33	-25	-23	.	9	-15	25	15	15	1	-9	23	1	1	-1
X.62	3465	585	105	-135	63	63	9	-33	15	-39	.	57	33	9	15	15	-15	-9	-1	1	1	7
X.63	3465	585	105	-135	-105	-105	57	-9	15	9	.	9	33	-39	-9	-9	15	15	-1	1	1	7
X.64	3465	585	105	-135	-63	-63	-39	33	25	-23	.	9	-15	25	-15	-15	1	9	-23	1	1	1
X.65	3465	585	105	-135	105	105	57	9	-15	9	.	9	33	-39	9	9	-15	-15	1	1	1	-7
X.66	3465	585	105	-135	-63	-63	9	33	-15	-39	.	57	33	9	-15	-15	-15	9	1	1	1	-7
X.67	4620	-180	-84	60	-112	112	44	16	.	-4	3	-52	4	28	16	16	-12	-16	.	4	-4	.
X.68	4620	-180	-84	60	112	-112	44	-16	.	-4	3	-52	4	28	-16	-16	-12	16	.	4	-4	.
X.69	4620	-180	-84	60	56	-52	-8	-40	28	3	44	4	-4	-8	-8	-12	8	-8	8	4	-4	8
X.70	4620	-180	-84	60	-56	-56	-52	8	40	28	3	44	4	-4	8	8	-12	-8	8	4	-4	-8
X.71	6930	1170	210	-270	-42	-42	-30	-42	30	2	.	-30	-30	2	6	6	2	-6	-2	2	2	14
X.72	6930	1170	210	-270	42	42	-30	42	-30	2	.	-30	-30	2	-6	-6	2	6	-2	2	2	-14
X.73	6930	-270	126	90	84	-252	18	12	20	-22	.	-30	6	-6	24	.	.	14	-12	4	-2	-4
X.74	6930	-270	126	90	-252	84	18	12	20	-22	.	-30	6	-6	.	24	.	14	-12	4	-2	-4
X.75	6930	-270	126	90	-84	252	18	-12	-20	-22	.	-30	6	-6	-24	.	.	14	12	-4	-2	4
X.76	6930	-270	126	90	252	-84	18	-12	-20	-22	.	-30	6	-6	.	-24	.	14	12	-4	-2	4
X.77	9240	-360	-168	120	56	56	-8	-8	40	24	-3	-8	8	24	-8	-8	-24	8	8	8	-8	-8
X.78	9240	-360	-168	120	-56	-56	-8	8	-40	24	-3	-8	8	24	8	8	-24	-8	-8	8	-8	8
X.79	13860	-540	-252	180	.	.	-60	.	.	-12	.	36	12	-44	.	.	28	.	.	-4	4	.

Character table of $E_3 = E(\mathrm{Co}_2)$ (continued)

2	8	10	10	10	10	8	9	9	8	8	8	3	7	5	5	6	5	5	5	4	2	2	8	9	9	8	8	8	8	8
3	.	.	.	.	.	1	.	.	.	.	.	.	2	2	2	1	1	1	1	1	.	.	1	.	.	.	.	.	.	.
5	1	.	.	.	.	.	.	.	.	.	.	1	.	.	.	.	.	.	.	.	.	.	.	.	.	.	.	.	.	.
7	.	.	.	.	.	.	.	.	.	.	.	.	.	.	.	.	.	.	.	.	1	1	.	.	.	.	.	.	.	.
11	.	.	.	.	.	.	.	.	.	.	.	.	.	.	.	.	.	.	.	.	.	.	.	.	.	.	.	.	.	.
	4l	4m	4n	4o	4p	4q	4r	4s	4t	4u	4v	5a	6a	6b	6c	6d	6e	6f	6g	6h	7a	7b	8a	8b	8c	8d	8e	8f	8g	8h
2P	2c	2f	2f	2f	2f	2f	2f	2f	2i	2f	2i	5a	3a	3a	3a	3a	3a	3a	3a	3a	7a	7b	4a	4c	4c	4f	4f	4a	4c	4c
3P	4l	4m	4n	4o	4p	4q	4r	4s	4t	4u	4v	5a	2a	2c	2a	2b	2d	2f	2e	2g	7b	7a	8a	8b	8c	8d	8e	8f	8g	8h
5P	4l	4m	4n	4o	4p	4q	4r	4s	4t	4u	4v	1a	6a	6b	6c	6d	6e	6f	6g	6h	7b	7a	8a	8b	8c	8d	8e	8f	8g	8h
7P	4l	4m	4n	4o	4p	4q	4r	4s	4t	4u	4v	5a	6a	6b	6c	6d	6e	6f	6g	6h	1a	1a	8a	8b	8c	8d	8e	8f	8g	8h
11P	4l	4m	4n	4o	4p	4q	4r	4s	4t	4u	4v	5a	6a	6b	6c	6d	6e	6f	6g	6h	7a	7b	8a	8b	8c	8d	8e	8f	8g	8h
X.1	1	1	1	1	1	1	1	1	1	1	1	1	1	1	1	1	1	1	1	1	1	1	1	1	1	1	1	1	1	1
X.2	−1	1	1	−1	−1	−1	1	−1	1	1	−1	1	1	1	1	1	−1	1	−1	−1	1	1	−1	1	1	−1	−1	−1	1	−1
X.3	1	1	1	−3	−3	1	1	−3	1	1	1	1	3	3	3	3	−1	−1	−1	−1	.	.	1	1	1	−3	−3	−3	1	1
X.4	−1	1	1	3	3	−1	1	3	1	1	−1	1	3	3	3	3	1	−1	1	1	.	.	−1	1	1	3	3	3	1	−1
X.5	.	2	2	−4	4	.	−2	.	−2	2	.	2	−4	−2	2	.	2	.	−2	.	1	1	.	2	−2	−2	2	.	2	.
X.6	.	2	2	4	−4	.	−2	.	−2	2	.	2	−4	−2	2	.	−2	.	2	.	1	1	.	2	−2	2	−2	.	2	.
X.7	5	1	1	1	1	−3	1	1	1	1	−3	.	.	.	.	.	.	.	.	.	A	$\bar{A}$	−3	1	1	1	1	1	1	−3
X.8	5	1	1	1	1	−3	1	1	1	1	−3	.	.	.	.	.	.	.	.	.	$\bar{A}$	A	−3	1	1	1	1	1	1	−3
X.9	−5	1	1	−1	−1	3	1	−1	1	1	3	.	.	.	.	.	.	.	.	.	A	$\bar{A}$	3	1	1	−1	−1	−1	1	3
X.10	−5	1	1	−1	−1	3	1	−1	1	1	3	.	.	.	.	.	.	.	.	.	$\bar{A}$	A	3	1	1	−1	−1	−1	1	3
X.11	5	3	3	1	1	1	3	1	−1	−1	1	.	1	1	1	1	1	1	1	1	−1	−1	1	3	3	1	1	1	−1	1
X.12	−5	3	3	−1	−1	−1	3	−1	−1	−1	−1	.	1	1	1	1	−1	1	−1	−1	−1	−1	−1	3	3	−1	−1	−1	−1	−1
X.13	−1	3	3	−1	−1	3	3	−1	−1	−1	3	−1	.	.	.	.	.	.	.	.	1	1	3	3	3	−1	−1	−1	−1	3
X.14	1	3	3	1	1	−3	3	1	−1	−1	−3	−1	.	.	.	.	.	.	.	.	1	1	−3	3	3	1	1	1	−1	−3
X.15	6	−2	−2	2	2	2	−2	2	2	2	2	−1	1	1	1	1	−1	1	−1	−1	.	.	2	−2	−2	2	2	2	2	2
X.16	−6	−2	−2	−2	−2	−2	−2	−2	2	2	−2	−1	1	1	1	1	1	1	1	1	.	.	−2	−2	−2	−2	−2	−2	2	−2
X.17	10	−2	−2	−2	−2	2	−2	−2	−2	−2	2	.	3	3	3	3	1	−1	1	1	.	.	2	−2	−2	−2	−2	−2	−2	2
X.18	−10	−2	−2	2	2	−2	−2	2	−2	−2	−2	.	3	3	3	3	−1	−1	−1	−1	.	.	−2	−2	−2	2	2	2	−2	−2
X.19	−1	−1	−1	−5	−5	3	−1	3	3	−1	−1	1	6	.	.	−2	.	−2	.	2	.	.	3	3	3	−1	−1	3	3	−1
X.20	1	3	3	−7	−7	−3	3	1	−1	3	1	1	6	.	.	−2	−2	2	−2	.	.	.	−3	−1	−1	1	1	1	−1	1
X.21	9	−1	−1	1	1	1	−1	1	−1	−1	1	1	−3	−3	−3	−3	−1	1	−1	−1	.	.	1	−1	−1	1	1	1	−1	1
X.22	−1	3	3	7	7	3	3	−1	−1	3	−1	1	6	.	.	−2	2	2	2	.	.	.	3	−1	−1	−1	−1	−1	−1	−1
X.23	1	−1	−1	5	5	−3	−1	−3	3	−1	1	1	6	.	.	−2	.	−2	.	−2	.	.	−3	3	3	1	1	−3	3	1
X.24	−9	−1	−1	−1	−1	−1	−1	−1	−1	−1	−1	1	−3	−3	−3	−3	1	1	1	1	.	.	−1	−1	−1	−1	−1	−1	−1	−1
X.25	5	1	1	−3	−3	−3	1	−3	1	1	−3	.	−2	−2	−2	−2	.	−2	.	.	.	.	−3	1	1	−3	−3	−3	1	−3
X.26	−5	1	1	3	3	3	1	3	1	1	3	.	−2	−2	−2	−2	.	−2	.	.	.	.	3	1	1	3	3	3	1	3
X.27	.	.	.	8	−8	.	.	.	.	.	.	.	−8	−4	4	.	.	.	.	.	−1	−1	.	.	.	4	−4	.	.	.
X.28	.	.	.	−8	8	.	.	.	.	.	.	.	−8	−4	4	.	.	.	.	.	−1	−1	.	.	.	−4	4	.	.	.
X.29	.	.	.	.	.	.	.	.	.	.	.	.	2	2	2	2	.	2	.	.	.	.	.	.	.	.	.	.	.	.
X.30	.	10	−6	−4	4	4	−2	.	2	2	.	.	−3	1	−3	1	3	1	−1	−1	.	.	−4	−2	2	2	−2	.	−2	.
X.31	.	6	6	4	−4	.	−6	.	2	−2	.	.	4	2	−2	.	2	.	−2	.	.	.	.	6	−6	2	−2	.	−2	.
X.32	.	−6	10	−4	4	−4	−2	.	2	2	.	.	−3	1	−3	1	1	1	−3	1	.	.	4	−2	2	2	−2	.	−2	.
X.33	.	10	−6	4	−4	−4	−2	.	2	2	.	.	−3	1	−3	1	−3	1	1	1	.	.	4	−2	2	−2	2	.	−2	.
X.34	.	−6	10	4	−4	4	−2	.	2	2	.	.	−3	1	−3	1	−1	1	3	−1	.	.	−4	−2	2	−2	2	.	−2	.
X.35	.	6	6	−4	4	.	−6	.	2	−2	.	.	4	2	−2	.	−2	.	2	.	.	.	.	6	−6	−2	2	.	−2	.
X.36	4	4	4	−4	−4	.	4	−4	.	.	.	−1	6	.	.	−2	.	−2	.	−2	.	.	.	−4	−4	.	.	4	.	.
X.37	−4	−4	−4	−4	−4	.	−4	−4	.	.	.	−1	6	.	.	−2	2	2	2	.	.	.	.	4	4	.	.	4	.	.
X.38	−4	4	4	4	4	.	4	4	.	.	.	−1	6	.	.	−2	.	−2	.	2	.	.	.	−4	−4	.	.	−4	.	.
X.39	4	−4	−4	4	4	.	−4	4	.	.	.	−1	6	.	.	−2	−2	2	−2	.	.	.	.	4	4	.	.	−4	.	.
X.40	.	2	2	−4	4	.	−2	.	−2	2	.	.	.	.	.	.	.	.	.	.	A	$\bar{A}$	.	2	−2	−2	2	.	2	.
X.41	.	2	2	−4	4	.	−2	.	−2	2	.	.	.	.	.	.	.	.	.	.	$\bar{A}$	A	.	2	−2	−2	2	.	2	.
X.42	.	2	2	4	−4	.	−2	.	−2	2	.	.	.	.	.	.	.	.	.	.	A	$\bar{A}$	.	2	−2	2	−2	.	2	.
X.43	.	2	2	4	−4	.	−2	.	−2	2	.	.	.	.	.	.	.	.	.	.	$\bar{A}$	A	.	2	−2	2	−2	.	2	.
X.44	−5	3	3	−1	−1	3	3	7	3	−1	−1	.	−6	.	.	2	.	2	.	−2	.	.	3	−1	−1	−1	−1	−1	3	−1
X.45	−5	−1	−1	3	3	3	−1	−5	−1	3	−1	.	−6	.	.	2	−2	−2	−2	.	.	.	3	3	3	−1	−1	3	−1	−1
X.46	5	−1	−1	−3	−3	−3	−1	5	−1	3	1	.	−6	.	.	2	2	−2	2	.	.	.	−3	3	3	1	1	−3	−1	1
X.47	5	3	3	1	1	−3	3	−7	3	−1	1	.	−6	.	.	2	.	2	.	2	.	.	−3	−1	−1	1	1	1	3	1
X.48	−6	−2	−2	−2	−2	−6	−2	−2	−2	−2	2	1	.	.	.	.	.	.	.	.	.	.	−6	−2	−2	2	2	−2	−2	2
X.49	6	−2	−2	2	2	6	−2	2	−2	−2	−2	1	.	.	.	.	.	.	.	.	.	.	6	−2	−2	−2	−2	2	−2	−2
X.50	.	.	.	.	.	.	.	.	.	.	.	−2	−4	−2	2	.	2	.	−2	.	1	1	.	.	.	.	.	.	.	.
X.51	.	.	.	.	.	.	.	.	.	.	.	−2	−4	−2	2	.	−2	.	2	.	1	1	.	.	.	.	.	.	.	.
X.52	.	−4	−4	.	.	.	4	.	−4	−4	.	.	−6	2	−6	2	.	2	.	.	.	.	.	4	−4	.	.	.	4	.
X.53	.	−4	−4	.	.	.	4	.	−4	4	.	.	8	4	−4	.	.	.	.	.	.	.	.	−4	4	.	.	.	4	.
X.54	.	−4	−4	.	.	.	4	.	4	−4	.	2	.	.	.	.	.	.	.	.	.	.	.	−4	4	.	.	.	−4	.
X.55	.	.	.	.	.	8	.	.	.	.	.	.	9	−5	3	1	−1	1	−1	1	.	.	−8	.	.	.	.	.	.	.
X.56	.	.	.	8	−8	.	.	.	.	.	.	.	−6	4	.	−2	2	−2	−2	.	.	.	.	.	.	−4	4	.	.	.
X.57	.	.	.	.	.	−8	.	.	.	.	.	.	9	−5	3	1	1	1	1	−1	.	.	8	.	.	.	.	.	.	.
X.58	.	.	.	−8	8	.	.	.	.	.	.	.	−6	4	.	−2	−2	−2	2	.	.	.	.	.	.	4	−4	.	.	.
X.59	5	1	1	−7	−7	−3	1	1	1	−3	1	.	.	.	.	.	.	.	.	.	.	.	−3	5	5	1	1	1	1	1
X.60	−5	1	1	7	7	3	1	−1	1	−3	−1	.	.	.	.	.	.	.	.	.	.	.	3	5	5	−1	−1	−1	1	−1
X.61	−5	5	5	−5	−5	3	5	3	−3	1	−1	.	.	.	.	.	.	.	.	.	.	.	3	1	1	−1	−1	3	−3	−1
X.62	−5	1	1	−1	−1	3	1	−9	1	−3	−1	.	.	.	.	.	.	.	.	.	.	.	3	−3	−3	3	3	−1	1	−1
X.63	−5	−3	−3	−5	−5	3	−3	3	−3	1	−1	.	.	.	.	.	.	.	.	.	.	.	3	1	1	3	3	−5	−3	−1
X.64	5	5	5	5	5	−3	5	−3	−3	1	1	.	.	.	.	.	.	.	.	.	.	.	−3	1	1	1	1	−3	−3	1
X.65	5	−3	−3	5	5	−3	−3	−3	−3	1	1	.	.	.	.	.	.	.	.	.	.	.	−3	1	1	−3	−3	5	−3	1
X.66	5	1	1	1	1	−3	1	9	1	−3	1	.	.	.	.	.	.	.	.	.	.	.	−3	−3	−3	−3	−3	1	1	1
X.67	.	−4	−4	.	.	4	4	.	.	.	−4	.	3	−3	−3	3	−1	−1	−1	1	.	.	−4	4	−4	.	.	.	.	4
X.68	.	−4	−4	.	.	−4	4	.	.	.	4	.	3	−3	−3	3	1	−1	1	−1	.	.	4	4	−4	.	.	.	.	−4
X.69	.	4	4	.	.	4	−4	.	.	.	4	.	3	−3	−3	3	−1	−1	−1	1	.	.	−4	−4	4	.	.	.	.	−4
X.70	.	4	4	.	.	−4	−4	.	.	.	−4	.	3	−3	−3	3	1	−1	1	−1	.	.	4	−4	4	.	.	.	.	4
X.71	−10	−2	−2	2	2	−6	−2	2	2	2	2	.	.	.	.	.	.	.	.	.	.	.	−6	−2	−2	−2	−2	2	2	2
X.72	10	−2	−2	−2	−2	6	−2	−2	2	2	−2	.	.	.	.	.	.	.	.	.	.	.	6	−2	−2	2	2	−2	2	−2
X.73	.	−6	10	4	−4	.	−2	.	−2	−2	4	.	.	.	.	.	.	.	.	.	.	.	.	−2	2	−2	2	.	2	−4
X.74	.	10	−6	−4	4	.	−2	.	−2	−2	4	.	.	.	.	.	.	.	.	.	.	.	.	−2	2	2	−2	.	2	−4
X.75	.	−6	10	−4	4	.	−2	.	−2	−2	−4	.	.	.	.	.	.	.	.	.	.	.	.	−2	2	2	−2	.	2	4
X.76	.	10	−6	4	−4	.	−2	.	−2	−2	−4	.	.	.	.	.	.	.	.	.	.	.	.	−2	2	−2	2	.	2	4
X.77	.	.	.	.	.	−8	.	.	.	.	.	.	−3	3	3	−3	−1	1	−1	1	.	.	8	.	.	.	.	.	.	.
X.78	.	.	.	.	.	8	.	.	.	.	.	.	−3	3	3	−3	1	1	1	−1	.	.	−8	.	.	.	.	.	.	.
X.79	.	−4	−4	.	.	.	4	.	4	4	.	.	.	.	.	.	.	.	.	.	.	.	.	4	−4	.	.	.	−4	.

Character table of $E_3 = E(Co_2)$ (continued)

2	8	7	6	6	6	6	6	3	2	2	.	5	5	5	4	4	3	2	2	2	2	2	2	5	5	2	3
3	.	.	.	.	.	.	.	.	.	.	.	1	1	1	1	1	1	.	.	.	.	.	.	.	.	.	1
5	.	.	.	.	.	.	.	1	1	1	.	.	.	.	.	.	.	.	.	.	.	.	.	.	.	1	.
7	.	.	.	.	.	.	.	.	.	.	.	.	.	.	.	.	.	1	1	1	1	1	1	.	.	.	.
11	.	.	.	.	.	.	.	.	.	.	1	.	.	.	.	.	.	.	.	.	.	.	.	.	.	.	.

	8i	8j	8k	8l	8m	8n	8o	10a	10b	10c	11a	12a	12b	12c	12d	12e	12f	14a	14b	14c	14d	14e	14f	16a	16b	20a	24a
2P	4a	4f	4m	4n	4m	4n	4i	5a	5a	5a	11a	6d	6d	6a	6d	6d	6f	7a	7b	7b	7a	7b	7a	8b	8c	10a	12c
3P	8i	8j	8k	8l	8m	8n	8o	10a	10b	10c	11a	4d	4e	4a	4g	4b	4q	14b	14a	14d	14c	14f	14e	16a	16b	20a	8a
5P	8i	8j	8k	8l	8m	8n	8o	2c	2a	2h	11a	12a	12b	12c	12d	12e	12f	2d	2d	2e	2e	2b	2b	16a	16b	4l	24a
7P	8i	8j	8k	8l	8m	8n	8o	10a	10b	10c	11a	12a	12b	12c	12d	12e	12f	2d	2d	2e	2e	2b	2b	16a	16b	20a	24a
11P	8i	8j	8k	8l	8m	8n	8o	10a	10b	10c	1a	12a	12b	12c	12d	12e	12f	14a	14b	14c	14d	14e	14f	16a	16b	20a	24a
X.1	1	1	1	1	1	1	1	1	1	1	1	1	1	1	1	1	1	1	1	1	1	1	1	1	1	1	1
X.2	1	-1	-1	1	1	1	1	1	1	-1	1	-1	-1	1	-1	1	-1	-1	-1	-1	-1	1	1	-1	1	-1	-1
X.3	1	-1	-1	-1	-1	1	1	1	1	-1	1	-1	-1	-1	-1	-1	1	.	.	.	.	.	.	-1	-1	1	1
X.4	1	1	1	1	-1	-1	1	1	1	-1	1	1	1	-1	1	-1	-1	.	.	.	.	.	.	1	-1	-1	-1
X.5	-2	.	2	-2	.	.	.	.	-2	.	.	2	-2	.	.	.	.	1	1	-1	-1	-1	-1	.	.	.	.
X.6	-2	.	-2	2	.	.	.	.	-2	.	.	-2	2	.	.	.	.	-1	-1	1	1	-1	-1	.	.	.	.
X.7	1	1	-1	-1	-1	-1	1	.	.	.	1	.	.	.	.	.	.	$-A$	$-\bar{A}$	$-\bar{A}$	$-A$	$\bar{A}$	A	-1	-1	.	.
X.8	1	1	-1	-1	-1	-1	1	.	.	.	1	.	.	.	.	.	.	$-\bar{A}$	$-A$	$-A$	$-\bar{A}$	A	$\bar{A}$	-1	-1	.	.
X.9	1	1	1	1	-1	-1	1	.	.	.	1	.	.	.	.	.	.	A	$\bar{A}$	$\bar{A}$	A	$\bar{A}$	A	1	-1	.	.
X.10	1	1	1	1	-1	-1	1	.	.	.	1	.	.	.	.	.	.	$\bar{A}$	A	A	$\bar{A}$	A	$\bar{A}$	1	-1	.	.
X.11	-1	3	-1	-1	1	1	-1	.	.	.	.	1	1	1	1	1	1	-1	-1	-1	-1	-1	-1	-1	1	.	1
X.12	-1	3	1	1	1	1	-1	.	.	.	.	-1	-1	1	-1	1	-1	1	1	1	1	-1	-1	-1	1	.	-1
X.13	-1	3	-1	-1	-1	-1	1	-1	-1	-1	.	.	.	.	.	.	.	1	1	1	1	1	1	-1	-1	-1	.
X.14	-1	3	1	1	-1	-1	1	-1	-1	1	.	.	.	.	.	.	.	-1	-1	-1	-1	1	1	1	-1	.	.
X.15	2	-2	.	.	.	.	.	2	-1	-1	.	-1	-1	1	-1	1	1	.	.	.	.	.	.	.	.	1	-1
X.16	2	-2	.	.	.	.	.	2	-1	-1	.	1	1	1	1	1	1	.	.	.	.	.	.	.	.	-1	1
X.17	-2	-2	.	.	.	.	.	-2	.	.	.	1	1	-1	1	-1	-1	1	.	.	.	.	.	.	.	.	-1
X.18	-2	-2	.	.	.	.	.	-2	.	.	.	1	-1	-1	-1	-1	-1	1	.	.	.	.	.	.	.	.	1
X.19	-1	-1	-1	-1	-1	-1	-1	1	1	1	.	-2	-2	2	.	.	.	.	.	.	.	.	.	1	1	-1	.
X.20	3	-1	-1	-1	1	1	-1	1	-1	-1	.	.	.	-2	2	.	.	.	.	.	.	.	.	1	-1	1	.
X.21	-1	-1	1	1	1	1	-1	1	1	-1	.	-1	-1	1	-1	1	1	.	.	.	.	.	.	1	-1	-1	1
X.22	3	-1	1	1	1	1	-1	1	-1	1	.	.	.	-2	-2	.	.	.	.	.	.	.	.	-1	-1	-1	.
X.23	-1	-1	1	1	-1	-1	1	1	1	-1	.	2	2	2	.	.	.	.	.	.	.	.	.	-1	1	1	.
X.24	-1	-1	-1	-1	-1	-1	-1	1	1	1	.	1	1	1	1	1	-1	.	.	.	.	.	.	-1	-1	1	-1
X.25	1	1	1	1	1	1	1	.	.	.	.	-2	.	-2	.	.	.	.	.	.	.	.	.	-1	1	.	.
X.26	1	1	-1	-1	1	1	1	.	.	.	.	-2	.	-2	.	.	.	.	.	.	.	.	.	-1	1	.	.
X.27	.	.	.	.	.	.	.	.	.	.	.	.	.	.	.	.	.	1	1	-1	-1	1	1	.	.	.	.
X.28	.	.	.	.	.	.	.	.	.	.	.	.	.	.	.	.	.	-1	-1	1	1	1	1	.	.	.	.
X.29	.	.	.	.	.	.	.	.	.	.	-1	.	2	.	2	.	.	.	.	.	.	.	.	.	.	.	.
X.30	-2	.	.	.	2	-2	.	.	.	.	.	-3	1	1	-1	1	.	.	.	.	.	.	.	.	.	.	-1
X.31	2	.	-2	2	.	.	.	.	.	.	.	2	-2	.	.	.	.	.	.	.	.	.	.	.	.	.	.
X.32	-2	.	.	.	-2	2	.	.	.	.	.	-1	3	1	-1	-1	-1	.	.	.	.	.	.	.	.	.	1
X.33	-2	.	.	.	2	-2	.	.	.	.	.	3	-1	1	-1	-1	-1	.	.	.	.	.	.	.	.	.	1
X.34	-2	.	.	.	-2	2	.	.	.	.	.	1	-3	1	1	-1	1	.	.	.	.	.	.	.	.	.	-1
X.35	2	.	2	-2	.	.	.	.	.	.	.	-2	2	.	.	.	.	.	.	.	.	.	.	.	.	.	.
X.36	.	.	.	.	.	.	.	-1	1	1	.	2	2	2	.	.	.	.	.	.	.	.	.	.	.	-1	.
X.37	.	.	.	.	.	.	.	-1	1	-1	.	.	-2	-2	2	.	.	.	.	.	.	.	.	.	.	1	.
X.38	.	.	.	.	.	.	.	-1	1	-1	.	-2	-2	2	.	.	.	.	.	.	.	.	.	.	.	1	.
X.39	.	.	.	.	.	.	.	-1	1	1	.	.	-2	2	2	.	.	.	.	.	.	.	.	.	.	-1	.
X.40	-2	.	-2	2	.	.	.	.	.	.	.	.	.	.	.	.	.	$-A$	$-\bar{A}$	$\bar{A}$	A	$-\bar{A}$	$-A$	.	.	.	.
X.41	-2	.	-2	2	.	.	.	.	.	.	.	.	.	.	.	.	.	$-\bar{A}$	$-A$	A	$\bar{A}$	A	$-\bar{A}$	.	.	.	.
X.42	-2	.	2	-2	.	.	.	.	.	.	.	.	.	.	.	.	.	A	$\bar{A}$	$-\bar{A}$	$-A$	$-\bar{A}$	$-A$	.	.	.	.
X.43	-2	.	2	-2	.	.	.	.	.	.	.	.	.	.	.	.	.	$\bar{A}$	A	$-A$	$-\bar{A}$	$-A$	$-\bar{A}$	.	.	.	.
X.44	-1	-1	-1	-1	-1	-1	-1	.	.	.	.	2	2	-2	.	.	.	.	.	.	.	.	.	1	1	.	.
X.45	3	-1	1	1	1	1	-1	.	.	.	.	.	.	2	2	.	.	.	.	.	.	.	.	-1	-1	.	.
X.46	3	-1	-1	-1	1	1	-1	.	.	.	.	.	.	2	-2	.	.	.	.	.	.	.	.	1	-1	.	.
X.47	-1	-1	1	1	-1	-1	-1	.	.	.	.	-2	-2	-2	.	.	.	.	.	.	.	.	.	-1	1	.	.
X.48	-2	2	.	.	.	.	.	2	1	-1	1	.	.	.	.	.	.	.	.	.	.	.	.	.	.	-1	.
X.49	-2	2	.	.	.	.	.	2	1	-1	-1	.	.	.	.	.	.	.	.	.	.	.	.	.	.	1	.
X.50	.	.	.	.	.	.	.	2	.	.	.	2	-2	.	.	.	.	-1	-1	1	1	-1	-1	.	.	.	.
X.51	.	.	.	.	.	.	.	2	.	.	.	-2	2	.	.	.	.	1	1	-1	-1	-1	-1	.	.	.	.
X.52	4	.	.	.	.	.	.	.	.	.	.	.	.	.	.	.	.	2	.	-2	.	.	.	.	.	.	.
X.53	-4	.	.	.	.	.	.	.	.	.	.	.	.	.	.	.	.	.	.	.	.	.	.	.	.	.	.
X.54	4	.	.	.	.	.	-2	.	.	.	.	.	.	.	.	.	.	.	.	.	.	.	.	.	.	.	.
X.55	.	.	.	.	.	.	.	.	.	.	.	1	1	1	-1	-1	-1	.	.	.	.	.	.	.	.	.	1
X.56	.	.	.	.	.	.	.	.	.	.	.	-2	2	-2	.	2	.	.	.	.	.	.	.	.	.	.	.
X.57	.	.	.	.	.	.	.	.	.	.	.	-1	-1	1	1	-1	1	.	.	.	.	.	.	.	.	.	-1
X.58	.	.	.	.	.	.	.	.	.	.	.	2	-2	-2	.	2	.	.	.	.	.	.	.	.	.	.	.
X.59	-3	-3	1	1	1	1	1	.	.	.	.	.	.	.	.	.	.	.	.	.	.	.	.	-1	-1	.	.
X.60	-3	-3	-1	-1	1	1	1	.	.	.	.	.	.	.	.	.	.	.	.	.	.	.	.	1	-1	.	.
X.61	1	-3	1	1	-1	-1	1	.	.	.	.	.	.	.	.	.	.	.	.	.	.	.	.	-1	1	.	.
X.62	-3	1	-1	-1	1	1	1	.	.	.	.	.	.	.	.	.	.	.	.	.	.	.	.	1	-1	.	.
X.63	1	1	1	1	-1	-1	1	.	.	.	.	.	.	.	.	.	.	.	.	.	.	.	.	-1	1	.	.
X.64	1	-3	-1	-1	-1	-1	1	.	.	.	.	.	.	.	.	.	.	.	.	.	.	.	.	1	1	.	.
X.65	1	1	-1	-1	-1	-1	1	.	.	.	.	.	.	.	.	.	.	.	.	.	.	.	.	1	1	.	.
X.66	-3	1	1	1	1	1	1	.	.	.	.	.	.	.	.	.	.	.	.	.	.	.	.	-1	-1	.	.
X.67	.	.	.	.	.	.	.	.	.	.	.	1	1	-1	-1	1	1	.	.	.	.	.	.	.	.	.	-1
X.68	.	.	.	.	.	.	.	.	.	.	.	-1	-1	-1	1	1	-1	.	.	.	.	.	.	.	.	.	1
X.69	.	.	.	.	.	.	.	.	.	.	.	1	1	-1	-1	1	1	.	.	.	.	.	.	.	.	.	-1
X.70	.	.	.	.	.	.	.	.	.	.	.	-1	-1	-1	1	1	1	.	.	.	.	.	.	.	.	.	1
X.71	2	2	.	.	.	.	-2	.	.	.	.	.	.	.	.	.	.	.	.	.	.	.	.	.	.	.	.
X.72	2	2	.	.	.	.	-2	.	.	.	.	.	.	.	.	.	.	.	.	.	.	.	.	.	.	.	.
X.73	2	.	.	.	2	-2	.	.	.	.	.	.	.	.	.	.	.	.	.	.	.	.	.	.	.	.	.
X.74	2	.	.	.	-2	2	.	.	.	.	.	.	.	.	.	.	.	.	.	.	.	.	.	.	.	.	.
X.75	2	.	.	.	2	-2	.	.	.	.	.	.	.	.	.	.	.	.	.	.	.	.	.	.	.	.	.
X.76	2	.	.	.	-2	2	.	.	.	.	.	.	.	.	.	.	.	.	.	.	.	.	.	.	.	.	.
X.77	.	.	.	.	.	.	.	.	.	.	.	1	1	1	-1	-1	1	.	.	.	.	.	.	.	.	.	-1
X.78	.	.	.	.	.	.	.	.	.	.	.	-1	-1	1	1	-1	-1	.	.	.	.	.	.	.	.	.	1
X.79	-4	.	.	.	.	.	.	.	.	.	.	.	.	.	.	.	.	.	.	.	.	.	.	.	.	.	.

where $A = \frac{1}{2}(-1 + i\sqrt{7})$.

B.4. *Character table of* $H(\mathrm{Fi}_{22}) = \langle x, y, h \rangle$

	1a	2a	2b	2c	2d	2e	2f	2g	2h	2i	2j	2k	2l	2m	3a	3b	3c	4_1	4_2	4_3	4_4	4_5
2	17	17	16	15	15	16	16	14	10	12	12	13	11	10	7	8	5	12	10	12	10	10
3	4	4	4	2	2	1	1	2	2	1	1	.	1	1	4	3	3	3	2	1	1	1
5	1	1	1	.	.	.	.	.	.	.	1	.	.	.	.	.	.	1	.	.	.	.
2P	1a	1a	1a	1a	1a	1a	1a	1a	1a	1a	1a	1a	1a	1a	3a	3b	3c	2a	2a	2a	2e	2a
3P	1a	2a	2b	2c	2d	2e	2f	2g	2h	2i	2j	2k	2l	2m	1a	1a	1a	4_1	4_2	4_3	4_4	4_5
5P	1a	2a	2b	2c	2d	2e	2f	2g	2h	2i	2j	2k	2l	2m	3a	3b	3c	4_1	4_2	4_3	4_4	4_5
X.1	1	1	1	1	1	1	1	1	1	1	1	1	1	1	1	1	1	1	1	1	1	1
X.2	1	1	1	1	1	1	1	1	-1	1	1	1	1	-1	1	1	1	1	-1	1	-1	-1
X.3	6	6	6	-2	-2	6	6	-2	-4	2	2	-2	2	.	-3	.	3	6	-4	-2	-4	.
X.4	6	6	6	-2	-2	6	6	-2	4	2	2	-2	2	.	-3	.	3	6	4	-2	4	.
X.5	10	10	10	-6	-6	10	10	-6	.	2	2	-6	2	.	1	4	-2	10	.	-6	.	.
X.6	15	15	15	-1	-1	15	15	-1	5	-1	-1	-1	-1	-3	6	.	3	15	5	-1	5	-3
X.7	15	15	15	-1	-1	15	15	-1	-5	-1	-1	-1	-1	3	6	.	3	15	-5	-1	-5	3
X.8	15	15	15	7	7	15	15	7	-5	3	3	7	3	-1	-3	3	.	15	-5	7	-5	-1
X.9	15	15	15	7	7	15	15	7	5	3	3	7	3	1	-3	3	.	15	5	7	5	1
X.10	20	20	20	4	4	20	20	4	.	-4	-4	4	-4	.	-7	2	2	20	.	4	.	.
X.11	20	20	20	4	4	20	20	4	-10	4	4	4	4	-2	2	-1	5	20	-10	4	-10	-2
X.12	20	20	20	4	4	20	20	4	10	4	4	4	4	2	2	-1	5	20	10	4	10	2
X.13	24	24	24	8	8	24	24	8	4	.	.	8	.	.	6	3	.	24	4	8	4	4
X.14	24	24	24	8	8	24	24	8	-4	.	.	8	.	.	6	3	.	24	-4	8	-4	-4
X.15	30	30	30	-10	-10	30	30	-10	10	2	2	-10	2	-2	3	3	3	30	10	-10	10	-2
X.16	30	30	30	-10	-10	30	30	-10	-10	2	2	-10	2	2	3	3	3	30	-10	-10	-10	2
X.17	32	-32	.	16	-16	.	.	.	.	8	-8	.	.	.	-4	8	2	.	.	.	.	.
X.18	40	40	-40	16	16	8	-8	-16	-10	4	4	.	-4	-6	4	7	1	.	10	.	2	6
X.19	40	40	-40	16	16	8	-8	-16	10	4	4	.	-4	6	4	7	1	.	-10	.	-2	-6
X.20	40	40	-40	-16	-16	8	-8	16	10	4	4	.	-4	-2	4	7	1	.	-10	.	-2	2
X.21	40	40	-40	-16	-16	8	-8	16	-10	4	4	.	-4	2	4	7	1	.	10	.	2	-2
X.22	60	60	60	-4	-4	60	60	-4	-10	4	4	4	4	2	6	-3	-3	60	-10	-4	-10	-2
X.23	60	60	60	-4	-4	60	60	-4	10	4	4	-4	4	2	6	-3	-3	60	10	-4	10	2
X.24	60	60	60	12	12	60	60	12	.	4	4	12	4	.	-3	.	-6	60	.	12	.	.
X.25	64	64	64	.	.	64	64	.	-16	.	.	.	.	.	-8	-2	4	64	-16	.	-16	.
X.26	64	64	64	.	.	64	64	.	16	.	.	.	.	.	-8	-2	4	64	16	.	16	.
X.27	80	80	-80	.	.	16	-16	.	20	8	8	.	-8	.	8	-4	2	.	-20	.	-4	-4
X.28	80	80	80	-16	-16	80	80	-16	.	.	.	.	.	-16	.	-10	2	-4	80	.	-16	.
X.29	80	80	-80	.	.	16	-16	.	-20	8	8	.	-8	-4	8	-4	2	.	20	.	4	4
X.30	81	81	81	9	9	81	81	9	9	-3	-3	9	-3	-3	.	.	.	81	9	9	9	-3
X.31	81	81	81	9	9	81	81	9	-9	-3	-3	9	-3	3	.	.	.	81	-9	9	-9	3
X.32	90	90	90	-6	-6	90	90	-6	.	-6	-6	-6	-6	.	9	.	.	90	.	-6	.	.
X.33	120	120	-120	16	16	24	-24	-16	10	-4	-4	.	4	-10	12	3	3	.	-10	.	-2	10
X.34	120	120	-120	16	16	24	-24	-16	-10	-4	-4	.	4	10	12	3	3	.	10	.	2	-10
X.35	120	120	120	24	24	-8	-8	24	.	.	.	.	.	-8	8	3	6	.	8	.	8	8
X.36	120	120	120	24	24	-8	-8	24	.	.	.	.	.	-8	8	3	6	8	.	8	.	8
X.37	120	120	-120	-16	-16	24	-24	16	10	-4	-4	.	4	2	12	3	3	.	10	.	2	-2
X.38	120	120	-120	-16	-16	24	-24	16	-10	-4	-4	.	4	-2	12	3	3	.	-10	.	-2	2
X.39	120	120	120	24	24	-8	-8	24	.	.	.	.	.	-8	8	3	6	.	8	.	8	-8
X.40	120	120	120	24	24	-8	-8	24	.	.	.	.	.	-8	8	3	6	8	.	8	.	-8
X.41	135	135	135	-33	-33	7	7	-33	-15	3	3	-1	3	5	.	9	.	-9	-15	15	1	5
X.42	135	135	135	39	39	7	7	39	-15	15	15	7	15	-7	.	9	.	-9	-15	-9	1	-7
X.43	135	135	135	39	39	7	7	39	15	15	15	7	15	7	.	9	.	-9	15	-9	-1	7
X.44	135	135	135	-33	-33	7	7	-33	15	3	3	-1	3	-5	.	9	.	-9	15	15	-1	-5
X.45	160	-160	.	-48	48	.	.	.	.	8	-8	.	.	.	-2	16	-2	.	.	.	.	.
X.46	160	-160	.	-48	48	.	.	.	.	8	-8	.	.	.	-2	16	-2	.	.	.	.	.
X.47	192	-192	.	-32	32	.	.	.	.	16	-16	.	.	.	12	.	6	.	.	.	.	.
X.48	216	216	-216	-48	48	-8	8	48	-6	12	12	.	-12	-2	9	.	.	6	.	.	-2	2
X.49	216	216	-216	48	48	-8	8	-48	6	12	12	.	-12	10	9	.	.	-6	.	.	2	-10
X.50	216	216	-216	48	48	-8	8	-48	6	12	12	.	-12	-10	9	.	.	-6	.	.	2	10
X.51	216	216	-216	-48	-48	-8	8	48	6	12	12	.	-12	2	9	.	.	-6	.	.	2	-2
X.52	240	240	-240	-32	-32	48	-48	32	40	8	8	.	-8	.	-12	.	3	.	-40	.	-8	.
X.53	240	240	-240	32	32	48	-48	-32	40	8	8	.	-8	.	-12	.	3	.	-40	.	-8	.
X.54	240	240	240	48	48	-16	-16	48	.	.	.	.	.	-16	6	12	.	16	.	16	.	.
X.55	240	240	-240	-32	-32	48	-48	32	-40	8	8	.	-8	.	-12	.	3	.	40	.	8	.
X.56	240	240	-240	32	32	48	-48	-32	-40	8	8	.	-8	.	-12	.	3	.	40	.	8	.
X.57	240	240	240	-48	-48	-16	-16	-48	.	.	.	.	.	16	6	12	.	16	.	-16	.	.
X.58	240	240	240	-48	-48	-16	-16	-48	.	.	.	.	.	16	6	12	.	16	.	-16	.	.
X.59	270	270	270	6	6	14	14	6	30	18	18	6	18	2	.	-9	.	-18	-30	6	-2	2
X.60	270	270	270	6	6	14	14	6	-30	18	18	6	18	-2	.	-9	.	-18	-30	6	2	-2
X.61	320	320	-320	-64	-64	64	-64	64	.	.	.	.	.	.	-4	14	-4	.	.	.	.	.
X.62	320	320	-320	64	64	64	-64	-64	.	.	.	.	.	.	-4	14	-4	.	.	.	.	.
X.63	320	-320	.	32	-32	.	.	.	.	-16	16	.	.	.	14	8	2	.	.	.	.	.
X.64	320	-320	.	32	-32	.	.	.	.	-16	16	.	.	.	14	8	2	.	.	.	.	.
X.65	405	405	405	45	45	21	21	45	45	9	9	13	9	9	.	.	.	-27	45	-3	-3	9
X.66	405	405	405	-27	-27	21	21	-27	-45	-3	-3	5	-3	3	.	.	.	-27	-45	21	3	3
X.67	405	405	405	-27	-27	21	21	-27	45	-3	-3	5	-3	-3	.	.	.	-27	45	21	-3	-3
X.68	405	405	405	45	45	21	21	45	-45	9	9	13	9	-9	.	.	.	-27	-45	-3	3	-9
X.69	480	480	-480	32	32	96	-96	-32	40	.	.	.	.	.	8	12	-3	.	-40	.	-8	-8
X.70	480	480	-480	.	.	96	-96	.	.	-16	-16	.	16	.	8	-24	6	.	.	.	.	.
X.71	480	480	-480	-32	-32	96	-96	32	-40	.	.	.	.	.	8	12	-3	.	40	.	8	-8
X.72	480	480	-480	-32	-32	96	-96	32	40	.	.	.	.	.	8	12	-3	.	-40	.	-8	8
X.73	480	480	-480	32	32	96	-96	-32	-40	.	.	.	.	.	8	12	-3	.	40	.	8	8
X.74	480	-480	.	-16	16	.	.	.	.	-8	8	.	.	.	-24	6	.	.	.	.	.	.
X.75	480	-480	.	112	-112	.	.	.	.	24	-24	.	.	.	12	24	.	.	.	.	.	.
X.76	540	540	540	60	60	28	28	60	-30	-12	-12	-4	-12	2	.	9	.	-36	30	-36	-2	-2
X.77	540	540	540	60	60	28	28	60	-30	-12	-12	-4	-12	2	.	9	.	-36	-30	-36	2	2
X.78	540	540	540	-84	-84	28	28	-84	-30	12	12	-20	12	-2	.	9	.	-36	30	12	-2	-2
X.79	540	540	540	-84	-84	28	28	-84	-30	12	12	-20	12	2	.	9	.	-36	-30	12	2	2
X.80	640	-640	.	64	-64	.	.	.	.	32	-32	.	.	.	-8	-8	10	.	.	.	.	.
X.81	640	640	-640	.	.	128	-128	.	.	.	.	.	.	.	-8	-8	-8	.	.	.	.	.
X.82	720	720	720	-48	-48	-48	-48	-48	.	.	.	.	.	16	.	-9	.	48	.	-16	.	.
X.83	720	720	720	-48	-48	-48	-48	-48	.	.	.	.	.	16	.	-9	.	48	.	-16	.	.
X.84	720	720	720	-48	-48	-48	-48	-48	.	.	.	.	.	16	.	-9	.	48	.	-16	.	.

Character table of $H(Fi_{22})$ (continued)

	4_6	4_7	4_8	4_9	4_{10}	4_{11}	4_{12}	4_{13}	4_{14}	4_{15}	4_{16}	4_{17}	4_{18}	4_{19}	4_{20}	4_{21}	4_{22}	4_{23}	4_{24}	4_{25}	4_{26}	4_{27}	$5a$	$6a$	$6b$
2	10	10	10	10	10	11	11	.	9	9	9	9	8	8	9	9	9	9	8	8	8	8	7	3	.
3	1	1	1	1	1	.	.	.	1	1	1	1	1	1	.	.	.	.	.	.	.	.	.	4	4
5	.	.	.	.	.	.	.	.	.	.	.	.	.	.	.	.	.	.	.	.	.	.	1	.	.
2P	2e	2e	2e	2e	2e	2e	2e	2c	2c	2g	2g	2i	2j	2e	2e	2f	2e	2i	2i	2j	2i	2i	5a	3a	3a
3P	4_6	4_7	4_8	4_9	4_{10}	4_{11}	4_{12}	4_{13}	4_{14}	4_{15}	4_{16}	4_{17}	4_{18}	4_{19}	4_{20}	4_{21}	4_{22}	4_{23}	4_{24}	4_{25}	4_{26}	4_{27}	5a	2a	2b
5P	4_6	4_7	4_8	4_9	4_{10}	4_{11}	4_{12}	4_{13}	4_{14}	4_{15}	4_{16}	4_{17}	4_{18}	4_{19}	4_{20}	4_{21}	4_{22}	4_{23}	4_{24}	4_{25}	4_{26}	4_1	1a	6a	6c
X.1	1	1	1	1	1	1	1	1	1	1	1	1	1	1	1	1	1	1	1	1	1	1	1	1	1
X.2	1	-1	-1	-1	1	1	1	1	1	1	1	-1	-1	-1	-1	1	1	-1	1	-1	1	1	1	1	1
X.3	-2	.	-4	-2	2	2	2	2	2	2	2	2	.	.	2	2	-2	.	-2	.	.	.	1	-3	-3
X.4	-2	.	4	-2	2	2	2	2	2	2	2	-2	-2	.	.	2	2	2	.	2	.	.	1	-3	-3
X.5	-6	.	.	-6	2	2	2	2	2	2	2	.	.	.	.	2	2	-2	.	-2	-2	.	1	1	1
X.6	-1	-3	-3	5	-1	-1	-1	3	3	3	3	1	1	-3	-3	-1	-1	1	-1	1	-1	-1	.	6	6
X.7	-1	3	3	-5	-1	-1	-1	3	3	3	3	-1	-1	3	3	-1	-1	-1	-1	-1	-1	-1	.	6	6
X.8	7	-1	-1	-5	7	3	3	-1	-1	-1	-1	-3	-3	-1	-1	3	3	-1	1	-1	1	1	.	-3	-3
X.9	7	1	1	5	7	3	3	-1	-1	-1	-1	3	3	1	1	3	3	-1	1	-1	1	1	.	-3	-3
X.10	4	.	.	.	4	-4	-4	4	4	4	4	.	.	.	.	-4	-4	.	.	.	.	.	.	-7	-7
X.11	4	-2	-2	-10	4	4	4	.	.	.	.	-2	-2	-2	-2	4	4	-2	.	-2	.	.	.	2	2
X.12	4	2	2	10	4	4	4	.	.	.	.	2	2	2	2	4	4	2	.	2	.	.	.	2	2
X.13	8	4	4	4	8	.	.	.	.	.	.	.	.	.	.	4	4	.	.	.	.	.	-1	6	6
X.14	8	-4	-4	-4	8	.	.	.	.	.	.	.	.	.	.	-4	-4	.	.	.	.	.	-1	6	6
X.15	-10	-2	-2	10	-10	2	2	-2	-2	-2	-2	-4	-4	-2	-2	2	2	.	.	.	.	.	.	3	3
X.16	-10	2	2	-10	-10	2	2	-2	-2	-2	-2	4	4	2	2	2	2	.	.	.	.	.	.	3	3
X.17	4	.	.	.	.	.	.	.	.	.	.	4	-4	.	.	.	.	.	.	4	.	-4	.	2	4
X.18	4	-2	-2	-2	-4	4	4	.	.	.	.	.	-4	-4	-2	-2	.	.	2	.	2	-2	.	4	-4
X.19	4	2	-2	2	-4	4	-4	.	.	.	.	.	4	-4	2	-2	.	.	2	.	2	-2	.	4	-4
X.20	-4	-6	6	2	4	4	-4	.	.	.	.	.	4	4	2	-2	.	.	-2	.	-2	2	.	4	-4
X.21	-4	6	-6	-2	4	4	4	.	.	.	.	.	4	-4	-2	-2	.	.	-2	.	-2	2	.	4	-4
X.22	-4	-2	-2	-10	-4	4	4	.	.	.	.	.	2	2	-2	-2	4	4	2	.	2	.	.	6	6
X.23	-4	2	2	10	-4	4	4	.	.	.	.	.	-2	-2	2	2	4	4	-2	.	-2	.	.	6	6
X.24	12	.	.	.	12	4	4	4	4	4	4	.	.	.	.	4	4	.	.	.	.	.	.	-3	-3
X.25	.	.	.	-16	.	.	.	.	.	.	.	.	.	.	.	.	.	.	.	.	.	.	-1	-8	-8
X.26	.	.	.	16	.	.	.	.	.	.	.	.	.	.	.	.	.	.	.	.	.	.	-1	-8	-8
X.27	.	-4	4	4	.	8	-8	.	.	.	.	.	4	-4	.	.	.	.	.	.	.	.	.	8	-8
X.28	-16	.	.	.	-16	.	.	.	.	.	.	.	.	.	.	.	.	.	.	.	.	.	.	-10	-10
X.29	.	-4	-4	-4	.	8	-8	.	.	.	.	.	-4	4	.	.	.	.	.	.	.	.	.	8	-8
X.30	9	-3	-3	9	9	-3	-3	-3	-3	-3	-3	3	3	-3	-3	-3	-3	-1	-1	-1	-1	-1	1	.	.
X.31	9	3	3	-9	9	-3	-3	-3	-3	-3	-3	-3	-3	3	3	-3	-3	1	-1	1	-1	-1	1	.	.
X.32	-6	.	.	-6	-6	-6	-6	2	2	2	2	.	.	.	.	-6	-6	.	2	.	2	2	.	9	9
X.33	4	2	-2	2	-4	-4	4	.	.	.	.	4	-4	-6	6	.	.	.	-2	.	-2	2	.	12	-12
X.34	4	-2	2	-2	-4	-4	4	.	.	.	.	-4	4	6	-6	.	.	.	-2	.	-2	2	.	12	-12
X.35	.	-8	-8	.	.	.	.	-4	12	-4	-4	.	.	.	.	.	.	.	.	.	.	.	.	3	3
X.36	.	-8	-8	.	.	.	.	12	-4	-4	-4	.	.	.	.	.	.	.	.	.	.	.	.	3	3
X.37	-4	-10	10	-2	4	-4	4	.	.	.	.	4	-4	6	-6	.	.	.	2	.	2	-2	.	12	-12
X.38	-4	10	-10	2	4	-4	4	.	.	.	.	-4	4	6	-6	.	.	.	2	.	2	-2	.	12	-12
X.39	.	8	8	.	.	.	.	-4	12	-4	-4	.	.	.	.	.	.	.	.	.	.	.	.	3	3
X.40	.	8	8	.	.	.	.	12	-4	-4	-4	.	.	.	.	.	.	.	.	.	.	.	.	3	3
X.41	-1	5	5	1	-1	3	3	3	3	3	3	3	3	-3	-3	3	-5	-1	-3	-1	-3	-3	.	.	.
X.42	-1	-7	-7	1	-1	-1	-1	3	3	3	3	-3	-3	1	1	-1	-1	-3	3	-3	3	3	.	.	.
X.43	-1	7	7	-1	-1	-1	-1	3	3	3	3	3	3	-1	-1	-1	-1	3	3	3	3	3	.	.	.
X.44	-1	-5	-5	-1	-1	3	3	3	3	3	3	-3	-3	-3	3	3	-5	1	-3	1	-3	-3	.	.	.
X.45	.	.	.	.	.	.	.	.	.	.	.	4	-4	.	.	.	.	.	.	-4	.	4	.	2	A
X.46	.	.	.	.	.	.	.	.	.	.	.	4	-4	.	.	.	.	.	.	-4	.	4	.	2	$\bar{A}$
X.47	.	.	.	.	.	.	.	.	.	.	.	8	-8	.	.	.	.	.	.	.	.	.	.	2	-12
X.48	4	10	-10	2	-4	-4	4	.	.	.	.	.	2	-2	.	.	.	4	-2	-4	-2	2	1	.	.
X.49	-4	-2	2	-2	4	4	4	.	.	.	.	.	-2	2	.	.	.	4	2	-4	2	-2	1	.	.
X.50	-4	2	-2	2	4	-4	4	.	.	.	.	.	2	-2	.	.	.	-4	2	4	2	-2	1	.	.
X.51	4	-10	10	-2	-4	-4	4	.	.	.	.	.	-2	2	.	.	.	-4	-2	4	-2	2	1	.	.
X.52	-8	.	.	8	8	8	-8	.	.	.	.	.	-8	8	.	.	.	.	.	.	.	.	.	-12	12
X.53	8	.	.	8	-8	8	-8	.	.	.	.	.	8	-8	.	.	.	.	.	.	.	.	.	-12	12
X.54	.	.	.	.	.	.	.	-8	-8	8	8	.	.	.	.	.	.	.	.	.	.	.	.	6	6
X.55	-8	.	.	-8	8	8	-8	.	.	.	.	.	8	-8	.	.	.	.	.	.	.	.	.	-12	12
X.56	8	.	.	-8	-8	8	-8	.	.	.	.	.	-8	8	.	.	.	.	.	.	.	.	.	-12	12
X.57	.	.	.	.	.	.	.	.	.	.	.	.	.	.	.	.	.	.	.	.	.	.	.	6	6
X.58	.	.	.	.	.	.	.	.	.	.	.	.	.	.	.	.	.	.	.	.	.	.	.	6	6
X.59	-2	2	2	-2	-2	2	2	6	6	6	6	.	.	2	2	2	-6	4	.	4	.	.	.	.	.
X.60	-2	-2	-2	2	-2	2	2	6	6	6	6	.	.	-2	-2	2	-6	-4	.	-4	.	.	.	.	.
X.61	-16	.	.	.	16	.	.	.	.	.	.	.	.	.	.	.	.	.	.	.	.	.	.	-4	4
X.62	16	.	.	.	-16	.	.	.	.	.	.	.	.	.	.	.	.	.	.	.	.	.	.	-4	4
X.63	.	.	.	.	.	.	.	.	.	.	8	-8	.	.	.	.	.	.	.	.	.	.	.	-14	$\bar{A}$
X.64	.	.	.	.	.	.	.	.	.	.	8	-8	.	.	.	.	.	.	.	.	.	.	.	-14	A
X.65	-3	9	9	-3	-3	9	9	-3	-3	-3	-3	3	3	1	1	-7	1	-1	-1	-1	-1	-1	.	.	.
X.66	-3	3	3	3	-3	13	13	-3	-3	-3	-3	3	-5	-5	-3	-3	3	1	3	1	-3	1	.	.	.
X.67	-3	-3	-3	-3	-3	13	13	-3	-3	-3	-3	-3	-3	5	5	-3	-3	-3	1	-3	1	1	.	.	.
X.68	-3	-9	-9	3	-3	9	9	-3	-3	-3	-3	-3	-3	-1	-1	-7	1	1	-1	1	-1	-1	.	.	.
X.69	8	8	-8	8	-8	.	.	.	.	.	.	.	.	.	.	.	.	.	.	.	.	.	.	12	-12
X.70	.	.	.	.	.	-16	16	.	.	.	.	.	.	.	.	.	.	.	.	.	.	.	.	-24	24
X.71	-8	8	-8	-8	8	.	.	.	.	.	.	.	.	.	.	.	.	.	.	.	.	.	.	12	-12
X.72	-8	-8	8	8	8	.	.	.	.	.	.	.	.	.	.	.	.	.	.	.	.	.	.	12	-12
X.73	8	-8	8	-8	-8	.	.	.	.	.	.	.	.	.	.	.	.	.	.	.	.	.	.	12	-12
X.74	.	.	.	.	.	.	.	.	.	.	12	-12	.	.	.	.	.	.	-4	.	4	.	.	24	.
X.75	.	.	.	.	.	.	.	.	.	.	-4	4	.	.	.	.	.	.	4	.	-4	.	.	-12	.
X.76	4	-2	-2	-2	4	4	4	.	.	.	.	.	6	6	-2	-2	4	-4	-2	.	-2	.	.	.	.
X.77	4	2	2	2	4	4	4	.	.	.	.	.	-6	-6	2	2	4	-4	2	.	2	.	.	.	.
X.78	4	-2	-2	-2	4	-4	-4	.	.	.	.	.	-6	-6	-2	-2	-4	4	2	.	2	.	.	.	.
X.79	4	2	2	2	4	-4	-4	.	.	.	.	.	6	6	2	2	-4	4	-2	.	-2	.	.	.	.
X.80	.	.	.	.	.	.	.	.	.	.	.	.	.	.	.	.	.	.	.	.	.	.	.	8	.
X.81	.	.	.	.	.	.	.	.	.	.	.	.	.	.	.	.	.	.	.	.	.	.	.	-8	8
X.82	.	.	.	.	.	24	-8	-8	-8	.	.	.	.	.	.	.	.	.	.	.	.	.	.	-9	-9
X.83	.	.	.	.	.	-8	24	-8	-8	.	.	.	.	.	.	.	.	.	.	.	.	.	.	-9	-9
X.84	.	.	.	.	.	-8	-8	8	8	.	.	.	.	.	.	.	.	.	.	.	.	.	.	-9	-9

Character table of $H(\mathrm{Fi}_{22})$ (continued)

	6c	6d	6e	6f	6g	6h	6i	6j	6k	6l	6m	6n	6o	6p	6q	6r	6s	6t	6u	6v	6w	6x	6y	6z	8a	8b	8c	8d	8e
2	7	8	7	7	7	7	7	5	6	6	4	7	7	5	5	4	4	4	4	5	5	5	3	4	7	7	7	7	7
3	4	3	3	2	2	2	2	3	2	2	3	1	1	2	2	2	2	2	2	1	1	1	2	1	1	1	.	.	.
5	.	.	.	.	.	.	.	.	.	.	.	.	.	.	.	.	.	.	.	.	.	.	.	.	.	.	.	.	.
2P	3a	3b	3b	3a	3a	3a	3a	3c	3b	3b	3c	3b	3b	3b	3b	3c	3c	3c	3c	3c	3c	3b	3c	3c	4_1	4_1	4_{11}	4_{11}	4_{11}
3P	2b	2a	2b	2c	2g	2g	2d	2a	2c	2d	2b	2e	2f	2h	2g	2g	2g	2d	2c	2i	2j	2m	2h	2l	8a	8b	8c	8d	8e
5P	6b	6d	6e	6f	6h	6g	6i	6j	6k	6l	6m	6n	6o	6p	6q	6s	6r	6t	6u	6v	6w	6x	6y	6z	8a	8b	8c	8d	8e
X.1	1	1	1	1	1	1	1	1	1	1	1	1	1	1	1	1	1	1	1	1	1	1	1	1	1	1	1	1	1
X.2	1	1	1	1	1	1	1	1	1	1	1	1	1	−1	1	1	1	1	1	1	1	−1	−1	1	−1	−1	−1	1	1
X.3	−3	.	.	1	1	1	1	3	−2	−2	3	.	.	2	−2	1	1	1	1	−1	−1	.	−1	−1	.	.	2	.	.
X.4	−3	.	.	1	1	1	1	3	−2	−2	3	.	.	−2	−2	1	1	1	1	−1	−1	.	1	−1	.	.	−2	.	.
X.5	1	4	4	−3	−3	−3	−3	−2	.	.	−2	4	4	.	.	.	.	.	.	2	2	.	.	2	.	.	.	−2	−2
X.6	6	.	.	2	2	2	2	3	2	2	3	.	.	2	2	−1	−1	−1	−1	−1	−1	.	−1	−1	−3	−3	1	−1	−1
X.7	6	.	.	2	2	2	2	3	2	2	3	.	.	−2	2	−1	−1	−1	−1	−1	−1	.	1	−1	3	3	−1	−1	−1
X.8	−3	3	3	1	1	1	1	.	1	1	.	3	3	1	1	−2	−2	−2	−2	.	.	−1	−2	.	−1	−1	−3	1	1
X.9	−3	3	3	1	1	1	1	.	1	1	.	3	3	−1	1	−2	−2	−2	−2	.	.	1	2	.	1	1	3	1	1
X.10	−7	2	2	1	1	1	1	2	−2	−2	2	2	2	.	−2	−2	−2	−2	−2	2	2	.	.	2	.	.	.	.	.
X.11	2	−1	−1	−2	−2	−2	−2	5	1	1	5	−1	−1	−1	1	1	1	1	1	1	1	1	−1	1	−2	−2	−2	.	.
X.12	2	−1	−1	−2	−2	−2	−2	5	1	1	5	−1	−1	1	1	1	1	1	1	1	1	−1	1	1	2	2	2	.	.
X.13	6	3	3	2	2	2	2	.	−1	−1	.	3	3	1	−1	2	2	2	2	.	.	1	−2	.	4	4	.	.	.
X.14	6	3	3	2	2	2	2	.	−1	−1	.	3	3	−1	−1	2	2	2	2	.	.	−1	2	.	−4	−4	.	.	.
X.15	3	3	3	−1	−1	−1	−1	3	−1	−1	3	3	3	1	−1	−1	−1	−1	−1	−1	−1	1	1	−1	−2	−2	−4	.	.
X.16	3	3	3	−1	−1	−1	−1	3	−1	−1	3	3	3	−1	−1	−1	−1	−1	−1	−1	−1	−1	−1	−1	2	2	4	.	.
X.17	.	−8	.	4	.	.	−4	−2	4	−4	.	.	.	.	.	.	.	2	−2	2	−2	.	.	.	.	.	.	.	.
X.18	−4	7	−7	4	−4	−4	4	1	1	1	−1	−1	1	−1	−1	−1	−1	1	1	1	1	−3	−1	−1	−2	2	.	2	−2
X.19	−4	7	−7	4	−4	−4	4	1	1	1	−1	−1	1	1	−1	−1	−1	1	1	1	1	3	1	−1	2	−2	.	2	−2
X.20	−4	7	−7	−4	4	4	−4	1	−1	−1	−1	−1	1	1	1	1	1	−1	−1	1	1	1	1	−1	−2	2	.	−2	2
X.21	−4	7	−7	−4	4	4	−4	1	−1	−1	−1	−1	1	−1	1	1	1	−1	−1	1	1	−1	−1	−1	2	−2	.	−2	2
X.22	6	−3	−3	2	2	2	2	−3	−1	−1	−3	−3	−3	−1	−1	−1	−1	−1	−1	1	1	1	−1	1	−2	−2	2	.	.
X.23	6	−3	−3	2	2	2	2	−3	−1	−1	−3	−3	−3	1	−1	−1	−1	−1	−1	1	1	−1	1	1	2	2	−2	.	.
X.24	−3	.	.	−3	−3	−3	−3	−6	.	.	−6	.	.	.	.	.	.	.	.	−2	−2	.	.	−2	.	.	.	.	.
X.25	−8	−2	−2	.	.	.	.	4	.	.	4	−2	−2	2	.	.	.	.	.	.	.	.	2	.	.	.	.	.	.
X.26	−8	−2	−2	.	.	.	.	4	.	.	4	−2	−2	−2	.	.	.	.	.	.	.	.	−2	.	.	.	.	.	.
X.27	−8	−4	4	.	.	.	.	2	.	.	−2	4	−4	2	.	.	.	.	.	2	2	−2	2	−2	.	.	.	.	.
X.28	−10	2	2	2	2	2	2	−4	2	2	−4	2	2	.	2	2	2	2	2	.	.	.	.	.	.	.	.	.	.
X.29	−8	−4	4	.	.	.	.	2	.	.	−2	4	−4	−2	.	.	.	.	.	2	2	2	−2	−2	.	.	.	.	.
X.30	.	.	.	.	.	.	.	.	.	.	.	.	.	.	.	.	.	.	.	.	.	.	.	.	−3	−3	3	−1	−1
X.31	.	.	.	.	.	.	.	.	.	.	.	.	.	.	.	.	.	.	.	.	.	.	.	.	3	3	−3	−1	−1
X.32	9	.	.	−3	−3	−3	−3	.	.	.	.	.	.	.	.	.	.	.	.	.	.	.	.	.	.	.	.	2	2
X.33	−12	3	−3	4	−4	−4	4	3	1	1	−3	3	−3	1	−1	−1	−1	1	1	−1	−1	−1	1	1	−2	2	.	−2	2
X.34	−12	3	−3	4	−4	−4	4	3	1	1	−3	3	−3	−1	−1	−1	−1	1	1	−1	−1	1	−1	1	2	−2	.	−2	2
X.35	3	6	6	3	3	3	3	.	.	.	.	−2	−2	.	.	.	.	.	.	.	.	2	.	.	.	.	.	.	.
X.36	3	6	6	3	3	3	3	.	.	.	.	−2	−2	.	.	.	.	.	.	.	.	2	.	.	.	.	.	.	.
X.37	−12	3	−3	−4	4	4	−4	3	−1	−1	−3	3	−3	−1	1	1	1	−1	−1	−1	−1	−1	−1	1	−2	2	.	2	−2
X.38	−12	3	−3	−4	4	4	−4	3	−1	−1	−3	3	−3	1	1	1	1	−1	−1	−1	−1	1	1	1	2	−2	.	2	−2
X.39	3	6	6	3	3	3	3	.	.	.	.	−2	−2	.	.	.	.	.	.	.	.	−2	.	.	.	.	.	.	.
X.40	3	6	6	3	3	3	3	.	.	.	.	−2	−2	.	.	.	.	.	.	.	.	−2	.	.	.	.	.	.	.
X.41	.	9	9	.	.	.	.	.	3	3	.	1	1	−3	3	.	.	.	.	.	.	−1	.	.	1	1	−1	1	1
X.42	.	9	9	.	.	.	.	.	3	3	.	1	1	−3	3	.	.	.	.	.	.	−1	.	.	1	1	1	−1	−1
X.43	.	9	9	.	.	.	.	.	3	3	.	1	1	3	3	.	.	.	.	.	.	1	.	.	−1	−1	−1	−1	−1
X.44	.	9	9	.	.	.	.	.	3	3	.	1	1	3	3	.	.	.	.	.	.	1	.	.	−1	−1	1	1	1
X.45	$\bar{A}$	−16	.	−6	B	$\bar{B}$	6	2	.	.	.	.	.	.	.	$\bar{B}$	B	.	.	2	−2	.	.	.	.	.	.	.	.
X.46	A	−16	.	−6	$\bar{B}$	B	6	2	.	.	.	.	.	.	.	B	$\bar{B}$	.	.	2	−2	.	.	.	.	.	.	.	.
X.47	.	.	.	4	.	.	−4	−6	−8	8	.	.	.	.	.	.	.	2	−2	−2	2	.	.	.	.	.	.	.	.
X.48	.	9	−9	.	.	.	.	.	−3	−3	.	1	−1	−3	3	.	.	.	.	.	.	1	.	.	−2	2	.	2	−2
X.49	.	9	−9	.	.	.	.	.	3	3	.	1	−1	3	−3	.	.	.	.	.	.	1	.	.	−2	2	.	−2	2
X.50	.	9	−9	.	.	.	.	.	3	3	.	1	−1	−3	−3	.	.	.	.	.	.	−1	.	.	2	−2	.	−2	2
X.51	.	9	−9	.	.	.	.	.	−3	−3	.	1	−1	3	3	.	.	.	.	.	.	−1	.	.	2	−2	.	2	−2
X.52	12	.	.	4	−4	−4	4	3	−2	−2	−3	.	.	−2	2	−1	−1	1	1	−1	−1	.	1	1	.	.	.	.	.
X.53	12	.	.	−4	4	4	−4	3	2	2	−3	.	.	−2	−2	1	1	−1	−1	−1	−1	.	1	1	.	.	.	.	.
X.54	6	12	12	6	6	6	6	.	.	.	.	−4	−4	.	.	.	.	.	.	.	.	.	.	.	.	.	.	.	.
X.55	12	.	.	4	−4	−4	4	3	−2	−2	−3	.	.	2	2	−1	−1	1	1	−1	−1	.	−1	1	.	.	.	.	.
X.56	12	.	.	−4	4	4	−4	3	2	2	−3	.	.	2	−2	1	1	−1	−1	−1	−1	.	−1	1	.	.	.	.	.
X.57	6	12	12	−6	−6	−6	−6	.	.	.	.	−4	−4	.	.	.	.	.	.	.	.	.	.	.	.	.	.	.	.
X.58	6	12	12	−6	−6	−6	−6	.	.	.	.	−4	−4	.	.	.	.	.	.	.	.	.	.	.	.	.	.	.	.
X.59	.	−9	−9	.	.	.	.	.	−3	−3	.	−1	−1	−3	−3	.	.	.	.	.	.	−1	.	.	−2	−2	.	.	.
X.60	.	−9	−9	.	.	.	.	.	−3	−3	.	−1	−1	3	−3	.	.	.	.	.	.	1	.	.	2	2	.	.	.
X.61	4	14	−14	−4	4	4	−4	−4	2	2	4	−2	2	.	−2	−2	−2	2	2	.	.	.	.	.	.	.	.	.	.
X.62	4	14	−14	4	−4	−4	4	−4	−2	−2	4	−2	2	.	2	2	2	−2	−2	.	.	.	.	.	.	.	.	.	.
X.63	A	−8	.	2	$\bar{A}$	A	−2	−2	−4	4	.	.	.	.	.	.	.	−2	2	2	−2	.	.	.	.	.	.	.	.
X.64	$\bar{A}$	−8	.	2	A	$\bar{A}$	−2	−2	−4	4	.	.	.	.	.	.	.	−2	2	2	−2	.	.	.	.	.	.	.	.
X.65	.	.	.	.	.	.	.	.	.	.	.	.	.	.	.	.	.	.	.	.	.	.	.	.	−3	−3	−1	3	3
X.66	.	.	.	.	.	.	.	.	.	.	.	.	.	.	.	.	.	.	.	.	.	.	.	.	3	3	−1	−3	−3
X.67	.	.	.	.	.	.	.	.	.	.	.	.	.	.	.	.	.	.	.	.	.	.	.	.	−3	−3	1	−3	−3
X.68	.	.	.	.	.	.	.	.	.	.	.	.	.	.	.	.	.	.	.	.	.	.	.	.	3	3	1	3	3
X.69	−12	−3	3	−4	4	4	−4	.	−1	−1	.	−3	3	1	1	−2	−2	2	2	.	.	−1	−2	.	4	−4	.	.	.
X.70	24	.	.	.	.	.	.	6	.	.	−6	.	.	.	.	.	.	.	.	2	2	.	.	−2	.	.	.	.	.
X.71	−12	−3	3	4	−4	−4	4	.	1	1	.	−3	3	−1	−1	2	2	−2	−2	.	.	−1	2	.	4	−4	.	.	.
X.72	−12	−3	3	4	−4	−4	4	.	1	1	.	−3	3	1	−1	2	2	−2	−2	.	.	1	−2	.	−4	4	.	.	.
X.73	−12	−3	3	−4	4	4	−4	.	−1	−1	.	−3	3	−1	1	−2	−2	2	2	.	.	1	2	.	−4	4	.	.	.
X.74	.	.	.	8	.	.	−8	−6	8	−8	.	.	.	.	.	.	.	−2	2	−2	2	.	.	.	.	.	.	.	.
X.75	.	−24	.	4	.	.	−4	.	4	−4	.	.	.	.	.	.	.	−4	4	.	.	.	.	.	.	.	.	.	.
X.76	.	9	9	.	.	.	.	.	−3	−3	.	1	1	−3	−3	.	.	.	.	.	.	1	.	.	2	2	−2	.	.
X.77	.	9	9	.	.	.	.	.	−3	−3	.	1	1	3	−3	.	.	.	.	.	.	−1	.	.	−2	−2	2	.	.
X.78	.	9	9	.	.	.	.	.	−3	−3	.	1	1	−3	−3	.	.	.	.	.	.	1	.	.	2	2	2	.	.
X.79	.	9	9	.	.	.	.	.	−3	−3	.	1	1	3	−3	.	.	.	.	.	.	−1	.	.	−2	−2	−2	.	.
X.80	.	8	.	−8	.	.	8	−10	4	−4	.	.	.	.	.	.	.	2	−2	2	−2	.	.	.	.	.	.	.	.
X.81	8	−8	8	.	.	.	.	−8	.	.	8	8	−8	.	.	.	.	.	.	.	.	.	.	.	.	.	.	.	.
X.82	−9	.	.	3	3	3	3	.	.	.	.	.	.	.	.	.	.	.	.	.	.	.	.	.	.	.	.	.	.
X.83	−9	.	.	3	3	3	3	.	.	.	.	.	.	.	.	.	.	.	.	.	.	.	.	.	.	.	.	.	.
X.84	−9	.	.	3	3	3	3	.	.	.	.	.	.	.	.	.	.	.	.	.	.	.	.	.	.	.	.	.	.

Character table of $H(\mathrm{Fi}_{22})$ (continued)

	8f	8g	8h	8i	8j	8k	9a	10a	10b	10c	12a	12b	12c	12d	12e	12f	12g	12h	12i	12j	12k	12l	12m	12n	12o	12p
2	7	7	6	5	5	5	2	3	2	2	5	6	6	5	5	5	5	5	5	5	5	5	5	5	5	5
3	.	.	.	.	.	.	2	.	.	.	3	2	2	2	1	1	1	1	1	1	1	1	1	1	1	1
5	.	.	.	.	.	.	.	1	1	1	.	.	.	.	.	.	.	.	.	.	.	.	.	.	.	.
2P	4_3	4_{11}	4_{12}	4_{13}	4_{14}	4_{21}	9a	5a	5a	5a	6a	6d	6d	6d	6n	6d	6n	6n	6n	6g	6h	6a	6g	6h	6n	6n
3P	8f	8g	8h	8i	8j	8k	3a	10a	10b	10c	4_1	4_1	4_1	4_2	4_4	4_5	4_6	4_7	4_8	4_{15}	4_{15}	4_3	4_{16}	4_{16}	4_{10}	4_9
5P	8f	8g	8h	8i	8j	8k	9a	2a	2h	2b	12a	12b	12c	12d	12e	12f	12g	12h	12i	12j	12l	12n	12m	12o	12p	
X.1	1	1	1	1	1	1	1	1	1	1	1	1	1	1	1	1	1	1	1	1	1	1	1	1	1	1
X.2	1	−1	−1	−1	−1	1	1	1	−1	1	1	1	1	−1	−1	−1	1	−1	−1	1	1	1	1	1	1	−1
X.3	2	−2	−2	.	.	.	.	1	1	1	−3	.	.	2	2	.	−2	.	−1	−1	1	−1	−1	−2	2	.
X.4	2	2	2	.	.	.	.	1	−1	1	−3	.	.	−2	−2	.	−2	.	−1	−1	1	−1	−1	−2	−2	.
X.5	2	.	.	.	.	−2	1	.	.	.	1	4	4	.	.	.	.	.	2	.	.	.	.	−3	−1	−1	.	.	2	2
X.6	3	1	1	−1	−1	−1	.	.	.	.	6	.	.	2	2	.	2	.	.	.	.	2	.	.	2	2
X.7	3	−1	−1	1	1	1	−1	.	.	.	6	.	.	−2	−2	.	2	.	.	.	.	2	.	.	2	2
X.8	−1	1	1	1	1	1	.	.	.	.	−3	3	3	1	1	−1	1	−1	−1	−1	−1	1	−1	−1	1	1
X.9	−1	−1	−1	−1	−1	1	.	.	.	.	−3	3	3	−1	−1	1	1	1	1	1	1	−1	−1	−1	1	−1
X.10	4	.	.	.	.	.	−1	.	.	.	−7	2	2	.	.	.	−2	.	.	.	1	1	1	1	−2	.
X.11	.	−2	−2	.	.	.	−1	.	.	.	2	−1	−1	−1	−1	1	1	1	1	.	.	−2	.	.	1	−1
X.12	.	2	2	.	.	.	−1	.	.	.	2	−1	−1	1	1	−1	−1	−1	−1	.	.	−2	.	.	1	−1
X.13	.	.	.	.	.	.	.	−1	−1	−1	6	3	3	1	1	1	−1	−1	.	.	.	2	.	.	−1	1
X.14	.	.	.	.	.	.	.	−1	1	−1	6	3	3	−1	−1	−1	−1	−1	−1	−1	.	.	2	.	.	−1	−1
X.15	−2	.	.	.	.	.	.	.	.	.	3	3	3	1	1	1	1	1	1	1	1	−1	1	1	−1	1
X.16	−2	.	.	.	.	.	.	.	.	.	3	3	3	−1	−1	−1	−1	−1	−1	1	1	−1	1	1	−1	−1
X.17	.	.	.	.	2	−2	.	.	.	.	.	.	.	.	.	.	.	.	.	−2	−2	.	2	2	.	.
X.18	.	.	.	.	1	.	.	.	.	.	3	−3	1	−1	3	1	1	−1	.	.	.	.	.	.	−1	−1
X.19	.	.	.	.	1	.	.	.	.	.	3	−3	−1	1	−3	1	−1	1	.	.	.	.	.	.	−1	−1
X.20	.	.	.	.	1	.	.	.	.	.	3	−3	−1	1	−1	−1	−3	3	.	.	.	.	.	.	1	1
X.21	.	.	.	.	1	.	.	.	.	.	3	−3	1	−1	1	−1	3	−3	.	.	.	.	.	.	1	1
X.22	.	2	2	.	.	.	.	.	.	.	6	−3	−3	−1	−1	1	−1	1	1	.	.	2	.	.	−1	−1
X.23	.	−2	−2	.	.	.	.	.	.	.	6	−3	−3	1	1	−1	−1	−1	−1	.	.	2	.	.	−1	1
X.24	4	.	.	.	.	.	.	.	.	.	−3	.	.	.	.	.	.	.	.	1	1	−3	1	1	.	.
X.25	.	.	.	.	1	−1	−1	−1	.	.	−8	−2	−2	2	2	.	.	.	.	.	.	.	.	.	.	2
X.26	.	.	.	.	1	−1	1	−1	.	.	−8	−2	−2	−2	−2	.	.	.	.	.	.	.	.	.	.	−2
X.27	.	.	.	.	.	−1	.	.	.	.	.	.	.	−2	2	2	.	2	−2	.	.	.	.	.	.	−2
X.28	.	.	.	.	.	−1	.	.	.	.	−10	2	2	.	.	.	2	.	.	.	2	.	.	2	.	2
X.29	.	.	.	.	.	−1	.	.	.	.	.	.	.	2	−2	−2	.	−2	2	.	.	.	.	.	.	2
X.30	−3	−1	−1	1	1	−1	.	1	−1	1	.	.	.	.	.	.	.	.	.	.	.	.	.	.	.	.
X.31	−3	1	1	−1	−1	−1	.	1	1	1	.	.	.	.	.	.	.	.	.	.	.	.	.	.	.	.
X.32	2	.	.	.	.	2	.	.	.	.	9	.	.	.	.	.	.	.	.	−1	−1	−3	−1	−1	.	.
X.33	.	.	.	.	.	.	.	.	.	.	3	−3	−1	1	1	1	−1	1	.	.	.	.	.	.	−1	−1
X.34	.	.	.	.	.	.	.	.	.	.	3	−3	1	−1	1	1	1	−1	.	.	.	.	.	.	−1	1
X.35	.	.	.	−2	2	.	.	.	.	.	−1	2	2	.	.	.	2	.	.	−2	−2	−1	−1	−1	−1	−1
X.36	.	.	.	2	−2	.	.	.	.	.	−1	2	2	.	.	.	2	.	.	−2	−2	−1	−1	−1	−1	−1
X.37	.	.	.	.	.	.	.	.	.	.	3	−3	1	−1	1	−1	−1	1	.	.	.	.	.	.	1	1
X.38	.	.	.	.	.	.	.	.	.	.	3	−3	−1	1	−1	−1	1	−1	.	.	.	.	.	.	1	−1
X.39	.	.	.	2	−2	.	.	.	.	.	−1	2	2	.	.	.	−2	.	.	2	2	−1	−1	−1	−1	−1
X.40	.	.	.	−2	2	.	.	.	.	.	−1	2	2	.	.	.	−2	.	.	2	2	−1	−1	−1	−1	−1
X.41	−1	3	−1	1	1	1	.	.	.	.	−3	−3	−3	1	−1	−1	−1	−1	.	.	.	.	.	.	−1	1
X.42	−1	1	1	−1	−1	−1	.	.	.	.	−3	−3	−3	1	−1	−1	−1	−1	.	.	.	.	.	.	−1	1
X.43	−1	−1	−1	1	1	−1	.	.	.	.	−3	−3	3	−1	1	−1	1	1	.	.	.	.	.	.	−1	−1
X.44	−1	−3	1	−1	−1	1	.	.	.	.	−3	−3	3	−1	1	−1	1	1	.	.	.	.	.	.	−1	1
X.45	.	.	.	.	.	1	.	.	.	.	.	.	.	.	.	.	.	.	.	C	C̄	.	−C	−C̄	.	.
X.46	.	.	.	.	.	1	.	.	.	.	.	.	.	.	.	.	.	.	.	C̄	C	.	−C̄	−C	.	.
X.47	.	.	.	.	.	.	−2	.	.	.	.	.	.	.	.	.	.	.	.	2	2	.	−2	−2	.	.
X.48	.	.	.	.	.	.	1	−1	−1	.	−3	3	3	1	−1	1	1	−1	.	.	.	.	.	.	−1	−1
X.49	.	.	.	.	.	.	1	1	−1	.	−3	3	−1	−1	1	−1	1	1	.	.	.	.	.	.	1	1
X.50	.	.	.	.	.	.	1	−1	−1	.	−3	3	3	1	1	−1	−1	1	.	.	.	.	.	.	1	1
X.51	.	.	.	.	.	.	1	1	−1	.	−3	3	−1	1	−1	1	−1	1	.	.	.	.	.	.	−1	1
X.52	.	.	.	.	.	.	.	.	.	.	2	−2	.	−2	.	.	.	.	.	.	.	.	.	.	2	2
X.53	.	.	.	.	.	.	.	.	.	.	2	−2	.	2	.	.	.	.	.	.	.	.	.	.	−2	2
X.54	.	.	.	.	.	.	.	−2	4	4	.	.	.	.	.	.	.	.	.	2	2	−2	2	2	.	.
X.55	.	.	.	.	.	.	.	.	.	.	−2	2	.	−2	.	.	.	.	.	.	.	.	.	.	2	−2
X.56	.	.	.	.	.	.	.	.	.	.	−2	2	.	2	.	.	.	.	.	.	.	.	.	.	−2	−2
X.57	.	.	.	.	.	.	.	−2	4	4	.	.	.	.	.	.	.	.	.	2	.	.	.	.	.	.
X.58	.	.	.	.	.	.	.	−2	4	4	.	.	.	.	.	.	.	.	.	2	.	.	.	.	.	.
X.59	−2	−4	.	.	.	.	.	.	.	.	3	3	−3	1	−1	1	−1	−1	.	.	.	.	.	.	1	1
X.60	−2	4	.	.	.	.	.	.	.	.	3	3	−3	1	1	1	1	1	.	.	.	.	.	.	1	−1
X.61	.	.	.	.	.	.	−1	.	.	.	6	−6	.	.	.	.	2	.	.	.	.	.	.	.	−2	.
X.62	.	.	.	.	.	.	−1	.	.	.	6	−6	.	.	.	.	−2	.	.	.	.	.	.	.	2	.
X.63	.	.	.	.	.	.	−1	.	.	.	.	.	.	.	.	.	.	.	.	−Ċ	−C	.	Ċ	C	.	.
X.64	.	.	.	.	.	.	−1	.	.	.	.	.	.	.	.	.	.	.	.	−C	−C̄	.	C	C̄	.	.
X.65	1	3	−1	−1	−1	−1	.	.	.	.	.	.	.	.	.	.	.	.	.	.	.	.	.	.	.	.
X.66	1	−1	−1	−1	−1	1	.	.	.	.	.	.	.	.	.	.	.	.	.	.	.	.	.	.	.	.
X.67	1	1	1	1	1	1	.	.	.	.	.	.	.	.	.	.	.	.	.	.	.	.	.	.	.	.
X.68	1	−3	1	1	1	−1	.	.	.	.	.	.	.	.	.	.	.	.	.	.	.	.	.	.	.	.
X.69	.	.	.	.	.	.	.	.	.	.	−3	3	−1	1	1	−1	−1	1	.	.	.	.	.	.	1	−1
X.70	.	.	.	.	.	.	.	.	.	.	.	.	.	.	.	.	.	.	.	.	.	.	.	.	.	.
X.71	.	.	.	.	.	.	.	.	.	.	−3	3	1	−1	1	1	1	−1	.	.	.	.	.	.	−1	1
X.72	.	.	.	.	.	.	.	.	.	.	−3	3	−1	1	−1	1	1	−1	.	.	.	.	.	.	−1	−1
X.73	.	.	.	.	.	.	.	.	.	.	−3	3	1	−1	−1	−1	1	−1	.	.	.	.	.	.	1	1
X.74	.	.	.	.	.	.	.	.	.	.	.	.	.	.	.	.	.	.	.	2	2	.	−2	−2	.	.
X.75	.	.	.	.	.	.	.	.	.	.	−3	−3	−3	1	1	1	1	1	.	.	.	.	.	.	1	1
X.76	.	−2	2	.	.	.	.	.	.	.	−3	−3	3	−1	1	1	−1	−1	.	.	.	.	.	.	1	1
X.77	.	2	−2	.	.	.	.	.	.	.	−3	−3	3	−1	1	1	−1	−1	.	.	.	.	.	.	1	−1
X.78	.	2	−2	.	.	.	.	.	.	.	−3	−3	3	1	1	1	1	1	.	.	.	.	.	.	1	1
X.79	.	−2	2	.	.	.	.	.	.	.	−3	−3	3	−1	−1	1	−1	−1	.	.	.	.	.	.	1	−1
X.80	.	.	.	.	.	−2	.	.	.	.	.	.	.	.	.	.	.	.	.	.	.	.	.	.	.	.
X.81	.	.	.	.	.	1	.	.	.	.	.	.	.	.	.	.	.	.	.	.	.	.	.	.	.	.
X.82	.	.	.	.	.	.	.	.	.	.	3	.	.	.	.	.	.	.	.	1	1	−1	1	1	.	.
X.83	.	.	.	.	.	.	.	.	.	.	3	.	.	.	.	.	.	.	.	1	1	−1	1	1	.	.
X.84	.	.	.	.	.	.	.	.	.	.	3	.	.	.	.	.	.	.	.	F	F̄	−1	F	F̄	.	.

Character table of $H(\mathrm{Fi}_{22})$ (continued)

	12q	12r	12s	12t	12u	16a	16b	18a	18b	18c	20a	24a	24b
2	3	4	4	3	3	5	5	2	2	2	2	4	4
3	2	1	1	1	1	.	.	2	2	2	.	1	1
5	.	.	.	.	.	.	.	.	.	.	1	.	.
2P	6j	6f	6f	6v	6w	8f	8f	9a	9a	9a	10a	12b	12b
3P	4_2	4_{13}	4_{14}	4_{17}	4_{18}	16a	16b	6a	6b	6c	20a	8a	8b
5P	12q	12r	12s	12t	12u	16b	16a	18a	18c	18b	4_2	24a	24b
X.1	1	1	1	1	1	1	1	1	1	1	1	1	1
X.2	-1	1	1	-1	-1	-1	-1	1	1	1	-1	-1	-1
X.3	-1	-1	-1	-1	-1	.	.	.	.	.	1	.	.
X.4	1	-1	-1	1	1	.	.	.	.	.	-1	.	.
X.5	.	-1	-1	.	.	.	.	1	1	1	-1	.	.
X.6	-1	.	.	.	.	1	1	-1	.	-1	.	.	.
X.7	1	.	.	.	.	-1	-1	1	1	.	.	.	.
X.8	-2	-1	-1	.	.	1	1	.	.	.	.	-1	-1
X.9	2	-1	-1	.	.	-1	-1	.	.	.	.	1	1
X.10	.	1	1	.	.	.	.	-1	-1	-1	.	.	.
X.11	-1	.	.	1	1	.	.	-1	-1	-1	.	1	1
X.12	1	.	.	-1	-1	.	.	-1	-1	-1	.	-1	-1
X.13	-2	.	.	.	.	.	.	.	.	.	-1	1	1
X.14	2	.	.	.	.	.	.	.	.	.	1	-1	-1
X.15	1	1	1	-1	-1	.	.	.	.	.	.	1	1
X.16	-1	1	1	1	1	.	.	.	.	.	.	-1	-1
X.17	.	.	.	.	.	.	.	-2	.	.	.	.	.
X.18	1	.	.	-1	1	.	.	1	-1	-1	.	1	-1
X.19	-1	.	.	1	-1	.	.	1	-1	-1	.	-1	1
X.20	-1	.	.	-1	1	.	.	1	-1	-1	.	1	1
X.21	1	.	.	1	-1	.	.	1	-1	-1	.	-1	-1
X.22	-1	.	.	-1	-1	.	.	.	.	.	.	1	1
X.23	1	.	.	1	1	.	.	.	.	.	.	-1	-1
X.24	.	1	1	.	.	.	.	.	.	.	.	.	.
X.25	2	.	.	.	.	.	.	1	1	1	-1	.	.
X.26	-2	.	.	.	.	.	.	1	1	1	1	.	.
X.27	-2	.	.	.	.	.	.	-1	1	1	.	.	.
X.28	.	.	.	.	.	.	.	-1	-1	-1	.	.	.
X.29	2	.	.	.	.	.	.	-1	1	1	.	.	.
X.30	.	.	.	.	.	1	1	.	.	.	-1	.	.
X.31	.	.	.	.	.	-1	-1	.	.	.	1	.	.
X.32	.	-1	-1	.	.	.	.	.	.	.	.	.	.
X.33	-1	.	.	1	-1	.	.	.	.	.	.	1	-1
X.34	1	.	.	-1	1	.	.	.	.	.	.	-1	1
X.35	.	-1	3	.	.	.	.	.	.	.	.	.	.
X.36	.	3	-1	.	.	.	.	.	.	.	.	.	.
X.37	1	.	.	1	-1	.	.	.	.	.	.	1	-1
X.38	-1	.	.	-1	1	.	.	.	.	.	.	-1	1
X.39	.	-1	3	.	.	.	.	.	.	.	.	.	.
X.40	.	3	-1	.	.	.	.	.	.	.	.	.	.
X.41	.	.	.	.	.	-1	-1	.	.	.	.	1	1
X.42	.	.	.	.	.	1	1	.	.	.	.	1	1
X.43	.	.	.	.	.	-1	-1	.	.	.	.	-1	-1
X.44	.	.	.	.	.	1	1	.	.	.	.	-1	-1
X.45	.	.	.	.	.	.	.	-1	D	$\bar{D}$	.	.	.
X.46	.	.	.	.	.	.	.	-1	$\bar{D}$	D	.	.	.
X.47	.	.	.	.	.	.	.	.	.	.	.	.	.
X.48	.	.	.	.	.	.	.	.	.	.	1	1	-1
X.49	.	.	.	.	.	.	.	.	.	.	-1	1	-1
X.50	.	.	.	.	.	.	.	.	.	.	1	-1	1
X.51	.	.	.	.	.	.	.	.	.	.	-1	-1	1
X.52	-1	.	.	1	-1	.	.	.	.	.	.	.	.
X.53	-1	.	.	-1	1	.	.	.	.	.	.	.	.
X.54	.	-2	-2	.	.	.	.	.	.	.	.	.	.
X.55	1	.	.	-1	1	.	.	.	.	.	.	.	.
X.56	1	.	.	1	-1	.	.	.	.	.	.	.	.
X.57	.	.	.	.	.	E	$\bar{E}$	.	.	.	.	.	.
X.58	.	.	.	.	.	$\bar{E}$	E	.	.	.	.	.	.
X.59	.	.	.	.	.	.	.	.	.	.	.	1	1
X.60	.	.	.	.	.	.	.	.	.	.	.	-1	-1
X.61	.	.	.	.	.	.	.	-1	1	1	.	.	.
X.62	.	.	.	.	.	.	.	-1	1	1	.	.	.
X.63	.	.	.	.	.	.	.	1	$\bar{D}$	D	.	.	.
X.64	.	.	.	.	.	.	.	1	D	$\bar{D}$	.	.	.
X.65	.	.	.	.	.	1	1	.	.	.	.	.	.
X.66	.	.	.	.	.	1	1	.	.	.	.	.	.
X.67	.	.	.	.	.	-1	-1	.	.	.	.	.	.
X.68	.	.	.	.	.	-1	-1	.	.	.	.	.	.
X.69	2	.	.	.	.	.	.	.	.	.	.	1	-1
X.70	.	.	.	.	.	.	.	.	.	.	.	.	.
X.71	-2	.	.	.	.	.	.	.	.	.	.	1	-1
X.72	2	.	.	.	.	.	.	.	.	.	.	-1	1
X.73	-2	.	.	.	.	.	.	.	.	.	.	-1	1
X.74	.	.	.	.	.	.	.	.	.	.	.	.	.
X.75	.	.	.	.	.	.	.	.	.	.	.	.	.
X.76	.	.	.	.	.	.	.	.	.	.	.	-1	-1
X.77	.	.	.	.	.	.	.	.	.	.	.	1	1
X.78	.	.	.	.	.	.	.	.	.	.	.	-1	-1
X.79	.	.	.	.	.	.	.	.	.	.	.	1	1
X.80	.	.	.	.	.	.	.	2	.	.	.	.	.
X.81	.	.	.	.	.	.	.	1	-1	-1	.	.	.
X.82	.	-3	1	.	.	.	.	.	.	.	.	.	.
X.83	.	1	-3	.	.	.	.	.	.	.	.	.	.
X.84	.	1	1	.	.	.	.	.	.	.	.	.	.

Character table of $H(Fi_{22})$ (continued)

	$1a$	$2a$	$2b$	$2c$	$2d$	$2e$	$2f$	$2g$	$2h$	$2i$	$2j$	$2k$	$2l$	$2m$	$3a$	$3b$	$3c$	4_1	4_2	4_3	4_4	4_5
2	17	17	16	15	15	16	16	14	10	12	12	13	11	10	7	8	5	12	10	12	10	10
3	4	4	4	2	2	1	1	2	2	1	1	.	1	1	4	3	3	3	2	1	1	1
5	1	1	1	.	.	.	.	.	1	.	.	.	.	.	.	.	.	1	.	.	.	.
2P	$1a$	$1a$	$1a$	$1a$	$1a$	$1a$	$1a$	$1a$	$1a$	$1a$	$1a$	$1a$	$1a$	$1a$	$3a$	$3b$	$3c$	$2a$	$2a$	$2a$	$2e$	$2a$
3P	$1a$	$2a$	$2b$	$2c$	$2d$	$2e$	$2f$	$2g$	$2h$	$2i$	$2j$	$2k$	$2l$	$2m$	$1a$	$1a$	$1a$	4_1	4_2	4_3	4_4	4_5
5P	$1a$	$2a$	$2b$	$2c$	$2d$	$2e$	$2f$	$2g$	$2h$	$2i$	$2j$	$2k$	$2l$	$2m$	$3a$	$3b$	$3c$	4_1	4_2	4_3	4_4	4_5
X.85	720	720	720	−48	−48	−48	−48	−48	.	.	.	16	.	.	−9	.	.	48	.	−16	.	.
X.86	768	−768	.	128	−128	.	.	.	.	.	.	.	.	.	.	−24	24	.	.	.	.	.
X.87	810	810	810	90	90	42	42	90	.	18	18	26	18	.	.	.	.	−54	.	−6	.	.
X.88	810	810	810	−54	−54	42	42	−54	.	−6	−6	10	−6	.	.	.	.	−54	.	42	.	.
X.89	810	810	810	18	18	42	42	18	.	−18	−18	18	−18	12	.	.	.	−54	.	18	.	12
X.90	810	810	810	18	18	42	42	18	.	−18	−18	18	−18	−12	.	.	.	−54	.	18	.	−12
X.91	864	864	−864	96	96	−32	32	−96	24	.	.	.	8	.	9	.	.	−24	.	8	.	−8
X.92	864	864	−864	−96	−96	−32	32	96	−24	.	.	.	8	.	9	.	.	24	.	−8	.	−8
X.93	864	864	−864	−96	−96	−32	32	96	24	.	.	.	−8	.	9	.	.	−24	.	8	.	8
X.94	864	864	−864	96	96	−32	32	−96	−24	.	.	.	−8	.	9	.	.	24	.	−8	.	8
X.95	960	960	960	.	−64	−64	.	.	.	.	.	.	.	.	16	24	−6	64	.	.	.	16
X.96	960	960	960	.	−64	−64	.	.	.	.	.	.	.	.	−16	24	−6	64	.	.	.	−16
X.97	960	−960	.	−160	160	.	.	.	.	16	−16	.	.	.	−12	24	6	.	.	.	.	.
X.98	960	−960	.	96	−96	.	.	.	.	16	−16	.	.	.	6	.	−6	.	.	.	.	.
X.99	960	−960	.	96	−96	.	.	.	.	16	−16	.	.	.	6	.	−6	.	.	.	.	.
X.100	1080	1080	−1080	48	48	−40	40	−48	30	12	12	.	−12	10	.	−9	.	.	−30	.	10	−10
X.101	1080	1080	−1080	−48	−48	−40	40	48	−30	12	12	.	−12	−2	.	−9	.	.	30	.	−10	2
X.102	1080	1080	1080	−24	−24	56	56	−24	60	.	.	−24	.	−4	.	−9	.	−72	60	−24	−4	−4
X.103	1080	1080	1080	−24	−24	56	56	−24	−60	.	.	−24	.	4	.	−9	.	−72	−60	−24	4	4
X.104	1080	1080	−1080	−48	−48	−40	40	48	30	12	12	.	−12	2	.	−9	.	.	−30	.	10	−2
X.105	1080	1080	−1080	48	48	−40	40	−48	−30	12	12	.	−12	−10	.	−9	.	.	30	.	−10	10
X.106	1280	−1280	.	−128	128	.	.	.	.	.	.	.	.	.	20	8	−4	.	.	.	.	.
X.107	1280	1280	.	−128	128	.	.	.	.	.	.	.	.	.	20	8	−4	.	.	.	.	.
X.108	1296	1296	−1296	.	.	−48	48	.	36	−24	−24	.	24	−12	.	.	.	−36	.	12	.	12
X.109	1296	1296	−1296	.	.	−48	48	.	−36	−24	−24	.	24	12	.	.	.	36	.	−12	.	−12
X.110	1440	1440	1440	96	96	−96	−96	96	.	.	.	−32	.	.	−18	.	.	96	.	32	.	.
X.111	1440	−1440	.	−48	48	.	.	.	.	−24	24	.	.	.	−18	.	.	.	.	.	.	.
X.112	1440	−1440	.	−48	48	.	.	.	.	−24	24	.	.	.	−18	.	.	.	.	.	.	.
X.113	1920	−1920	.	−64	64	.	.	.	.	32	−32	.	.	.	−24	−24	−6	.	.	.	.	.
X.114	2048	−2048	.	.	.	.	.	.	.	.	.	.	.	.	32	−16	8	.	.	.	.	.
X.115	2592	−2592	.	144	−144	.	.	.	.	−24	24	.	.	.	.	.	.	.	.	.	.	.

Character table of $H(Fi_{22})$ (continued)

	4_6	4_7	4_8	4_9	4_{10}	4_{11}	4_{12}	4_{13}	4_{14}	4_{15}	4_{16}	4_{17}	4_{18}	4_{19}	4_{20}	4_{21}	4_{22}	4_{23}	4_{24}	4_{25}	4_{26}	4_{27}	$5a$	$6a$	$6b$
2	10	10	10	10	10	10	11	11	9	9	9	9	9	8	8	9	9	9	9	8	8	8	7	3	7
3	1	1	1	1	1	1	.	.	1	1	1	1	1	1	.	.	.	.	.	.	.	.	.	4	4
5	.	.	.	.	.	.	.	.	.	.	.	.	.	.	.	.	.	.	.	.	.	.	1	.	.
2P	$2e$	$2e$	$2e$	$2e$	$2e$	$2e$	$2e$	$2c$	$2c$	$2g$	$2g$	$2i$	$2j$	$2e$	$2e$	$2f$	$2e$	$2i$	$2i$	$2j$	$2i$	$2i$	$5a$	$3a$	$3a$
3P	4_6	4_7	4_8	4_9	4_{10}	4_{11}	4_{12}	4_{13}	4_{14}	4_{15}	4_{16}	4_{17}	4_{18}	4_{19}	4_{20}	4_{21}	4_{22}	4_{23}	4_{24}	4_{25}	4_{26}	4_{27}	$5a$	$2a$	$2b$
5P	4_6	4_7	4_8	4_9	4_{10}	4_{11}	4_{12}	4_{13}	4_{14}	4_{15}	4_{16}	4_{17}	4_{18}	4_{19}	4_{20}	4_{21}	4_{22}	4_{23}	4_{24}	4_{25}	4_{26}	4_1	$1a$	$6a$	$6c$
X.85	.	.	.	.	.	.	.	.	−8	−8	8	8	.	.	.	.	.	.	.	.	.	.	.	−9	−9
X.86	.	.	.	.	.	.	.	.	.	.	.	.	.	.	.	.	.	.	.	.	.	.	.	−2	24
X.87	−6	.	.	.	−6	−14	−14	−6	−6	−6	−6	.	.	.	.	2	2	.	−2	.	−2	−2	.	.	.
X.88	−6	.	.	.	−6	−6	−6	−6	−6	−6	.	.	.	10	−6	2	2	.	2	.	2	2	.	.	.
X.89	−6	12	12	.	−6	−2	−2	6	6	6	6	−6	−6	−4	−4	−2	6	−2	.	−2	.	.	.	.	.
X.90	−6	−12	−12	.	−6	−2	−2	6	6	6	6	6	6	4	4	−2	6	2	.	2	.	.	.	.	.
X.91	−8	8	−8	−8	8	.	.	.	.	.	.	.	.	.	.	.	.	.	.	.	.	.	.	−1	.
X.92	8	8	−8	8	−8	.	.	.	.	.	.	.	.	.	.	.	.	.	.	.	.	.	.	−1	.
X.93	8	−8	8	−8	−8	.	.	.	.	.	.	.	.	.	.	.	.	.	.	.	.	.	.	−1	.
X.94	−8	−8	8	8	8	.	.	.	.	.	.	.	.	.	.	.	.	.	.	.	.	.	.	−1	.
X.95	.	−16	−16	.	.	.	.	.	.	.	.	.	.	.	.	.	.	.	.	.	.	.	.	24	24
X.96	.	16	16	.	.	.	.	.	.	.	.	.	.	.	.	.	.	.	.	.	.	.	.	24	24
X.97	.	.	.	.	.	.	.	.	.	.	.	.	.	−8	8	.	.	.	.	.	.	.	12	.	.
X.98	.	.	.	.	.	.	.	.	.	.	.	.	.	8	−8	.	.	.	.	.	.	.	.	−6	G
X.99	.	.	.	.	.	.	.	.	.	.	.	.	.	8	−8	.	.	.	.	.	.	.	.	−6	$\bar{G}$
X.100	−4	−2	2	−10	4	−4	4	.	.	.	.	.	.	.	−2	2	.	−4	−2	4	−2	2	.	.	.
X.101	4	10	−10	10	−4	4	4	.	.	.	.	.	.	.	2	−2	.	−4	2	4	2	−2	.	.	.
X.102	8	−4	−4	−4	8	.	.	.	.	.	.	.	.	.	−4	−4	.	.	.	.	.	.	.	.	.
X.103	8	4	4	4	8	.	.	.	.	.	.	.	.	.	4	4	.	.	.	.	.	.	.	.	.
X.104	4	−10	10	−10	−4	4	4	.	.	.	.	.	.	.	−2	2	.	4	2	−4	2	−2	.	.	.
X.105	−4	2	−2	10	4	−4	4	.	.	.	.	.	.	.	2	−2	.	4	−2	−4	−2	2	.	.	.
X.106	.	.	.	.	.	.	.	.	.	.	.	.	.	.	.	.	.	.	.	.	.	.	.	−20	H
X.107	.	.	.	.	.	.	.	.	.	.	.	.	.	.	.	.	.	.	.	.	.	.	.	−20	$\bar{H}$
X.108	.	12	−12	−12	.	8	−8	.	.	.	.	4	−4	.	.	.	.	.	.	.	.	.	.	1	.
X.109	.	−12	12	12	.	8	−8	.	.	.	.	−4	4	.	.	.	.	.	.	.	.	.	.	1	.
X.110	.	.	.	.	.	.	.	.	.	.	.	.	.	.	.	.	.	.	.	.	.	.	.	−18	−18
X.111	.	.	.	.	.	.	.	4	−4	.	.	.	.	.	.	.	4	.	−4	.	.	.	.	18	G
X.112	.	.	.	.	.	.	.	4	−4	.	.	.	.	.	.	.	4	.	−4	.	.	.	.	18	$\bar{G}$
X.113	.	.	.	.	.	.	.	.	.	.	.	.	.	.	.	.	.	.	.	.	.	.	.	24	.
X.114	.	.	.	.	.	.	.	.	.	.	.	.	.	.	.	.	.	.	.	.	.	.	−2	−32	.
X.115	.	.	.	.	.	.	.	−12	12	.	.	.	.	.	.	.	−4	.	4	.	.	.	2	.	.

Character table of $H(\mathrm{Fi}_{22})$ (continued)

	6c	6d	6e	6f	6g	6h	6i	6j	6k	6l	6m	6n	6o	6p	6q	6r	6s	6t	6u	6v	6w	6x	6y	6z	8a	8b	8c	8d	8e
2	7	8	7	7	7	7	7	5	6	6	4	7	7	5	5	4	4	4	4	5	5	5	3	4	7	7	7	7	7
3	4	3	3	2	2	2	2	3	2	2	3	1	1	2	2	2	2	2	2	1	1	1	2	1	1	1	1	.	.
5	.	.	.	.	.	.	.	.	.	.	.	.	.	.	.	.	.	.	.	.	.	.	.	.	.	.	.	.	.
2P	3a	3b	3b	3a	3a	3a	3c	3b	3b	3c	3b	3b	3b	3b	3c	3c	3c	3c	3c	3c	3b	3c	3c	4_1	4_1	4_{11}	4_{11}	4_{11}	
3P	2b	2a	2b	2c	2g	2g	2d	2a	2c	2d	2b	2e	2f	2h	2g	2g	2g	2d	2c	2i	2j	2m	2h	2l	8a	8b	8c	8d	8e
5P	6b	6d	6e	6f	6h	6g	6i	6j	6k	6l	6m	6n	6o	6p	6q	6s	6r	6t	6u	6v	6w	6x	6y	6z	8a	8b	8c	8d	8e
X.85	-9	.	.	3	3	3	.	.	.	.	.	.	.	.	.	.	.	.	.	.	.	.	.	.	.	.	.	.	.
X.86	.	-24	.	8	.	.	-8	.	-4	4	.	.	.	.	.	.	.	.	4	-4	.	.	.	.	.	.	.	.	.
X.87	.	.	.	.	.	.	.	.	.	.	.	.	.	.	.	.	.	.	.	.	.	.	.	.	.	.	-2	-2	.
X.88	.	.	.	.	.	.	.	.	.	.	.	.	.	.	.	.	.	.	.	.	.	.	.	.	.	.	2	2	.
X.89	.	.	.	.	.	.	.	.	.	.	.	.	.	.	.	.	.	.	.	.	.	.	.	.	.	2	.	.	.
X.90	.	.	.	.	.	.	.	.	.	.	.	.	.	.	.	.	.	.	.	.	.	.	.	.	.	-2	.	.	.
X.91	.	9	-9	.	.	.	.	.	-3	-3	.	1	-1	3	3	.	.	.	.	.	.	.	-1	.	-4	4	.	.	.
X.92	.	9	-9	.	.	.	.	.	3	3	.	1	-1	-3	-3	.	.	.	.	.	.	.	-1	.	-4	4	.	.	.
X.93	.	9	-9	.	.	.	.	.	3	3	.	1	-1	3	-3	.	.	.	.	.	.	.	1	.	-4	-4	.	.	.
X.94	.	9	-9	.	.	.	.	.	-3	-3	.	1	-1	3	3	.	.	.	.	.	.	.	1	.	4	-4	.	.	.
X.95	24	-6	-6	.	.	.	.	.	.	.	.	2	2	.	.	.	.	.	.	.	.	-2	.	.	.	.	.	.	.
X.96	24	-6	-6	.	.	.	.	.	.	.	.	2	2	.	.	.	.	.	.	.	.	2	.	.	.	.	.	.	.
X.97	.	-24	.	-4	.	.	4	-6	-4	4	.	.	.	.	.	.	.	.	-2	2	-2	2	.	.	.	.	.	.	.
X.98	$\bar G$	.	-6	$\bar B$	$\bar B$	6	6	.	.	.	.	.	.	.	.	B	$\bar B$	.	-2	2	.	.	.	.	.	.	.	.	.
X.99	G	.	-6	B	$\bar B$	6	6	.	.	.	.	.	.	.	.	$\bar B$	B	.	-2	2	.	.	.	.	.	.	.	.	.
X.100	.	-9	9	.	.	.	.	.	3	3	.	-1	1	-3	-3	.	.	.	.	.	.	.	1	.	-2	2	.	2	-2
X.101	.	-9	9	.	.	.	.	.	-3	-3	.	-1	1	3	3	.	.	.	.	.	.	.	1	.	-2	2	.	-2	2
X.102	.	-9	-9	.	.	.	.	.	3	3	.	-1	-1	3	3	.	.	.	.	.	.	.	-1	.	4	4	.	.	.
X.103	.	-9	-9	.	.	.	.	.	3	3	.	-1	-1	-3	3	.	.	.	.	.	.	.	1	.	-4	-4	.	.	.
X.104	.	-9	9	.	.	.	.	.	-3	-3	.	-1	1	-3	3	.	.	.	.	.	.	.	-1	.	2	-2	.	-2	2
X.105	.	-9	9	.	.	.	.	.	3	3	.	-1	1	3	-3	.	.	.	.	.	.	.	-1	.	2	-2	.	2	-2
X.106	$\bar H$	-8	.	4	I	$\bar I$	-4	4	4	-4	.	.	.	.	.	.	.	.	B	$\bar B$	2	-2	.	.	.	.	.	.	.
X.107	H	-8	.	4	$\bar I$	I	-4	4	4	-4	.	.	.	.	.	.	.	.	$\bar B$	B	2	-2	.	.	.	.	.	.	.
X.108	.	.	.	.	.	.	.	.	.	.	.	.	.	.	.	.	.	.	.	.	.	.	.	.	.	.	.	.	.
X.109	.	.	.	.	.	.	.	.	.	.	.	.	.	.	.	.	.	.	.	.	.	.	.	.	.	.	.	.	.
X.110	-18	.	.	-6	-6	-6	-6	.	.	.	.	.	.	.	.	.	.	.	.	.	.	.	.	.	.	.	.	.	.
X.111	$\bar G$	.	.	-6	$\bar A$	A	6	.	.	.	.	.	.	.	.	.	.	.	.	.	.	.	.	.	.	.	.	.	.
X.112	G	.	.	-6	A	$\bar A$	6	.	.	.	.	.	.	.	.	.	.	.	.	.	.	.	.	.	.	.	.	.	.
X.113	.	24	.	8	.	.	-8	6	-4	4	.	.	.	.	.	.	.	.	-2	2	2	-2	.	.	.	.	.	.	.
X.114	.	16	.	.	.	.	.	-8	.	.	.	.	.	.	.	.	.	.	.	.	.	.	.	.	.	.	.	.	.
X.115	.	.	.	.	.	.	.	.	.	.	.	.	.	.	.	.	.	.	.	.	.	.	.	.	.	.	.	.	.

Character table of $H(\mathrm{Fi}_{22})$ (continued)

	8f	8g	8h	8i	8j	8k	9a	10a	10b	10c	12a	12b	12c	12d	12e	12f	12g	12h	12i	12j	12k	12l	12m	12n	12o	12p
2	7	7	6	5	5	5	2	3	2	2	5	6	6	5	5	5	5	5	5	5	5	5	5	5	5	5
3	.	.	.	.	.	.	2	.	.	.	3	2	2	2	1	1	1	1	1	1	1	1	.	1	1	1
5	.	.	.	.	.	.	.	1	1	1	.	.	.	.	.	.	.	.	.	.	.	.	.	.	.	.
2P	4_3	4_{11}	4_{12}	4_{13}	4_{14}	4_{21}	9a	5a	5a	5a	6a	6d	6d	6d	6n	6s	6n	6n	6n	6g	6h	6a	6g	6h	6n	6n
3P	8f	8g	8h	8i	8j	8k	3a	10a	10b	10c	4_1	4_1	4_1	4_2	4_4	4_5	4_6	4_7	4_8	4_{15}	4_{15}	4_3	4_{16}	4_{16}	4_{10}	4_9
5P	8f	8g	8h	8i	8j	8k	9a	2a	2h	2b	12a	12b	12c	12d	12e	12f	12g	12h	12i	12j	12k	12l	12n	12m	12o	12p
X.85	.	.	.	.	.	.	2	.	.	.	3	.	.	.	.	.	.	.	.	.	F	$F-1$	F	F	.	.
X.86	.	.	.	.	.	.	.	2	.	.	.	.	.	.	.	.	.	.	.	.	.	.	.	.	.	.
X.87	2	.	.	.	2	.	.	.	-2	.	.	.	.	.	.	.	.	.	.	.	.	.	.	.	.	.
X.88	2	.	.	.	.	.	.	.	-2	.	.	.	.	.	.	.	.	.	.	.	.	.	.	.	.	.
X.89	-2	-2	2	.	.	.	.	.	.	.	.	.	.	.	.	.	.	.	.	.	.	.	.	.	.	.
X.90	-2	2	-2	.	.	.	.	.	.	.	.	.	.	.	.	.	.	.	.	.	.	.	.	.	.	.
X.91	.	.	.	.	.	.	.	-1	-1	1	.	-3	3	-3	-1	1	1	-1	1	.	.	.	.	.	-1	1
X.92	.	.	.	.	.	.	.	-1	1	1	.	-3	3	3	1	1	-1	-1	1	.	.	.	.	.	1	-1
X.93	.	.	.	.	.	.	.	-1	-1	1	.	-3	3	-3	-1	-1	1	-1	1	.	.	.	.	.	1	1
X.94	.	.	.	.	.	.	.	-1	1	1	.	-3	3	3	1	-1	1	1	-1	.	.	.	.	.	-1	-1
X.95	.	.	.	.	.	.	.	.	.	.	.	-8	-2	-2	.	.	-2	.	2	2	.	.	.	.	.	.
X.96	.	.	.	.	.	.	.	.	.	.	.	-8	-2	-2	.	.	2	.	-2	-2	.	.	.	.	.	.
X.97	.	.	.	.	.	.	.	.	.	.	.	.	.	.	.	.	.	.	.	.	-2	-2	.	2	2	.
X.98	.	.	.	.	.	.	.	.	.	.	.	.	.	.	.	.	.	.	.	.	$-\bar C$	$-C$	.	$\bar C$	C	.
X.99	.	.	.	.	.	.	.	.	.	.	.	.	.	.	.	.	.	.	.	.	$-C$	$-\bar C$	.	C	$\bar C$	.
X.100	.	.	.	.	.	.	.	.	.	.	.	3	-3	3	1	-1	-1	1	-1	.	.	.	.	.	1	1
X.101	.	.	.	.	.	.	.	.	.	.	.	3	-3	-3	-1	-1	1	1	-1	.	.	.	.	.	-1	-1
X.102	.	.	.	.	.	.	.	.	.	.	.	3	3	3	-1	-1	-1	-1	1	.	.	.	.	.	-1	-1
X.103	.	.	.	.	.	.	.	.	.	.	.	3	3	-3	1	1	-1	-1	1	.	.	.	.	.	-1	1
X.104	.	.	.	.	.	.	.	.	.	.	.	3	-3	3	1	1	1	-1	1	.	.	.	.	.	-1	-1
X.105	.	.	.	.	.	.	.	.	.	.	.	3	-3	-3	-1	1	-1	-1	1	.	.	.	.	.	1	1
X.106	.	.	.	.	.	.	.	.	-1	.	.	.	.	.	.	.	.	.	.	.	.	.	.	.	.	.
X.107	.	.	.	.	.	.	.	.	-1	.	.	.	.	.	.	.	.	.	.	.	.	.	.	.	.	.
X.108	.	.	.	.	.	.	.	.	.	.	1	1	-1	.	.	.	.	.	.	.	.	.	.	.	.	.
X.109	.	.	.	.	.	.	.	.	.	.	1	-1	-1	.	.	.	.	.	.	.	.	.	.	.	.	.
X.110	.	.	.	.	.	.	.	.	.	.	6	.	.	.	.	.	.	.	.	.	.	.	2	.	.	.
X.111	.	.	.	.	.	.	.	.	.	.	.	.	.	.	.	.	.	.	.	.	$\bar C$	C	.	$-\bar C$	$-C$	.
X.112	.	.	.	.	.	.	.	.	.	.	.	.	.	.	.	.	.	.	.	.	C	$\bar C$	.	$-C$	$-\bar C$	.
X.113	.	.	.	.	.	.	.	.	.	.	.	.	.	.	.	.	.	.	.	.	.	.	.	.	.	.
X.114	.	.	.	.	.	.	2	2	.	.	.	.	.	.	.	.	.	.	.	.	.	.	.	.	.	.
X.115	.	.	.	.	.	.	-2	.	.	.	.	.	.	.	.	.	.	.	.	.	.	.	.	.	.	.

Character table of $H(\mathrm{Fi}_{22})$ *(continued)*

		2 3 4 4 3 3 5 5 2 2 2 2 4 4												
		3 2 1 1 1 1 . . 2 2 2 . 1 1												
		5 . . . . 1 . .												
	12q	12r	12s	12t	12u	16a	16b	18a	18b	18c	20a	24a	24b
2P	6j	6f	6f	6v	6w	8f	8f	9a	9a	9a	10a	12b	12b
3P	4$_2$	4$_{13}$	4$_{14}$	4$_{17}$	4$_{18}$	16a	16b	6a	6b	6c	20a	8a	8b
5P	12q	12r	12s	12t	12u	16b	16a	18a	18c	18b	4$_2$	24a	24b
X.85	.	1	1	.	.	.	.	.	.	.	.	.	.
X.86	.	.	.	.	.	.	.	.	.	.	.	.	.
X.87	.	.	.	.	.	.	.	.	.	.	.	.	.
X.88	.	.	.	.	.	.	.	.	.	.	.	.	.
X.89	.	.	.	.	.	.	.	.	.	.	.	.	.
X.90	.	.	.	.	.	.	.	.	.	.	.	.	.
X.91	.	.	.	.	.	.	.	.	.	.	1	-1	1
X.92	.	.	.	.	.	.	.	.	.	.	-1	-1	1
X.93	.	.	.	.	.	.	.	.	.	.	1	1	-1
X.94	.	.	.	.	.	.	.	.	.	.	-1	1	-1
X.95	.	.	.	.	.	.	.	.	.	.	.	.	.
X.96	.	.	.	.	.	.	.	.	.	.	.	.	.
X.97	.	.	.	.	.	.	.	.	.	.	.	.	.
X.98	.	.	.	.	.	.	.	.	.	.	.	.	.
X.99	.	.	.	.	.	.	.	.	.	.	.	.	.
X.100	.	.	.	.	.	.	.	.	.	.	.	1	-1
X.101	.	.	.	.	.	.	.	.	.	.	.	1	-1
X.102	.	.	.	.	.	.	.	.	.	.	.	1	1
X.103	.	.	.	.	.	.	.	.	.	.	.	-1	-1
X.104	.	.	.	.	.	.	.	.	.	.	.	-1	1
X.105	.	.	.	.	.	.	.	.	.	.	.	-1	1
X.106	.	.	.	.	.	.	.	1	D	$\bar{D}$	.	.	.
X.107	.	.	.	.	.	.	.	1	$\bar{D}$	D	.	.	.
X.108	.	.	.	.	.	.	.	.	.	.	-1	.	.
X.109	.	.	.	.	.	.	.	.	.	.	1	.	.
X.110	.	.	.	.	.	.	.	.	.	.	.	.	.
X.111	.	.	.	.	.	.	.	.	.	.	.	.	.
X.112	.	.	.	.	.	.	.	.	.	.	.	.	.
X.113	.	.	.	.	.	.	.	.	.	.	.	.	.
X.114	.	.	.	.	.	.	-2	.	.	.	.	.	.
X.115	.	.	.	.	.	.	.	.	.	.	.	.	.

where $A = -6i\sqrt{3}$; $B = 2i\sqrt{3}$; $C = 1 + i\sqrt{3}$; $D = -i\sqrt{3}$; $E = -2(\zeta_8^3 + \zeta_8)$; $F = -1 + 2i\sqrt{3}$; $G = -18i\sqrt{3}$; $H = 12i\sqrt{3}$; *and* $I = 4i\sqrt{3}$.

B.5. Character table of $D(\mathrm{Fi}_{22}) = \langle x, y \rangle$

	1a	2a	2b	2c	2d	2e	2f	2g	2h	2i	2j	2k	2l	2m	2n	2o	2p	2q	2r	2s	3a	4_1	4_2	4_3
2	17	17	16	15	15	16	16	14	12	12	13	13	13	11	12	10	11	11	10	10	5	12	12	10
3	1	1	1	1	1	.	.	1	1	1	.	.	.	1	.	1	.	.	.	.	1	.	.	1
5	1	1	1	.	.	.	.	.	.	.	.	.	.	.	.	.	.	.	.	.	1	.	.	1
2P	1a	1a	1a	1a	1a	1a	1a	1a	1a	1a	1a	1a	1a	1a	1a	1a	1a	1a	1a	1a	3a	2a	2a	2a
3P	1a	2a	2b	2c	2d	2e	2f	2g	2h	2i	2j	2k	2l	2m	2n	2o	2p	2q	2r	2s	1a	4_1	4_2	4_3
5P	1a	2a	2b	2c	2d	2e	2f	2g	2h	2i	2j	2k	2l	2m	2n	2o	2p	2q	2r	2s	3a	4_1	4_2	4_3
X.1	1	1	1	1	1	1	1	1	1	1	1	1	1	1	1	1	1	1	1	1	1	1	1	1
X.2	1	1	1	1	1	1	1	1	1	1	1	1	1	1	1	−1	1	1	−1	1	1	1	1	−1
X.3	4	4	4	4	4	4	4	4	4	4	4	4	4	4	4	−2	.	.	−2	.	1	4	4	−2
X.4	4	4	4	4	4	4	4	4	4	4	4	4	4	4	2	.	.	.	2	.	1	4	4	2
X.5	5	5	5	5	5	5	5	5	5	5	5	5	5	5	5	1	1	1	1	1	−1	5	5	1
X.6	5	5	5	5	−3	−3	5	5	−3	1	1	5	5	5	−3	1	−3	−3	1	1	2	−3	5	−3
X.7	5	5	5	5	5	5	5	5	5	5	5	5	5	5	5	−1	1	1	−1	1	−1	5	5	−1
X.8	5	5	5	5	−3	−3	5	5	−3	1	1	5	5	5	−3	1	−3	−3	1	1	2	−3	5	3
X.9	6	6	6	6	6	6	6	6	6	6	6	6	6	6	6	.	−2	−2	.	−2	.	6	6	.
X.10	10	10	10	−6	−6	10	10	−6	2	2	10	10	−6	2	−6	.	2	2	.	2	−2	−6	10	.
X.11	10	10	10	2	2	10	10	2	−2	−2	10	10	2	−2	2	4	2	2	.	2	1	2	10	4
X.12	10	10	10	2	2	10	10	2	−2	−2	10	10	2	−2	2	−2	−2	−2	2	−2	1	2	10	−2
X.13	10	10	10	2	2	10	10	2	−2	−2	10	10	2	−2	2	2	−2	−2	−2	−2	1	2	10	2
X.14	10	10	10	2	2	10	10	2	−2	−2	10	10	2	−2	2	−4	2	2	.	2	1	2	10	−4
X.15	15	15	15	15	15	15	15	15	15	15	−1	−1	15	15	−1	3	3	3	3	3	.	−1	−1	3
X.16	15	15	15	15	15	15	15	15	15	15	−1	−1	15	15	−1	3	−1	−1	3	−1	.	−1	−1	3
X.17	15	15	15	−9	−9	15	15	−9	3	3	−1	−1	−9	3	7	3	3	3	−1	3	.	7	−1	3
X.18	15	15	15	−9	−9	15	15	−9	3	3	−1	−1	−9	3	7	3	−1	−1	−1	−1	.	7	−1	3
X.19	15	15	15	−9	−9	15	15	−9	3	3	15	15	−9	3	−9	3	−1	−1	−1	−1	.	−9	15	3
X.20	15	15	15	15	15	15	15	15	15	15	−1	−1	15	15	−1	−3	−1	−1	−3	−1	.	−1	−1	−3
X.21	15	15	15	15	15	15	15	15	15	15	−1	−1	15	15	−1	−3	3	3	−3	3	.	−1	−1	−3
X.22	15	15	15	−9	−9	15	15	−9	3	3	−1	−1	−9	3	7	−3	−1	−1	1	−1	.	7	−1	−3
X.23	15	15	15	−9	−9	15	15	−9	3	3	15	15	−9	3	−9	−3	−1	−1	1	−1	.	−9	15	−3
X.24	15	15	15	−9	−9	15	15	−9	3	3	−1	−1	−9	3	7	−3	3	3	1	3	.	7	−1	−3
X.25	20	20	20	4	4	20	20	4	−4	−4	20	20	4	−4	4	2	.	.	−2	.	−1	4	20	−2
X.26	20	20	20	4	4	20	20	4	−4	−4	20	20	4	−4	4	2	.	.	2	.	−1	4	20	2
X.27	30	30	30	6	6	30	30	6	−6	−6	−2	−2	6	−6	6	.	−2	−2	4	−2	.	6	−2	.
X.28	30	30	30	6	6	30	30	6	−6	−6	−2	−2	6	−6	6	.	−2	−2	−4	−2	.	6	−2	.
X.29	30	30	30	30	30	30	30	30	30	30	−2	−2	30	30	−2	.	−2	−2	.	−2	.	−2	−2	.
X.30	30	30	30	−18	−18	30	30	−18	6	6	−2	−2	−18	6	14	.	−2	−2	.	−2	.	14	−2	.
X.31	30	30	30	6	6	30	30	6	−6	−6	−2	−2	6	−6	6	6	2	2	2	2	.	6	−2	6
X.32	30	30	30	6	6	30	30	6	−6	−6	−2	−2	6	−6	6	−6	2	2	−2	2	.	6	−2	−6
X.33	32	−32	.	16	−16	.	.	.	8	−8	.	.	.	.	.	.	8	−8	.	2	.	.	.	.
X.34	40	40	−40	16	16	8	−8	−16	4	4	8	−8	.	.	−4	.	−2	−4	−4	6	4	1	.	2
X.35	40	40	−40	−16	−16	8	−8	16	4	4	8	−8	.	.	−4	.	2	−4	−4	−2	4	1	.	−2
X.36	40	40	−40	−16	−16	8	−8	16	4	4	8	−8	.	.	−4	.	−10	4	4	2	−4	1	.	10
X.37	40	40	−40	16	16	8	−8	−16	4	4	8	−8	.	.	−4	.	10	4	4	−6	4	1	.	−10
X.38	40	40	−40	−16	−16	8	−8	16	4	4	8	−8	.	.	−4	.	−2	−4	−4	2	4	1	.	2
X.39	40	40	−40	16	16	8	−8	−16	4	4	8	−8	.	.	−4	.	2	−4	−4	6	4	1	.	−2
X.40	40	40	−40	16	16	8	−8	−16	4	4	8	−8	.	.	−4	.	−10	4	4	−6	−4	1	.	10
X.41	40	40	−40	−16	−16	8	−8	16	4	4	8	−8	.	.	−4	.	10	4	4	2	−4	1	.	−10
X.42	60	60	60	−36	−36	60	60	−36	12	12	−4	−4	−36	12	−4	−6	.	2	.	.	.	−4	−4	−6
X.43	60	60	60	−36	−36	60	60	−36	12	12	−4	−4	−36	12	−4	6	.	−2	.	.	.	−4	−4	6
X.44	60	60	60	12	12	60	60	12	−12	−12	−4	−4	12	−12	12	.	4	4	.	4	.	12	−4	.
X.45	60	60	60	12	12	60	60	12	−12	−12	−4	−4	12	−12	−20	−6	.	2	.	.	.	−20	−4	−6
X.46	60	60	60	12	12	60	60	12	−12	−12	−4	−4	12	−12	−20	6	.	−2	.	.	.	−20	−4	6
X.47	60	60	60	12	12	60	60	12	−12	−12	−4	−4	12	−12	12	.	−4	−4	.	−4	.	12	−4	.
X.48	80	80	−80	−32	−32	16	−16	32	8	8	16	−16	.	.	−8	.	8	.	.	.	−1	.	.	−8
X.49	80	80	−80	−32	−32	16	−16	32	8	8	16	−16	.	.	−8	.	−8	.	.	.	−1	.	.	8
X.50	80	80	−80	.	.	16	−16	.	−8	−8	16	−16	.	.	8	.	12	.	−4	.	2	.	.	−12
X.51	80	80	−80	32	32	16	−16	−32	8	8	16	−16	.	.	−8	.	−8	.	.	.	−1	.	.	8
X.52	80	80	−80	32	32	16	−16	−32	8	8	16	−16	.	.	−8	.	8	.	.	.	−1	.	.	−8
X.53	80	80	−80	.	.	16	−16	.	−8	−8	16	−16	.	.	8	.	−12	.	4	.	2	.	.	12
X.54	96	96	−96	.	.	−32	32	.	.	.	.	.	.	.	.	.	8	8	.	−8	.	.	.	.
X.55	96	96	−96	.	.	−32	32	.	.	.	.	.	.	.	.	.	8	8	.	−8	.	.	.	.
X.56	96	96	−96	.	.	−32	32	.	.	.	.	.	.	.	.	.	−8	−8	.	8	.	.	.	.
X.57	96	96	−96	.	.	−32	32	.	.	.	.	.	.	.	.	.	−8	−8	.	8	.	.	.	.
X.58	96	−96	.	48	−48	.	.	.	24	−24	.	.	.	.	.	.	−8	8	.	.	.	.	.	.
X.59	96	−96	.	48	−48	.	.	.	24	−24	.	.	.	.	.	.	−8	8	.	.	.	.	.	.
X.60	120	120	120	24	24	−8	−8	24	.	.	8	8	−8	.	.	8	−12	12	12	−4	12	.	−8	−8	−12
X.61	120	120	120	24	24	−8	−8	24	.	.	−8	−8	−8	.	−8	.	.	.	.	−8	.	.	8	8
X.62	120	120	−48	−48	24	−24	48	12	12	−8	8	.	−12	.	6	4	4	2	−4	.	8	8	.	−6
X.63	120	120	120	24	24	−8	−8	24	.	.	−8	−8	8	.	−8	.	.	.	−8	.	.	8	8	.
X.64	120	120	120	−24	−24	−8	−8	−24	.	.	8	8	8	.	−8	−12	.	.	4	.	8	−8	−12	.
X.65	120	120	120	24	24	−8	−8	24	.	.	−8	−8	−8	.	−8	.	.	.	8	.	8	8	.	.
X.66	120	120	120	48	48	24	−24	−48	12	12	−8	8	.	−12	.	−6	4	4	−10	−4	.	.	.	6
X.67	120	120	−120	48	48	24	−24	−48	12	12	−8	8	.	−12	.	6	−4	−4	2	4	.	.	.	−6
X.68	120	120	−120	−48	−48	24	−24	48	12	12	−8	8	.	−12	.	−6	−4	−4	6	4	.	.	.	6
X.69	120	120	120	24	24	−8	−8	24	.	.	8	8	−8	.	8	12	−4	−4	4	−4	.	−8	−8	12
X.70	120	120	120	24	24	−8	−8	24	.	.	−8	−8	−8	.	−8	.	12	.	−4	.	8	8	.
X.71	120	120	120	−24	−24	−8	−8	−24	.	.	8	8	8	.	−8	12	.	.	−4	.	8	−8	12	.
X.72	120	120	120	−24	−24	−8	−8	−24	.	.	8	8	8	.	−8	12	.	.	−4	.	8	−8	12
X.73	120	120	120	24	24	−8	−8	24	.	.	8	8	−8	.	8	12	12	12	4	12	.	−8	−8	12
X.74	120	120	−120	−48	−48	24	−24	48	12	12	−8	8	.	−12	.	6	−4	−4	−6	4	.	.	.	−6
X.75	120	120	−120	48	48	24	−24	−48	12	12	−8	8	.	−12	.	−6	−4	−4	2	4	.	.	.	6
X.76	120	120	−120	48	48	24	−24	−48	12	12	−8	8	.	−12	.	6	4	4	10	−4	.	.	.	−6
X.77	120	120	−120	−48	−48	24	−24	48	12	12	−8	8	.	−12	.	−6	4	4	−2	−4	.	.	.	6
X.78	120	120	120	−24	−24	−8	−8	−24	.	.	8	8	8	.	−8	−12	.	.	4	.	8	−8	−12
X.79	120	120	120	24	24	−8	−8	24	.	.	8	8	8	−8	.	8	−12	−4	−4	−4	−4	.	−8	−8	−12
X.80	128	−128	.	64	−64	.	.	.	32	−32	.	.	.	.	.	.	.	.	2	.	.	.	.	.
X.81	160	−160	.	80	−80	.	.	.	40	−40	.	.	.	.	.	.	.	.	−2	.	.	.	.	.
X.82	160	−160	.	−48	48	.	.	.	8	−8	.	.	.	.	.	.	8	−8	.	4	.	.	.	.
X.83	160	160	−160	.	.	32	−32	.	−16	−16	32	−32	.	16	.	.	.	.	−2	.	.	.	.	.
X.84	160	−160	.	−48	48	.	.	.	8	−8	.	.	.	.	.	.	8	−8	.	−2	.	.	.	.

Character table of $D(\mathrm{Fi}_{22})$ *(continued)*

	4_4	4_5	4_6	4_7	4_8	4_9	4_{10}	4_{11}	4_{12}	4_{13}	4_{14}	4_{15}	4_{16}	4_{17}	4_{18}	4_{19}	4_{20}	4_{21}	4_{22}	4_{23}	4_{24}	4_{25}	4_{26}	4_{27}	4_{28}	4_{29}
2	11	11	10	10	10	10	10	10	10	10	10	8	8	9	9	9	9	9	9	9	9	9	9	8	8	8
3	.	.	.	.	.	.	.	.	.	.	.	1	1	.	.	.	.	.	.	.	.	.	.	.	.	.
5	.	.	.	.	.	.	.	.	.	.	.	.	.	.	.	.	.	.	.	.	.	.	.	.	.	.
2P	2_e	2_e	2_e	2_e	2_e	2_e	2_e	2_e	2_a	2_e	2_e	2_h	2_i	2_c	2_c	2_j	2_g	2_f	2_j	2_e	2_e	2_e	2_g	2_j	2_h	2_j
3P	4_4	4_5	4_6	4_7	4_8	4_9	4_{10}	4_{11}	4_{12}	4_{13}	4_{14}	4_{15}	4_{16}	4_{17}	4_{18}	4_{19}	4_{20}	4_{21}	4_{22}	4_{23}	4_{24}	4_{25}	4_{26}	4_{27}	4_{28}	4_{29}
5P	4_4	4_5	4_6	4_7	4_8	4_9	4_{10}	4_{11}	4_{12}	4_{13}	4_{14}	4_{15}	4_{16}	4_{17}	4_{18}	4_{19}	4_{20}	4_{21}	4_{22}	4_{23}	4_{24}	4_{25}	4_{26}	4_{27}	4_{28}	4_{29}
X.1	1	1	1	1	1	1	1	1	1	1	1	1	1	1	1	1	1	1	1	1	1	1	1	1	1	1
X.2	1	1	-1	1	-1	1	1	-1	-1	1	-1	-1	-1	1	1	1	1	1	1	-1	1	-1	1	-1	-1	-1
X.3	4	4	-2	.	-2	.	4	-2	-2	4	-2	-2	-2	.	.	.	.	4	.	-2	4	-2	.	-2	-2	-2
X.4	4	4	2	.	2	.	4	2	2	4	2	2	2	.	.	.	.	4	.	2	4	2	.	2	2	2
X.5	5	5	1	1	1	1	5	1	1	5	1	1	1	i	1	i	1	5	1	1	5	1	1	1	1	1
X.6	1	1	1	-3	1	-3	1	1	-3	-3	3	3	1	1	1	1	1	1	1	1	1	1	1	1	1	-3
X.7	5	5	-1	1	-1	1	5	-1	-1	5	-1	-1	-1	1	1	1	1	5	1	-1	5	-1	1	-1	-1	-1
X.8	1	1	-1	1	3	-3	1	-1	-1	3	-3	-3	-3	1	1	1	1	1	1	-1	1	-1	1	-1	1	3
X.9	6	6	.	-2	.	-2	6	.	.	6	.	.	.	.	.	.	.	-2	-2	-2	6	-2	.	6	.	-2
X.10	2	2	.	2	.	2	-6	.	.	.	-6	.	.	2	2	2	2	2	2	.	2	.	2	.	.	.
X.11	-2	-2	.	2	4	2	2	.	2	4	2	2	2	-2	-2	2	-2	-2	2	.	-2	.	-2	.	-2	4
X.12	-2	-2	2	-2	-2	2	2	2	-2	-4	-4	2	2	-2	2	-2	-2	2	2	-2	2	2	2	.	.	-2
X.13	-2	-2	-2	-2	2	-2	2	-2	-2	2	2	4	4	2	2	-2	2	-2	-2	-2	-2	-2	2	-2	.	2
X.14	-2	-2	.	2	-4	2	2	.	2	-4	-2	-2	-2	-2	2	-2	-2	2	.	-2	.	-2	.	2	.	-4
X.15	-1	-1	3	3	3	3	-1	3	3	-1	3	3	3	3	3	-1	3	-1	-1	3	-1	3	3	-1	3	-1
X.16	-1	-1	3	-1	3	-1	-1	3	-1	-1	3	3	3	3	-1	-1	3	-1	-1	3	-1	3	-1	-1	3	-1
X.17	3	3	-1	3	3	-1	-1	-1	-1	3	-3	-3	3	3	-1	3	3	-1	-1	-5	-1	3	3	3	1	-1
X.18	3	3	-1	-1	3	-1	-1	-1	-1	3	-3	-3	-1	-1	3	-1	3	-1	3	-5	-1	-1	3	3	1	-1
X.19	3	3	-1	-1	3	-1	-9	-1	-1	-9	-3	-3	-3	-1	-1	-1	-1	3	-1	-1	3	-1	-1	-1	-1	3
X.20	-1	-1	-3	-1	-3	-1	-1	-3	-3	-1	-3	-3	-3	-1	-1	3	-1	-1	3	-3	-1	-3	-1	1	-3	1
X.21	-1	-1	-3	-3	-3	-3	-1	-3	-3	-3	3	3	-3	-1	-1	3	-1	3	-1	-3	-1	-3	-3	1	-3	1
X.22	3	3	1	-1	-3	-1	-1	1	1	-1	-3	3	3	-1	-1	3	1	3	3	1	-5	1	-1	-3	-1	1
X.23	3	3	1	-1	-3	-1	-9	1	1	-9	3	3	3	-1	-1	-1	1	3	1	3	1	1	-1	-1	-1	-3
X.24	3	3	1	3	-3	3	-1	1	1	-1	-3	3	3	3	-1	3	3	-1	1	-5	1	3	-3	-1	1	1
X.25	-4	-4	-2	.	-2	.	4	-2	-2	4	-2	2	2	.	.	.	.	-4	.	-2	-4	-2	.	-2	-2	-2
X.26	-4	-4	2	.	2	.	4	2	2	4	-2	-2	.	.	.	.	-4	.	-2	4	-2	2	.	-2	-2	2
X.27	10	10	4	-2	.	-2	-2	4	4	-2	.	-6	-6	2	2	-2	2	-6	-2	4	2	-4	2	-4	-2	.
X.28	10	10	-4	-2	.	-2	-2	-4	-4	-2	.	6	6	2	2	-2	2	-6	-2	-4	2	-4	2	4	2	.
X.29	-2	-2	.	-2	.	-2	.	.	-2	.	.	.	.	-2	-2	-2	-2	-2	-2	.	-2	.	.	-2	.	.
X.30	6	6	.	-2	.	-2	-2	.	.	.	-2	.	.	.	.	.	.	-2	-2	-2	6	-2	.	-10	.	-2
X.31	10	10	2	2	6	2	-2	2	2	-2	6	.	.	-2	-2	2	-2	-6	2	2	2	2	-2	2	-4	-2
X.32	10	10	-2	2	-6	2	-2	-2	-2	-2	-6	.	.	-2	-2	2	-2	-6	2	-2	2	2	-2	-2	4	2
X.33	.	.	.	.	.	.	.	.	.	.	.	.	.	.	.	.	.	4	.	.	.	.	-4	.	.	.
X.34	4	-4	2	4	6	-4	-2	6	-4	6	4	-4	.	.	.	-4	.	.	4	-2	.	2	.	2	.	2
X.35	4	-4	6	4	-6	-4	-4	-6	2	4	6	4	-4	.	.	-4	.	.	4	2	.	-2	.	-2	.	-2
X.36	4	-4	-6	-4	-2	4	-4	-6	-2	4	4	-4	.	.	.	4	.	.	-4	-2	.	2	.	2	.	2
X.37	4	-4	-2	-4	2	4	4	2	-6	-4	2	4	-4	.	.	.	4	.	.	-4	-2	.	-2	.	-2	-2
X.38	4	-4	-6	4	-6	-4	-4	-6	2	4	-6	-4	4	.	.	-4	.	.	4	2	.	2	.	2	.	2
X.39	4	-4	-2	4	-6	-4	4	2	-6	4	6	-4	4	.	.	.	4	.	.	4	2	.	-2	.	-2	-2
X.40	4	-4	2	-4	-2	4	-2	6	-4	-2	4	4	.	.	.	4	.	.	-4	-2	.	2	.	2	.	2
X.41	4	-4	6	-4	2	4	-4	-6	2	4	-2	-4	4	.	.	.	4	.	.	-4	2	.	-2	.	-2	-2
X.42	-4	-4	2	.	-6	.	4	2	2	4	-6	6	6	.	.	.	.	-4	.	-2	4	2	.	2	-2	2
X.43	-4	-4	-2	.	6	.	4	-2	-2	4	6	-6	-6	.	.	.	.	-4	.	-2	4	-2	.	-2	2	-2
X.44	-12	-12	.	4	.	4	-4	.	.	.	-4	.	.	-4	-4	-4	-4	4	-4	.	4	-4	.	4	.	.
X.45	4	4	2	.	-6	.	4	2	2	4	-6	-6	-6	.	.	.	.	4	.	-2	4	2	.	2	2	2
X.46	4	4	-2	.	6	.	4	-2	-2	4	6	6	6	.	.	.	.	4	.	-2	4	-2	.	-2	-2	-2
X.47	-12	-12	.	-4	.	-4	-4	.	.	.	-4	.	.	4	4	4	4	4	4	.	4	4	.	4	.	.
X.48	8	-8	.	.	8	.	-8	.	.	.	-8	.	.	8	-8	-8	8	.	.	.	.	.	.	.	.	.
X.49	8	-8	.	.	-8	.	-8	.	.	.	8	.	.	8	8	8	-8	.	.	.	.	.	.	.	.	.
X.50	-8	8	-4	.	-4	.	.	4	4	.	4	.	.	.	.	.	.	.	.	.	-4	.	4	.	4	.	-4
X.51	8	-8	.	.	-8	.	8	.	.	.	-8	.	.	-8	8	8	-8	.	.	.	.	.	.	.	.	.
X.52	8	-8	.	.	8	.	8	.	.	.	8	.	.	-8	-8	-8	8	.	.	.	.	.	.	.	.	.
X.53	-8	8	4	.	4	.	.	-4	-4	.	-4	.	.	.	.	.	.	.	.	.	4	.	-4	.	-4	.	4
X.54	.	.	.	8	.	-8	.	.	.	.	8	.	.	.	.	.	.	.	.	.	.	.	.	.	.	.
X.55	.	.	.	8	.	-8	.	.	.	.	-8	.	.	.	.	.	.	.	.	.	.	.	.	.	.	.
X.56	.	.	.	-8	.	8	.	.	.	.	8	.	.	.	.	.	.	.	.	.	.	.	.	.	.	.
X.57	.	.	.	-8	.	8	.	.	.	.	-8	.	.	.	.	.	.	.	.	.	.	.	.	.	.	.
X.58	.	.	.	.	.	.	.	.	.	.	.	.	.	.	.	.	.	-4	.	.	.	.	4	.	.	.
X.59	.	.	.	.	.	.	.	.	.	.	.	.	.	.	.	.	.	-4	.	.	.	.	4	.	.	.
X.60	.	.	-4	-4	4	-4	.	-4	-4	.	4	.	.	.	.	.	.	.	.	.	4	.	4	.	.	.
X.61	.	.	8	.	.	.	.	8	-8	.	.	.	.	-4	12	.	-4	.	.	.	.	.	-4	.	.	.
X.62	-4	4	10	-4	-2	4	4	-10	-2	-4	2	.	.	.	.	.	-4	.	.	4	-2	.	2	.	2	-4	2
X.63	.	.	8	.	.	.	.	8	-8	.	.	.	.	12	-4	.	-4	.	.	.	.	.	-4	.	.	.
X.64	.	.	4	.	4	.	.	4	4	.	4	.	.	-4	12	.	4	.	.	4	-4	.	-4	.	.	.
X.65	.	.	-8	.	.	.	.	-8	8	.	.	.	.	-4	12	.	-4	.	.	.	.	.	-4	.	.	.
X.66	-4	4	-2	-4	2	4	2	10	4	-2	.	.	.	.	.	.	-4	.	.	4	2	.	-2	.	-2	-4	-2
X.67	-4	4	-6	4	-2	-4	-4	6	-2	4	2	.	.	.	.	.	4	.	.	-4	6	.	-6	.	2	-4	2
X.68	-4	4	2	4	2	-4	4	2	-6	-4	2	.	.	.	.	.	4	.	.	-4	-6	.	6	.	-2	-4	-2
X.69	.	.	4	12	-4	12	.	4	4	.	-4	.	.	.	.	.	.	.	.	.	-4	.	-4	.	.	.
X.70	.	.	-8	.	.	.	.	-8	8	.	.	.	.	12	-4	.	-4	.	.	.	.	.	-4	.	.	.
X.71	.	.	-4	.	-4	.	.	-4	-4	.	-4	.	.	.	.	.	.	4	.	.	4	.	4	.	.	.
X.72	.	.	-4	.	-4	.	.	-4	-4	.	-4	.	.	.	.	.	.	4	.	.	4	.	4	.	.	.
X.73	.	.	4	-4	-4	-4	.	4	4	.	.	.	.	.	.	.	.	-4	.	.	-4	.	-4	.	.	.
X.74	-4	4	2	4	-2	-4	-2	6	-4	2	.	.	.	.	.	.	4	.	.	-4	6	.	-6	.	2	4	2
X.75	-4	4	6	4	2	-4	-6	2	4	-2	.	.	.	.	.	.	4	.	.	-4	-6	.	6	.	-2	4	-2
X.76	-4	4	2	-4	-2	4	-4	-2	-10	4	2	.	.	.	.	.	-4	.	.	4	-2	.	2	.	2	4	-2
X.77	-4	4	-10	-4	2	4	4	10	2	-4	-2	.	.	.	.	.	-4	.	.	4	2	.	-2	.	-2	4	-2
X.78	.	.	4	.	4	.	.	4	4	.	-4	.	.	.	.	.	.	4	.	.	4	-4	.	-4	.	.
X.79	.	.	-4	12	4	12	.	-4	-4	.	4	.	.	.	.	.	.	4	.	.	4	.	.	.	.	.
X.80	.	.	.	.	.	.	.	.	.	.	.	.	.	.	.	.	.	.	.	.	.	.	.	.	.	.
X.81	.	.	.	.	.	.	.	.	.	.	.	.	.	.	.	.	.	4	.	.	.	.	-4	.	.	.
X.82	.	.	.	.	.	.	.	.	.	.	.	.	.	.	.	.	.	4	.	.	.	.	-4	.	.	.
X.83	-16	16	.	.	.	.	.	.	.	.	.	.	.	.	.	.	.	.	.	.	.	.	.	.	.	.
X.84	.	.	.	.	.	.	.	.	.	.	.	.	.	.	.	.	.	4	.	.	.	.	-4	.	.	.

Character table of $D(\mathrm{Fi}_{22})$ *(continued)*

2	8	8	8	8	8	8	8	7	7	7	6	6	2	5	5	5	4	4	4	4	4	4	3	7	7	7
3	.	.	.	.	.	.	.	.	.	.	.	.	.	1	1	1	1	1	1	1	1	1	1	.	.	.
5	.	.	.	.	.	.	.	.	.	.	.	.	1	.												

	4_{30}	4_{31}	4_{32}	4_{33}	4_{34}	4_{35}	4_{36}	4_{37}	4_{38}	4_{39}	4_{40}	4_{41}	$5a$	$6a$	$6b$	$6c$	$6d$	$6e$	$6f$	$6g$	$6h$	$6i$	$6j$	$8a$	$8b$	$8c$
$2P$	$2f$	$2h$	$2j$	$2j$	$2j$	$2i$	$2h$	$2k$	$2h$	$2j$	$2p$	$2q$	$5a$	$3a$	$3a$	$3a$	$3a$	$3a$	$3a$	$3a$	$3a$	$3a$	$3a$	4_4	4_4	4_1
$3P$	4_{30}	4_{31}	4_{32}	4_{33}	4_{34}	4_{35}	4_{36}	4_{37}	4_{38}	4_{39}	4_{40}	4_{41}	$5a$	$2h$	$2i$	$2a$	$2d$	$2g$	$2g$	$2b$	$2c$	$2m$	$2o$	$8a$	$8b$	$8c$
$5P$	4_{30}	4_{31}	4_{32}	4_{33}	4_{34}	4_{35}	4_{36}	4_{37}	4_{38}	4_{39}	4_{40}	4_{41}	$1a$	$6a$	$6b$	$6c$	$6d$	$6f$	$6e$	$6g$	$6h$	$6i$	$6j$	$8a$	$8b$	$8c$
X.1	1	1	1	1	1	1	1	1	1	1	1	1	1	1	1	1	1	1	1	1	1	1	1	1	1	1
X.2	1	1	-1	-1	1	-1	1	1	1	1	-1	-1	1	1	1	1	1	1	1	1	1	1	1	-1	-1	1
X.3	.	.	-2	-2	.	-2	.	.	.	.	.	.	-1	1	1	1	1	1	1	1	1	1	1	-2	.	.
X.4	.	.	2	2	.	2	.	.	.	.	.	.	-1	1	1	1	1	1	1	1	1	1	1	2	.	.
X.5	1	1	1	1	1	1	1	1	1	1	-1	-1	.	-1	-1	-1	-1	-1	-1	-1	-1	-1	-1	1	1	1
X.6	1	-1	-3	1	1	-1	-1	1	-1	1	-1	-1	.	-2	-2	2	.	.	.	2	.	-2	.	-1	-1	1
X.7	1	1	-1	-1	1	-1	1	1	1	1	1	1	.	-1	-1	-1	-1	-1	-1	-1	-1	-1	-1	-1	-1	1
X.8	1	-1	3	-1	1	-1	1	-1	1	1	1	1	.	-2	-2	2	.	.	.	2	.	-2	.	-1	1	1
X.9	-2	-2	.	.	.	-2	.	.	-2	-2	-2	-2	.	1	.	.	.	.	.	.	.	.	.	-2	.	-2
X.10	2	-2	.	.	.	2	.	.	-2	-2	-2	-2	.	2	2	-2	.	.	.	-2	.	2	.	-2	.	2
X.11	2	.	4	.	.	2	-2	.	2	.	.	2	.	1	1	1	-1	-1	-1	1	-1	1	1	.	-2	-2
X.12	-2	.	-2	-2	.	.	.	-2	.	.	-2	.	.	1	1	1	-1	-1	-1	1	-1	1	1	.	.	2
X.13	-2	.	2	-2	-2	.	.	2	.	.	-2	.	.	1	1	1	-1	-1	-1	1	-1	1	-1	.	.	2
X.14	2	.	-4	.	.	2	2	.	2	.	.	2	.	1	1	1	-1	-1	-1	1	-1	1	-1	.	2	-2
X.15	3	3	-1	-1	-1	3	3	-1	3	-1	1	1	.	.	.	.	.	.	.	.	.	.	.	-1	-1	-1
X.16	-1	-1	-1	-1	3	3	-1	-1	-1	-1	-1	-1	.	.	.	.	.	.	.	.	.	.	.	3	-1	3
X.17	3	-3	-1	3	-1	1	-3	-1	3	-1	1	1	.	.	.	.	.	.	.	.	.	.	.	1	-3	-1
X.18	-1	1	-1	3	3	1	1	-1	1	-1	-1	-1	.	.	.	.	.	.	.	.	.	.	.	-3	-3	3
X.19	-1	1	3	-1	-1	1	1	-1	1	-1	-1	-1	.	.	.	.	.	.	.	.	.	.	.	1	1	-1
X.20	-1	-1	1	1	3	-3	-1	-1	-1	-1	1	1	.	.	.	.	.	.	.	.	.	.	.	3	1	3
X.21	3	3	1	1	-1	-3	3	-1	3	-1	-1	-1	.	.	.	.	.	.	.	.	.	.	.	-1	1	-1
X.22	-1	1	1	-3	3	-1	1	-1	1	-1	1	1	.	.	.	.	.	.	.	.	.	.	.	-3	3	3
X.23	-1	1	-3	-1	-1	1	1	-1	1	-1	1	1	.	.	.	.	.	.	.	.	.	.	.	1	1	-1
X.24	3	-3	1	-3	-1	-1	-3	-1	-3	-1	-1	-1	.	.	.	.	.	.	.	.	.	.	.	1	3	-1
X.25	.	.	-2	-2	.	2	.	-2	.	.	.	.	.	-1	-1	-1	1	1	1	-1	1	1	1	.	2	.
X.26	.	.	2	2	.	-2	.	.	.	.	.	.	.	-1	-1	-1	1	1	1	-1	1	-1	-1	.	-2	.
X.27	-2	.	.	-4	-2	-2	.	2	.	.	2	.	.	.	.	.	.	.	.	.	.	.	.	.	-2	2
X.28	-2	.	.	4	-2	2	.	2	.	.	2	.	.	.	.	.	.	.	.	.	.	.	.	.	2	2
X.29	-2	-2	.	.	-2	.	.	-2	2	-2	2	.	.	.	.	.	.	.	.	.	.	.	.	-2	.	-2
X.30	-2	2	.	.	-2	.	.	-2	2	2	2	.	.	.	.	.	.	.	.	.	.	.	.	2	.	-2
X.31	2	.	-2	2	2	-4	.	-2	.	.	-2	.	.	.	.	.	.	.	.	.	.	.	.	4	-2	.
X.32	2	.	2	-2	2	4	.	-2	.	.	-2	.	.	.	.	.	.	.	.	.	.	.	.	-4	-2	.
X.33	.	4	.	.	.	.	-4	.	.	2	.	.	2	2	-2	-2	2	.	.	.	-2	.	.	.	.	.
X.34	.	-2	-2	-2	.	.	-2	.	2	.	.	.	.	1	1	1	-1	-1	-1	1	-1	1	2	.	.	.
X.35	.	2	2	2	.	.	2	.	-2	.	.	.	.	1	1	1	-1	1	1	-1	-1	-1	-2	.	.	.
X.36	.	-2	-2	-2	.	.	-2	.	2	.	.	.	.	1	1	1	-1	1	1	-1	-1	-1	2	.	.	.
X.37	.	2	2	2	.	.	2	.	-2	.	.	.	.	1	1	1	-1	1	1	-1	-1	-1	1	-2	.	.
X.38	.	2	-2	-2	.	.	2	.	-2	.	.	.	.	1	1	1	-1	1	1	-1	-1	-1	1	-2	.	.
X.39	.	-2	2	2	.	.	-2	.	2	.	.	.	.	1	1	1	-1	-1	-1	1	-1	1	-2	.	.	.
X.40	.	2	-2	-2	.	.	2	.	-2	.	.	.	.	1	1	1	-1	-1	-1	1	-1	1	-2	.	.	.
X.41	.	-2	2	2	.	.	-2	.	2	.	.	.	.	1	1	1	-1	1	1	-1	-1	-1	1	2	.	.
X.42	.	.	2	2	.	-2	.	.	.	.	.	.	.	.	.	.	.	.	.	.	.	.	.	.	-2	.
X.43	.	.	-2	-2	.	2	.	.	.	.	.	.	.	.	.	.	.	.	.	.	.	.	.	.	2	.
X.44	4	.	.	.	-4	.	.	.	.	.	.	.	.	.	.	.	.	.	.	.	.	.	.	.	.	4
X.45	.	.	2	2	.	-2	.	.	.	.	.	.	.	.	.	.	.	.	.	.	.	.	.	.	2	.
X.46	.	.	-2	-2	.	-2	.	.	.	.	.	.	.	.	.	.	.	.	.	.	.	.	.	.	-2	.
X.47	-4	.	.	.	4	.	.	.	.	.	.	.	.	.	.	.	.	.	.	.	.	.	.	.	.	-4
X.48	.	.	.	.	.	.	.	.	.	.	.	.	.	-1	-1	-1	1	-1	-1	1	1	1	-1	.	.	.
X.49	.	.	.	.	.	.	.	.	.	.	.	.	.	-1	-1	-1	1	-1	-1	1	1	1	1	.	.	.
X.50	.	.	4	-4	.	.	.	.	.	.	.	.	.	-2	-2	2	.	.	.	-2	.	2	.	.	.	.
X.51	.	.	.	.	.	.	.	.	.	.	.	.	.	-1	-1	-1	1	1	1	-1	1	1	1	.	.	.
X.52	.	.	.	.	.	.	.	.	.	.	.	.	.	-1	-1	-1	1	1	1	-1	1	-1	-1	.	.	.
X.53	.	.	-4	4	.	.	.	.	.	.	.	.	.	-2	-2	2	.	.	.	-2	.	2	.	.	.	.
X.54	.	.	.	.	.	.	.	.	.	.	.	.	4	-4	1	.	.	.	.	.	.	.	.	.	.	.
X.55	.	.	.	.	.	.	.	.	.	.	.	.	-4	4	1	.	.	.	.	.	.	.	.	.	.	.
X.56	.	.	.	.	.	.	.	.	.	.	.	.	.	.	1	.	.	.	.	.	.	.	.	.	.	.
X.57	.	.	.	.	.	.	.	.	.	.	.	.	.	.	1	.	.	.	.	.	.	.	.	.	.	.
X.58	.	-4	.	.	.	.	4	.	.	.	.	.	.	.	1	.	.	.	.	.	.	.	.	.	.	.
X.59	.	-4	.	.	.	.	4	.	.	.	.	.	.	.	1	.	.	.	.	.	.	.	.	.	.	.
X.60	-4	.	.	.	.	.	.	.	.	.	.	.	-2	-2	.	.	.	.	.	.	.	.	.	.	.	.
X.61	.	.	.	.	.	.	.	.	.	.	.	.	.	.	.	.	.	.	.	.	.	.	.	.	.	.
X.62	.	-2	-2	-2	.	4	-2	.	2	.	.	.	.	.	.	.	.	.	.	.	.	.	.	.	-2	.
X.63	.	.	.	.	.	.	.	.	.	.	.	.	.	.	.	.	.	.	.	.	.	.	.	.	.	.
X.64	.	.	.	.	-4	.	.	-4	.	4	.	.	.	.	.	.	.	.	.	.	.	.	.	.	.	.
X.65	.	.	.	.	.	.	.	.	.	.	.	.	.	.	.	.	.	.	.	.	.	.	.	.	.	.
X.66	.	2	2	2	.	4	2	.	-2	.	.	.	.	.	.	.	.	.	.	.	.	.	.	2	.	.
X.67	.	-2	-2	-2	.	4	-2	.	2	.	.	.	.	.	.	.	.	.	.	.	.	.	.	-2	.	.
X.68	.	2	2	2	.	4	2	.	-2	.	.	.	.	.	.	.	.	.	.	.	.	.	.	2	.	.
X.69	-4	.	.	.	.	.	.	.	.	.	.	.	-2	-2	.	.	.	.	.	.	.	.	.	.	.	.
X.70	.	.	.	.	.	.	.	.	.	.	.	.	.	.	.	.	.	.	.	.	.	.	.	.	.	.
X.71	.	.	.	.	-4	.	.	-4	.	4	.	.	.	.	.	.	.	.	.	.	.	.	.	.	.	.
X.72	.	.	.	.	-4	.	.	4	.	-4	.	.	.	.	.	.	.	.	.	.	.	.	.	.	.	.
X.73	-4	.	.	.	.	.	.	.	.	.	.	.	2	2	.	.	.	.	.	.	.	.	.	.	.	.
X.74	.	2	-2	-2	.	-4	2	.	-2	.	.	.	.	.	.	.	.	.	.	.	.	.	.	2	.	.
X.75	.	-2	2	2	.	-4	-2	.	2	.	.	.	.	.	.	.	.	.	.	.	.	.	.	-2	.	.
X.76	.	2	-2	-2	.	-4	2	.	-2	.	.	.	.	.	.	.	.	.	.	.	.	.	.	2	.	.
X.77	.	-2	2	2	.	-4	-2	.	2	.	.	.	.	.	.	.	.	.	.	.	.	.	.	-2	.	.
X.78	.	.	.	.	-4	.	.	4	.	-4	.	.	.	.	.	.	.	.	.	.	.	.	.	.	.	.
X.79	-4	.	.	.	.	.	.	.	.	.	.	.	2	2	.	.	.	.	.	.	.	.	.	.	.	.
X.80	.	.	.	.	.	.	.	.	.	.	.	.	-2	2	-2	-2	2	.	.	.	-2	.	.	.	.	.
X.81	.	4	.	.	.	.	-4	.	.	.	.	.	-2	2	2	2	-2	.	.	.	2	.	.	.	.	.
X.82	.	-4	.	.	.	.	4	.	.	.	.	.	-4	4	-4	.	.	.	.	.	.	.	.	.	.	.
X.83	.	.	.	.	.	.	.	.	.	.	.	.	2	2	-2	.	.	.	2	.	-2	.	.	.	.	.
X.84	.	-4	.	.	.	.	4	.	.	.	.	.	2	-2	2	.	C	$\bar{C}$	.	.	.	.	.	.	.	.

Character table of $D(\mathrm{Fi}_{22})$ (continued)

	8d	8e	8f	8g	8h	8i	8j	8k	8l	8m	8n	8o	8p	10a	10b	10c	12a	12b	12c	16a	16b
2	7	7	7	7	6	6	6	6	5	5	5	5	5	2	2	2	3	3	3	5	5
3	.	.	.	.	.	.	.	.	.	.	.	.	.	.	.	.	1	1	1	.	.
5	.	.	.	.	.	.	.	.	.	.	.	.	.	1	1	1	.	.	.	.	.
2P	4_2	4_4	4_4	4_2	4_1	4_5	4_7	4_7	4_{17}	4_{18}	4_{19}	4_{21}	4_{22}	$5a$	$5a$	$5a$	$6a$	$6b$	$6c$	$8c$	$8c$
3P	$8d$	$8e$	$8f$	$8g$	$8h$	$8i$	$8k$	$8j$	$8l$	$8m$	$8n$	$8o$	$8p$	$10a$	$10c$	$10b$	4_{15}	4_{16}	4_3	$16a$	$16b$
5P	$8d$	$8e$	$8f$	$8g$	$8h$	$8i$	$8j$	$8k$	$8l$	$8m$	$8n$	$8o$	$8p$	$2a$	$2b$	$2b$	$12a$	$12b$	$12c$	$16b$	$16a$
X.1	1	1	1	1	1	1	1	1	1	1	1	1	1	1	1	1	1	1	1	1	1
X.2	-1	1	-1	-1	1	-1	-1	-1	-1	-1	-1	1	-1	1	1	1	-1	-1	-1	-1	-1
X.3	-2	.	-2	-2	.	-2	.	.	.	.	.	.	.	-1	-1	-1	1	1	1	.	.
X.4	2	.	2	2	.	2	.	.	.	.	.	.	.	-1	-1	-1	-1	-1	-1	.	.
X.5	1	1	1	1	1	1	-1	-1	-1	-1	-1	1	-1	.	.	.	1	1	1	-1	-1
X.6	1	-1	3	1	1	-1	-1	-1	1	1	-1	-1	-1	.	.	.	.	.	.	1	1
X.7	-1	1	-1	-1	1	1	1	1	1	1	1	1	1	.	.	.	-1	-1	-1	1	1
X.8	-1	-1	-3	-1	1	1	1	1	-1	-1	1	-1	1	.	.	.	.	.	.	-1	-1
X.9	.	-2	.	.	-2	.	.	.	.	.	.	-2	.	1	1	1	.	.	.	.	.
X.10	.	-2	.	.	2	.	.	.	.	.	.	-2	.	.	.	.	.	.	.	.	.
X.11	.	.	2	.	-2	-2	.	.	.	.	.	.	.	.	.	.	-1	-1	1	.	.
X.12	2	.	-4	2	2	.	.	.	.	.	.	.	.	.	.	.	-1	-1	1	.	.
X.13	-2	.	4	-2	2	.	.	.	.	.	.	.	.	.	.	.	1	1	-1	.	.
X.14	.	.	-2	.	-2	2	2	.	.	.	.	.	.	.	.	.	1	1	-1	.	.
X.15	-1	-1	-1	-1	-1	-1	1	1	1	1	-1	-1	-1	.	.	.	.	.	.	-1	-1
X.16	-1	3	-1	-1	-1	-1	-1	-1	1	1	-1	1	-1	.	.	.	.	.	.	1	1
X.17	-1	1	1	-1	-1	1	1	1	-1	-1	-1	1	-1	.	.	.	.	.	.	1	1
X.18	-1	-3	1	-1	-1	1	-1	-1	1	1	1	1	1	.	.	.	.	.	.	-1	-1
X.19	-1	1	-3	-1	-1	1	-1	-1	1	1	-1	1	-1	.	.	.	.	.	.	1	1
X.20	1	3	1	1	-1	1	-1	-1	1	1	-1	-1	-1	.	.	.	.	.	.	-1	-1
X.21	1	-1	1	1	-1	1	-1	-1	-1	-1	-1	1	-1	.	.	.	.	.	.	1	1
X.22	1	-3	-1	1	-1	1	1	1	-1	-1	-1	1	-1	.	.	.	.	.	.	1	1
X.23	1	1	3	1	-1	1	-1	-1	1	1	-1	1	1	.	.	.	.	.	.	-1	-1
X.24	1	1	-1	1	-1	1	-1	-1	-1	1	1	1	1	.	.	.	.	.	.	-1	-1
X.25	-2	.	2	-2	.	2	.	.	.	.	.	.	.	.	.	.	-1	-1	1	.	.
X.26	2	.	-2	2	.	-2	.	.	.	.	.	.	.	.	.	.	1	1	-1	.	.
X.27	.	.	2	.	-2	2	.	.	.	.	.	.	.	.	.	.	.	.	.	.	.
X.28	.	.	-2	.	-2	-2	.	.	.	.	.	.	.	.	.	.	.	.	.	.	.
X.29	.	-2	.	.	2	.	.	.	.	.	2	.	.	.	.	.	.	.	.	.	.
X.30	.	2	.	.	2	.	.	.	.	.	-2	.	.	.	.	.	.	.	.	.	.
X.31	-2	.	.	-2	2	2	.	.	.	.	.	.	.	.	.	.	.	.	.	.	.
X.32	2	.	.	2	2	2	.	.	.	.	.	.	.	.	.	.	.	.	.	.	.
X.33	.	.	.	.	.	.	.	.	.	.	.	.	-2	.	.	.	.	.	.	.	.
X.34	-2	-2	.	2	.	.	.	.	.	.	.	.	.	.	.	.	1	-1	-1	.	.
X.35	-2	2	.	2	.	.	.	.	.	.	.	.	.	.	.	.	1	-1	1	.	.
X.36	2	-2	.	-2	.	.	.	.	.	.	.	.	.	.	.	.	1	-1	1	.	.
X.37	2	2	.	-2	.	.	.	.	.	.	.	.	.	.	.	.	1	-1	-1	.	.
X.38	2	2	.	-2	.	.	.	.	.	.	.	.	.	.	.	.	-1	1	1	.	.
X.39	2	-2	.	-2	.	.	.	.	.	.	.	.	.	.	.	.	-1	1	1	.	.
X.40	-2	2	.	2	.	.	.	.	.	.	.	.	.	.	.	.	-1	1	-1	.	.
X.41	-2	-2	.	2	.	.	.	.	.	.	.	.	.	.	.	.	-1	1	-1	.	.
X.42	-2	.	-2	-2	2	.	.	.	.	.	.	.	.	.	.	.	.	.	.	.	.
X.43	2	.	2	2	-2	.	.	.	.	.	.	.	.	.	.	.	.	.	.	.	.
X.44	.	.	.	.	.	.	.	.	.	.	.	.	.	.	.	.	.	.	.	.	.
X.45	-2	.	2	-2	-2	.	.	.	.	.	.	.	.	.	.	.	.	.	.	.	.
X.46	2	.	-2	2	2	.	.	.	.	.	.	.	.	.	.	.	.	.	.	.	.
X.47	.	.	.	.	.	.	.	.	.	.	.	.	.	.	.	.	.	.	.	.	.
X.48	.	.	.	.	.	.	.	.	.	.	.	.	.	.	.	.	1	-1	1	.	.
X.49	.	.	.	.	.	.	.	.	.	.	.	.	.	.	.	.	-1	1	-1	.	.
X.50	.	.	.	.	.	.	.	.	.	.	.	.	.	.	.	.	.	.	.	.	.
X.51	.	.	.	.	.	.	.	.	.	.	.	.	.	.	.	.	1	-1	-1	.	.
X.52	.	.	.	.	.	.	.	.	.	.	.	.	.	.	.	.	-1	1	1	.	.
X.53	.	.	.	.	.	.	.	.	.	.	.	.	.	.	.	.	.	.	.	.	.
X.54	.	.	.	.	.	.	.	.	.	.	.	.	.	1	-1	-1	.	.	.	.	.
X.55	.	.	.	.	.	.	.	.	.	.	.	.	.	1	-1	-1	.	.	.	.	.
X.56	.	.	.	.	A	$\bar{A}$	.	.	.	.	.	.	.	1	-1	-1	.	.	.	.	.
X.57	.	.	.	.	$\bar{A}$	A	.	.	.	.	.	.	.	1	-1	-1	.	.	.	.	.
X.58	.	.	.	.	.	.	.	.	.	.	.	.	.	-1	B	$-B$	.	.	.	.	.
X.59	.	.	.	.	.	.	.	.	.	.	.	.	.	-1	$-B$	B	.	.	.	.	.
X.60	.	.	.	.	2	2	.	.	.	.	.	.	.	.	.	.	.	.	.	.	.
X.61	.	.	.	.	.	.	.	.	2	-2	.	.	.	.	.	.	.	.	.	.	.
X.62	2	2	.	-2	.	.	.	.	.	.	.	.	.	.	.	.	.	.	.	.	.
X.63	.	.	.	.	.	.	.	.	-2	2	.	.	.	.	.	.	.	.	.	.	.
X.64	.	.	.	.	.	.	.	.	.	.	-2	2	.	.	.	.	.	.	.	.	.
X.65	.	.	.	.	.	.	.	.	-2	2	.	.	.	.	.	.	.	.	.	.	.
X.66	2	-2	.	-2	.	.	.	.	.	.	.	.	.	.	.	.	.	.	.	.	.
X.67	-2	2	.	2	.	.	.	.	.	.	.	.	.	.	.	.	.	.	.	.	.
X.68	-2	-2	.	2	.	.	.	.	.	.	.	.	.	.	.	.	.	.	.	.	.
X.69	.	.	.	.	2	2	.	.	.	.	.	.	.	.	.	.	.	.	.	.	.
X.70	.	.	.	.	.	.	.	.	2	-2	.	.	.	.	.	.	.	.	.	.	.
X.71	.	.	.	.	.	.	.	.	.	.	2	-2	.	.	.	.	.	.	.	.	.
X.72	.	.	.	.	.	.	.	.	.	.	-2	2	.	.	.	.	.	.	.	.	.
X.73	.	.	.	.	-2	-2	.	.	.	.	.	.	.	.	.	.	.	.	.	.	.
X.74	2	-2	.	-2	.	.	.	.	.	.	.	.	.	.	.	.	.	.	.	.	.
X.75	2	2	.	-2	.	.	.	.	.	.	.	.	.	.	.	.	.	.	.	.	.
X.76	-2	-2	.	2	.	.	.	.	.	.	.	.	.	.	.	.	.	.	.	.	.
X.77	-2	2	.	2	.	.	.	.	.	.	.	.	.	.	.	.	.	.	.	.	.
X.78	.	.	.	.	.	.	.	.	.	.	2	.	-2	.	.	.	.	.	.	.	.
X.79	.	.	.	.	-2	-2	.	.	.	.	.	.	.	.	.	.	.	.	.	.	.
X.80	.	.	.	.	.	.	.	.	.	.	.	.	.	2	.	.	.	.	.	.	.
X.81	.	.	.	.	.	.	.	.	.	.	.	.	.	.	.	.	.	.	.	.	.
X.82	.	.	.	.	.	.	.	.	.	.	.	.	.	.	.	.	.	.	.	.	.
X.83	.	.	.	.	.	.	.	.	.	.	.	.	.	.	.	.	.	.	.	.	.
X.84	.	.	.	.	.	.	.	.	.	.	.	.	.	.	.	.	.	.	.	.	.

Character table of $D(\mathrm{Fi}_{22})$ (continued)

2	17	17	16	15	15	16	16	14	12	12	13	13	13	11	12	10	11	11	10	10	5	12	12	10
3	1	1	1	1	1	.	.	1	1	1	.	.	.	1	.	1	.	.	.	.	.	1	.	.	1
5	1	1	1	.	.	.	.	.	.	.	.	.	.	.	.	.	.	.	.	.	.	.	.	.
	$1a$	$2a$	$2b$	$2c$	$2d$	$2e$	$2f$	$2g$	$2h$	$2i$	$2j$	$2k$	$2l$	$2m$	$2n$	$2o$	$2p$	$2q$	$2r$	$2s$	$3a$	4_1	4_2	4_3
2P	$1a$	$1a$	$1a$	$1a$	$1a$	$1a$	$1a$	$1a$	$1a$	$1a$	$1a$	$1a$	$1a$	$1a$	$1a$	$1a$	$1a$	$1a$	$1a$	$1a$	$3a$	$2a$	$2a$	$2a$
3P	$1a$	$2a$	$2b$	$2c$	$2d$	$2e$	$2f$	$2g$	$2h$	$2i$	$2j$	$2k$	$2l$	$2m$	$2n$	$2o$	$2p$	$2q$	$2r$	$2s$	$1a$	4_1	4_2	4_3
5P	$1a$	$2a$	$2b$	$2c$	$2d$	$2e$	$2f$	$2g$	$2h$	$2i$	$2j$	$2k$	$2l$	$2m$	$2n$	$2o$	$2p$	$2q$	$2r$	$2s$	$3a$	4_1	4_2	4_3
X.85	160	−160	.	−48	48	.	.	8	−8	.	.	.	.	.	.	8	−8	.	−2	.	.	.	.	.
X.86	240	240	240	48	48	−16	−16	48	.	16	16	−16	.	16	.	−8	−8	.	−8	.	.	−16	−16	.
X.87	240	240	240	48	48	−16	−16	48	.	−16	−16	−16	.	−16	.	.	.	.	.	.	.	16	16	.
X.88	240	240	−240	.	.	48	−48	.	−24	−24	−16	16	.	24	.	−12	.	.	4	.	.	.	.	12
X.89	240	240	240	−48	−48	−16	−16	−48	.	16	16	16	.	−16	.	.	.	.	.	.	.	16	−16	.
X.90	240	240	−240	.	.	48	−48	.	−24	−24	−16	16	.	24	.	12	.	.	−4	.	.	.	.	−12
X.91	240	240	240	−48	−48	−16	−16	−48	.	−16	−16	16	.	16	.	.	.	.	.	.	.	−16	16	.
X.92	240	240	240	−48	−48	−16	−16	−48	.	−16	−16	16	.	16	.	.	.	.	.	.	.	−16	16	.
X.93	320	−320	.	32	−32	.	.	.	−16	16	.	.	.	.	.	16	−16	.	2	.	.	.	.	.
X.94	320	−320	.	32	−32	.	.	.	−16	16	.	.	.	.	.	−16	16	.	2	.	.	.	.	.
X.95	384	384	−384	.	.	−128	128	.	.	.	.	.	.	.	.	.	.	.	.	.	.	.	.	.
X.96	480	−480	.	−144	144	.	.	.	24	−24	.	.	.	.	.	.	−8	8	.	.	.	.	.	.
X.97	640	−640	.	64	−64	.	.	.	−32	32	.	.	.	.	.	.	.	.	−2	.	.	.	.	.

Character table of $D(\mathrm{Fi}_{22})$ (continued)

2	11	11	10	10	10	10	10	10	10	10	10	8	8	9	9	9	9	9	9	9	9	9	9	8	8	8
3	.	.	.	.	.	.	.	.	.	.	.	1	1	.	.	.	.	.	.	.	.	.	.	.	.	.
5																										
	4_4	4_5	4_6	4_7	4_8	4_9	4_{10}	4_{11}	4_{12}	4_{13}	4_{14}	4_{15}	4_{16}	4_{17}	4_{18}	4_{19}	4_{20}	4_{21}	4_{22}	4_{23}	4_{24}	4_{25}	4_{26}	4_{27}	4_{28}	4_{29}
2P	$2e$	$2e$	$2e$	$2e$	$2e$	$2e$	$2e$	$2e$	$2a$	$2e$	$2e$	$2h$	$2i$	$2c$	$2c$	$2j$	$2g$	$2f$	$2j$	$2e$	$2e$	$2e$	$2g$	$2j$	$2h$	$2j$
3P	4_4	4_5	4_6	4_7	4_8	4_9	4_{10}	4_{11}	4_{12}	4_{13}	4_{14}	4_{15}	4_{16}	4_{17}	4_{18}	4_{19}	4_{20}	4_{21}	4_{22}	4_{23}	4_{24}	4_{25}	4_{26}	4_{27}	4_{28}	4_{29}
5P	4_4	4_5	4_6	4_7	4_8	4_9	4_{10}	4_{11}	4_{12}	4_{13}	4_{14}	4_{15}	4_{16}	4_{17}	4_{18}	4_{19}	4_{20}	4_{21}	4_{22}	4_{23}	4_{24}	4_{25}	4_{26}	4_{27}	4_{28}	4_{29}
X.85	.	.	.	.	.	.	.	.	.	.	.	.	.	.	.	.	4	.	.	.	.	−4	.	.	.	.
X.86	.	.	−8	.	−8	.	.	.	.	.	.	.	.	.	.	.	.	.	.	.	.	8	.	.	.	.
X.87	.	.	.	.	.	.	.	.	.	.	.	.	−8	−8	.	8	.	.	.	.	.	8	.	.	.	.
X.88	8	−8	4	.	4	.	.	−4	−4	.	−4	.	.	.	.	.	.	4	.	−4	.	4	.	−4	.	.
X.89	.	.	.	.	.	.	.	.	.	.	.	.	.	−8	.	.	−8	.	.	.	.	.	.	.	.	.
X.90	8	−8	−4	.	−4	.	.	4	4	.	4	.	.	.	.	.	.	−4	.	4	.	−4	.	4	.	.
X.91	.	.	.	.	.	.	.	.	.	.	.	.	.	.	.	.	.	.	.	.	.	.	.	.	.	.
X.92	.	.	.	.	.	.	.	.	.	.	.	.	.	.	.	.	.	.	.	.	.	.	.	.	.	.
X.93	.	.	.	.	.	.	.	.	.	.	.	.	.	.	.	.	−8	.	.	.	.	8	.	.	.	.
X.94	.	.	.	.	.	.	.	.	.	.	.	.	.	.	.	.	8	.	.	.	.	−8	.	.	.	.
X.95	.	.	.	.	.	.	.	.	.	.	.	.	.	.	.	.	.	.	.	.	.	.	.	.	.	.
X.96	.	.	.	.	.	.	.	.	.	.	.	.	.	.	.	.	−4	.	.	.	.	4	.	.	.	.
X.97	.	.	.	.	.	.	.	.	.	.	.	.	.	.	.	.	.	.	.	.	.	.	.	.	.	.

Character table of $D(\mathrm{Fi}_{22})$ (continued)

2	8	8	8	8	8	8	8	7	7	7	6	6	2	5	5	5	4	4	4	4	4	3	7	7	7
3	.	.	.	.	.	.	.	.	.	.	.	.	1	.	.	.	.	.	.	.	.	.	.	.	.
5													1	1	1	1	1	1	1	1	1				
	4_{30}	4_{31}	4_{32}	4_{33}	4_{34}	4_{35}	4_{36}	4_{37}	4_{38}	4_{39}	4_{40}	4_{41}	$5a$	$6a$	$6b$	$6c$	$6d$	$6e$	$6f$	$6g$	$6h$	$6i$	$6j$	$8a$	$8b$	$8c$
2P	$2f$	$2h$	$2j$	$2j$	$2j$	$2i$	$2h$	$2k$	$2h$	$2j$	$2p$	$2q$	$5a$	$3a$	$3a$	$3a$	$3a$	$3a$	$3a$	$3a$	$3a$	$3a$	$3a$	4_4	4_4	4_1
3P	4_{30}	4_{31}	4_{32}	4_{33}	4_{34}	4_{35}	4_{36}	4_{37}	4_{38}	4_{39}	4_{40}	4_{41}	$5a$	$2h$	$2i$	$2a$	$2d$	$2g$	$2g$	$2b$	$2c$	$2m$	$2o$	$8a$	$8b$	$8c$
5P	4_{30}	4_{31}	4_{32}	4_{33}	4_{34}	4_{35}	4_{36}	4_{37}	4_{38}	4_{39}	4_{40}	4_{41}	$1a$	$6a$	$6b$	$6c$	$6d$	$6f$	$6e$	$6g$	$6h$	$6i$	$6j$	$8a$	$8b$	$8c$
X.85	.	−4	.	.	.	.	4	.	.	.	.	.	.	2	−2	2	.	C	C	.	.	.	.	.	.	.
X.86	8	.	.	.	.	.	.	.	.	.	.	.	.	.	.	.	.	.	.	.	.	.	.	.	.	.
X.87	.	.	.	.	.	.	.	.	.	.	.	.	.	.	.	.	.	.	.	.	.	.	.	.	.	.
X.88	.	.	4	−4	.	.	.	.	.	.	.	.	.	.	.	.	.	.	.	.	.	.	.	.	.	.
X.89	.	.	.	.	.	8	.	.	.	.	.	.	.	.	.	.	.	.	.	.	.	.	.	.	.	.
X.90	.	.	−4	4	.	.	.	.	.	.	.	.	.	.	.	.	.	.	.	.	.	.	.	.	.	.
X.91	.	.	.	.	.	.	.	.	.	.	.	.	.	.	.	.	.	.	.	.	.	.	.	.	.	.
X.92	.	.	.	.	.	.	.	.	.	.	.	.	.	.	.	.	.	.	.	.	.	.	.	.	.	.
X.93	.	.	.	.	.	.	.	.	.	.	.	.	.	2	−2	−2	−2	.	.	.	.	2	.	.	.	.
X.94	.	.	.	.	.	.	.	.	.	.	.	.	.	2	−2	−2	−2	.	.	.	.	2	.	.	.	.
X.95	.	.	.	.	.	.	.	.	.	.	.	.	−1	.	.	.	.	.	.	.	.	.	.	.	.	.
X.96	.	4	.	.	.	.	.	−4	.	.	.	.	.	.	.	.	.	.	.	.	.	.	.	.	.	.
X.97	.	.	.	.	.	.	.	.	.	.	.	.	.	−2	2	2	2	.	.	.	.	−2	.	.	.	.

Character table of $D(\mathrm{Fi}_{22})$ (continued)

2	7	7	7	7	6	6	6	6	5	5	5	5	5	2	2	2	3	3	3	5	5
3	.	.	.	.	.	.	.	.	.	.	.	.	.	1	1	1	1	1	1	.	.
5																					
	$8d$	$8e$	$8f$	$8g$	$8h$	$8i$	$8j$	$8k$	$8l$	$8m$	$8n$	$8o$	$8p$	$10a$	$10b$	$10c$	$12a$	$12b$	$12c$	$16a$	$16b$
2P	4_2	4_4	4_4	4_2	4_1	4_5	4_7	4_7	4_{17}	4_{18}	4_{19}	4_{21}	4_{22}	$5a$	$5a$	$6a$	$6b$	$6c$	$8c$	$8c$
3P	$8d$	$8e$	$8f$	$8g$	$8h$	$8i$	$8j$	$8k$	$8j$	$8m$	$8n$	$8o$	$8p$	$10a$	$10c$	$10b$	4_{15}	4_{16}	4_3	$16a$	$16b$
5P	$8d$	$8e$	$8f$	$8g$	$8h$	$8i$	$8j$	$8k$	$8l$	$8m$	$8n$	$8o$	$8p$	$2a$	$2b$	$2b$	$12a$	$12b$	$12c$	$16b$	$16a$
X.85	.	.	.	.	.	.	.	.	.	.	.	.	.	.	.	.	.	.	.	.	.
X.86	.	.	.	.	.	.	.	.	.	.	.	.	.	.	.	.	.	.	.	.	.
X.87	.	.	.	.	.	.	.	.	.	.	.	.	.	.	.	.	.	.	.	.	.
X.88	.	.	.	.	.	.	.	.	.	.	.	.	.	.	.	.	.	.	.	.	.
X.89	.	.	.	.	.	.	.	.	.	.	.	.	.	.	.	.	.	.	.	.	.
X.90	.	.	.	.	.	.	.	.	.	.	.	.	.	.	.	.	.	.	.	.	.
X.91	.	.	.	.	.	.	.	.	.	.	.	.	.	.	.	.	.	.	.	D	$\bar{D}$
X.92	.	.	.	.	.	.	.	.	.	.	.	.	.	.	.	.	.	.	.	$\bar{D}$	D
X.93	.	.	.	.	.	.	.	.	.	.	.	.	.	.	.	.	.	.	.	.	.
X.94	.	.	.	.	.	.	.	.	.	.	.	.	.	.	.	.	.	.	.	.	.
X.95	.	.	.	.	.	.	.	.	.	.	.	.	−1	1	1	.	.	.	.	.	.
X.96	.	.	.	.	.	.	.	.	.	.	.	.	.	.	.	.	.	.	.	.	.
X.97	.	.	.	.	.	.	.	.	.	.	.	.	.	.	.	.	.	.	.	.	.

where $A = 4i$, $B = i\sqrt{5}$, $C = -2i\sqrt{3}$, and $D = -2(\zeta_8^3 + \zeta_8)$.

B.6. *Character table of* $E_2 = E(Fi_{22}) = \langle x, y, e \rangle$

2	17	16	17	16	13	13	6	12	12	10	9	10	9	9	8	7	7	2	6	4	4	5	4	1	1	7
3	2	2	1	2	1	.	2	.	.	1	1	.	.	.	.	.	.	.	2	2	2	1	1	.	.	.
5	1	1	1	.	.	.	.	.	.	.	.	.	.	.	.	.	.	1	.	.	.	.	.	.	.	.
7	1	1	.	.	.	.	.	.	.	.	.	.	.	.	.	.	.	.	.	.	.	.	.	1	1	.
11	1	.	.	.	.	.	.	.	.	.	.	.	.	.	.	.	.	.	.	.	.	.	.	.	.	.

	1a	2a	2b	2c	2d	2e	3a	4a	4b	4c	4d	4e	4f	4g	4h	4i	4j	5a	6a	6b	6c	6d	6e	7a	7b	8a
2P	1a	1a	1a	1a	1a	1a	3a	2b	2b	2b	2c	2b	2d	2d	2d	2d	2e	5a	3a	3a	3a	3a	3a	7a	7b	4a
3P	1a	2a	2b	2c	2d	2e	1a	4a	4b	4c	4d	4e	4f	4g	4h	4i	4j	5a	2c	2a	2c	2b	2d	7b	7a	8a
5P	1a	2a	2b	2c	2d	2e	3a	4a	4b	4c	4d	4e	4f	4g	4h	4i	4j	1a	6a	6b	6c	6d	6e	7b	7a	8a
7P	1a	2a	2b	2c	2d	2e	3a	4a	4b	4c	4d	4e	4f	4g	4h	4i	4j	5a	6a	6b	6c	6d	6e	1a	1a	8a
11P	1a	2a	2b	2c	2d	2e	3a	4a	4b	4c	4d	4e	4f	4g	4h	4i	4j	5a	6a	6b	6c	6d	6e	7a	7b	8a
X.1	1	1	1	1	1	1	1	1	1	1	1	1	1	1	1	1	1	1	1	1	1	1	1	1	1	1
X.2	21	21	21	21	5	5	3	5	5	5	5	1	1	1	1	1	1	1	3	3	3	3	−1	.	.	1
X.3	45	45	45	45	−3	−3	.	−3	−3	−3	−3	1	1	1	1	1	1	.	.	.	.	.	.	A	Ā	1
X.4	45	45	45	45	−3	−3	.	−3	−3	−3	−3	1	1	1	1	1	1	.	.	.	.	.	.	Ā	A	1
X.5	55	55	55	55	7	7	1	7	7	7	7	7	3	3	3	−1	−1	.	1	1	1	1	1	−1	−1	3
X.6	77	−35	13	−3	13	−3	5	−3	5	−7	1	1	−3	5	1	1	1	2	−3	1	−3	1	1	.	.	1
X.7	99	99	99	99	3	3	.	3	3	3	3	3	3	3	3	−1	−1	−1	.	.	.	.	.	1	1	3
X.8	154	154	154	154	10	10	1	10	10	10	10	10	−2	−2	−2	2	2	−1	1	1	1	1	1	.	.	−2
X.9	210	210	210	210	2	2	3	2	2	2	2	2	−2	−2	−2	−2	−2	.	3	3	3	3	−1	.	.	−2
X.10	231	231	231	231	7	7	−3	7	7	7	7	7	−1	−1	−1	−1	−1	1	−3	−3	−3	−3	1	.	.	−1
X.11	280	280	280	280	−8	−8	1	−8	−8	−8	−8	−8	.	.	.	.	.	.	1	1	1	1	1	.	.	.
X.12	280	280	280	280	−8	−8	1	−8	−8	−8	−8	−8	.	.	.	.	.	.	1	1	1	1	1	.	.	.
X.13	330	90	10	−6	26	10	6	−6	2	6	−2	−2	−2	6	2	2	2	.	6	.	−2	2	1	1	1	−2
X.14	385	385	385	385	1	1	−2	1	1	1	1	1	1	1	1	1	1	.	−2	−2	−2	−2	−2	.	.	1
X.15	385	−175	65	−15	17	1	−2	1	9	−11	5	−3	5	−3	1	1	1	.	6	−4	.	2	2	.	.	1
X.16	385	−175	65	−15	17	1	7	1	9	−11	5	−3	5	−3	1	1	1	.	−9	5	−3	−1	−1	.	.	1
X.17	616	−56	−24	8	24	−8	4	8	−8	.	.	.	4	4	−4	4	−4	1	−4	−2	2	.	.	.	.	.
X.18	616	−56	−24	8	24	−8	4	8	−8	.	.	.	4	4	−4	−4	4	1	−4	−2	2	.	.	.	.	.
X.19	616	−280	104	−24	8	8	−5	8	8	−8	8	−8	.	.	.	.	.	1	3	−1	3	−1	−1	.	.	.
X.20	616	−280	104	−24	8	8	−5	8	8	−8	8	−8	.	.	.	.	.	1	3	−1	3	−1	−1	.	.	.
X.21	693	−315	117	−27	21	5	.	5	13	−15	9	−7	−3	5	1	1	1	−2	.	.	.	.	.	.	.	1
X.22	770	−350	130	−30	−14	18	5	18	2	2	10	−14	−2	−2	−2	−2	−2	.	−3	1	−3	1	1	.	.	−2
X.23	990	270	30	−18	−18	−2	.	14	6	−6	−6	2	−6	2	−2	2	2	.	.	.	.	.	.	A	Ā	2
X.24	990	270	30	−18	−18	−2	.	14	6	−6	−6	2	−6	2	−2	2	2	.	.	.	.	.	.	A	Ā	2
X.25	1155	−525	195	−45	35	−13	3	−13	11	−17	−1	7	−5	3	−1	−1	−1	.	3	−3	−3	3	−1	.	.	−1
X.26	1155	−525	195	−45	−13	3	3	−5	7	−1	−1	7	−1	3	−1	−1	.	.	3	−3	−3	3	−1	.	.	3
X.27	1980	540	60	−36	60	28	.	−4	12	12	−12	−4	4	4	4	.	.	.	.	.	.	.	.	−1	−1	−4
X.28	2310	−1050	390	−90	22	−10	−3	−10	6	−10	−2	6	2	2	2	−2	−2	.	−3	3	3	−3	1	.	.	2
X.29	2310	630	70	−42	−10	6	6	22	14	−6	−14	2	2	−6	−2	−2	−2	.	6	.	.	−2	2	.	.	2
X.30	2310	630	70	−42	22	38	6	−10	−18	−6	2	2	2	−6	−2	2	2	.	6	.	.	−2	−2	.	.	2
X.31	2310	630	70	−42	−22	−26	6	−10	14	18	2	−6	−6	2	−2	−2	−2	.	6	.	.	−2	−2	.	.	2
X.32	2640	720	80	−48	16	16	−6	16	16	.	−16	.	.	.	.	.	.	.	−6	.	.	2	−2	1	1	.
X.33	3465	−1575	585	−135	−39	9	.	9	−15	21	−3	−3	−3	5	1	1	1	.	.	.	.	.	.	.	.	1
X.34	3465	−1575	585	−135	9	−7	.	−7	1	−3	−3	5	1	−7	−3	1	1	.	.	.	.	.	.	.	.	−3
X.35	4620	1260	140	−84	44	12	−6	−20	−4	12	4	−4	−4	−4	−4	.	.	.	−6	.	.	2	2	.	.	4
X.36	5544	−504	−216	72	24	−8	.	8	−8	.	.	.	4	4	−4	4	−4	−1	.	.	.	.	.	.	.	.
X.37	5544	−504	−216	72	24	−8	.	8	−8	.	.	.	4	4	−4	−4	4	−1	.	.	.	.	.	.	.	.
X.38	6160	−560	−240	80	48	−16	4	16	−16	.	.	.	−8	−8	8	.	.	.	−4	−2	2	.	.	.	.	.
X.39	6160	−560	−240	80	−48	16	4	−16	16	.	.	.	.	.	.	.	.	.	−4	−2	2	.	.	.	.	.
X.40	6160	−560	−240	80	−48	16	4	−16	16	.	.	.	.	.	.	.	.	.	−4	−2	2	.	.	.	.	.
X.41	6930	1890	210	−126	−30	18	.	2	−22	−18	6	6	−2	6	2	−2	−2	.	.	.	.	.	.	.	.	−2
X.42	6930	1890	210	−126	−30	−46	.	2	10	6	6	−2	6	−2	2	2	2	.	.	.	.	.	.	.	.	−2
X.43	9856	−896	−384	128	.	.	−8	.	.	.	.	.	.	.	.	.	.	.	1	8	4	−4	.	.	.	.

Character table of $E_2 = E(\mathrm{Fi}_{22})$ (continued)

	8b	8c	8d	8e	8f	8g	10a	10b	10c	11a	11b	12a	12b	14a	14b	16a	16b
2	7	7	6	5	5	5	2	2	2	.	.	4	3	1	1	5	5
3	.	.	.	.	.	.	.	.	.	.	.	1	1	.	.	.	.
5	.	.	.	.	.	.	1	1	1	.	.	.	.	.	.	.	.
7	.	.	.	.	.	.	.	.	.	.	.	.	.	1	1	.	.
11	.	.	.	.	.	.	.	.	.	1	1	.	.	.	.	.	.
2P	4b	4b	4a	4d	4f	4g	5a	5a	5a	11b	11a	6a	6d	7a	7b	8a	8a
3P	8b	8c	8d	8e	8f	8g	10b	10a	10c	11a	11b	4d	4c	14b	14a	16a	16b
5P	8b	8c	8d	8e	8f	8g	2a	2a	2b	11a	11b	12a	12b	14b	14a	16b	16a
7P	8b	8c	8d	8e	8f	8g	10b	10a	10c	11b	11a	12a	12b	2a	2a	16b	16a
11P	8b	8c	8d	8e	8f	8g	10a	10b	10c	1a	1a	12a	12b	14a	14b	16a	16b
X.1	1	1	1	1	1	1	1	1	1	1	1	1	1	1	1	1	1
X.2	1	1	1	1	-1	-1	1	1	1	-1	-1	-1	-1	.	.	-1	-1
X.3	1	1	1	1	-1	-1	.	.	.	1	1	.	.	A	$\bar{A}$	-1	-1
X.4	1	1	1	1	-1	-1	.	.	.	1	1	.	.	$\bar{A}$	A	-1	-1
X.5	3	3	-1	-1	1	1	.	.	.	.	.	1	1	-1	-1	1	1
X.6	1	-3	1	-1	1	1	.	.	-2	.	.	1	-1	.	.	-1	-1
X.7	3	3	-1	-1	-1	-1	-1	-1	-1	.	.	1	1	-1	-1	-1	-1
X.8	-2	-2	2	2	.	.	-1	-1	-1	.	.	1	1	.	.	.	.
X.9	-2	-2	-2	-2	.	.	.	.	.	1	1	-1	-1	.	.	.	.
X.10	-1	-1	-1	-1	-1	-1	1	1	1	.	.	1	1	.	.	-1	-1
X.11	.	.	.	.	.	.	.	.	.	B	$\bar{B}$	1	1	.	.	.	.
X.12	.	.	.	.	.	.	.	.	.	$\bar{B}$	B	1	1	.	.	.	.
X.13	-2	2	-2	.	.	.	.	.	.	.	.	-2	.	-1	-1	.	.
X.14	1	1	1	1	1	1	.	.	.	.	.	-2	-2	.	.	1	1
X.15	-3	1	1	-1	-1	-1	.	.	.	.	.	2	-2	.	.	1	1
X.16	-3	1	1	-1	-1	-1	.	.	.	.	.	-1	1	.	.	1	1
X.17	.	.	.	.	-2	2	-1	-1	1	.	.	.	.	.	.	.	.
X.18	.	.	.	.	2	-2	-1	-1	1	.	.	.	.	.	.	.	.
X.19	.	.	.	.	.	.	C	$-C$	-1	.	.	-1	1	.	.	.	.
X.20	.	.	.	.	.	.	$-C$	C	-1	.	.	-1	1	.	.	.	.
X.21	1	-3	1	-1	1	1	.	.	2	.	.	.	.	.	.	-1	-1
X.22	2	2	-2	2	.	.	.	.	.	.	.	1	-1	.	.	.	.
X.23	-2	2	-2	.	.	.	.	.	.	.	.	.	.	$-\bar{A}$	$-A$	.	.
X.24	-2	2	-2	.	.	.	.	.	.	.	.	.	.	$-A$	$-\bar{A}$	.	.
X.25	3	-1	-1	1	-1	-1	.	.	.	.	.	-1	1	.	.	1	1
X.26	-5	-1	-1	1	1	1	.	.	.	.	.	-1	1	.	.	-1	-1
X.27	.	.	.	.	.	.	.	.	.	.	.	.	.	1	1	.	.
X.28	-2	-2	-2	2	.	.	.	.	.	.	.	1	-1	.	.	.	.
X.29	2	-2	2	.	.	.	.	.	.	.	.	-2	.	.	.	.	.
X.30	2	-2	-2	.	.	.	.	.	.	.	.	2	.	.	.	.	.
X.31	-2	2	2	.	.	.	.	.	.	.	.	2	.	.	.	.	.
X.32	.	.	.	.	.	.	.	.	.	.	.	2	.	-1	-1	.	.
X.33	1	-3	1	-1	-1	-1	.	.	.	.	.	.	.	.	.	1	1
X.34	1	5	1	-1	1	1	.	.	.	.	.	.	.	.	.	-1	-1
X.35	.	.	.	.	.	.	.	.	.	.	.	-2	.	.	.	.	.
X.36	.	.	.	.	2	-2	1	1	-1	.	.	.	.	.	.	.	.
X.37	.	.	.	.	-2	2	1	1	-1	.	.	.	.	.	.	.	.
X.38	.	.	.	.	.	.	.	.	.	.	.	.	.	.	.	.	.
X.39	.	.	.	.	.	.	.	.	.	.	.	.	.	.	.	D	$\bar{D}$
X.40	.	.	.	.	.	.	.	.	.	.	.	.	.	.	.	$\bar{D}$	D
X.41	-2	2	2	.	.	.	.	.	.	.	.	.	.	.	.	.	.
X.42	2	-2	-2	.	.	.	.	.	.	.	.	.	.	.	.	.	.
X.43	.	.	.	.	.	.	-1	-1	1	.	.	.	.	.	.	.	.

where $A = \frac{1}{2}(-1 + i\sqrt{7})$, $B = \frac{1}{2}(-1 + i\sqrt{11})$, $C = -i\sqrt{5}$, and $D = -2(\zeta_8^3 + \zeta_8)$.

Appendix C. Generating matrices of Fischer group $\mathfrak{G}_2$

The generating matrices are block matrices:

$$\mathcal{H} = \left(\begin{array}{c|c} \mathcal{H}_1 & 0 \\ \hline 0 & \mathcal{H}_2 \end{array}\right), \quad \mathcal{X} = \left(\begin{array}{c|c} \mathcal{X}_1 & 0 \\ \hline 0 & \mathcal{X}_2 \end{array}\right), \quad \mathcal{Y} = \left(\begin{array}{c|c} \mathcal{Y}_1 & 0 \\ \hline 0 & \mathcal{Y}_2 \end{array}\right), \quad \mathcal{E} = \left(\begin{array}{c|c} \mathcal{E}_1 & \mathcal{E}_2 \\ \hline \mathcal{E}_3 & \mathcal{E}_4 \end{array}\right),$$

where $\mathcal{H}_1, \mathcal{X}_1, \mathcal{Y}_1 \in \mathrm{GL}_{46}(13)$ and $\mathcal{H}_2, \mathcal{X}_2, \mathcal{Y}_2 \in \mathrm{GL}_{32}(13)$; while $\mathcal{E}_1 \in \mathrm{Mat}_{38,38}(13)$, $\mathcal{E}_4 \in \mathrm{Mat}_{40,40}(13)$, $\mathcal{E}_2 \in \mathrm{Mat}_{38,40}(13)$, and $\mathcal{E}_3 \in \mathrm{Mat}_{40,38}(13)$. Below, the element 0 of GF(13) is replaced by a dot.

Matrix $\mathcal{H}_1$:

Matrix $\mathcal{H}_2$:

```
1 7 .  .  .  1  1  8  8  .  12  .  11  6  1  .  .  .  7  .  6  6  8  12  6  6  2  7  1  7  8  1  7
 . 4 .  .  .  .  6  7  .  12  .  .  8  .  1  .  5  11  7  6  6  12  5  5  .  7  .  7  7  1  6
 . 5 .  . 12  8  2  11 12  .  12  .  12  2  6  6  7  5  3  4  10  3  8  5  5  .  1  9  12  10  1  6
 . 7 1  .  . 12  5  5 12  3  .  2  6  11  .  .  6  2  7  7  6  2  8  8  11  6  12  6  5  11  6
 . 6 . 12  . 12 12  6  6  .  .  7  .  .  .  6  12  6  7  6  .  6  6  .  7  11  6  6  .  6
 . 9 .  .  .  1  2  8  8  .  12  .  12  5  .  12  1  8  2  6  7  7  .  8  8  2  5  3  8  7  1  9
 . 6 .  .  . 12  6  5  .  1  .  .  7  .  .  6  12  6  6  7  12  6  6  .  8  11  5  7  12  5
 . 2 .  .  . 6  . 10 10  .  12  .  .  4  8  8  5  3  11  4  9  4  6  1  2  1  5  7  9  10  1  2
 . 9 .  .  . 11 11 11 11  .  1  2  12  .  . 11  11  1  12  11  .  12  11  9  1  9  11  12  12  11
 . 4 .  .  .  3  2  2  2  .  . 12  10  1  .  2  3  12  1  2  .  2  2  4  12  3  1  1  .  1
 . 2 .  .  .  5  12  9  9  .  .  4  7  7  6  1  10  3  10  3  5  1  2  .  4  5  9  10  .  1
 . 7 .  .  .  .  7  7  .  .  .  6  .  .  7  1  7  7  6  1  7  7  .  6  .  7  6  .  7
 . 9 .  .  .  7  1  4  4  .  12  .  12 10  7  6  7  10  .  10  2  10  6  10  9  .  10  7  3  4  1  10
 . 8 .  .  . 11 11 11 11  .  .  1  3  .  1  12  10  9  1  12  12  12  10  10  10  2  9  11  .  10
 . 6 .  .  . 12  .  7  7  .  12  .  12  7  .  12  1  6  12  6  6  6  6  12  6  6  12  6  .  7  1  7
 . 7 .  .  .  5  11  2  2  .  1  .  1  11  7  8  5  8  10  10  3  10  5  7  8  .  12  5  2  3  .  7
 . 1 .  .  .  1  .  .  .  .  1  .  12  .  .  .  1  1  12  .  1  .  1  .  1  .  12
 . 9 .  .  .  7  .  3  3  .  .  .  11  6  6  7  9  .  10  3  9  7  10  9  12  10  5  3  3  .  10
 . 5 .  .  .  7  . 10 10  .  1  .  .  2  5  5  7  4  2  3  10  3  7  5  5  12  2  7  10  9  12  4
 . 12 .  .  .  5  .  9 10  .  11  .  .  6  7  8  6  1  10  5  9  2  6  .  11  4  5  10  10  2  2
 . 2 .  .  .  6  .  9  9  .  .  1  4  6  7  6  3  .  4  9  3  7  3  2  12  4  5  9  10  .  3
 . . .  .  .  5  12  8  8  .  .  1  5  7  8  5  2  10  4  9  3  6  1  1  12  5  4  8  10  .  1
 . 2 .  .  .  9  1  4  3  12  2  1  1  7  6  6  7  12  4  8  5  11  8  .  3  8  9  4  3  11  11
 . 12 .  .  .  5  12  9  9  .  12  .  6  8  9  5  1  9  5  9  3  6  12  .  12  6  4  9  10  1  1
 . 6 .  .  . 12  6  5  .  1  .  .  7  .  12  1  7  .  6  6  7  .  7  6  12  7  11  5  6  12  6
 . 7 .  .  .  .  .  6  6  12  1  .  1  6  12  .  6  1  7  7  6  1  7  7  12  6  .  7  6  12  6
 . 9 .  .  .  2  1  7  7  .  1  .  .  5  12  12  1  8  3  6  7  7  1  9  8  1  5  1  7  7  12  8
 . 2 .  .  .  1  1  1  1  .  .  .  12  .  .  .  1  1  12  1  .  1  2  12  2  1  1  .  1
 . 5 .  .  .  8  1  11  11  .  .  12  2  7  6  7  4  1  3  10  4  7  4  4  1  3  7  10  10  .  4
 . 4 .  .  .  7  . 10 10  .  .  .  3  6  6  7  4  1  3  10  3  7  4  4  .  2  7  10  9  .  4
 . 2 .  .  .  1  .  .  .  .  1  .  12  .  .  .  1  .  1  12  1  1  .  1  1  2  .  1  .  12
 . 11 .  .  . 11  .  .  .  .  12  .  .  1  .  .  .  12  11  1  12  12  12  11  12  12  .  .  .  .  1  12
```

Matrix $\mathcal{X}_1$:

```
1. . . . . . . . . . . . . . . . . . . . . . . . . . . . . . . . .
. . 3 10 6 11 . . . . . . . . . . . . . . . . . . . . . . . . . . .
. 6 1 3 12 . . . . . . . . . . . . . . . . . . . . . . . . . . . .
. 5 . 6 5 7 . . . . . . . . . . . . . . . . . . . . . . . . . . .
. . . . 1 . . . . . . . . . . . . . . . . . . . . . . . . . . . .
. 5 9 12 . 1 . . . . . . . . . . . . . . . . . . . . . . . . . . .
. . . . . . 1 2 10 6 2 9 12 . 3 . . 12 9 9 1 5 2 5 1 1 8 1 10 2 12 2 1
. . . . . 2 11 12 6 2 8 7 12 . 12 3 10 7 3 10 5 3 9 11 9 6 10 2 4 . 4 4
. . . . . 7 5 6 9 5 7 11 8 6 7 8 . 1 12 1 1 5 9 . 4 11 3 6 . 11 5 8
. . . . . 1 . 8 12 3 1 10 7 6 7 12 8 4 12 11 2 9 11 5 4 5 . 2 . 12 . 8
. . . . . 2 7 6 5 5 2 1 1 7 3 12 6 9 7 3 5 10 4 9 . 1 8 12 1 5 9 9
. . . . . 2 11 4 1 12 5 9 2 1 10 1 1 1 8 . 7 . 11 4 9 . 6 10 7 12 12 12
. . . . . 3 8 2 11 12 6 5 1 6 6 10 1 2 2 8 11 4 1 1 11 10 11 3 3 8
. . . . . 10 8 8 11 11 5 3 1 9 10 10 9 6 7 9 8 6 10 1 2 . 4 4 9 6
. . . . . 10 11 11 0 9 . 9 5 1 3 9 2 5 7 6 5 . 9 7 11 12 5 1 7
. . . . . 11 12 12 7 2 11 7 6 6 4 5 8 7 7 . 3 5 9 5 5 . 7 8 10
. . . . . 6 11 4 2 6 10 6 3 4 9 8 8 7 2 8 . 2 1 12 11 4 . 12
. . . . . 4 8 5 7 8 . 6 6 12 . 6 6 9 10 2 2 5 5 10 5 3 11
. . . . . 1 2 12 6 6 11 10 6 6 11 8 4 8 12 6 9 8 3 6 4 12
. . . . . 6 . 9 11 5 1 5 11 2 8 10 7 12 1 10 1 5 8 12 3
. . . . . 5 7 5 1 3 2 4 1 5 2 8 2 4 5 12 4 3 10 7 4
. . . . . 11 1 . 2 2 4 6 8 5 3 5 2 12 11 6 1 8 10
. . . . . 11 3 3 10 10 6 5 11 4 8 7 9 6 8 6 . 6 1
. . . . . 9 6 10 12 10 5 1 5 8 9 . 6 4 12 12 3 6 9
. . . . . 8 10 1 5 9 9 10 1 2 5 11 8 2 1 7 2 5
. . . . . 1 11 4 1 4 2 10 . 8 4 6 4 6 7 6 4 3
. . . . . 2 12 . 9 8 . 1 11 10 12 9 11 5 1 7 2
. . . . . 4 . 7 3 10 12 12 1 11 8 3 12 2 10 1 4
. . . . . 9 8 . 8 2 2 10 2 . 2 2 3 5 . 2
. . . . . 7 12 6 . 7 7 4 6 4 4 6 6 10 8
. . . . . 12 1 11 6 4 1 8 12 11 2 8 10 10
. . . . . 11 9 . 11 5 12 5 7 7 . 12 3
. . . . . 11 3 1 . 10 1 6 9 11 12 3
. . . . . 1 5 . 1 6 9 11 8 8 2
. . . . . 7 6 2 . 12 6 5 6 12
. . . . . 10 2 7 1 7 8 7 10
. . . . . 8 . 6 10 5 10 1
. . . . . 11 7 12 4 7 7
. . . . . 9 11 . 5 3
. . . . . 10 7 9 8 4
. . . . . 4 12 3 7 8
. . . . . 12 11 9 6
. . . . . . 12 7 3
. . . . . 12 11 9
. . . . . 11 2 5
. . . . . 12 1 9
```

Matrix $\mathcal{X}_2$:

```
 6  7  .  6  7  .  .  . 12  .  .  6  .  .  .  .  8  .  . 12  .  7 12  7  .  1  8 12  7  7  7  8
 7  7  .  8  7  .  1  . 12  .  7  .  .  .  .  .  .  5  .  3  2  .  8  .  6  . 11  6  2  7  6  7  6
 2  2  .  8  2  .  7  8  .  6  . 11 12  7  . 12 12  . 10 12  6  9  4  2  .  9  5  4  3  6 12 11
 7  6  .  5  5  .  .  .  3  .  .  7  . 12  .  .  5  . 12  .  .  6  1  6  .  1  5  .  6  6  7  5
 7  7  .  7  7  .  .  .  .  .  6  .  .  .  .  6  .  1  1  .  6  2  7  . 12  6  .  6  5  5  6
 4  5  .  5  6  1  .  . 12  .  .  6  .  1  . 10  . 10 11  .  5 10  7  .  3  8 11  7  9  8  8
 8  8  .  7  7  .  .  .  1  .  6  . 12  .  .  5  .  1  .  .  7  2  7  . 12  6  .  6  5  5  6
 2  4  1 11  5  .  7  5 12  8  .  8  1  6  .  1  9  .  2  .  7 11  6  5  .  4  4  8  3  4 10 11
 3  1  .  2  .  .  1  1  1 12  .  2  .  .  .  .  9  .  4  4 12  2  4 11  . 10 11  3  .  9 11 10
11 12  . 11  .  . 11  .  .  1  . 11  .  .  .  4  . 10  9  1 11 10  2  .  3  2  9  .  3  2  3
 4  5  . 12  5  .  6  6 12  7  1  9  1  6  .  8  .  3  1  7 11  8  4  .  4  3  8  3  1  8  9
 6  6  .  6  6  .  .  .  .  .  7  .  .  .  .  7  .  .  .  .  6  .  7  .  7 12  6  7  7  7  7
10 10  .  4 10  .  7  7 12  6  .  4  .  7  .  .  4  .  .  6  4  6  9  .  7 10  7 11 10  3  4
 3  2  .  3  1  .  1  .  .  .  1  .  .  .  .  8  .  6  4  .  3  4 12  .  8 11  4  .  9 11 11
 7  7 12  8  7  .  1  1 11 12  .  8  .  1  . 12  7  .  1 12  7  .  5 12  .  7  .  7  6  5  6
10 11  .  5 11  .  7  5  .  8  .  2  1  5  .  1  .  4  2  7  5  8 11  .  3  9  9  9  8  2  2
 .  .  . 12  . 12  .  1  . 12  . 12  .  .  .  1  . 12  .  .  .  1  .  1  .  1  . 12  .  .  .
11 10  .  3  9  .  7  7  1  6  .  4 12  6  .  2  1  1  1  6  4  8  9  .  6  9  7 10  9  3  3
 4  3 12  9  2  .  6  8  1  6  . 11 12  7  . 12 11  . 10 12  6  8  6  2  .  9  3  5  3  4  9  9
 4  4  1 12  5  .  8  6 11  6  . 11  1  7  .  1  7  .  5  4  6  .  8  2  .  3  3 10  4  1 10  9
 4  3  1  9  3  .  7  6  1  6  . 10  .  6  .  1  8  .  1  1  6 11  8  .  6  3  7  3  3 10  9
 4  4  1 11  4  .  7  5  .  7  .  9  1  6  .  1  7  .  4  2  7 12  9  4  .  3  2  9  3  1  9  9
 8  8  .  .  9  .  4  6  2  7  .  1  .  6  .  1  7  .  8  8  8  .  4 12  . 11 10  3 10  .  5  5
 4  5  1 12  5  .  8  5 12  7  . 10  1  6  .  1  6  .  6  3  7  .  9  4  .  1  3 10  3  1 10 10
 8  7  .  6  6  .  .  .  1 12  .  7  .  .  . 12  .  6  .  .  .  .  6  2  6  .  5  .  7  5  5  6
 7  6  .  6  6  .  .  .  1  .  .  7  .  .  .  .  6  . 12  .  .  6  .  6  .  1  6  .  6  7  7  6
 7  5  .  4  5  . 12  1  2 12  .  7 12  .  .  .  8  . 10 11  .  5 12  6  .  3  7 11  7  8  8  7
12 12  . 12  .  . 12  .  .  .  . 12  .  .  .  .  2  . 11 11  . 12 11  1  .  2  1 12  .  2  1  1
 3  3  .  9  3  .  6  7  .  6  . 10  .  7  .  . 11  . 12 12  7  9  6  3  .  7  4  5  4  4 11 11
 3  3  .  9  3  .  6  7  .  6  . 10  .  7  .  . 11  . 12 12  7  9  6  3  .  8  3  6  4  4 10 10
12  .  . 12  . 12 12  1  1  . 11  . 12  . 12 12  2  .  1  . 12 11  1 12 12  2  .  1  .  1  1  1
 .  .  .  2  1  .  1  . 11  .  .  1  1  1  .  . 12  .  1  2 12  1  . 12  . 12  . 12  .  2  . 12 12 12
```

Matrix $\mathcal{Y}_1$:

```
1  .  .  .  .  .  .  .  .  .  .  .  .  .  .  .  .  .  .  .  .  .  .  .  .  .  .  .  .  .  .  .  .  .
.  9  6  7  2  1  .  .  .  .  .  .  .  .  .  .  .  .  .  .  .  .  .  .  .  .  .  .  .  .  .  .  .
.  .  7 12  7 10  .  .  .  .  .  .  .  .  .  .  .  .  .  .  .  .  .  .  .  .  .  .  .  .  .  .  .
.  1 11 11  3  6  .  .  .  .  .  .  .  .  .  .  .  .  .  .  .  .  .  .  .  .  .  .  .  .  .  .  .
.  2  9 10 12  .  .  .  .  .  .  .  .  .  .  .  .  .  .  .  .  .  .  .  .  .  .  .  .  .  .  .  .
.  .  .  1  6  .  .  .  .  .  .  .  .  .  .  .  .  .  .  .  .  .  .  .  .  .  .  .  .  .  .  .  .
.  .  .  . 12 12  1  7 10  2  3  2  2 10  6  4  9 11  3 10  7  3  8 12 11  5  5  9  8  2  9  5  3  5  8  1  6  8  . 12  4  4  5 12
.  .  .  . 11  7  3  8  4  6  .  7  .  . 12  5  9  9  4  .  7  .  1  2  8  3  3  9  8 12  .  3  4  4  8  7  4 10  4  8  4  5  2  7 11  4
.  .  .  .  6  4  3 12 11  6  5  3  4  5  6  4  . 12  3  7  2  8  3 11  9  3  .  6  4  2  7  7  . 10  4  8  7  2  8  2  7  2 11  2
.  .  .  . 12  3  7 11  .  8  9 10 11  7  2 11  1 11  8 10  . 11 12  3  6 10  3  .  5  8  8 10  7  . 10  2 11 13 10  4  3  4  2  0
.  .  .  . 11  6  4  9  6 10  5  7  6  3  4  .  6  3  4  5  3  4  3  6  3 10  8  6  8  1  2  2  8  3  1  1  4  .  1  . 12  1  5  4
.  .  .  . 11  4  8  2  4  4 11  5  6 12 12  7  1  8  5  1  6  1 11  3  8  .  2  9  4  9  2 11  5  2  .  5  . 10  . 11  .  2  1  4
.  .  .  . 10  4  4  2  4  9  7  3  3  8  6  1  9  7 11  9  .  9  8  9  6 12 10 11  6  .  4 10  5  6 12  6  4 12 11  6 10  1  7 10
.  .  .  .  3 11  5  4  5  3  4  7  8  3  4  7 11  3  3 11  8  6 11  1  . 11  7  7  7  . 10  3  8  3  6  2 12  9  . 10 10  1  6 12
.  .  .  .  3  8  2  5  1  3  3  2  2  7  . 10 12 10  6  .  1  5  8 10  4  9  2  1  4  3 12  1  9  2 12  9 12 12  6  1  .  7 12  1
.  .  .  .  2 12  2  5  7  1  6  7  .  2  5 12  6  4  4  2  6  6  6  2  2  5  .  2 12  4  4  1  2  7  .  4  8  2  3  9  4  9  8 11
.  .  .  .  7  3  .  6  9  .  7 11  1 12  8 12  3  8  7  7  5  6  8  7 12  7  1 11 11  9  1  1  8 11  8  5 11  8  3  9  1  6 11  9
.  .  .  .  9  3  8  4  4  4  9  7  3 11 11 12  5  1  1 12  3 11  7  6  5  3  1  8 12  9  2  4  6  .  4  8  9  5  5 11  5  7  7  2
.  .  .  . 12  .  .  8  8  1 11  5  6  .  3  6  4  5  7 12  .  6  3  7  2  4  3  .  6  3 10  6 11 12  4 11  .  3  6  8  4  .  2 12
.  .  .  .  7  6  9  2  8  1  2  2  4  7  6  6  4  .  9 10  1  .  4  9 10  .  2  8  6  5 11  6 12 11  8 10 12 11 11  1  3  5 12  5
.  .  .  .  8  1 10  8  .  1  4  4  7  2  3  7  7  6  6  2  .  1 10 10 10  1  1  1  3  6  .  8 11  8  4  .  3  8  7  7  9  4 12  3
.  .  .  .  2  2 10  5  4  1  5  1  5  .  3 11  3  2  5  5  1  8  7  3  8  6  2  .  6  3  1  5  9  8  7 10 12  9 12  .  2 12  5
.  .  .  .  2 11  .  3  9  .  7 10 11  5 10  9  2 12 11  3  4  4 11  .  2  4  1  7  .  2  3  4  9  3  9  . 10  2 10  2  7  2  3  3
.  .  .  .  4  6  2  9  3  5  2  2 10  7  8  1  8 10 12  3  9 10  6 11  .  7  2  1  3  3  5  1  2 10  1  8  4  .  1  5  1 10  5  4
.  .  .  .  5 12  9 10  3  3  .  7  3  3  5  .  9 12  9  2 11 11  3  6  7 12 12  4  9  5 10  2  .  8  . 11 12  3 10  .  4 10  7  8
.  .  .  . 12  2  1 12  1  3  4  5  9  4 11 10  2  2  9  7 12  6  8  2 12 10  3  7  1 12  3 12  7 11  4  1  3  6  1 11  6  5  9  8
.  .  .  . 11  2 11  9 12 10 12  3  5  9  7  5 12 11  4  7  .  1 12  .  2  9  3  9  2  5 11  6  2  7  5 10 10 10  9  9  4  5  1
.  .  .  .  9  7  3  4  7  3  2 11  1  1  7  9 10  1  .  6  9 12 12  .  4  8 11 10  1  2  8  6  7  5  2  6 10  9  6 10  1 12  6
.  .  .  .  4  8  7  8  7 11  6  7  .  5  .  3  3 10  5  3 12 12 10 12  9  3  8  8 11  5  1  6  .  8  9 11  2 12  5  2  4  7  5
.  .  .  .  6  3  4  4  1  3  1  7  4  3  4 11  7 10  7 12  . 12  1 11  3 10  1  .  1  2  2  .  1  2  .  9  2  4  .  4  6  9 11  4  3  7  7
.  .  .  .  1 11  6  5  8  1  2  5  4  1  4  5 11 12  5  9 11  8 12 12  9  5  3  5 10 11  8  2 11 11  6  7  . 11  8  9 11  9  .  2
.  .  .  .  2 10 10  5  6 11 10  8  7  5 10  2 12  4  7  4 10  3 10  8  7  3  6 11  3  7  8  .  4 11  9  1  5  5  .  9  .  6  1
.  .  .  .  2 10 10  5  1  7  9  2  9  4  4  6  2  4  6  1  8  .  8  8  1  5  .  8 11  6  .  9  1  . 12 10  2  7 10 11  4  1 10
.  .  .  . 12  . 11  5  6  9  2  6 12  4  9  .  .  3 11 12 12  9  3  1  5  2  8  2  9 10  7  2 10 10  6  1  6  4  6  3  1  4  .  6
.  .  .  .  6  9 11  9  7 12  .  7  .  8  1 10 12  2  9  8  6 10  4  .  7  5  7  .  1 11 11  1 10  3  4  3  9  1  4 11  6 10  8 11
.  .  .  .  3  3 12 11 11  .  5  3  9 11  7  5  1  1  1  4  9  6 12  8  7 10  4  4 10  1  5  6 12  .  5  8 10 12  6 12  . 12 10  3
.  .  .  .  5 12  7  5  8  5  7  4  8 10  .  8  1 11  9 12  .  5  3  5  9  3  2  1 10  3  .  4  5 12  4  1  4  4  6  .  2 11  2
.  .  .  .  2 11  8  .  4  8  7  2  2  6 12  2  2  3  9  .  3  6  1  3 11  .  4  2  9 11  5  1  .  1  8  5  5  3  1 11 10  1 12  2
.  .  .  .  4  9  3  .  1  2  1  7  9  5  .  5  6  9  .  5  6 12 12  6  8 11  1 10 12 12 10  5  6  3  2  4  5  3  .  1  5 12  6 12
.  .  .  .  3  7 12  6  6  5  2 10  4  2  8  9 10  2  9  1  . 10  4 12 11  6  . 12  6  7 12  2  2  .  5  4 10  . 12 11 11 11  7 10
.  .  .  .  9  4 10  .  4  5  5 12  3  7 10  8 11  8  1 10  2  4  4 11  5  9  2  3 10 11  3  7  5 11 11  2 10  1  9  .  8  8  5 10
.  .  .  .  1 12  4  .  3 11  6 10  2  .  4  .  8  .  8 12  8  .  .  8  8 12  8 10  4  3  2 10 11  8  .  5 11 11 10  5 10  9  7 10  3  4  9  8 10  7
.  .  .  .  .  .  2  7 11 11  .  12  9  .  6  4 11 11  2 12  8  .  9 12  6 12  7  5  1  4  7  4  3  3  2  7  9  2  1  6  1  .
.  .  .  .  1 12  4  .  1  2  3  3  8  .  8 10  . 11 11 12  1  7  3  1 11  9  9  8  4  .  4  5  .  3  7  .  2  5 10  8  6  2  7  3
.  .  .  .  2 11  8  .  6  8 11  3  .  7  1  8 12  4  9  5  9  .  4  2  9 10  3 12  6  .  8 10  6  3  6  2 11  8  6  5  6  .  4  4
.  .  .  .  1 12  4  . 11  1 12  4  6  . 12  1  9  6  6  8  5 10  2  8 11 10 11 10  5  2 11  3  4  3  2  1 12 12  6  2  2  4  3  5
```

Matrix $\mathcal{Y}_2$:

```
6  7  5  .  12  8 12  6  6  7  .  6  5  6  7 12  8  .  7  5  8  . 12  2  7  .  .  6  6  7 12  7
8  7  6  1  .  6  .  7  7  7  .  8  6  6  8  .  5  .  5  5  7  . 12  .  6  .  1  7  6  8  .  7
8 11  8  8 11 12  1  8  7  3  8  4  9  9  1  5 12  . 11 11  9  6 12  1  6  . 10 11 10  9  3  2
7  6  8 12  .  5  .  6  7  6  .  7  8  7  6  2  5  .  7  9  6  .  3 11  7  . 12  6  7  5  1  6
7  6  7  .  1  5  .  7  7  7 12  6  7  7  7  .  .  6  6  6  .  .  .  .  5  .  1  6  6  7  1  7
5  8  5  1 12  9  .  6  4  6  1  6  5  6  5 11 10  .  8  7  7  . 12  1  9  . 11  7  7  5 10  6
7  5  7 11  .  5 12  6  8  7 12  6  7  7  7  1  5  .  6  6  7  .  1  .  5  .  1  5  6  7  2  7
10 10  9  6  . 11  . 10 11  3  6  3  8  9  5  7  9  .  9  8 11  7 12  1  3  .  2 11 11 10  7  4
2 11  2  .  2  9  1  2  2  1 11  2  3  2  1  2  9  . 11  1 11 12  1 10 10  .  2 12 12  2  4  1
11  1 11 12 10  4  . 11 12 12  1 11 10 11 12 11  4  .  3  .  2  1  .  3  2  . 10  .  1 11 10 12
12  9 10  6  1  8 12 11 11  3  6  4 10 10  5  7  8  .  9  8 10  7  .  .  1  .  3  9  9 11  8  4
6  7  6  . 12  7  1  6  6  6  .  7  6  6  6  .  7  .  7  7  6  .  .  7  . 12  7  7  6 12  6
3  3  3  7  .  4  .  4  3 10  6 10  3  4 10  6  4  .  3  3  3  6 12  . 10  .  3  3  4  7 10
3 11  1  .  2  9  .  2  3  1 11  2  2  1  3  2  8  . 10 12 12  .  . 11  9  .  4  . 12  3  4  1
7  6  7  1  1  6  .  8  6  7  .  7  7  8  6 12  7  .  6  6  5 12 12  .  6  .  1  6  5  8  .  7
5  3  4  5  1  2  .  4  6 10  5 10  3  3 12  8  .  .  2  2  4  7  1  .  8  .  3  3  3  4  9 11
12  .  .  . 12 12  1  . 12  .  .  .  .  .  . 12  .  .  1  .  1  1  1  .  1  . 12 12  . 12  .  .
4  3  3  7  .  2  .  3  3 10  6 11  4  4 10  7  2  .  2  3  3  6  . 12  9  .  .  3  4  .  7 10
9  9 11  6  . 10  .  9  9  3  7  3 11 11  1  6 11  . 11 11  9  6  1  .  4  . 11  9  9  9  7  3
12 10  9  9  1  8  . 12 10  4  6  6 10 10  6  7  7  .  7  8  9  6 11 12  2  .  3 11  9 12  7  4
10 10  9  7  .  9  . 10 10  4  6  4 10 10  4  7  8  .  9 10 10  6  . 12  3  .  . 10 10 10  7  3
11  9 10  6  1  8  . 11 12  4  5  4 10 10  6  8  7  .  8  9 10  7  . 12  1  .  3 10 10 11  9  4
1  4  3  5 11  6  .  1  2  8  8  7  2  2  8  6  7  .  6  5  5  7  2  1 12  .  9  3  5  .  4  8
12 10  9  7  1  8  . 11 12  4  5  5  9  9  9  7  8  .  7  7 10  7 12  .  1  .  4 11 10 12  8  5
6  5  8 12  1  5  .  7  7  7 12  6  8  8  7  1  6  .  6  7  6  .  1 12  5  .  1  5  6  7  2  7
7  6  7  .  .  6  .  6  6  6  .  7  7  7  7  .  6  .  7  7  6  .  1  .  6  . 12  7  7  6  .  6
5  7  6  . 12  8  .  5  5  6  1  6  7  6  5  .  8  .  8  8  7  1  .  8  . 10  6  7  5 11  5
12  1 12  . 12  2  . 12 12 12  1 12 12 12 12 12  2  .  1  .  1  .  . 11  .  1  . 11  1 12 11 12
9 10  9  7 12 11  .  9  9  3  7  3  9  9  3  6 11  . 11 10 10  7  .  1  4 12 12 10 10  9  5  3
9 10 10  7  . 10  . 10  9  3  7  3 10 10  3  6 11  . 10 10 10  6  .  4  . 12 10 10  9  6  3
.  1  . 12 12  1 12 12  . 12  . 12  1 12 12 12  .  1  .  1 12  2  1  1  1  . 12  .  1 12 12  .
1  .  1  1  2 12  .  2  .  .  . 12  1  1  1  .  . 12 12 12  . 12 12 12 12  .  2  . 12  1  1  .
```

Matrix $\mathcal{E}_1$:

```
1  .  .
.  .  .  4 11  4  5  8  7  .  3  5  8  7  4  .  8  5  5  4  4  9  2 12  7  4  9  7 12  4  1  9  9  7  6 10  1  6
.  .  . 11  1 11  .  .  .  .  .  .  .  .  .  .  5  . 12 11  8  2 11  9  7  8  5  3  .  7  5 12  .  6  6
.  .  .  2 12  2 12  4  9  7  5  2  5  5  .  2  3  7 11  9  4  6 10  3  8  .  5  3  .  4  6  .  2 10 10  4  6  8
.  .  .  5  4  5  4  9  3  .  5  4  9  3 11  .  9  4  4 11 11  2 12  7  3 11  2  3 11  8  6  8  7  6 12 12  6  2
.  .  .  7  3  7 10  6  1  7  9  .  7 10  1  2  5  5 10  5 11  3 10  1  .  2  5  5  9 11  9  2  3 10 11  6 10
. 12 11 12  4  1  3  . 12  7 10 10 10  5  4  7  4  8  .  3 12  9  1  4  9  9  8  1 10  9  8  4  4 12  9  1  1 11
.  8  1  9  2  4  6 12 11  3  7  9  2  9  4  4  1  8  7  4  8  7  8 10 10  7  3  7  6  3  4 12  6  2  9 10 10  3
.  7  4 11  5  2  4  5  3  8 11  1  8 10  .  1  4  1 12  9  .  4  7  8 12 12 10 12  1  7  3 11  5  9  4  2  6  9
.  .  4 10  3  3  9  2 10  4  4  .  5  4 12  2  4  5 12  9  5  .  8  2  4  2  7  . 11  .  4  3  9  4 10  1 11
.  6 12  5  5  5  8  5  8  7  .  9  6  1  9  3  4  7  3  3 12  4  6  1 10  2  7  1 12  1  2  7  .  2  3 10  9  6  1
.  1  9 12  8  1  8  1  6  8 12 10  3  2  5  . 11 11  2 10  .  8  9  7  5 11  3 10  6  7  9 11  2  2  2  2  9
.  8  3 10 12  3  4  1  3  3  2  .  4  7  1  9  7  4  6  7  3 11  .  8  1  5  6 10  6  5  2  5  4  4  8  1  2  2
.  7  7  4 12  9 12  2  9 11  2 11  .  6  3  8  4  8 12  4  9  3  4  9 11  5 10 12 11 12  6 11  3  6  8  9  7  9
. 10  7  6  9  7  5  9  7 11  7  2  5  3  4  3  3  1  8  .  9  2 11  8  2  9  3  1  2 11  2 11 12 12 10 12  3
.  7  2  2  9 11  6  4 11  6  5 10 12  7 10  4 10  2  9 10  5  5  3  5  9 11  9  5  5 12  3  6  . 11  9  9 12
.  4  6  5 12  8  .  .  . 11 11  3  6  2  2  6 10  1  5  6  9  2  2 10  2  1 11  6  1 12 12  3  4 12  .  4  2
. 11  8  2 10 11 10 11  1 10  .  4  6  2  .  2  3  . 12  8  .  1  5  4 12  . 12  3  .  1  6  2  6  .  8  5  7 10
. 12 11  3  3 10  .  .  .  2  2  8  7  2 11  4  5  4  7  8 10  8  6  5  9  3  9  5  4  6  3  5 11  9 10  4 11
.  7  9  8  8  5  3  2 12  3  1 10  .  8 10 10  2  9  9  4 11  4  .  1 10  1  3  6 11  9 10 11  8  1 12 11  .  .
.  1  3  9  8  4  7  9  2  7  3  8  8  6  2  9  7 10  6  3  9  .  1  1  9  5  2  6  .  6  2  3  7  4  4  4  .
.  . 12  8  7  5  3  2 12  3  8  4  4  .  3  6  7  4 11  7  2 12  9  .  5  1 11  4 11  .  3  .  9  .  5  1  4
.  2  3  7 12  6 12  8  9 12 12  9 12  8 10  6  7  1  5  3  7  9  9 10  7  6  8 11  5  4  7  8  9  3  8  7  9  6
.  1  . 10  7  3 11  3  5 11  8  2 12 12  4  2  8  1  8  2  2  2  .  .  2 10  7  1  3  2  . 11  3  5 11  8  4
.  3  6  5  1  8  7  9  2  7  2 10  .  3 11  3  2  6  .  5  7  9 11  7  1  5  9 10  5  4  1 10  8  7 12  1 11  .
.  8 11  3 11 10  2 10  8  2  2  8  2 10  6  1 12 11  8  7 12  8  2 10  9  6  4  1 12 11  3  9  5  2  4 12 10 11
.  6  5  1 12 12  9  6 10  9 11 12  .  . 11  9  .  3  6 12 11  4  5 10  6  .  1 11  .  .  1  .  9 11  .  6 10  9
.  7  6 11  1  2  8  1  6  8  4  2  .  .  4  8  .  7  4  2  4 11  4  .  3  6 11  2 11  1  9  4  3 12  9  8  4  7
.  1  1 12  9  1  .  .  .  .  .  .  .  .  .  .  .  .  2  .  .  5  8 12  4  2  8  4  7  8 12  2  3 12  2  9  9  4
.  6  8  3  4 10  .  .  .  .  .  .  .  .  .  .  .  3  .  .  1  8  1  1 10 10  9  .  4  6 11  5  6 11 10  1  3
.  2  .  2  3 11  .  .  .  .  .  .  .  .  .  .  .  .  1  9  9  8  7 11  3  2 12  1  2  2  5  1  2 11  8
.  2  .  2  3 11  .  .  .  .  .  .  .  .  .  .  6  . 11  6  5  9  5  2  6  9 10  8  1  6 12  1  4 10  4
.  2  4  7  7  6  .  .  .  .  .  .  .  .  .  .  4  .  8 10  5  7  8  . 12  6 12  3  6  1 11  2  .  5  9  7
.  2  .  2  3 11  .  .  .  .  .  .  .  .  .  .  5  . 12 11  8  2 11  9  7  2  1  3  1  9  . 12  9  3  5
.  4 11  8  4  5  .  .  .  .  .  .  .  .  .  .  .  9  3  3  7 11  8  1 12  6  9  9  1 12 12 12 10  4
.  . 12  2 12 11  .  .  .  .  .  .  .  .  .  . 11  . 12  9  5  1  4  7  6  7  7  .  3  4 10  7  7  4  2
.  9  9  4  3  9  .  .  .  .  .  .  .  .  .  .  4  . 10  2 10 10  9  9 12 11  3  8  9  9 10  4  5  7  5  4
.  4  3 11  9  2  .  .  .  .  .  .  .  .  .  .  3  .  2  4 10  9  4  8 12 12  1  7  4  3 10  9  1  6  9
```

Matrix $\mathcal{E}_2$:

```
 .  8  3  2  5  3  9 10  .  .  .  .  .  .  .  .  .  .  .  .  .  .  .  .  .  .  .  .  .  .  .  .  .  .  .  .  .  .  .  .  .
 .  4  8  6  1 10  7 10  .  .  .  .  .  .  .  .  .  .  .  .  .  .  .  .  .  .  .  .  .  .  .  .  .  .  .  .  .  .  .  .  .
 1  9  6 12  6  1 10  .  .  .  .  .  .  .  .  .  .  .  .  .  .  .  .  .  .  .  .  .  .  .  .  .  .  .  .  .  .  .  .  .  .
 9 12  4  6 12  4  1  8  .  .  .  .  .  .  .  .  .  .  .  .  .  .  .  .  .  .  .  .  .  .  .  .  .  .  .  .  .  .  .  .  .
 3  5  8 12  7  4  .  1  .  .  .  .  .  .  .  .  .  .  .  .  .  .  .  .  .  .  .  .  .  .  .  .  .  .  .  .  .  .  .  .  .
 4 10  5  8  .  5  5  3  3  .  .  8  5  4 11  9  3  8  .  4  4  4  5  8 11  1  .  3 11  . 10  1  2  6  .  2 11  8  2  5
 .  6 10  9  9 10  5  9  8  5  8  9  5 11  .  6  8  9  8  2  7 11  7  9  4  9  .  9  .  7  .  4 12  3  7  6  1  6  .
 3  3  4  7  .  6  3  .  4  8  .  3  4  2  8  6  4  3  5 10  7  2  5  7  .  5  . 10  7  . 12  5  6 11  7  6  7  6  9 11
 3  5  8 11  .  2  1 12  4  8 10  7  1 12  3  6  4  7  5  4  7 12 11  2  4  . 10 10 12 10  9  3  1  7  5  9  4  6 12  9
 2  9  7 10  2  6  3  .  9 11  1 11  .  5 11  6  9 11  2  4  7  5 10  1 11  . 10 12  9 10  5  7  4  3  . 10  3  8 10  1
 5 12  8 12  5 12  .  4 10 10  2 12  9  6  5  8 10 12  3 10  5  6  9 11  6 12 11  7 10 11  4  7  3 10 11  2 11  3 11  9
11  2  6 11  .  4  6  3  . 10  4  9  . 11  6  5  .  9  3 11  8 11  9  7  1 11  .  8  1  .  7  1 12  2 10 12  1  3  3  7
 1  . 11  .  5  1  1  1  2  .  3  .  5  4 10  6  2  .  .  7  4 12  3  4 11  8  9  2  8  6 11 11  2  6 12  1  5  2  4
 9  1  9  7 10  7  5  4  4  8  2 11  2  3  6  .  4 11  5  1  .  3  9  .  4 10  7  3  5  7  5  4  8  5  1  6  7  6  6  6
11 12 12  9 10  2  9  8  9  1 10  .  6  2  4  5  9  . 12 11  8  2 11 11  8  2  1  4  5  1  .  4  8  5  1 10  3  5 12  6
 9  4  7  6  6  3  6  1  8  . 12  8 10  5  2  4  8  8  .  3  9  5  .  4  2  9 11  5  2 11  1  6 11  6  2  .  .  2  2  2
 8  4  6  7  2 10  9 10  .  8  .  2  2  1  .  2  .  2  5  . 11  1  .  4  8  .  6  . 10  6 11  1  3  8  8  2 11 12  1  3
 8  2  2  9  6  .  8  1  2  2  .  3  8  8  2  2  . 11  3 11  8  8  1 10  4 11  6  4 11  9  7  9  9  7  4  9  6  1  9
 6  4  2  6  4  5  1 11  6  2  3 10  9  8  8  .  6 10 11  9  .  8  .  7  3 11 11  2  4 11  8 11  9  6  9  2 11  8  1 11
 .  5 10 12  3  9  2  6 10 10  9  4 11  2  2  5 10  4  3 11  8  2  4  6  6  6 11  9  3 11 11  4 10  4  6 10  3  7  2 12
 6 10  9  2 12  9  2  4  2 12 11  2  7  8  8  6  2  2  1  7  7  8 12 10 12  4  9  2  5  9  7 11  8  3  1  3 10 12  8  7
 2  7  8  .  2  9  4 12  4  . 10  3  9  .  . 10  4  3  .  9  3  . 10 11  8  2 10 11 10 10  3  2  3  4  . 11  2  2  9  .
10  3 10  .  6 11  6  4  9  2  3 10 11 12 12  3  9 10 11 11 10 12 11  6  1  4  .  8 10  .  8  5  3  6  5  9  4  . 11  4
 4 12  5  8  4 12  8  8  3  8 10  3  6  .  9  3  3  5  6  4  .  4  4  3  1  2  1  7  2 11 12  6  .  6 11  2  4  6  8
 3  6  1 10  7  8  4  4  .  6  . 12  6  6  .  .  7 12  .  6 10 10  3  1  5  5  5  1  8  8  1  9  1 12  9  6  5
 6  2  4  4  9  . 12 10  .  9  .  5  4  4  .  4  5  .  4  2 10  3 10  2 10 10  2  4  3  3 10  8  5  8  1  1  6
 7  .  9  7  8  7 12 10  . 11  .  5  9  9  .  2  5  .  9  7  9  4 12  2  4  .  2  8  9  .  3  5  1 12 12  9  6
 7  5  5  7  .  .  1  8  . 11  .  3 11 11  .  2  3  . 11  1  7  6  6  6 12  6  6  7  1  7  3  9  3 10  .  5 12
11  3  7  3  7  6  8  1  . 10  .  8 10 10  .  3  8  . 10  5  4  9  7  6  5  6  6  2  8  7  3  4  3 10  2 11  3
 3 12  5  3  4  7  5  6  .  5  .  .  8  5  5  .  8  8  .  5 10 12  1  .  1  3  1  1  . 10 12  7  7  7  6 12  3  3
11  . 11 11  7 11  . 10  .  1  .  . 10  1  1  . 12 10  .  1  1  6  7  9  7 12  7  7  5  1  6  9  7  9  4  1  9  1
 1  2  3  6 10 11 10  .  .  8  .  .  7  8  8  .  5  7  .  8 12  8  5  5  2 11  2  2  8  2 11 11 12 11  2  . 12 10
 8  5  9 10  9  7  4  .  7  .  .  4  7  7  .  6  4  .  7  2 12  1 10  4  1  4  4 12 12  9  5  1  5  8  . 10  6
 2  1 11 12 10  1 12 10  .  6  .  . 11  6  6  .  7 11  .  6  3  4  9 11  6  7  6  6  4  6  7  3  9  3 10  .  5 12
 4  4  4  9  .  8 12  .  .  2  .  . 11  2  2  . 11 11  .  2  6 12 11 10  9  2  9  9 12 11  4  5  8  5  8  .  9  .
12  9  5  4  1  5  9 12  . 10  .  7 10 10  .  3  7  . 10 11 11  2  1  6  6  6  6 11  7  7  9  8  9  4  . 10  4
 9  .  8  1  3  9 11  8  .  4  .  .  2  4  4  .  9  2  .  4  1  7  6  3  4 10  4  4  7  3  9  2  5  2 11  . 11  7
```

Matrix $\mathcal{E}_3$:

```
. 4 2 . 8 . . . . . . . . . . . . . . . . . 5 6 6 1 9 3 2 3 . 5 3 7 8 6 1 10 4
. . 11 4 11 9 . . . . . . . . . . . . . . 2 . 6 10 1 . 5 9 9 7 12 5 1 . 12 12 8 8 3
. . 8 10 8 3 . . . . . . . . . . . . . 12 . 4 6 4 4 1 1 10 9 . 11 6 3 5 3 1 1 5
. . 5 3 5 10 . . . . . . . . . . . . 3 . . 4 9 2 12 5 4 5 6 9 9 10 2 . 11 6 9 7
. . 3 7 3 6 . . . . . . . . . . . 10 . . 4 11 5 . 12 6 6 4 2 12 4 10 1 12 5 10 4
. . 5 3 5 10 . . . . . . . . . . 2 . . 6 10 1 . 5 9 9 6 3 5 6 2 8 5 1 2 6
. . . . . . . . . . . . . . . 1 . . 3 5 7 . 9 11 11 3 8 9 3 1 4 9 7 1 3
. . . . . . . . . . . . . . 5 . . 12 11 8 2 11 9 7 8 3 . 7 5 12 . 6 6
. . . . . . . 10 9 11 3 12 1 6 7 4 4 7 . 9 10 9 11 11 9 2 5 2 10 . 5 6 2 5 3 5
. . . . . . 5 8 11 3 12 9 . 6 1 . 5 8 8 5 7 10 5 1 5 11 10 11 6 5 8 . . 4 4 4 10
. . . . . 2 11 2 5 3 10 9 7 5 12 . 11 11 5 1 . 7 7 . 3 2 4 12 4 . 3 2 . 11 1
. . . . 4 9 11 3 11 8 8 6 6 . 8 9 12 10 6 4 7 7 1 3 1 8 2 1 3 . 10 . 12 . 10
. . . . 4 9 2 2 9 9 9 9 11 . 10 7 7 10 9 11 4 8 2 11 7 2 . 11 . 10 12 5 5 7 4 8
. . . . . 4 5 4 10 1 5 2 6 12 8 4 4 5 2 . 9 4 4 4 12 12 8 6 7 . 1 8 12 .
. . . . . 1 1 9 1 4 9 1 11 3 3 2 8 1 6 3 11 9 6 1 8 10 8 10 2 1 10 5 10 .
. . . . 2 11 3 4 5 11 12 12 4 1 3 2 12 1 1 3 . 8 8 9 6 9 9 7 1 8 9 4 10 7
. . . . 9 4 2 8 1 11 12 1 2 4 5 11 6 1 9 11 11 8 1 1 7 8 5 6 4 4 5 8 5 4
. . . . 8 5 8 9 3 3 10 4 8 11 . 2 12 9 5 6 8 9 4 1 7 1 6 7 10 9 12 3 10 10 12
. . . . 2 11 9 7 10 9 11 6 11 3 8 5 . 11 . 3 3 3 11 2 2 . 9 9 2 6 4 . 1 3 5
. . . . 4 9 9 11 9 5 . 7 8 3 2 8 6 7 7 6 8 4 8 4 6 10 3 12 9 12 7 6 5 10 5
. . . . 7 6 2 2 5 1 1 1 10 . 11 10 3 11 3 3 4 . 8 9 7 6 . 10 6 9 3 9 4
. . . . 5 8 7 3 . 5 2 11 1 1 6 2 3 8 11 5 10 2 12 7 10 9 7 5 3 . 10 11 11 3 3
. . . . 9 4 4 7 8 12 8 10 8 8 11 12 12 10 . 4 8 2 2 10 2 11 8 7 . 1 9 5 11 7 1
. . . . 2 11 10 11 10 . 4 1 7 9 11 6 2 7 11 12 8 12 6 3 9 11 . 10 4 9 . 5 1 10 10
. . . . 4 9 8 5 8 8 12 8 . 10 12 6 . 7 5 . 3 12 9 2 7 6 7 3 9 8 8 . 7 3
. . . . 11 2 5 4 3 10 6 10 4 . 8 9 10 12 2 7 7 5 8 7 . 4 10 6 8 2 11 6 3 10
. . . . 2 11 6 6 3 3 6 3 5 4 12 10 2 2 3 9 7 . 4 . 10 11 5 3 12 9 2 9 4 8 11
. . . . 7 6 6 1 5 10 10 . 12 10 5 4 9 2 11 10 3 5 7 2 9 12 7 6 2 10 7 1 . 2
. . . . 11 2 2 2 3 8 8 8 3 12 8 2 1 10 10 2 7 7 7 11 8 1 12 3 4 3 5 7 1
. . . . 11 2 10 1 2 5 8 9 6 . 10 12 4 10 11 5 3 10 10 1 7 9 11 6 . 5 . 8 6 4 7
. . . . 6 7 7 . 3 9 . 10 11 6 3 1 12 9 9 2 10 2 4 9 4 11 11 3 12 12 12 11 1 3 5
. . . . 11 2 3 10 1 7 3 8 4 9 2 10 10 12 3 2 4 5 4 10 4 10 5 12 4 3 4 3 12 4 12
. . . . . 3 7 5 8 1 12 11 4 9 9 6 12 1 1 10 10 1 1 4 8 3 10 . 12 5 3 5 12
. . . . 12 9 1 11 2 11 3 11 3 10 3 5 8 1 5 4 4 3 10 3 7 2 4 10 10 1 9 5 2 6
. . . . 4 10 9 3 4 2 4 . 12 6 1 12 7 2 2 8 5 3 12 5 3 12 11 10 9 11 5 6 1
. . . . 5 1 8 5 2 4 1 10 2 3 9 4 10 11 4 2 7 . 4 4 4 6 5 11
. . . . 2 11 8 9 10 3 7 3 9 . 5 4 3 1 11 6 6 8 5 6 1 9 4 8 7 1 6 12 10 7
. . . . 2 11 12 8 2 4 1 9 3 12 4 9 9 10 2 10 7 10 5 12 4 12 3 7 1 3 3 6 1 7 .
. . . . 4 9 8 10 . 4 9 11 7 5 12 8 1 6 . 7 11 3 2 3 9 7 4 8 12 3 5 5 . 8 1
. . . . 9 4 9 5 8 11 5 1 9 6 . 1 11 . . 10 12 3 6 6 12 6 8 1 11 9 9 12 2 . .
```

Matrix $\mathcal{E}_4$:

```
.  4  5  3  8 1211 12  .  2  .  .  1  2  2  .  .  .  .11  1  .  2  7 10  3  8 11  1 1111 1012  2  5  9  5  8  . 12  1
8 1012 12  5  4  2  2  .  2  .  .  1  2  2  .  .  .  .11  1  .  2  5  8  5  6  5  8  5  5  8  5  8  .  7  .  .  . 12  7
2  7  7  6  6  9  9  .  .  9  .  .10  9  9  .  .  .  4 10  .  9  9  4  9  8  9  4  9  9  4  9  4  .  5  .  .  .  1  5
12  5  3  5  4 11  3 11  . 12  .  . 1112 12  .  .  .  1 11  . 12  .  .  .  1  .  .  .  .  .  .  .  .  . 12  .  .  . 1212
10  8  1  5  8  3  4 11  .  1  .  .  2  1  1  .  .  . 12  2  .  1  .  .  . 12  .  .  .  .  .  .  .  .  1  .  .  .  1  1
2 12  7  .  4 12  5  8  .  5  .  .  2  5  5  .  .  .  8  2  .  5  9  4  9 12  9  4  9  9  4  9  4  .  1  .  .  . 10  1
1  6 10  .  2  6  9  4  .  2  .  .  4  2  2  .  .  . 11  4  .  2  .  .  . 11  .  .  .  .  .  .  .  .  2  .  .  .  2  2
.  4  .  3  1  5  2  9  .  .  .  .  .  .  .  .  .  .  .  .  .  .  .  .  .  .  .  .  .  .  .  .  .  .  .  .  .  .  .  .
7  9  . 10 12  8  8  2  .  5  3  .  6 1212  6  .  .  .  5  .  6  8  .  3 1011 11  5  8 11  3  5  8  5  1  8  8  7 11  6
3  . 1010  3  4 1112  6 12 10  .  . 12 12  .  7  . 12  6  .  1 1210  9 11  6  5  2 11  3  5  2  4  8  2  2  6  5  .
10  9  .  .  6  7  3  3 11  .  9  8  8  2 10  8  2  5  . 11  8  6 12  3 11  1  4  1  5  5  .  1  5 10  9  2  2  2  4  1
. 11  5  .  3 11  8  8  7 1110  7  7 1011  7  6  6 11  7  7  2  7 10  3  5 11  5  8  5  4 11  8 11  9  8  8  6  5  7
3  . 10  .  3  4 10  9  3  8  7  3 10  8  2 1010 10  8  3 10  5  3  .  6  8  8  8  5  8  6  8  5  2  5  5  5 10  9 10
9 10  3  1  9  2  3 10  .  1  9  6  6  2  1  6  .  7  1  .  6 12  7  9 11  2  7  8 11  2  3  8 11  9  4 1111  1  8  6
4 12  .  8  1 11  3  8 10  2  7 10 10  1  2 10  3  3  2 1010 11 10  7  6  8  8  8  5  8  .  8  5  2 12  5  5  3  8 10
10  4  3  1 1011  4  6 11  5 10  8  8 11  1  8  2  5  5 11  8  2 11 10 10  4  8  4  3  .  . 11  3  8  5  .  .  1  7  8
5 11  1  .  7  4 10  8  .  7  7  7  7  .  7  7  .  6  7  .  7  6 12  7  5  6  8  .  7  .  .  .  7  5  1  . 12  7  7
8 12  9  6  6 12 12  2  7  6  6  6  6  6  .  6  6  7  6  7  6  7  6  7  1  6  1  7 11  .  6  .  .  6  6  2 12  6  6  8 12  6
5  9  3  1 10  2  2  7  8  5  7  5  5  4  8  5  5  8  5  8  5  8  5  8  1  7 12  4  8  4  3  7  3  4  3  1 12  .  . 11  7  5
1 11  5  9  7  .  1 12  3 11  .  3 10 12  5 10 10 10 11  3 10  2 10  .  7  5  5  5  8  5  .  5  8 12  8  8  8 10  5  3
.  7  5  8  4  4  6  4  8  6  3  5 11  .  9 11  5  8  6  8 11  7  8  3 10 12  9 12  1  9  .  6  1  2 10  4  4  5  2 11
5  1  3  7  7  9  6  6  6 11  1  7  7 10 11  7  7  6 11  6  7  2  5  1 11  9 12 10  4  9  . 10  4  8  4  4  4  5 10  7
4  3  9  9  4  .  8  7  6  5  3  .  .  6  5  .  7  .  5  6  .  1  6  3  3  5 12  5  8 12  3  5  8  4  8  8  8  .  5  6
. 10  1  1 11  7  1 11  8  3  4  5  5  2  .  5  5  8  3  8  5  4  1  4  9  9  .  9 11 11  .  9 11  6  7  2  2  4  6 12
12  2  5 10  1  8  1  4  7  3  .  .  .  2 10  .  6  .  3  7  . 10  7  .  .  3  2 10  3  3  .  9  3  4 10  3  3  7  3  .
1  7  .  5  6  7  7  6  8  9 10  5  5  9  6  5  5  8  9  8  5  4  8 10  3  2  6  9  4 12  .  9  4  5  8  7  7 11  6  5
1  3 11  .  4 10  7  .  2 10  6  5  5 10  7  5 11  8 10  2  5  9  2  6  . 11  7  4  8  2 10  4  8  7  .  5  5 12  7  5
10 10  .  8 12  9  6  2  4  8 10  8  8  9 11  8  9  5  8  4  8  5 10  4  9  .  .  8  7  3  .  8  7  9  2  3  3  . 11  8
8  9  5  4  1  3  3  1  8  1  7  5  5  1  4  5  8  1  8  5  6  8  7  .  . 11  1  .  3 10  1  . 10  9  9  9  4 11  5
.  5  .  .  9  4  3  2  5  5  4  2  2  4  1  2  8 11  5  5  2  9  4  4  9 10  2  4 10  6  .  4 10  7 12  .  .  7  1  2
8  2  8  . 12  1  5 12  6  8  6  8  8  7  5  8  7  5  8  6  8  5  .  6  1  3  3  9 10  6 10  8 10  8  7  7  7  3  5  8
4  3  5  3  8 11  3  7  4  2  4  8  8  2  5  8  9  5  2  4  8  5 10  4  9  .  .  1  7  9  .  2  7 10  2 10 10  . 11  2
3  3  6  7  9  7  5  1  7  2 10  7  7  1  8  7  6  6  2  7  7  5  .  3  3  1  8  8  5  8 10  8  5  8 1211 11  .  8  .
. 12  5 12  7  5  1  7 10 11  . 10 10 11 11 10  3  3 11 10 10  2 10  .  5  .  5  8  5  5  7  5  8 11  2  8  8  3  5 10
5  6  2 12  9  5  . 12  7  1  3  .  .  1  1  .  6  .  1  7  . 12  1  3  4  8  7  8 11  2 10  8 11  2 12  4  4  7  8  .
12  7  4  5  2  5 12  1 10  3  3  3  3  3  3  3 10  3 10  3 10  4  3  4 10  2  3  3 10  3  3 11  3  3  3  4  3  3
10  9 10  1 12 10  9  1  5  4  3  8  4  7  8  5  4  5  8  9  5  3 10  4  7  4  9  1  .  4  9  1 12 12 12  2  7  8
7  5  9  4  5  4  6  2  5  4  3  8  8  4  7  8  8  5  4  5  8  9  5  3 10  4  7  4  9  1  .  4  9  1 12 12 12  2  7  8
1  6  1  3  7  1  9 11  7  3  .  .  .  2 10  .  6  .  3  7  . 10  7  .  . 10  2 10  3  3  .  9  3 11  3 10 10  7  3  .
10 11  6 12 12  8  7  3  6 10  .  . 11  3  .  7  . 10  6  .  3  6  .  . 10 11  3 10 10  .  4 10  9  3 10 10  6 10  .
```

References

[1] M. Aschbacher. *3-transposition groups*. Cambridge University Press, Cambridge, 1997.

[2] J. Cannon and C. Playoust. *An Introduction to* MAGMA. School of Mathematics and Statistics, University of Sydney, 1993.

[3] R.W. Carter. *Simple groups of Lie type*. John Wiley and Sons, London, 1972.

[4] J.H. Conway. A group of order 8,315,553,613,086,720,000. *Bull. London Math. Soc.*, **1**:79–88, 1969.

[5] J.H. Conway. Three lectures on exceptional groups, in M.B. Powel, G. Higman (eds.) *Finite simple groups*. Academic Press, London, 1971, pp. 215–247.

[6] J.H. Conway, R.T. Curtis, S.P. Norton, R.A. Parker, and R.A. Wilson. *Atlas of finite groups*. Clarendon Press, Oxford, 1985.

[7] B. Fischer. Finite groups generated by 3-transpositions. *Inventiones Math.*, **13**:232–246, 1971.

[8] D.F. Holt. The mechanical computation of first and second cohomology groups. *J. Symbolic Computation*, **1**:351–361, 1985.

[9] G. James. The modular characters of the Mathieu groups. *J. Algebra*, **27**:57–111, 1973.

[10] G. O. Michler. Constructing finite simple groups from irreducible subgroups of $GL_n(2)$, this volume.

[11] G. O. Michler. Theory of finite simple groups. Cambridge University Press, Cambridge, 2006.

[12] C. E. Praeger, L.H. Soicher. Low rank representations and graphs for sporadic groups. Australian Math. Soc. Lecture Series **8**, Cambridge University Press, Cambridge, 1997.

DEPARTMENT OF MATHEMATICS, CORNELL UNIVERSITY, ITHACA, N.Y. 14853, USA

DEPARTMENT OF MATHEMATICS, CORNELL UNIVERSITY, ITHACA, N.Y. 14853, USA

Contemporary Mathematics
Volume **470**, 2008

.

Constructing Simple Groups from Irreducible Subgroups of $GL_n(2)$

Gerhard O. Michler

ABSTRACT. In this article we describe a new relation between the structure of the Sylow 2-subgroups of certain finite simple groups and some 2-modular representations of indecomposable subgroups T of $GL_n(2)$. It is the theoretical basis for a new algorithm constructing finite simple groups from irreducible or indecomposable subgroups T of $GL_n(2)$. Since each simple group is an irreducible subgroup of finitely many linear groups $GL_n(2)$ applications of this algorithm can be iterated. In particular, new uniform existence proofs for all known sporadic simple groups can be given this way by starting from fairly small indecomposable subgroups of $GL_n(2)$. But it is not restricted to sporadic groups. In this article we apply it to the irreducible subgroup $T = GL_4(2)$ of $GL_4(2)$. Thus we obtain a new self contained existence proof of Conway's sporadic group Co_3 (up to isomorphism). Other applications are mentioned in the introduction.

1. Introduction

The celebrated Brauer-Fowler Theorem asserts that there are only finitely many non isomorphic simple groups G having a 2-central involution z such that its centralizer $C_G(z)$ is isomorphic to a given group H of even order. But it does not give any hint how to find such a group H without knowing one of the simple groups G.

It is the purpose of this article to provide an algorithm solving this problem for certain simple groups. This new algorithm is described in Section 2. It has been applied in [21] to show that all known finite sporadic simple groups can be constructed uniformly from irreducible or indecomposable subgroups T of fairly small general linear groups $GL_n(2)$. But it also constructs many alternating groups and simple groups of Lie type, and yields new relations between all these classes of simple groups. Here a subgroup U of any $GL_n(2)$ over the field $F = GF(2)$ is called *indecomposable* and *irreducible* if the canonical n-dimensional F-vector space $V = F^n$ is an indecomposable and an irreducible FU-module, respectively.

In this article we only consider finite simple groups G whose Sylow 2-subgroups S have a non-cyclic elementary abelian characteristic subgroup. In view of the classification of the simple groups with dihedral [13] or semi-dihedral Sylow 2-subgroups [1] and the Bender-Suzuki Theorem [2] and Suzuki [23] this is not a serious restriction by Theorem 4.8.5 of [20]. In particular, a fixed Sylow 2-subgroup

2000 *Mathematics Subject Classification.* 20D08, 20D05, 20C40.
Key words and phrases. construction of simple groups, sporadic simple groups, Conway sporadic simple group Co_3.
This research has also been supported by the grant NFS/SCREMS DMS-0532/06. The author thanks the referees and the editors for helpful suggestions and corrections, and help in improving the presentation.

S of G contains a maximal non-cyclic characteristic elementary abelian subgroup A. In many examples dealt with in [20] such a characteristic subgroup A is the unique maximal elementary abelian normal subgroup of S. In all other cases A is the intersection of finitely many maximal elementary abelian normal subgroups B_i of S, $i = 1, \ldots, k$, which are conjugate in the automorphism group $\mathrm{Aut}(S)$ of S. In particular, they all have the same order.

In Section 3 we apply the algorithm to the irreducible subgroup $T = GL_4(2)$ of $GL_4(2)$ in order to demonstrate its practicability. Thus we obtain a new self contained abstract existence proof for Conway's third sporadic group Co_3, see Theorem 4.2. It realizes this simple group as a simple subgroup $\mathfrak{G}$ of the matrix group $GL_{23}(13)$ and provides a system of representatives of all its conjugacy classes of as short words in four generating matrices.

Conway constructed his simple group Co_3 in [6] as the automorphism group of a 23-dimensional sublattice of the 24-dimensional Leech lattice using completely different methods, see [7]. Our Theorem 4.2 will be used in [21] to give a self contained uniqueness proof for Conway's sporadic group Co_3 by means of the author's uniqueness criterion stated as Theorem 7.5.1 in [20]. The original uniqueness proof due to D. Fendel [11] depends on Feit's classification theorem of the 23-dimensional irreducible integral lattices [10] which according to D. Gorenstein [14, p. 121] "is one of the deepest papers ever written in the representation theory of finite groups."

In the Appendices we state the character tables, systems of representatives of the conjugacy classes of the local subgroups of Co_3 constructed in Sections 3 and 4. The character table of the constructed simple group $\mathfrak{G}$ is stated there; the table was constructed using MAGMA and the faithful permutation representation documented in this work.

Algorithm 2.5 also reveals a new relation between the structure of the Sylow 2-subgroups of finite simple groups G and the indecomposable modular representations over $F = GF(2)$ of certain subgroups T of some $GL_n(2)$. In particular, whenever G_1 is a finite simple group and V is an irreducible or indecomposable FG_1-module, Algorithm 2.5 can be applied to find out whether there is a larger simple group G_2 having a Sylow 2-subgroup S with a maximal non-cyclic elementary abelian normal subgroup A isomorphic to V such that $N_{G_2}(A)$ is isomorphic to an extension of G_1 by V. In the course of the proof of Proposition 3.3, Algorithm 2.5 is applied to an indecomposable multiplicity free 6-dimensional module V of $T \cong GL_4(2)$. Here it returns the simple symplectic group $\mathrm{Sp}_6(2)$.

It will be shown elsewhere that the sporadic simple groups of Fischer [12], Conway [6] and the largest Janko group J_4 [17] can be constructed uniformly by means of Algorithm 2.5 by choosing G_1 to be one of the simple Mathieu groups M_{22}, M_{23}, M_{24}, which also can be constructed from the irreducible subgroups S_5 of $GL_4(2)$, $(3 \times A_5) : 2$ of $GL_4(2)$, and $3S_6$ of $GL_6(2)$, respectively.

But there are many more successful applications of this algorithm. By Theorem 9.3.2 of [20] it is easy to see that it returns Janko's simple sporadic group J_1 of order 175560 when it is applied to the irreducible Frobenius subgroup of order 21 in $GL_3(2)$. From [25] follows that the application of Algorithm 2.5 to $T = GL_5(2)$ yields Thompson's sporadic simple group Th, see [24]. Choosing $T = GL_3(2)$ as irreducible subgroup the algorithm establishes simultaneously the simple groups of Lie type $G_1 = G_2(3)$ and $G_2 = G_2(5)$, see [21]. In particular, Algorithm 2.5 constructs in this application two non isomorphic centralizers $H_i = C_{G_i}(z)$, $i = 1, 2$ from the same irreducible subgroup $T = GL_3(2)$. Finally, it has to be mentioned that there are also several examples where the algorithm fails, see [18].

Concerning our notation and terminology we refer to the books [5] and [20]. The computer algebra system MAGMA is described in Cannon-Playoust [3]. The author gratefully acknowledges computational help by his student Hyun Kyu Kim. He owes also thanks to L. Wang (Peking University) for providing the matrices of example 2.1 and his critical remarks on an earlier version of Lemma 2.1.

2. The Algorithm

In this article we only consider finite simple groups G whose Sylow 2-subgroups S have a non-cyclic elementary abelian characteristic subgroup. In view of the classification of the simple groups with dihedral [13] or semi-dihedral Sylow 2-subgroups [1] and the Bender-Suzuki Theorem [2] and Suzuki [23] this is not a serious restriction by Theorem 4.8.5 of [20]. It asserts that a fixed Sylow 2-subgroup S of any other simple group G contains a maximal non-cyclic characteristic elementary abelian subgroup A such that

$$(*) \qquad G = \langle N_G(A), C_G(x) \mid x \in I(N) \rangle,$$

where $I(N)$ denotes a set of representatives of the conjugacy classes of involutions of $N = N_G(A)$. This result provides the theoretical background for Algorithm 7.4.6 of [20] constructing the simple groups G having a 2 central involution z with a given centralizer $H = C_G(z)$. This algorithm constructs first the subgroup $U = \langle H, N_G(A) \rangle$ of G. Often U itself turns out to be a simple group. If U has only one conjugacy class of involutions then $G = U$ by Corollary 4.8.10 of [20]. But U may be a proper simple subgroup of G, see Theorem 8.6.6 of [20].

In many examples dealt with in [20] such a characteristic subgroup A is the unique maximal elementary abelian normal subgroup of S. In all other cases A is the intersection of finitely many maximal elementary abelian normal subgroups B_i of S, $i = 1, \ldots, k$, which are conjugate under the automorphism group Aut(S) of S. Such an example is dealt with in Sections 3 and 4. Therefore we prove an analogue of $(*)$ also for this situation.

LEMMA 2.1. *Let G be a finite simple group having no strongly embedded subgroups. Suppose that $A \neq 1$ is a maximal non-cyclic elementary abelian characteristic subgroup of the Sylow 2-subgroup S of G such that A is not a maximal elementary abelian normal subgroup of S. Then A is the intersection of finitely many maximal elementary abelian normal subgroups B_i of S which form an Aut(S)-orbit, $i = 1, 2, .., k$. Let $E = N_G(B_1)$ be of maximal order among the normalizers of the B_i. If $N_G(S) \leq E$ then*

$$G = \langle N_G(B_1), C_G(x) \mid x \in I(E) \rangle$$

where $I(E)$ denotes a set of representatives of the conjugacy classes of involutions of $E = N_G(B_1)$.

PROOF. Let B be a maximal elementary abelian normal subgroup of S containing A. Let D be the intersection of all B^α, $\alpha \in \mathrm{Aut}(S)$. Hence $D = A$ because A is a maximal elementary abelian characteristic subgroup of S.

Let $U = \langle E, C_G(x) \mid x \in I(E) \rangle$, and let a be any involution of U. Then there is a Sylow 2-subgroup T of U such that $a \in T$. By Sylow's Theorem $T^u = S \leq E$ for some $u \in U$. Hence there is an $e \in E$ such that $a^{ue} = x \in I(E)$. Thus $C_G(a) = [C_G(x)]^{(ue)^{-1}} \leq U$. As $N_G(S) \leq E$ by hypothesis, Theorem 4.8.4 of [20] implies that $U = G$, because G does not have any strongly embedded subgroup. $\square$

If a maximal non-cyclic elementary abelian normal subgroup of S is also a maximal elementary abelian characteristic subgroup of S, the condition $N_G(S) \leq N_G(A)$ is automatically satisfied. Otherwise it has to be checked. In the application below S is self normalizing in the target group G. So the conditions of Lemma 2.1 hold again.

In view of the above remarks and Lemma 2.1 we now let A be a maximal non-cyclic elementary abelian normal subgroup of S of a finite simple group G without strongly embedded subgroups such that $N_G(A)$ is maximal among the normalizers of maximal elementary abelian normal subgroups of S and $N_G(S) \leq N_G(A)$. Let n be its 2-rank, $C = C_G(A)$, $E = N_G(A)$, $T = E/C$, and let $F = GF(2)$ be the field with 2 elements. Then T is isomorphic to a subgroup of $\mathrm{GL}_n(2)$. Furthermore, the FT-module structure of A is uniquely determined by the matrices of the generators t_i of T with respect to a given basis B of the vector space A over F.

The FT-module structure of $A = F^n$ varies. There are many finite simple groups G having a non-cyclic elementary abelian normal subgroup A which is an irreducible FT-module, e.g. the alternating groups $G = A_{4k}, k \geq 3$. However, there are also finite simple groups G having a Sylow 2-subgroup S such that S has a maximal non-cyclic elementary abelian characteristic subgroup A which is not an irreducible FE-module, where $E = N_G(A)$. In [22] L. Wang and the author have given a self-contained existence and uniqueness proof for the Tits simple group $G = {}^2F_4(2)'$. It provides the following example.

EXAMPLE 2.2. By [22] the Tits simple group $G = {}^2F_4(2)'$ has a 2-central involution z belonging to a Sylow 2-subgroup S of $H = C_G(z)$ such that the unique maximal elementary abelian normal subgroup A of S is an indecomposable, but not irreducible FE-module, where $E = N_G(A)$. In fact, $T = E/C_G(A) = E/A \leq \mathrm{GL}_5(2)$ is generated by the following matrices:

$$a = \begin{pmatrix} 1&0&0&0&0 \\ 0&1&0&0&0 \\ 1&0&1&0&0 \\ 1&0&1&1&0 \\ 0&0&0&0&1 \end{pmatrix}, \quad b = \begin{pmatrix} 0&1&0&0&0 \\ 1&0&0&0&0 \\ 0&0&1&0&0 \\ 0&0&0&1&0 \\ 0&0&0&1&1 \end{pmatrix}, \quad c = \begin{pmatrix} 0&1&0&0&0 \\ 1&1&0&0&0 \\ 0&1&1&0&0 \\ 1&1&0&0&1 \\ 0&1&0&1&1 \end{pmatrix}$$

of orders 2, 4, and 3, respectively. It follows that the indecomposable FT-module A has a composition series of length 3 with irreducible composition factors of dimensions 2, 1, and 2.

REMARK 2.3. The explicit construction of the centralizer $H = C_G(z)$ from the indecomposable subgroup U of $\mathrm{GL}_5(2)$ defined in Example 2.2 is performed in [21, Chapter 10]. For completeness, we provide a brief sketch:

Using the MAGMA command FU:=FPGroup(U) we determine a presentation of $U = \langle a, b, c \rangle$. Letting V be the natural 5-dimensional vector space over $F = GF(2)$, we use Holt's Algorithm 7.4.5 of [20] to compute the second cohomological dimension $\dim_F[H^2(U,V)]$ and build all eight extensions E_i of U by V. For each E_i, we use MAGMA to obtain a faithful permutation representation PE_i with trivial stabilizer, and apply the first four steps of Algorithm 2.5 described below. In two of the extensions, the normal subgroup B is not a maximal elementary abelian normal subgroup. In the remaining six cases, there is exactly one conjugacy class of 2-central involutions z_i, and $D_i = C_{E_i}(z_i)$ is a Sylow 2-subgroup of the corresponding E_i.

Each D_i has exactly three normal subgroups with cyclic center, and in all but one of the cases the automorphism group of these subgroups turns out to be a 2-group. The only exception is for a non-split extension of U by V, where one of the normal subgroups has an automorphism of order 5. Letting E be this non-split extension, W the corresponding normal subgroup of order 2^9, and D the centralizer in E of the corresponding 2-central involution z, it is easy to verify using MAGMA that $C_D(W) = Z(W)$ and that W has a cyclic complement $C = \langle v \rangle$ in D of order 4.

Letting PW be the restriction of the faithful permutation representation PE to W, we obtain (again using MAGMA) a faithful permutation representation PA of $\mathrm{Aut}(PW)$, and a homomorphism $\pi\colon D \to PA$ with kernel $\langle z \rangle$. We then search in PA for an automorphism β of W of order 5 that satisfies the condition $\beta^{\pi(v)} = \beta^3$, and thus find a Frobenius subgroup $PF = \langle \beta, \pi(v) \rangle$ of order 20. The subgroup H is then uniquely determined as the semi-direct product of PF by W, and z is in the center of H. □

The main purpose of this article is to describe an algorithm which constructs the centralizer $H = C_G(z)$ of a 2-central involution z of some finite simple groups G from a given indecomposable subgroup U of some general linear group $\mathrm{GL}_n(2)$ provided all of the algorithm's conditions can be satisfied in the course of the construction. For its statement we need the following notations and definitions from [20].

The set of all faithful characters of a finite group U is denoted by $f\mathrm{char}_{\mathbb{C}}(U)$, and $mf\mathrm{char}_{\mathbb{C}}(U)$ denotes the set of all multiplicity-free faithful characters of U.

DEFINITION 2.4. *Let U_1, U_2 be a pair of finite groups intersecting in D. Then*

$$\Sigma = \left\{ (\nu, \omega) \in mf\mathrm{char}_{\mathbb{C}}(U_1) \times mf\mathrm{char}_{\mathbb{C}}(U_2) \mid \nu_{|D} = \omega_{|D} \right\}$$

is called the set of compatible pairs of multiplicity-free faithful characters of U_1 and U_2.

For each $(\nu, \omega) \in \Sigma$ the integer $n = \nu(1) = \omega(1)$ is called the degree *of the compatible pair (ν, ω).*

ALGORITHM 2.5. *Let T be an indecomposable subgroup of $\mathrm{GL}_n(2)$ acting on $V = F^n$ by matrix multiplication.*

- Step 1: *Calculate a faithful permutation representation PT of T and a finite presentation $T = \langle t_i \mid 1 \le i \le r \rangle$ with set $\mathcal{R}(T)$ of defining relations.*
- Step 2: *Compute all extension groups E of T by V by means of Holt's Algorithm [15]. Determine a complete set $\mathfrak{S}$ of non isomorphic extension groups E by means of the Cannon-Holt Algorithm [4].*

- Step 3: *Let $E \in \mathfrak{S}$. From the given presentation of E determine a faithful permutation representation PE of E. Using it and Kratzer's Algorithm 5.3.18 of [20] calculate a complete system of representatives of all the conjugacy classes of E.*
- Step 4: *Let $z \neq 1$ be a 2-central involution of E. Calculate $D = C_E(z)$ and fix a Sylow 2-subgroup S of D. Check that the elementary abelian normal subgroup V of E is a maximal elementary abelian normal subgroup of S. If it is not maximal, then the algorithm terminates.*
- Step 5: *Construct a group $H > D$ with the following properties:*
  - (a) *z belongs to the center $Z(H)$ of H.*
  - (b) *The index $|H : D|$ is odd.*
  - (c) *The normalizer $N_H(V) = D = C_E(z)$.*

  If no such H exists the algorithm terminates.

  Otherwise, apply for each constructed group H the following steps of Algorithm 7.4.6 of [20]. By step 5(c) it may be assumed from now on that $D = H \cap E$.
- Step 6: *Compute the character tables of the groups D, H, and E, and the fusion patterns of the conjugacy classes of D in H and in E, respectively. Using Kratzer's Algorithm 7.3.10 of [20] calculate the finite set of compatible pairs*
  $$\Sigma = \{(\nu, \omega) \in \mathrm{char}_{\mathbb{C}}(H) \times \mathrm{char}_{\mathbb{C}}(E) \mid \nu_{|D} = \omega_{|D} \in mf\mathrm{char}_{\mathbb{C}}(D)\}.$$
- Step 7: *Let $(\nu, \omega) \in \Sigma$ be a compatible pair of faithful characters of H and E of minimal degree $\nu(1) = n$ not yet dealt with. Let $p > 0$ be a prime not dividing $|H||E|$ which is minimal among the primes p not yet dealt with. Let F be a finite splitting field of characteristic $p > 0$ for the irreducible constituents of the characters of the compatible pair (ν, ω). Then do the following steps:*
  - (a) *Construct the faithful multiplicity-free semisimple FH-module $\mathfrak{V}$ corresponding to ν and the faithful FE-module $\mathfrak{W}$ corresponding to ω.*
  - (b) *Identify H and E with their isomorphic images in $\mathrm{GL}_n(F)$ under their representations afforded by $\mathfrak{V}$ and $\mathfrak{W}$, respectively. Determine then by means of Theorem 7.2.2 of [20] a double coset decomposition*
    $$C_{\mathrm{GL}_n(F)} = \bigcup_{i=1}^{s} C_{\mathrm{GL}_n(F)}(H) T_i C_{\mathrm{GL}_n(F)}(E)$$
    of the centralizer $C_{\mathrm{GL}_n(F)}(D)$ of D in $\mathrm{GL}_n(F)$. For each double coset representative T_i, let $G_i = \langle H, T_i^{-1} E T_i \rangle$, $1 \leq i \leq s$. Then $C_{G_i}(z) \geq H$ for all i.
  - (c) *For each $i \in \{1, 2, \ldots s\}$ compute the orders of some suitable elements of G_i and use this information to check whether a Sylow 2-subgroup of G_i may be isomorphic to S.*

    If the Sylow 2-subgroups of all constructed groups G_i are not isomorphic to S, then the algorithm ends.

  Otherwise, let $G = \langle H, T^{-1} E T \rangle$ be any of the groups G_i having fulfilled the Sylow 2-subgroup test of Step 6(c). Then the canonical n-dimensional vector space $M = F^n$ is an irreducible FG-module with multiplicity-free restriction $M_{|H} \cong \mathfrak{V}$.

- Step 8: *Since $M_{|H}$ has a proper non-zero FH-submodule U one can construct a permutation representation $\pi : G \longrightarrow S_m$ into the symmetric group on m letters with stabilizer $\tilde{H} \geq H$ and a strong base and generating set for $\pi(G)$ by Theorem 6.2.1 and Remark 5.2.12 of [20], respectively.*
- Step 9: *Check by means of Proposition 5.2.14 of [20] that π has stabilizer $\tilde{H} = H$. If so, then $\pi : G \longrightarrow S_m$ is faithful, and $|G| = |H|m$. In particular, G has a 2-central involution t with $C_G(t) \cong H$ and one can calculate a presentation of G by means of Theorem 5.2.18 of [20].*
- Step 10: *Using the faithful permutation representation $\pi : G \longrightarrow S_m$ of G and Kratzer's Algorithm 5.3.18 of [20] compute a complete set of representatives of all the conjugacy classes of G.*
- Step 11: *Compute a character table of G by means of MAGMA or the algorithms described in [20]. If no conjugacy class of G is in the kernel of some irreducible character then G is a simple group. Otherwise, the concrete character table of G provides matrix generators of a proper normal subgroup of G.*

In the following two sections we apply this algorithm to the irreducible subgroup $T = GL_4(2)$ of $GL_4(2)$ in order to provide a new abstract existence proof for Conway's sporadic group Co_3 (up to isomorphism), see [6].

3. Construction of the involution centralizer of Conway's sporadic group Co_3 from $GL_4(2)$

It is well known that the general linear groups $\Gamma = GL_n(2)$ can be generated by all elementary matrices $e_{i,j}$. Furthermore, if $n \geq 3$ a special case of a classical theorem of R. Steinberg yields that Γ has the following set $\mathcal{R}(\Gamma)$ of defining relations:

$$e_{i,j}^2 = 1 \qquad \text{for all } 1 \leq i, j \leq n, \, i \neq j,$$
$$[e_{i,j}, e_{j,l}] = e_{i,l} \quad \text{for } i \neq l,$$
$$[e_{i,j}, e_{k,l}] = 1 \qquad \text{for } j \neq k \text{ and } i \neq l,$$

see [19, p. 40]. In [9] Dennis and Stein reduced Steinberg's sets of defining relations of the classical groups substantially. In the case $n = 4$ dealt with here; their presentation is given in part (a) of the following result.

LEMMA 3.1. *Let $T = GL_4(2)$ and $F = GF(2)$. Let X be the set of the 10 elementary matrices $\{e_{1,2}, e_{1,3}, e_{2,1}, e_{2,3}, e_{2,4}, e_{3,1}, e_{3,2}, e_{3,4}, e_{4,2}, e_{4,3}\}$. Then the following assertions hold:*

(a) *$T = \langle X \rangle$ has the following set $\mathcal{R}(T)$ of defining relations:*

$$e_{1,2}^2 = e_{1,3}^2 = e_{2,1}^2 = e_{2,3}^2 = e_{2,4}^2 = e_{3,1}^2 = e_{3,2}^2 = e_{3,4}^2 = e_{4,2}^2 = e_{4,3}^2 = 1,$$
$$[e_{1,2}, e_{1,3}] = [e_{1,2}, e_{3,2}] = [e_{1,2}, e_{3,4}] = [e_{1,2}, e_{4,2}] = [e_{1,2}, e_{4,3}] = 1,$$
$$[e_{1,3}, e_{1,2}] = [e_{1,3}, e_{2,3}] = [e_{1,3}, e_{2,4}] = [e_{1,3}, e_{4,2}] = [e_{1,3}, e_{4,3}] = 1,$$
$$[e_{2,1}, e_{2,3}] = [e_{2,1}, e_{2,4}] = [e_{2,1}, e_{3,1}] = [e_{2,1}, e_{3,4}] = [e_{2,1}, e_{4,3}] = 1,$$
$$[e_{2,3}, e_{1,3}] = [e_{2,3}, e_{2,1}] = [e_{2,3}, e_{2,4}] = [e_{2,3}, e_{4,3}] = 1,$$

$$[e_{2,4}, e_{1,3}] = [e_{2,4}, e_{2,1}] = [e_{2,4}, e_{2,3}] = [e_{2,4}, e_{3,1}] = [e_{2,4}, e_{3,4}] = 1,$$
$$[e_{3,1}, e_{2,1}] = [e_{3,1}, e_{2,4}] = [e_{3,1}, e_{3,2}] = [e_{3,1}, e_{3,4}] = [e_{3,1}, e_{4,2}] = 1,$$
$$[e_{3,2}, e_{1,2}] = [e_{3,2}, e_{3,1}] = [e_{3,2}, e_{3,4}] = [e_{3,2}, e_{4,2}] = 1,$$
$$[e_{3,4}, e_{1,2}] = [e_{3,4}, e_{2,1}] = [e_{3,4}, e_{2,4}] = [e_{3,4}, e_{3,1}] = [e_{3,4}, e_{3,2}] = 1,$$
$$[e_{4,2}, e_{1,2}] = [e_{4,2}, e_{1,3}] = [e_{4,2}, e_{3,1}] = [e_{4,2}, e_{3,2}] = [e_{4,2}, e_{4,3}] = 1,$$
$$[e_{4,3}, e_{1,2}] = [e_{4,3}, e_{1,3}] = [e_{4,3}, e_{2,1}] = [e_{4,3}, e_{2,3}] = [e_{4,3}, e_{4,2}] = 1,$$
$$[e_{1,2}, e_{2,3}] = e_{1,3}, \quad [e_{1,3}, e_{3,2}] = e_{1,2}, \quad [e_{2,1}, e_{1,3}] = e_{2,3}, \quad [e_{2,3}, e_{3,1}] = e_{2,1},$$
$$[e_{2,3}, e_{3,4}] = e_{2,4}, \quad [e_{2,4}, e_{4,3}] = e_{2,3}, \quad [e_{3,1}, e_{1,2}] = e_{3,2}, \quad [e_{3,2}, e_{2,1}] = e_{3,1},$$
$$[e_{3,2}, e_{2,4}] = e_{3,4}, \quad [e_{3,4}, e_{4,2}] = e_{3,2}, \quad [e_{4,2}, e_{2,3}] = e_{4,3}, \quad [e_{4,3}, e_{3,2}] = e_{4,2},$$
$$(e_{1,2} e_{2,1}^{-1} e_{1,2})^4 = 1.$$

(b) *T has a faithful permutation representation PT of degree 15 with stabilizer*
$$U = \langle e_{1,2}, e_{1,3}, e_{2,1}, e_{2,3}, e_{3,1}, e_{3,2}, e_{4,3} \rangle.$$

(c) *The simple group T has two non-isomorphic simple modules V_1 and V_2 of dimension 4 over $F = GF(2)$. They are dual to each other, and we may assume without loss of generality that A_1 is isomorphic to the 4-dimensional F-vector space F^4, and that T acts on V_1 by matrix multiplication.*

(d) *The second irreducible representation V_2 of T is described by the transpose inverse matrices of the generating matrices $e_{i,j}$ of T defining V_1.*

(e) $\dim_F[H^2(T, V_1)] = 1 = \dim_F[H^2(T, V_2)]$.

(f) *The two non-split extensions E_1 and E_2 of $T = GL_4(2)$ by V_1 and V_2 are isomorphic groups.*

(g) *Let E be E_1 and let the last four generators of E be generators of V_1. Then the first ten generators e_i of E are chosen so that they map onto the given generators $\{e_{1,2}, e_{1,3}, e_{2,1}, e_{2,3}, e_{2,4}, e_{3,1}, e_{3,2}, e_{3,4}, e_{4,2}, e_{4,3}\}$ of T in this order.*

The non-split extension $E = \langle e_i \mid 1 \le i \le 14 \rangle$ of T by V_1 has the following set $\mathcal{R}(E)$ of defining relations:

$$e_1^2 = e_2^2 = e_5^2 = e_{11}^2 = e_{12}^2 = e_{13}^2 = e_{14}^2 = 1,$$
$$[e_r, e_s] = 1 \quad for \; 11 \le s < r \le 14$$
$$e_1 e_{11} e_1 e_{11} e_{12} = e_2 e_{11} e_2 e_{11} e_{13} = e_3^{-1} e_{12} e_3 e_{11} e_{12} = 1,$$
$$[e_1, e_{12}] = [e_1, e_{13}] = [e_1, e_{14}] = [e_2, e_{12}] = [e_2, e_{13}] = [e_2, e_{14}] = 1,$$
$$[e_3, e_{11}] = [e_3, e_{13}] = [e_3, e_{14}] = [e_4, e_{11}] = [e_4, e_{13}] = [e_4, e_{14}] = 1,$$
$$[e_5, e_{11}] = [e_5, e_{13}] = [e_5, e_{14}] = [e_6, e_{11}] = [e_6, e_{12}] = [e_6, e_{14}] = 1,$$
$$[e_7, e_{11}] = [e_7, e_{12}] = [e_7, e_{14}] = [e_8, e_{11}] = [e_8, e_{12}] = [e_8, e_{14}] = 1,$$
$$[e_9, e_{11}] = [e_9, e_{12}] = [e_9, e_{13}] = [e_{10}, e_{11}] = [e_{10}, e_{12}] = [e_{10}, e_{13}] = 1,$$
$$[e_1, e_7] = [e_1, e_9] = [e_2, e_4] = [e_2, e_{10}] = [e_3, e_4] = [e_3, e_{10}] = 1,$$
$$[e_4, e_2] = [e_4, e_3] = [e_4, e_{10}] = [e_5, e_8] = [e_6, e_7] = [e_6, e_8] = [e_7, e_1] = 1,$$
$$[e_7, e_6] = [e_7, e_8] = [e_7, e_9] = [e_8, e_5] = [e_8, e_6] = [e_8, e_7] = [e_9, e_1] = 1,$$
$$[e_9, e_7] = [e_9, e_{10}] = [e_{10}, e_2] = [e_{10}, e_3] = [e_{10}, e_4] = [e_{10}, e_9] = 1,$$
$$e_4^{-1} e_{12} e_4 e_{12} e_{13} = e_5 e_{12} e_5 e_{12} e_{14} = e_6^{-1} e_{13} e_6 e_{11} e_{13} = 1,$$

$$e_7^{-1}e_{13}e_7e_{12}e_{13} = e_8^{-1}e_{13}e_8e_{13}e_{14} = e_9^{-1}e_{14}e_9e_{12}e_{14} = 1,$$

$$e_{10}^{-1}e_{14}e_{10}e_{13}e_{14} = e_1e_2e_1e_2e_{12}e_{13} = 1,$$

$$e_1e_4^{-1}e_1e_4e_2e_{11}e_{12}e_{13}e_{14} = e_1e_8^{-1}e_1e_8e_{12}e_{14} = 1,$$

$$e_1e_{10}^{-1}e_1e_{10}e_{12} = e_2e_1e_2e_1e_{12}e_{13} = e_2e_5e_2e_5e_{13} = 1,$$

$$e_2e_7^{-1}e_2e_7e_1e_{11}e_{13}e_{14} = e_2e_9^{-1}e_2e_9e_{13} = e_3^{-1}e_2e_3e_2e_4^{-1}e_{11} = 1,$$

$$e_3^2e_{11} = e_4^2e_{13} = e_6^2e_{11} = e_7^2e_{12} = e_8^2e_{14} = e_9^2e_{12} = e_{10}^2e_{13} = 1,$$

$$e_3^{-1}e_5e_3e_5e_{11} = e_3^{-1}e_6^{-1}e_3e_6e_{11} = e_3^{-1}e_8^{-1}e_3e_8e_{14} = 1,$$

$$e_4^{-1}e_5e_4e_5e_{13} = e_4^{-1}e_6^{-1}e_4e_6e_3^{-1}e_{11}e_{14} = 1,$$

$$e_4^{-1}e_8^{-1}e_4e_8e_5e_{12}e_{13} = e_5e_2e_5e_2e_{13} = e_5e_3^{-1}e_5e_3e_{11} = 1,$$

$$e_5e_4^{-1}e_5e_4e_{13} = e_5e_6^{-1}e_5e_6e_{11}e_{14} = e_5e_{10}^{-1}e_5e_{10}e_4^{-1}e_{14} = 1,$$

$$e_6^{-1}e_1e_6e_1e_7^{-1}e_{11}e_{14} = e_6^{-1}e_3^{-1}e_6e_3e_{11} = e_6^{-1}e_5e_6e_5e_{11}e_{14} = 1,$$

$$e_6^{-1}e_9^{-1}e_6e_9e_{11} = e_7^{-1}e_3^{-1}e_7e_3e_6^{-1}e_{14} = 1,$$

$$e_7^{-1}e_5e_7e_5e_8^{-1}e_{12}e_{14} = e_8^{-1}e_1e_8e_1e_{12}e_{14} = e_8^{-1}e_3^{-1}e_8e_3e_{14} = 1,$$

$$e_8^{-1}e_9^{-1}e_8e_9e_7^{-1}e_{11}e_{12}e_{14} = e_9^{-1}e_2e_9e_2e_{13} = 1,$$

$$e_9^{-1}e_4^{-1}e_9e_4e_{10}^{-1}e_{12} = e_9^{-1}e_6^{-1}e_9e_6e_{11} = e_{10}^{-1}e_1e_{10}e_1e_{12} = 1,$$

$$e_{10}^{-1}e_7^{-1}e_{10}e_7e_9^{-1}e_{12}e_{13} = (e_1e_3^{-1}e_1)^4 = 1.$$

PROOF. By the introductory remarks we know that (a) is a presentation of the group T.

(b) As $T = GL_4(2)$ is a matrix group we use the parabolic subgroup U of the statement as stabilizer in order to get a transitive permutation representation PT of T of degree $2^4 - 1$. The MAGMA command h,PT:=CosetAction(T,sub<T|U>) provides also the induced isomorphism h between T and PT.

The assertions in (c) and (d) are obvious.

For (e), (f), and (g), let X denote the set of the ten generating elementary matrices $e_{i,j}$ of T. The MAGMA command FEalg:=MatrixAlgebra<F,4|X> constructs the algebra of all 4 by 4 matrices over F. MAGMA uses the isomorphism h to define on the natural 4-dimensional vector space $V_1 = F^4$ a PT-module structure by means of its command nmodQ:=GModule(PT,FEalg). Now Holt's algorithm [16] implemented in MAGMA can be applied. Thus the command CohomologicalDimension(PT,nmodQ,2) calculates the dimension of the second cohomology group $H^2(T, V_1)$, which is 1. Using the presentation of T given in (a) Holt's algorithm also delivers for each of the two 2-cocycles a presentation of the extension groups E_i, $i = 0,1$ by means of the MAGMA command P:=ExtensionProcess(PT,nmodQ,T). The presentation given in the statement (g) is the non-split extension E_1:=Extension(P,[1]) of T by V_1. The same commands applied to the dual module V_2 yield a similar presentation for the non-split extension E_2 of T by V_2. The isomorphism between the two extensions has been verified by means of MAGMA using permutation representations of E_1 and E_2 and the algorithms of Cannon and Holt described in [4].   □

REMARK 3.2. The construction of a group H with center $Z(H)$ of order 2 which will later be shown to be isomorphic to the centralizer of a 2-central involution of Conway's sporadic group Co_3 requires two applications of Algorithm 2.5. For completeness, we now give an outline of the intermediate application.

Let $E = \langle e_i \,|\, 1 \leq i \leq 14 \rangle$ be the finitely presented group constructed in Lemma 3.1. We will show $z = e_3^3$ is a 2-central involution whose centralizer $D = C_E(z)$ has center $Z(D) = \langle z \rangle$ and a Fitting subgroup Q of order 2^7.

Let $E_1 = D/Z(D)$; this group has an elementary abelian Fitting subgroup A of order 2^6 with complement $K \cong GL_3(2)$. It follows from [20, Prop. 8.6.5] that E_1 is isomorphic to the normalizer of a maximal elementary abelian normal subgroup of a Sylow 2-subgroup S_1 of the simple symplectic group $Sp_6(2)$. Thus, E_1 has a conjugacy class of 2-central involutions with centralizer $D_1 = C_{E_1}(z_1)$ of order $2^9 \cdot 3$. The Fitting subgroup A has a complement $C \cong S_4$ in D_1. By [20, Prop. 8.6.5] there is a monomorphism $\phi \colon E_1 \to GL_6(2)$ and subgroups $D_2 = \phi(D_1) \cong A \colon C$ and $E_1 = \phi(E_1) \cong A \colon K$ such that there is a subgroup $G_2 \geq D_2$ of $GL_6(2)$ with $H_2 \cong B \colon S_6$, B a normal subgroup of $\phi(S_1)$ of order 2^5, and $Z(H_2) = \langle \phi(z_1) \rangle$.

The subgroup $P = \langle H_2, E_2 \rangle \cong Sp_6(2)$ of $GL_6(2)$ is constructed using the following amalgam:

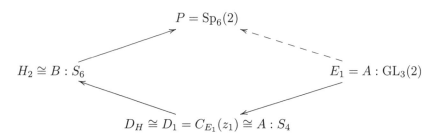

In particular, $|H_2 : \phi(D_1)| = 15$ and all conditions of Step 5 of Algorithm 2.5 hold.

In order to find H_2 and then $P \cong Sp_6(2)$ inside $GL_6(2)$ in steps (e) to (l) of Proposition 3.3 below, we let K and C act on A and obtain an injection into $GL_6(2)$, which must be modified using inner automorphisms of $GL_6(2)$. Suitable centralizers are then calculated so we can retrieve the normal subgroup A and all data of the diagram above. The construction we give below is shorter than an application of all steps 6 through 11 of Algorithm 2.5.

PROPOSITION 3.3. *Keep the notation of Lemma 3.1. Let* $E = \langle e_i \,|\, 1 \leq i \leq 14 \rangle$ *be the non-split extension of* $T = GL_4(2)$ *by its simple module* V_1 *of dimension 4 over* $F = GF(2)$. *Then the following statements hold:*

(a) E *has a faithful permutation representation of degree 168 with stabilizer* $\langle e_2 e_9 e_6^{-1} e_8 e_3, e_9 e_8^{-1} e_4 e_{10} \rangle$.

(b) $z := e_3^2$ *is a 2-central involution of* E *with centralizer* $D = C_E(z)$ *of order* $2^{10} \cdot 3 \cdot 7$. *Furthermore,* D *has unique normal subgroup* Q *of order* 2^7. *It is extra-special with center* $Z(Q) = \langle z \rangle$ *and* Q *has a complement* $C \cong PSL_3(2)$.

(c) *Let* $x_1 := e_6^{e_{10}}$, $x_2 := e_5^{e_9}$, $c_1 := e_{12} e_3 e_8 e_6 e_{14} e_9$ *and* $c_2 := e_8 e_{10} e_{13} e_7 e_3 e_5 e_4$. *Then* $S = \langle e_3, e_7, e_8, x_1, x_2 \rangle$ *is a Sylow 2-subgroup of* E *with elementary abelian normal subgroup* $A = \langle e_{11}, e_{12}, e_{13}, e_{14} \rangle$. *Furthermore,* $D = \langle S, e_4 \rangle$ *and* $C = \langle c_1, c_2 \rangle$.

(d) *The normal subgroup Q of D is generated by:*

$$f_2 = e_8^2, \quad f_2 = x_1, \quad f_3 = e_7^2, \quad f_4 = e_3, \quad f_5 = x_2 e_3 x_2, \quad \text{and} \quad f_6 = e_7 e_3 e_7.$$

Let $V = Q/Z(Q)$ and $v_i = f_i Z(Q) \in V$ for $i = 1, \ldots, 6$. Then $\mathcal{B} = \{v_i \mid 1 \leq i \leq 6\}$ is a basis of the elementary abelian group V. The conjugate actions of c_1 and c_2 on Q induce endomorphisms on V described by the following matrices with respect to the basis $\mathcal{B}$:

$$Mc_1 = \begin{pmatrix} 1&1&1&0&0&1 \\ 0&1&1&0&1&0 \\ 0&0&1&0&0&0 \\ 1&1&1&1&0&0 \\ 0&1&0&0&1&1 \\ 0&0&1&0&1&1 \end{pmatrix} \quad \text{and} \quad Mc_2 = \begin{pmatrix} 1&1&1&0&1&1 \\ 0&0&1&0&1&1 \\ 0&1&1&1&0&0 \\ 0&1&0&0&0&0 \\ 0&1&1&0&1&0 \\ 1&0&1&0&0&1 \end{pmatrix}.$$

(e) $E_1 = D/Z(Q) = \langle c_1, c_2, v_i \mid 1 \leq i \leq 6 \rangle$ *has the following set $\mathcal{R}(E_1)$ of defining relations:*

$$c_1^4 = c_2^4 = 1, \quad v_i^2 = 1, \quad \text{for } 1 \leq i \leq 6,$$

$$(c_1^{-1} c_2^{-1})^4 = c_2 c_1^{-1} c_2^{-1} c_1^{-2} c_2^{-1} c_1^{-1} c_2 c_1^{-1} = c_1 c_2^{-1} c_1 c_2^{-1} c_1^{-1} c_2^2 c_1^{-1} c_2^{-1} = 1,$$

$$[v_i, v_j] = 1 \quad \text{for } 1 \leq i, j \leq 6,$$

$$c_1^{-1} v_1 c_1 (v_1 v_3 v_4 v_5)^{-1} = c_1^{-1} v_2 c_1 (v_1 v_2 v_3 v_4 v_6)^{-1} = c_1^{-1} v_3 c_1 (v_3)^{-1} = 1,$$

$$c_1^{-1} v_4 c_1 (v_3 v_4)^{-1} = c_1^{-1} v_5 c_1 (v_1 v_5 v_6)^{-1} = c_1^{-1} v_6 c_1 (v_4 v_5 v_6)^{-1},$$

$$c_2^{-1} v_1 c_2 (v_1 v_6)^{-1} = c_2^{-1} v_2 c_2 (v_1 v_3 v_4 v_5)^{-1} = c_2^{-1} v_3 c_2 (v_1 v_2 v_3 v_5 v_6)^{-1} = 1,$$

$$c_2^{-1} v_4 c_2 (v_3)^{-1} = c_2^{-1} v_5 c_2 (v_1 v_2 v_5)^{-1} = c_2^{-1} v_6 c_2 (v_1 v_2 v_6)^{-1} = 1.$$

(f) *The element v_4 is a 2-central involution of E_1 and the Fitting subgroup V of $D_1 = C_{E_1}(v_4)$ has a complement $C_1 = \langle t_1, t_2, c_1^2 \rangle \cong S_4$, where both $t_1 = c_1^2 c_2 c_1 c_2 c_1^{-1} c_2^{-1} c_1^{-1} c_2$ and $t_2 = c_1^{-1} c_2 c_1 c_2$ have order 3.*

The conjugate actions of t_1 and t_2 on V induce endomorphisms given by the following matrices with respect to the basis $\mathcal{B}_1 = \{v_4, v_1, v_2, v_3, v_5, v_6\}$:

$$Mt_1 = \begin{pmatrix} 1&0&0&0&0&0 \\ 0&1&1&1&0&0 \\ 0&0&0&1&1&1 \\ 0&1&0&1&1&1 \\ 0&0&1&1&0&1 \\ 0&1&1&1&1&1 \end{pmatrix} \quad \text{and} \quad Mt_2 = \begin{pmatrix} 1&1&0&1&1&1 \\ 0&1&0&1&1&0 \\ 0&0&0&1&1&1 \\ 0&1&0&1&1&1 \\ 0&0&1&1&0&1 \\ 0&1&1&1&1&1 \end{pmatrix}.$$

(g) *Let $MC_1 = \langle Mt_1, Mt_2, (Mc_1)^2 \rangle$. Then $C_{\mathrm{GL}_6(2)}(MC_1)$ has order 2 and contains the involution*

$$Ma = \begin{pmatrix} 1&0&1&0&1&0 \\ 0&1&0&0&0&0 \\ 0&0&1&0&0&0 \\ 0&0&0&1&0&0 \\ 0&0&0&0&1&0 \\ 0&0&0&0&0&1 \end{pmatrix}.$$

(h) *The matrix*

$$Mw = \begin{pmatrix} 1&1&1&0&1&1 \\ 1&0&1&0&1&0 \\ 0&1&1&0&1&0 \\ 0&0&1&0&1&0 \\ 0&0&0&1&1&0 \\ 0&1&0&0&1&1 \end{pmatrix}$$

of order 8 satisfies the following conditions:

(1) *The group $MP = \langle MC_1, Mc_2, Ma^{Mw} \rangle$ is a subgroup of $\mathrm{GL}_6(2)$ and has order 1451520.*

(2) $(Mt_1)^{Mw}, (Mt_2)^{Mw} \in MP$ *and* $(Mc_1^2)^{Mw} \in MP$.

(i) $Mt = Ma^{Mw}$ *is 2-central involution of MP, and the Fitting subgroup MF_1 of $MH_1 = C_{MP}(Mt)$ is elementary abelian of order 2^5. It has a complement MC_2 isomorphic to the symmetric group S_6.*

(j) *A Sylow 2-subgroup MT of MP has a unique elementary abelian normal subgroup MA of order 2^6 and $N_{MP}(MA) \cong E_1$.*

(k) *The matrices* $Mp_1 = (Mt_1)^{Mw}$ *and* $Mp_2 = Mc_2Mt(Mc_2)^3MtMc_2$ *of respective orders 3 and 6 generate* MH_1.

(l) $MP = \langle MH_1, Mc_2 \rangle$ *is a simple group isomorphic to the symplectic group* $\mathrm{Sp}_6(2)$.

(m) *The simple group* MP *is isomorphic to the finitely presented group* $P = \langle p_1, p_2, c_2 \rangle$ *having the following set* $\mathcal{R}(P)$ *of defining relations:*

$$p_1^3 = p_2^6 = c_2^4 = 1,$$

$$(p_2c_2p_2)^2 = (c_2^{-1}p_2)^3 = [p_2^{-1}, c_2^{-1}]^2 = 1,$$

$$p_1c_2^{-1}p_2p_1^{-1}p_2^{-1}p_1^{-1}p_2^{-1}c_2p_1p_2 = p_1^{-1}p_2c_2p_2^{-2}p_1p_2^2c_2^{-1}p_2^{-1},$$

$$(p_1p_2^{-1})^6 = (p_1^{-1}p_2^{-2})^4 = p_1^{-1}c_2p_2^{-1}p_1p_2^{-2}p_1c_2p_2^{-1}p_1^{-1}p_2^2,$$

$$(p_1^{-1}p_2^{-2}p_1p_2^{-1}p_1^{-1}p_2^{-1})^2 = 1,$$

$$c_2p_2^{-2}p_1^{-1}c_2^{-1}p_1^{-1}p_2^{-1}p_1^{-1}c_2p_1c_2^2p_2p_1 = 1.$$

(n) *The twofold cover* $H = \langle h_1, h_2, h_3, h_4 \rangle$ *of* P *has the following set* $\mathcal{R}(H)$ *of defining relations:*

$$h_1^3 = h_4^2 = 1, \quad h_2^6 = h_3^4 = h_4,$$

$$[h_1, h_4] = [h_2, h_4] = [h_3, h_4] = 1,$$

$$h_2h_3h_2^2h_3h_2h_4^{-1} = (h_3^{-1}h_2)^3 = h_2h_3h_2^{-1}h_3^{-1}h_2h_3h_2^{-1}h_3^{-1}h_4^{-1} = 1,$$

$$h_1h_3^{-1}h_2h_1^{-1}h_2^{-1}h_1^{-1}h_2^{-1}h_3h_1h_2 = h_1^{-1}h_2h_3h_2^{-2}h_1h_2^2h_3^{-1}h_2^{-1} = 1,$$

$$(h_1h_2^{-1})^6 = h_1^{-1}h_2^{-2}h_1^{-1}h_2^{-2}h_1^{-1}h_2^{-2}h_1^{-1}h_2^{-2}h_4^{-1} = 1,$$

$$h_1^{-1}h_3h_2^{-1}h_1h_2^{-2}h_1h_3h_2^{-1}h_1^{-1}h_2^2 = (h_1^{-1}h_2^{-2}h_1h_2^{-1}h_1^{-1}h_2^{-1})^2 = 1,$$

$$h_3h_2^{-2}h_1^{-1}h_3^{-1}h_1^{-1}h_2^{-1}h_1^{-1}h_3h_1h_3^2h_2h_1h_4^{-1} = 1.$$

(o) H *has a faithful permutation representation of degree* 1920 *with stabilizer* $\langle (h_4^2h_1^{-1})^2, (h_2h_1^{-1}k_4^{-1}h_3^{-1})^8 \rangle$

(p) *The Sylow 2-subgroups of* H *and* E *are isomorphic.*

PROOF. (a) The faithful permutation representation PE of degree 168 of $E = \langle e_i \mid 1 \le i \le 14 \rangle$ has been found by means of the presentation of E given in Lemma 3.1 and MAGMA.

(b) Using the faithful permutation representation PE of E and MAGMA it has been checked that $z = e_3^2$ is a 2-central involution of E, $D = C_E(z)$ has order 2^{10}. Furthermore, an application of the MAGMA command NormalSubgroups(PD) yields that D has a unique normal subgroup Q of order 2^7. It is extra-special with its center $Z(Q) = \langle z \rangle$, and $V = Q/Z(Q)$ is elementary abelian. Applying then the MAGMA command C:=HasComplemnet(D,Q) it follows that Q has a complement C. An obvious isomorphism test with MAGMA shows that $C \cong \mathrm{PSL}_3(2)$.

(c) The five generators of the Sylow 2-subgroup $S = \langle e_3, e_7, e_8, x_1, x_2 \rangle$ have been constructed by means of the permutation representation PE and MAGMA. As $A = \langle e_{11}, e_{12}, e_{13}, e_{14} \rangle$ is an elementary abelian normal subgroup of E contained in D it is also normal in S. Using now the MAGMA command

    subs:=Subgroups(S : Al:=Normal, IsElementaryAbelian:=true)

we see that A is the maximal elementary abelian normal subgroup of S. Therefore condition (2) of Algorithm 2.5 is satisfied by E.

The generators of $D = \langle S, e_4 \rangle$ and $C = \langle c_1, c_2 \rangle$ given in the statement have been determined by means of PE and MAGMA.

(d) Similarly it has been verified that $Q = \langle f_i \mid 1 \leq i \leq 6 \rangle$. Since Q is extra-special of order 2^7 the set $\mathcal{B} = \{v_i = f_i Z(Q) \mid 1 \leq i \leq 6\}$ is a basis of the elementary abelian group $V = Q/Z(Q)$. Let $F = GF(2)$. Since Q is normal in D the conjugate actions of the generators c_i of C induce endomorphisms of the F-vector space V described with respect to $\mathcal{B}$ by the matrices Mc_i, $i = 1, 2$, as has been established by means of MAGMA.

(e) Using the faithful permutation representation PE of E and the MAGMA command FPGroup(C) to the permutation subgroup of PE generated by the permutations of the 2 generators of $C = \langle c_1, c_2 \rangle$ we get the following set $\mathcal{R}C$ of defining relations of C:

$$c_1^4 = c_2^4 = 1,$$
$$(c_1^{-1}c_2^{-1})^4 = c_2 c_1^{-1} c_2^{-1} c_1^{-2} c_2^{-1} c_1^{-1} c_2 c_1^{-1} = 1,$$
$$c_1 c_2^{-1} c_1 c_2^{-1} c_1^{-1} c_2^2 c_1^{-1} c_2^{-1} = 1.$$

Since $Q = \langle f_i \mid 1 \leq i \leq 6 \rangle$ is extra-special with center $Z(Q) = \langle z \rangle$ and D is the semi-direct product of Q and C the actions of the generators c_j of C on the f_i can be explicitly determined by means of the matrices Mc_i and a MAGMA calculation in the permutation representation PE of E. It follows that $D = \langle c_1, c_2, f_1, f_2, \ldots, f_6 \rangle$ has set $\mathcal{R}D$ of defining relations consisting of $\mathcal{R}C$ and the following relations:

$$f_3^2 = f_4^2 = f_6^2 = z^2 = 1, \quad f_1^2 = f_2^2 = f_5^2 = z,$$
$$c_1^{-1} f_1 c_1 (f_1 f_3 f_4 f_5)^{-1} = c_1^{-1} f_2 c_1 (f_1 f_2 f_3 f_4 f_6)^{-1} = z,$$
$$c_1^{-1} f_3 c_1 (f_3)^{-1} = c_1^{-1} f_5 c_1 (f_1 f_5 f_6)^{-1} = c_1^{-1} f_6 c_1 (f_4 f_5 f_6)^{-1} = z,$$
$$c_1^{-1} f_4 c_1 (f_3 f_4)^{-1} = c_2^{-1} f_1 c_2 (f_1 f_6)^{-1} = 1,$$
$$c_2^{-1} f_2 c_2 (f_1 f_3 f_4 f_5)^{-1} = c_2^{-1} f_3 c_2 (f_1 f_2 f_3 f_5 f_6)^{-1} = 1,$$
$$c_2^{-1} f_4 c_2 (f_3)^{-1} = c_2^{-1} f_5 c_2 (f_1 f_2 f_5)^{-1} = c_2^{-1} f_6 c_2 (f_1 f_2 f_6)^{-1} = z,$$
$$[f_1, f_2] = [f_1, f_3] = [f_1, f_5] = [f_2, f_4] = [f_2, f_6] = z,$$
$$[f_1, f_4] = [f_1, f_6] = [f_2, f_3] = [f_2, f_5] = [f_3, f_4] = [f_3, f_5] = 1,$$
$$[f_3, f_6] = [f_4, f_5] = z, \quad \text{and} \quad [f_4, f_6] = [f_5, f_6] = 1.$$

Thus $E_1 = D/Z(Q) = V : C$ has the presentation given in the statement. Furthermore, it has a faithful permutation representation PE_1 of degree 64 with stabilizer C.

(f) We now apply Algorithm 2.5 to the extension E_1 of the F-module V by $C \cong \mathrm{PSL}_3(2)$. Using MAGMA it has been checked that V is a uniserial indecomposable FC-module with two non-isomorphic composition factors having both dimension 3 over F. By Proposition 8.6.5 of [20] we know that E_1 is isomorphic to $N_G(A)$, where A is a maximal elementary abelian normal subgroup of a Sylow 2-subgroup of the simple symplectic group $G = \mathrm{Sp}_6(2)$. Therefore we use the conjugate action of C on V to get an embedding of C into $GL_6(2)$. Using the permutation representation PE_1 of E_1 and MAGMA we see that $C_{E_1}(v_1)$ is a semidirect product of V and the subgroup $C_1 = \langle t_1, t_2, c_1^2 \rangle \cong S_4$ where $t_1 = c_1^2 c_2 c_1 c_2 c_1^{-1} c_2^{-1} c_1^{-1} c_2$ and $t_2 = c_1^{-1} c_2 c_1 c_2$ have both order 3. Furthermore, with respect to $\mathcal{B}_1$ the conjugate actions of t_1 and t_2 on V are described by the matrices Mt_1 and Mt_2 in $GL_6(2)$, respectively.

(g) Let $MC_1 = \langle Mt_1, Mt_2, (Mc_1)^2 \rangle$. Another MAGMA calculation now shows that $C_{GL_6(2)}(MC_1)$ is generated by the matrix Ma of order 2 given in the statement.

(h) In order to be able to do the following calculation finding the matrix Mw of the statement we use the faithful permutation representation $PL63$ of $GL_6(2)$ with degree 63. Its stabilizer MU is generated by the five elementary matrices $M_{2,1}$, $M_{3,1}$, $M_{4,1}$, $M_{5,1}$, $M_{6,1}$ of $GL_6(2)$ and the following two matrices:

$$Mg_1 = \begin{pmatrix} 100000 \\ 011000 \\ 001000 \\ 000100 \\ 000010 \\ 000001 \end{pmatrix} \quad \text{and} \quad Mg_2 = \begin{pmatrix} 100000 \\ 000001 \\ 010000 \\ 001000 \\ 000100 \\ 000010 \end{pmatrix}.$$

Let pt_1, pt_2, pc_1, pc_2, pa be the respective images of the matrices Mt_1, Mt_2, Mc_1, Mc_2, Ma under the isomorphism $h : GL_6(2) \to PL63$. Let $PC_1 = \langle pt_1, pt_2, pc_1^2 \rangle$. For each $x \in PL63$ let $W(x) = \langle PC_1, pc_2, pa^x \rangle$, $U(x) = \langle pt_1^x, pt_2^x, (pc_1^2)^x \rangle$ and $D(x) = U(x) \cap W(x)$. Using then the MAGMA command

```
exists(w){x: x in PL63|Order(W(x)) eq 1451520 and D(x) eq  U(x)}
```

we obtain a permutation pw of order 8 in $PL63$ such that $h(Mw) = pw$, where Mw is the matrix of the statement. It satisfies both conditions (1) and (2) of the statement by the definition of w.

(i) All assertions of this statement have been verified by means of the faithful permutation representation $PL63$ and MAGMA.

(j) Let $MP = \langle MC_1, Mc_2, Ma^{Mw} \rangle$. Let $PP = h(MP)$ be the image of the matrix group MP in $PL63$, and let $pt = h(Mt)$ where $Mt = Ma^{Mw}$. Let PT be any Sylow 2-subgroup of $PH_1 = C_{PP}(pt)$. Using the MAGMA command

```
subs:=Subgroups(PT : Al:=Normal, IsElementaryAbelian:=true)
```

we see that PT has a unique maximal elementary abelian normal subgroup PA of order 2^6. Furthermore, an isomorphism test shows that $N_{PP}(PA) \cong E_1$. Since h is an isomorphism the statement holds.

(k) Again this assertion has been verified by means of $PL63$ and MAGMA. The elements $Mp_1 = (Mt_1)^{Mw}$ and $Mp_2 = Mc_2Mt(Mc_2)^3MtMc_2$ generate MH_1 and have orders 3 and 6, respectively.

(l) Another application of MAGMA yields that $PP = \langle PH_1, pc_2 \rangle$ is a simple group isomorphic to $Sp_6(2)$, see also [20, Prop. 8.6.5].

(m) Since $MP = \langle Mp_1, Mp_2, Mc_2 \rangle \cong PP$ under the isomorphism h we apply the MAGMA command FPGroup(PP) to the permutation group $PP = \langle pp_1, pp_2, pc_2 \rangle$ of degree 63. This way we obtain the presentation of MP given in the statement.

(n) By (m) the finitely presented group $P = \langle p_1, p_2, c_2 \rangle$ satisfies all conditions of Holt's Algorithm 7.4.5 of [20] implemented in MAGMA [15] and the trivial matrix representation F of PP over F. It follows by means of MAGMA that the second cohomological dimension $\dim_F[H^2(H_1, F)] = 1$. Using then the extension process of MAGMA we found the presentation of the unique 2-fold cover H of P with center $Z(H) = \langle h_4 \rangle$ given in (n).

(o) This permutation representation PH of H has been established by means of MAGMA and a stand alone program due to P. Young. Using the Todd-Coxeter Algorithm implemented in MAGMA statement (o) can easily be checked.

(p) Using now the faithful permutation representation PE and PH of the groups E and H defined in (a) and (o) and MAGMA it can be verified that E and H have isomorphic Sylow 2-subgroups. This completes the proof. $\qquad\square$

4. Construction of Conway's simple group Co3

In Lemma 4.1 of this section the main condition of Algorithm 2.5 is verified for the amalgam $H \leftarrow D \rightarrow E$ constructed in Sections 2 and 3. Therefore steps 6 through 11 of the algorithm can be applied to give here a self-contained existence proof for the sporadic group Co3 of Conway [**7**].

LEMMA 4.1. *Let $H = H(\mathrm{Co}_3) = \langle h_i \mid 1 \le i \le 4 \rangle$ be the finitely presented group constructed in Proposition 3.3. Let $E = \langle e_i \mid 1 \le i \le 14 \rangle$ be the non-split extension of $T = GL_4(2)$ by its simple module V_1 of dimension 4 over $F = GF(2)$ constructed in Lemma 3.1. Then the following statements hold:*

(a) *Each Sylow 2-subgroup S of H has a maximal elementary abelian normal subgroup A of order 2^4 and $D = N_H(A) \cong D_1 = C_E(z)$, where $z = e_3^2$ is in E.*

(b) *There is a Sylow 2-subgroup S such that $D = \langle x, y \rangle$ and $H = \langle x, y, h \rangle$, where $x = h_3$ and $y = h_9 h_{14} h_{15}$ and $h = h_1$ have respective orders 8, 8 and 3.*

(c) *The Goldschmidt index of the amalgam $H \leftarrow D \rightarrow E$ is 1.*

(d) *A system of representatives r_i of the 43 conjugacy classes of H and the corresponding centralizers orders $|C_H(r_i)|$ are given in Table A.1.*

(e) *A system of representatives d_i of the 30 conjugacy classes of D and the corresponding centralizers orders $|C_D(d_i)|$ are given in Table A.3.*

(f) *Let $\sigma : N_H(A) \rightarrow D_1 = C_E(z)$ be the isomorphism given in (a). Then $E = \langle x_1 = \sigma(x), y_1 = \sigma(y), e \rangle$ for $e = e_1$ of order 2. A system of representatives s_i of the 25 conjugacy classes of E and the corresponding centralizers orders $|C_E(s_i)|$ are given in Table A.2.*

(g) *The character tables of H, D and E are stated in Tables B.3, B.2 and B.1, respectively.*

PROOF. (a) By Proposition 3.3(a) and (o) the groups E and H have faithful permutation representations PE and PH of degrees 168 and 1920, respectively. Let S be any Sylow 2-subgroup of H. By means of MAGMA one verifies that S has two non conjugate maximal elementary abelian normal subgroups A_i with normalizers $N_H(A_i)$ of orders $2^{10} \cdot 3$ and $2^{10} \cdot 3 \cdot 7$. Let A be the one having the larger order, and let $D = N_H(A)$. By Proposition 3.3 $z := e_3^2$ is a 2-central involution of E with centralizer $D_1 = C_E(z)$ of order $2^{10} \cdot 3 \cdot 7$. Using now both permutation representations PE and PH an isomorphism test of MAGMA shows that there is a group isomorphism $\sigma : D \rightarrow D_1$.

(b) The generators of the three groups D, H, and E have been found computationally using the permutations representations PH and PE and the explicit isomorphism σ between $D = \langle x, y \rangle$ and $D_1 = \langle x_1 = \sigma(x), y_1 = \sigma(y) \rangle$.

(c) The Goldschmidt index has been calculated by means of Kratzer's Algorithm 7.1.10 of [20].

(d) and (e) The systems of representatives of the conjugacy classes of H and D have been computed by means of PH, MAGMA and Kratzer's Algorithm 5.3.18 of [20].

(f) This assertion is proven in the same manner as (d), using the permutation representation PE.

(g) The character tables were obtained using PH and MAGMA.   □

THEOREM 4.2. *Keep the notation of Lemma 4.1 and Proposition 3.3. Using the notation of the three character tables B.3, B.2, and B.1 of the groups H, D, and E, respectively, the following statements hold:*

(a) *There is exactly one compatible pair $(\chi, \tau) \in mf\mathrm{char}_{\mathbb{C}}(H) \times mf\mathrm{char}_{\mathbb{C}}(E)$ of degree 23 of the groups $H = \langle D, h \rangle$ and $E = \langle D, e \rangle$:*

$$\chi_3 + \chi_4 = \tau_1 + \tau_2 + \tau_4$$

with common restriction

$$\tau_{|D} = \chi_{|D} = \psi_1 + \psi_6 + \psi_8 + \psi_{10},$$

where irreducible characters with bold face indices denote faithful irreducible characters.

(b) *Let $\mathfrak{V}$ and $\mathfrak{W}$ be the uniquely determined (up to isomorphism) faithful semisimple multiplicity-free 23-dimensional modules of H and E over $F = GF(23)$ corresponding to the compatible pair χ, τ, respectively.*
   Let $\kappa_{\mathfrak{V}} : H \to \mathrm{GL}_{23}(23)$ and $\kappa_{\mathfrak{W}} : E \to \mathrm{GL}_{23}(23)$ be the representations of H and E afforded by the modules $\mathfrak{V}$ and $\mathfrak{W}$, respectively.
   Let $\mathfrak{h} = \kappa_{\mathfrak{V}}(h)$, $\mathfrak{x} = \kappa_{\mathfrak{V}}(x)$, $\mathfrak{y} = \kappa_{\mathfrak{V}}(y)$ in $\kappa_{\mathfrak{V}}(H) \leq \mathrm{GL}_{23}(23)$. Then the following assertions hold:
   (1) $\mathfrak{V}_{|D} \cong \mathfrak{W}_{|D}$, *and there is a transformation matrix $T \in \mathrm{GL}_{23}(23)$ such that*

$$\mathfrak{x} = T^{-1}\kappa_{\mathfrak{W}}(x_1)T, \mathfrak{y} = T^{-1}\kappa_{\mathfrak{W}}(y_1)T.$$

   Let $\mathfrak{e} = T^{-1}\kappa_{\mathfrak{W}}(e_1)T \in \mathrm{GL}_{23}(23)$.
   (2) $\mathfrak{G} = \langle \mathfrak{h}, \mathfrak{x}, \mathfrak{y}, \mathfrak{e} \rangle$ *has a faithful permutation representation of degree 170775 with stabilizer $\mathfrak{E} = \langle \mathfrak{x}, \mathfrak{y}, \mathfrak{e} \rangle$.*

(3) *The generating matrices of $\mathfrak{G}$ are:*

$$\mathfrak{x} = \begin{pmatrix} & & & & & & & & \end{pmatrix}$$

$$\mathfrak{y} = \begin{pmatrix} & & & & & & & & \end{pmatrix}$$

$$\mathfrak{h} = \begin{pmatrix} & & & & & & & & \end{pmatrix}$$

$$\mathfrak{e} = \begin{pmatrix} & & & & & & & & \end{pmatrix}$$

(4) $\mathfrak{G} = \langle \mathfrak{h}, \mathfrak{x}, \mathfrak{y}, \mathfrak{e} \rangle$, *and $\mathfrak{G}$ has 42 conjugacy classes $\mathfrak{g}_i{}^{\mathfrak{G}}$ with representatives $\mathfrak{g}_i$ and centralizer orders $|C_{\mathfrak{G}}(\mathfrak{g}_i)|$ as given in Table A.4.*

(5) *The character table of $\mathfrak{G}$ given in Table B.4 coincides with that of
Co$_3$ in the Atlas [8, p. 154].*

(c) $\mathfrak{G}$ *is a finite simple group with 2-central involution $\kappa_{\mathfrak{V}}(z) = (\mathfrak{x})^4$ such that*

$$C_{\mathfrak{G}}(\kappa_{\mathfrak{V}}(z)) = \kappa_{\mathfrak{V}}(H), \;\; and \;\; |\mathfrak{G}| = 2^{10} \cdot 3^7 \cdot 5^3 \cdot 7 \cdot 11 \cdot 23.$$

(d) *The simple group $\mathfrak{G}$ is isomorphic to Conway's simple group Co$_3$.*

PROOF. Throughout we keep the notation of Lemma 4.1.

(a) The character tables of the groups $H = \langle x, y, h \rangle$, $D = \langle x, y \rangle$, $D_1 = \langle x_1, y_1 \rangle$ and E are stated in the appendix. In the following we use their notations. Using MAGMA, these character tables and the fusion of the classes of D in H and D_1 in E, the compatible pair stated in assertion (a) has been calculated by means of MAGMA and Kratzer's Algorithm 7.3.10 of [20].

(b) In order to construct the semisimple faithful representation $\mathfrak{V}$ corresponding to the character $\chi = \chi_3 + \chi_4$ we use the faithful permutation representation PH of H of degree 1920 constructed in Proposition 3.3. Since both irreducible characters χ_3, χ_4 of H are not contained in the permutation character of PH the MAGMA command LowIndexSubgroups(PH, 1000) was applied. MAGMA provided generating permutations of 34 subgroups U_i of H having index $|H : U_i| \le 1000$. Furthermore it calculated all their character tables and the fusion between the classes of each U_i and H. Thus it was easy to calculate the inner products $((1_{U_i})^H, \chi_k)$ for all $\{1 \le i \le 34\}$ and $k = 3, 4$. It follows that χ_3 and χ_4 are constituents of the permutation characters PU_i of the subgroups U_{10} and U_3 of indices 240 and 72 for $i = 10$ and $i = 3$, respectively. As $H = \langle x, y, h \rangle$, MAGMA provided the permutation matrices $P_i(x)$, $P_i(y)$ and $P_i(h)$ for $i = 10, 3$. Applying the the Meataxe algorithm of MAGMA over the field $GF(23)$ one obtains from PU_{10} the 8×8 matrices $M(x)$, $M(y)$, $M(h)$ corresponding to the faithful irreducible character χ_3. Similarly PU_3 yields the three 15×15 matrices $N(x), N(y), N(h) \in \mathrm{GL}_{23}(23)$ corresponding to χ_4. The three generating matrices of $\mathfrak{H}$ are obtained by defining:

$$
\begin{aligned}
\mathfrak{x} &:= & \mathrm{GL}(23, 23)!\mathrm{DiagonalJoin}(N(x), M(x)), \\
\mathfrak{y} &:= & \mathrm{GL}(23, 23)!\mathrm{DiagonalJoin}(N(y), M(y)), \;\; and \\
\mathfrak{h} &:= & \mathrm{GL}(23, 23)!\mathrm{DiagonalJoin}(N(h), M(h)).
\end{aligned}
$$

These three matrices describe the semisimple H-module $\mathfrak{W}$ over $F = GF(23)$.

The faithful semisimple representation $\mathfrak{W}$ corresponding to the character $\tau = \tau_1 + \tau_2 + \tau_4$ of $E = \langle x_1, y_1, e_1 \rangle$ has been obtained similarly from the permutation module PE of degree 168 directly, because all three characters τ_1, τ_2, and τ_4 are contained in the permutation character of PE. Let $\mathfrak{W}(x_1)$, $\mathfrak{W}(y_1)$ and $\mathfrak{W}(e_1)$ be the blocked diagonal matrices of the three generators x_1, y_1 and e_1 of E with respect to $\mathfrak{W}$.

Then the matrix groups $\mathfrak{D} = \langle \mathfrak{x}, \mathfrak{y} \rangle$ and $\mathfrak{D}_1 = \langle \mathfrak{W}(x_1), \mathfrak{W}(y_1) \rangle$ are isomorphic, and so are the semisimple representations $\mathfrak{V}_{|\mathfrak{D}}$ and $\mathfrak{W}_{|\mathfrak{D}_1}$ by the definition of the compatible pair $(\chi, \tau) \in mf\mathrm{char}_{\mathbb{C}}(H) \times mf\mathrm{char}_{\mathbb{C}}(E)$ and Maschke's Theorem which is applicable because the orders of H and E are coprime to 23. Let

$Y = \mathrm{GL}(23, 23) = \mathrm{GL}_{23}(23)$. Applying then Parker's isomorphism test of Proposition 6.1.6 of [20] by means of the MAGMA command

`IsIsomorphic(GModule(sub<Y|V(x),V(y)>),GModule(sub<Y|W(x1),W(y1)>))`

we obtain the transformation matrix

$$
\mathcal{T} =
\begin{pmatrix}
14 & 9 & 14 & 14 & 9 & 14 & \cdot & \overset{17}{9} & 14 & \cdot & 14 & \cdot & \cdot & 9 & \cdot & \cdot & \cdot & \cdot & \cdot & \cdot & \cdot & \cdot & \cdot \\
\cdot & 9 & \cdot & 14 & \cdot & \cdot & \cdot & 9 & \cdot & \cdot & 14 & 14 & \cdot & \cdot & 9 & \cdot & \cdot & \cdot & \cdot & \cdot & \cdot & \cdot & \cdot \\
14 & 9 & 14 & \cdot & 9 & \cdot & \cdot & 9 & 14 & 14 & \cdot & 14 & 9 & \cdot & \cdot & \cdot & \cdot & \cdot & \cdot & \cdot & \cdot & \cdot & \cdot \\
\cdot & 9 & \cdot & \cdot & 14 & \cdot & 9 & \cdot & 14 & \cdot & \cdot & 9 & 9 & 9 & \cdot & \cdot & \cdot & \cdot & \cdot & \cdot & \cdot & \cdot & \cdot \\
14 & \cdot & 14 & \cdot & 9 & 14 & \cdot & \cdot & 14 & \cdot & \cdot & 14 & \cdot & 9 & 9 & \cdot & \cdot & \cdot & \cdot & \cdot & \cdot & \cdot & \cdot \\
14 & \cdot & 14 & 14 & 9 & \cdot & \cdot & 14 & 14 & 14 & \cdot & 9 & \cdot & 9 & \cdot & \cdot & \cdot & \cdot & \cdot & \cdot & \cdot & \cdot & \cdot \\
\cdot & \cdot & 14 & \cdot & \cdot & 14 & 5 & \cdot & \cdot & 14 & 14 & 14 & 9 & 9 & \cdot & \cdot & \cdot & \cdot & \cdot & \cdot & \cdot & \cdot & \cdot \\
14 & 20 & 3 & 20 & 20 & 9 & \cdot & 20 & 3 & 9 & 20 & 20 & 3 & 3 & 9 & \cdot & \cdot & \cdot & 14 & 14 & \cdot & \cdot & \cdot \\
20 & 20 & 20 & 3 & 3 & 20 & \cdot & 20 & 9 & 3 & 3 & 20 & 20 & 3 & 3 & \cdot & \cdot & \cdot & \cdot & \cdot & 14 & 9 & \cdot \\
3 & 14 & 14 & 3 & 20 & 20 & \cdot & 3 & 3 & 20 & 14 & 3 & 3 & 3 & 3 & 9 & \cdot & 14 & 14 & 9 & 9 & \cdot & \cdot \\
3 & 14 & 14 & 9 & 20 & 14 & \cdot & 3 & 3 & 3 & 20 & 20 & 9 & 20 & 3 & \cdot & 14 & \cdot & 9 & \cdot & \cdot & 9 & \cdot \\
20 & 14 & 20 & 3 & 14 & 14 & \cdot & 3 & 20 & 20 & 3 & 9 & 3 & 20 & 20 & \cdot & \cdot & \cdot & 14 & \cdot & \cdot & 9 & \cdot \\
20 & 3 & 20 & 14 & 14 & 3 & \cdot & 14 & 20 & 20 & 3 & 20 & 3 & 20 & 9 & \cdot & \cdot & \cdot & 9 & \cdot & \cdot & 9 & \cdot \\
20 & 20 & 20 & 20 & 14 & 14 & \cdot & 20 & 20 & 20 & 9 & 3 & 3 & 20 & 3 & \cdot & 14 & 14 & 14 & \cdot & \cdot & 9 & 14 \\
3 & 20 & 14 & 3 & 20 & 14 & \cdot & 20 & 3 & 14 & 3 & 3 & 20 & 20 & 20 & \cdot & 9 & 14 & 14 & \cdot & 9 & 14 & \cdot \\
3 & 3 & 14 & 20 & 20 & 3 & \cdot & 14 & 3 & 14 & 20 & 20 & 20 & 20 & 3 & 14 & 14 & \cdot & 14 & 14 & \cdot & \cdot & \cdot \\
3 & 20 & 14 & 3 & 20 & 3 & \cdot & 20 & 3 & 3 & 14 & 14 & 9 & 20 & 9 & 14 & \cdot & 14 & \cdot & \cdot & 9 & 14 & \cdot \\
20 & 20 & 20 & 3 & 14 & 20 & \cdot & 20 & 20 & 3 & 3 & 20 & 9 & 14 & 3 & 9 & 9 & 14 & 14 & 9 & 18 & \cdot & \cdot \\
3 & 20 & 14 & 20 & 20 & 20 & \cdot & 20 & 3 & 20 & 20 & 20 & 3 & 3 & 9 & \cdot & \cdot & \cdot & \cdot & 9 & 9 & \cdot & \cdot \\
20 & 3 & 20 & 20 & 3 & 9 & \cdot & 14 & 9 & 3 & 20 & 14 & 20 & 3 & 9 & 14 & \cdot & 18 & 9 & 14 & 14 & \cdot & \cdot \\
3 & 3 & 14 & 3 & 20 & 20 & \cdot & 20 & 3 & 14 & 3 & 14 & 3 & 3 & \cdot & 9 & \cdot & 14 & 14 & \cdot & \cdot & 9 & \cdot \\
20 & 3 & 20 & 20 & 3 & 9 & \cdot & 14 & 9 & 3 & 20 & 14 & 20 & 3 & 9 & 14 & \cdot & 18 & 9 & 14 & 14 & \cdot & \cdot \\
14 & 3 & 3 & 14 & 20 & 9 & \cdot & 14 & 3 & 20 & 3 & 3 & 14 & 3 & 3 & \cdot & 14 & \cdot & \cdot & 14 & 9 & 14 & \cdot
\end{pmatrix}
$$

satisfying $\mathfrak{V}(x) = (\mathfrak{W}(x_1))^{\mathcal{T}}$ and $\mathfrak{V}(y) = (\mathfrak{W}(y_1))^{\mathcal{T}}$.

Let $\mathfrak{e} = (\mathfrak{W}(e_1))^{\mathcal{T}}$. Then $\mathfrak{E} = \langle \mathfrak{x}, \mathfrak{y}, \mathfrak{e} \rangle \cong E$.

Let $\mathfrak{G} = \langle \mathfrak{h}, \mathfrak{x}, \mathfrak{y}, \mathfrak{e} \rangle$. Using the MAGMA command `CosetAction(G, E)` we obtain a faithful permutation representation PG of $\mathfrak{G}$ of degree 170775 with stabilizer $\mathfrak{E}$. In particular $|\mathfrak{G}| = 2^{18} \cdot 3^6 \cdot 5^3 \cdot 7 \cdot 11 \cdot 23$.

Using the faithful permutation PG of degree 170775 and Kratzer's Algorithm 5.3.18 of [20] we calculated the representatives of all the conjugacy classes of $\mathfrak{G}$, see Table A.4.

Furthermore, the character table of $\mathfrak{G}$ has been calculated by means of the above permutation representation and MAGMA. It is stated in Table B.4 and coincides with the one of Co$_3$ in [8, pp. 154–155].

(c) Let $\mathfrak{z} = (\mathfrak{x})^4$. Then $C_{\mathfrak{G}_2}(\mathfrak{z})$ contains $\mathfrak{H} = \langle \mathfrak{h}, \mathfrak{x}, \mathfrak{y} \rangle$ which is isomorphic to H. By the Table A.4 we know that $|C_{\mathfrak{G}}(\mathfrak{z})| = |H|$. Hence $C_{\mathfrak{G}} = (\mathfrak{z}) \cong \mathfrak{H}$. The character table of $\mathfrak{G}$ implies that $\mathfrak{G}$ is a simple group, see Table B.4.

(d) This assertion follows from (c) and Fendel's Theorem of [11]. This completes the proof. $\qquad\square$

Appendix A. Representatives of conjugacy classes

A.1. *Conjugacy classes of* $H = \langle x, y, h \rangle$

Class	Representative	$|Class|$	$|Centralizer|$	2P	3P	5P	7P
1	1	1	$2^{10} \cdot 3^4 \cdot 5 \cdot 7$	1	1	1	1
2_1	$(x)^4$	1	$2^{10} \cdot 3^4 \cdot 5 \cdot 7$	1	2_1	2_1	2_1
2_2	$(x^2y)^4$	630	$2^9 \cdot 3^2$	1	2_2	2_2	2_2
2_3	$(x^2hy)^3$	7560	$2^7 \cdot 3$	1	2_3	2_3	2_3
3_1	$(yh)^4$	672	$2^5 \cdot 3^3 \cdot 5$	3_1	1	3_1	3_1
3_2	$(xh)^3$	2240	$2^4 \cdot 3^4$	3_2	1	3_2	3_2
3_3	h	13440	$2^3 \cdot 3^3$	3_3	1	3_3	3_3
4_1	$(xhy)^3$	126	$2^9 \cdot 3^2 \cdot 5$	2_1	4_1	4_1	4_1
4_2	$(x)^2$	1890	$2^9 \cdot 3$	2_1	4_2	4_2	4_2
4_3	$(xyh)^3$	3780	$2^8 \cdot 3$	2_2	4_3	4_3	4_3
4_4	$(xy^2hy)^3$	3780	$2^8 \cdot 3$	2_2	4_4	4_4	4_4
4_5	$(x^2y)^2$	22680	2^7	2_2	4_5	4_5	4_5
5	$(y^2h)^4$	48384	$2^2 \cdot 3 \cdot 5$	5	5	1	5
6_1	$(yh)^2$	672	$2^5 \cdot 3^3 \cdot 5$	3_1	2_1	6_1	6_1
6_2	$(xh^2)^3$	2240	$2^4 \cdot 3^4$	3_2	2_1	6_2	6_2
6_3	$(xhy)^2$	13440	$2^3 \cdot 3^3$	3_3	2_1	6_3	6_3
6_4	$(xyh)^2$	20160	$2^4 \cdot 3^2$	3_2	2_2	6_4	6_4
6_5	x^3hy	20160	$2^4 \cdot 3^2$	3_2	2_2	6_5	6_5
6_6	$xhxh^2y$	20160	$2^4 \cdot 3^2$	3_1	2_2	6_6	6_6
6_7	xy^2	80640	$2^2 \cdot 3^2$	3_3	2_2	6_7	6_7
6_8	x^2hy	241920	$2^2 \cdot 3$	3_3	2_3	6_8	6_8
7	$(xy)^2$	207360	$2 \cdot 7$	7	7	7	1
8_1	$(xy^3h)^3$	15120	$2^6 \cdot 3$	4_2	8_1	8_1	8_1
8_2	$(x^2yhyxh)^3$	15120	$2^6 \cdot 3$	4_2	8_2	8_2	8_2
8_3	x	90720	2^5	4_2	8_3	8_3	8_3
8_4	x^2y	90720	2^5	4_5	8_4	8_4	8_4
8_5	xhy^2	90720	2^5	4_5	8_5	8_5	8_5
8_6	x^2yhy	181440	2^4	4_3	8_6	8_6	8_6
9	xh	161280	$2 \cdot 3^2$	9	3_2	9	9
10	$(y^2h)^2$	48384	$2^2 \cdot 3 \cdot 5$	5	10	2_1	10
12_1	xy^2xyhxh	20160	$2^4 \cdot 3^2$	6_1	4_1	12_1	12_1
12_2	yh	60480	$2^4 \cdot 3$	6_1	4_2	12_2	12_2
12_3	xhy	80640	$2^2 \cdot 3^2$	6_3	4_1	12_3	12_3
12_4	xyh	120960	$2^3 \cdot 3$	6_4	4_3	12_4	12_4
12_5	xy^2hy	120960	$2^3 \cdot 3$	6_4	4_4	12_5	12_5
14	xy	207360	$2 \cdot 7$	7	14	14	2_1
15	$(x^3yh)^2$	96768	$2 \cdot 3 \cdot 5$	15	5	3_1	15
18	xh^2	161280	$2 \cdot 3^2$	9	6_2	18	18
20_1	y^2h	145152	$2^2 \cdot 5$	10	20_1	4_1	20_1
20_2	$(y^2h)^{11}$	145152	$2^2 \cdot 5$	10	20_2	4_1	20_2
24_1	xy^3h	120960	$2^3 \cdot 3$	12_2	8_1	24_1	24_1
24_2	x^2yhyxh	120960	$2^3 \cdot 3$	12_2	8_2	24_2	24_2
30	x^3yh	96768	$2 \cdot 3 \cdot 5$	15	10	6_1	30

A.2. Conjugacy classes of $E = \langle x, y, e \rangle$

Class	Representative	|Class|	|Centralizer|	2P	3P	5P	7P
1	1	1	$2^{10} \cdot 3^2 \cdot 5 \cdot 7$	1	1	1	1
2_1	$(x)^4$	15	$2^{10} \cdot 3 \cdot 7$	1	2_1	2_1	2_1
2_2	e	840	$2^7 \cdot 3$	1	2_2	2_2	2_2
2_3	$(x^3ey)^3$	840	$2^7 \cdot 3$	1	2_3	2_3	2_3
3_1	$(x^3ey)^2$	1792	$2^2 \cdot 3^2 \cdot 5$	3_1	1	3_1	3_1
3_2	$(ye)^2$	4480	$2^3 \cdot 3^2$	3_2	1	3_2	3_2
4_1	$(x^2yxy)^3$	210	$2^9 \cdot 3$	2_1	4_1	4_1	4_1
4_2	$(x)^2$	630	2^9	2_1	4_2	4_2	4_2
4_3	$(y)^2$	2520	2^7	2_1	4_3	4_3	4_3
4_4	xy^2ey	20160	2^4	2_3	4_4	4_4	4_4
5	xey	21504	$3 \cdot 5$	5	5	1	5
6_1	xy^2	13440	$2^3 \cdot 3$	3_2	2_1	6_1	6_1
6_2	ye	26880	$2^2 \cdot 3$	3_2	2_2	6_2	6_2
6_3	x^3ey	26880	$2^2 \cdot 3$	3_1	2_3	6_3	6_3
7_1	$(xy)^2$	23040	$2 \cdot 7$	7_1	7_2	7_2	1
7_2	$(xy)^6$	23040	$2 \cdot 7$	7_2	7_1	7_1	1
8_1	x^2y	5040	2^6	4_2	8_1	8_1	8_1
8_2	x^2ey	5040	2^6	4_2	8_2	8_2	8_2
8_3	x	10080	2^5	4_2	8_3	8_3	8_3
8_4	y	20160	2^4	4_3	8_4	8_4	8_4
12	x^2yxy	26880	$2^2 \cdot 3$	6_1	4_1	12	12
14_1	xy	23040	$2 \cdot 7$	7_1	14_2	14_2	2_1
14_2	$(xy)^3$	23040	$2 \cdot 7$	7_2	14_1	14_1	2_1
15_1	$xyxey$	21504	$3 \cdot 5$	15_1	5	3_1	15_2
15_2	$(xyxey)^7$	21504	$3 \cdot 5$	15_2	5	3_1	15_1

A.3. Conjugacy classes of $D = \langle x, y \rangle$

Class	Representative	|Class|	|Centralizer|	2P	3P	7P
1	1	1	$2^{10} \cdot 3 \cdot 7$	1	1	1
2_1	$(x)^4$	1	$2^{10} \cdot 3 \cdot 7$	1	2_1	2_1
2_2	$(x^2y)^4$	14	$2^9 \cdot 3$	1	2_2	2_2
2_3	$x^2yx^2y^3$	56	$2^7 \cdot 3$	1	2_3	2_3
2_4	$x^2y^2xy^2$	168	2^7	1	2_4	2_4
2_5	$x^2yx^2yxy^2$	336	2^6	1	2_5	2_5
3	$(xy^2)^2$	896	$2^3 \cdot 3$	3	1	3
4_1	$(x^2yxy)^3$	14	$2^9 \cdot 3$	2_1	4_1	4_1
4_2	$(x)^2$	42	2^9	2_1	4_2	4_2
4_3	x^2y^3xy	84	2^8	2_2	4_3	4_3
4_4	$(x^2yxyxy^2)^2$	84	2^8	2_2	4_4	4_4
4_5	$(y)^2$	168	2^7	2_1	4_5	4_5
4_6	$(x^2y)^2$	168	2^7	2_2	4_6	4_6
4_7	$x^2yxyx^2y^2$	336	2^6	2_2	4_7	4_7
4_8	$xyxy^3xy^2$	1344	2^4	2_4	4_8	4_8
6_1	$(x^2yxy)^2$	896	$2^3 \cdot 3$	3	2_1	6_1
6_2	xy^2	1792	$2^2 \cdot 3$	3	2_2	6_2
6_3	x^2yxy^2xy	1792	$2^2 \cdot 3$	3	2_3	6_3
7_1	$(xy)^2$	1536	$2 \cdot 7$	7_1	7_2	1
7_2	$(xy)^6$	1536	$2 \cdot 7$	7_2	7_1	1
8_1	x^2yxy^3	336	2^6	4_2	8_1	8_1
8_2	$x^2y^3xy^2xy$	336	2^6	4_2	8_2	8_2
8_3	x	672	2^5	4_2	8_3	8_3
8_4	x^2y	672	2^5	4_6	8_4	8_4
8_5	x^6y	672	2^5	4_6	8_5	8_5
8_6	y	1344	2^4	4_5	8_6	8_6
8_7	x^2yxyxy^2	1344	2^4	4_4	8_7	8_7
12	x^2yxy	1792	$2^2 \cdot 3$	6_1	4_1	12
14_1	xy	1536	$2 \cdot 7$	7_1	14_2	2_1
14_2	$(xy)^3$	1536	$2 \cdot 7$	7_2	14_1	2_1

A.4. *Conjugacy classes of simple group* $G = \langle x, y, h, e \rangle$

Class	Representative	$|Class|$	$|Centralizer|$	2P	3P	5P	7P	11P	23P
1	1	1	$2^{10} \cdot 3^7 \cdot 5^3 \cdot 7 \cdot 11 \cdot 23$	1	1	1	1	1	1
2_1	$(x)^4$	170775	$2^{10} \cdot 3^4 \cdot 5 \cdot 7$	1	2_1	2_1	2_1	2_1	2_1
2_2	e	2608200	$2^7 \cdot 3^3 \cdot 5 \cdot 11$	1	2_2	2_2	2_2	2_2	2_2
3_1	$(xh)^3$	1416800	$2^5 \cdot 3^7 \cdot 5$	3_1	1	3_1	3_1	3_1	3_1
3_2	h	17001600	$2^3 \cdot 3^6 \cdot 5$	3_2	1	3_2	3_2	3_2	3_2
3_3	$(xyeh)^7$	109296000	$2^3 \cdot 3^4 \cdot 7$	3_3	1	3_3	3_3	3_3	3_3
4_1	$(xhy)^3$	21517650	$2^9 \cdot 3^2 \cdot 5$	2_1	4_1	4_1	4_1	4_1	4_1
4_2	$(x)^2$	322764750	$2^9 \cdot 3$	2_1	4_2	4_2	4_2	4_2	4_2
5_1	$(y^2 h)^4$	330511104	$2^2 \cdot 3 \cdot 5^3$	5_1	5_1	1	5_1	5_1	5_1
5_2	xhe	1652555520	$2^2 \cdot 3 \cdot 5^2$	5_2	5_2	1	5_2	5_2	5_2
6_1	$(yh)^2$	114760800	$2^5 \cdot 3^3 \cdot 5$	3_1	2_1	6_1	6_1	6_1	6_1
6_2	$(xh^2)^3$	382536000	$2^4 \cdot 3^4$	3_1	2_1	6_2	6_2	6_2	6_2
6_3	xy^2	2295216000	$2^3 \cdot 3^3$	3_2	2_1	6_3	6_3	6_3	6_3
6_4	ye	4590432000	$2^2 \cdot 3^3$	3_2	2_2	6_4	6_4	6_4	6_4
6_5	$xehyh$	6885648000	$2^3 \cdot 3^2$	3_3	2_2	6_5	6_5	6_5	6_5
7	$(xy)^2$	11803968000	$2 \cdot 3 \cdot 7$	7	7	7	1	7	7
8_1	$x^2 y$	2582118000	$2^6 \cdot 3$	4_2	8_1	8_1	8_1	8_1	8_1
8_2	$x^2 ey$	2582118000	$2^6 \cdot 3$	4_2	8_2	8_2	8_2	8_2	8_2
8_3	x	15492708000	2^5	4_2	8_3	8_3	8_3	8_3	8_3
9_1	xh	3060288000	$2 \cdot 3^4$	9_1	3_1	9_1	9_1	9_1	9_1
9_2	$yeh^2 e$	6120576000	3^4	9_2	3_1	9_2	9_2	9_2	9_2
10_1	$(y^2 h)^2$	8262777600	$2^2 \cdot 3 \cdot 5$	5_1	10_1	2_1	10_1	10_1	10_1
10_2	$xyxeh$	24788332800	$2^2 \cdot 5$	5_2	10_2	2_2	10_2	10_2	10_2
11_1	$(he)^2$	22534848000	$2 \cdot 11$	11_2	11_1	11_1	11_1	1	11_1
11_2	$(he)^4$	22534848000	$2 \cdot 11$	11_1	11_2	11_2	11_1	1	11_2
12_1	$xy^2 hy$	3442824000	$2^4 \cdot 3^2$	6_1	4_1	12_1	12_1	12_1	12_1
12_2	yh	10328472000	$2^4 \cdot 3$	6_1	4_2	12_2	12_2	12_2	12_2
12_3	xhy	13771296000	$2^2 \cdot 3^2$	6_3	4_1	12_3	12_3	12_3	12_3
14	xy	35411904000	$2 \cdot 7$	7	14	14	2_1	14	14
15_1	$xehe$	16525555200	$2 \cdot 3 \cdot 5$	15_1	5_1	3_1	15_1	15_1	15_1
15_2	$x^2 he$	33051110400	$3 \cdot 5$	15_2	5_2	3_2	15_2	15_2	15_2
18	xh^2	27542592000	$2 \cdot 3^2$	9_1	6_2	18	18	18	18
20_1	$y^2 h$	24788332800	$2^2 \cdot 5$	10_1	20_1	4_1	20_1	20_2	20_1
20_2	$(y^2 h)^{11}$	24788332800	$2^2 \cdot 5$	10_1	20_2	4_1	20_2	20_1	20_2
21	$xyeh$	23607936000	$3 \cdot 7$	21	7	21	3_3	21	21
22_1	he	22534848000	$2 \cdot 11$	11_1	22_1	22_1	22_2	2_2	22_1
22_2	$(he)^7$	22534848000	$2 \cdot 11$	11_2	22_2	22_2	22_1	2_2	22_2
23_1	yeh	21555072000	23	23_1	23_1	23_2	23_2	23_2	1
23_2	$(yeh)^5$	21555072000	23	23_2	23_2	23_1	23_1	23_1	1
24_1	$x^2 heh$	20656944000	$2^3 \cdot 3$	12_2	8_1	24_1	24_1	24_1	24_1
24_2	$xy^3 h$	20656944000	$2^3 \cdot 3$	12_2	8_2	24_2	24_2	24_2	24_2
30	$x^3 yh$	16525555200	$2 \cdot 3 \cdot 5$	15_1	10_1	6_1	30	30	30

Appendix B. Character Tables of Local Subgroups of Co$_3$

B.1. *Character table of $E = \langle x, y, e \rangle$*

2	10	10	7	7	2	3	9	9	7	4	.	3	2	2	1	1	6	6	5	4	2	1	1	.	.
3	2	1	1	1	2	2	1	.	.	.	.	1	1	1	1	.	.	.	.	1	.	.	1	1
5	1	.	.	1	.	.	.	.	1	.	.	.	.	.	.	.	.	.	.	.	.	.	1	1
7	1	1	.	.	.	.	.	.	.	.	.	.	1	1	.	.	.	.	.	1	1	.	.

	1a	2a	2b	2c	3a	3b	4a	4b	4c	4d	5a	6a	6b	6c	7a	7b	8a	8b	8c	8d	12a	14a	14b	15a	15 b
2P	1a	1a	1a	1a	3a	3b	2a	2a	2a	2c	5a	3b	3b	3a	7a	7b	4b	4b	4b	4c	6a	7a	7b	15a	15b
3P	1a	2a	2b	2c	1a	1a	4a	4b	4c	4d	5a	2a	2b	2c	7b	7a	8a	8b	8c	8d	4a	14b	14a	5a	5a
5P	1a	2a	2b	2c	3a	3b	4a	4b	4c	4d	1a	6a	6b	6c	7b	7a	8a	8b	8c	8d	12a	14b	14a	3a	3 a
7P	1a	2a	2b	2c	3a	3b	4a	4b	4c	4d	5a	6a	6b	6c	1a	1a	8a	8b	8c	8d	12a	2a	2a	15b	15 a

X.1	1	1	1	1	1	1	1	1	1	1	1	1	1	1	1	1	1	1	1	1	1	1	1	1	1
X.2	7	7	−1	3	4	1	−1	−1	3	1	2	1	−1	.	.	−1	−1	−1	1	−1	.	.	.	−1	−1
X.3	14	14	6	2	−1	2	6	6	2	.	−1	2	.	−1	.	.	2	2	2	.	.	.	.	−1	−1
X.4	15	−1	−1	3	.	3	−5	3	−1	1	.	−1	−1	.	1	1	1	−3	1	−1	1	−1	−1	.	.
X.5	20	20	4	4	5	−1	4	4	4	.	.	−1	1	1	−1	−1	.	.	.	.	1	−1	−1	.	.
X.6	21	21	−3	1	6	.	−3	−3	1	−1	1	.	.	−2	.	.	1	1	1	−1	.	.	.	1	1
X.7	21	21	−3	1	−3	.	−3	−3	1	−1	1	.	.	1	.	.	1	1	1	−1	.	.	.	A	$\bar{A}$
X.8	21	21	−3	1	−3	.	−3	−3	1	−1	1	.	.	1	.	.	1	1	1	−1	.	.	.	$\bar{A}$	A
X.9	28	28	−4	4	1	1	−4	−4	4	.	−2	1	−1	1	.	.	.	.	.	−1	.	.	.	1	1
X.10	35	35	3	−5	5	2	3	3	−5	−1	.	2	.	1	.	.	−1	−1	−1	−1	.	.	.	.	.
X.11	45	45	−3	−3	.	.	−3	−3	−3	1	.	.	.	.	B	$\bar{B}$	1	1	1	1	.	$\bar{B}$	B	.	.
X.12	45	−3	−3	−3	.	.	9	1	1	1	.	.	.	.	$\bar{B}$	B	3	−1	−1	−1	.	−B	−$\bar{B}$	.	.
X.13	45	45	−3	−3	.	.	−3	−3	−3	1	.	.	.	.	$\bar{B}$	B	1	1	1	1	.	B	$\bar{B}$	.	.
X.14	45	−3	−3	−3	.	.	9	1	1	1	.	.	.	.	B	$\bar{B}$	3	−1	−1	−1	.	−$\bar{B}$	−B	.	.
X.15	56	56	8	.	−4	−1	8	8	.	.	1	−1	−1	.	.	.	.	.	.	.	−1	.	.	1	1
X.16	64	64	.	.	4	−2	.	.	.	.	−1	−2	.	.	1	1	.	.	.	.	1	1	−1	−1	.
X.17	70	70	−2	2	−5	1	−2	−2	2	.	.	1	1	−1	.	.	−2	−2	−2	−2	.	1	.	.	.
X.18	90	−6	−6	6	.	.	−6	10	−2	.	.	.	.	.	−1	−1	−2	−2	2	2	.	1	1	.	.
X.19	105	−7	−7	−3	.	3	13	5	1	−1	.	.	−1	−1	.	.	−1	3	−1	1	1	.	.	.	.
X.20	105	−7	1	9	.	3	5	−3	−3	1	.	.	−1	1	.	.	−1	3	−1	−1	−1	.	.	.	.
X.21	105	−7	1	−3	.	3	3	−19	5	−1	.	.	−1	1	.	.	3	−1	−1	1	−1	.	.	.	.
X.22	120	−8	−8	.	.	−3	8	8	.	.	.	.	.	1	1	.	1	1	1	.	.	−1	−1	−1	.
X.23	210	−14	2	6	.	−3	−14	2	−2	.	.	.	.	1	−1	.	.	2	2	−2	.	1	.	.	.
X.24	315	−21	3	3	.	.	15	−9	−1	−1	.	.	.	.	.	.	1	−3	1	1	.	.	.	.	.
X.25	315	−21	3	−9	.	.	−9	−1	3	1	.	.	.	.	.	.	−3	1	1	−1	.	.	.	.	.

where $A = -\frac{1}{2}(1 + i\sqrt{15})$, $B = -\frac{1}{2}(1 + i\sqrt{7})$.

B.2. *Character table of $D = \langle x, y \rangle$*

2	10	10	9	7	7	6	3	9	9	8	8	7	7	6	4	3	2	2	1	1	6	6	5	5	5	4	4	2	1	1
3	1	1	1	1	.	.	1	1	.	.	.	.	.	.	.	.	1	1	1	.	.	.	.	.	.	.	.	1	.	.
7	1	1	.	.	.	.	.	.	.	.	.	.	.	.	.	.	.	.	1	1	.	.	.	.	.	.	.	.	1	1

	1a	2a	2b	2c	2d	2e	3a	4a	4b	4c	4d	4e	4f	4g	4h	6a	6b	6c	7a	7b	8a	8b	8c	8d	8e	8f	8g	12a	14a	14b
2P	1a	1a	1a	1a	1a	1a	3a	2a	2a	2b	2b	2a	2b	2b	2d	3a	3a	3a	7a	7b	4b	4b	4b	4b	4f	4f	4e	4d	6a	7a	7b
3P	1a	2a	2b	2c	2d	2e	1a	4a	4b	4c	4d	4e	4f	4g	4h	2a	2b	2c	7b	7a	8a	8b	8c	8d	8e	8f	8g	4a	14b	14a
7P	1a	2a	2b	2c	2d	2e	3a	4a	4b	4c	4d	4e	4f	4g	4h	6a	6b	6c	1a	1a	8a	8b	8c	8d	8e	8f	8g	12a	2a	2a

X.1	1	1	1	1	1	1	1	1	1	1	1	1	1	1	1	1	1	1	1	1	1	1	1	1	1	1	1	1	1	1
X.2	3	3	3	3	−1	−1	.	3	3	−1	−1	−1	−1	−1	1	.	.	.	A	$\bar{A}$	−1	−1	1	−1	1	1	1	.	A	$\bar{A}$
X.3	3	3	3	3	−1	−1	.	3	3	−1	−1	−1	−1	−1	1	.	.	.	$\bar{A}$	A	−1	−1	1	−1	1	1	1	.	$\bar{A}$	A
X.4	6	6	6	6	2	2	.	6	6	2	2	2	2	2	2	.	.	.	−1	−1	2	2	.	2	.	.	.	.	−1	−1
X.5	7	7	−1	−1	3	−1	1	−5	3	−1	3	−1	−1	1	1	1	−1	−1	.	.	1	−3	1	1	−1	1	−1	1	.	.
X.6	7	7	−1	−1	−1	−1	1	−5	3	3	3	−1	3	−1	−1	1	−1	1	.	.	−3	1	1	1	1	1	1	.	.	.
X.7	7	7	7	−1	−1	3	1	−1	−1	3	3	3	−1	−1	1	1	−1	1	.	.	−1	−1	1	1	1	1	−1	−1	.	.
X.8	7	7	7	−1	−1	1	1	−1	−1	−1	−1	3	3	1	1	1	1	.	.	−1	−1	1	−1	−1	−1	−1	−1	.	.	.
X.9	7	7	7	7	−1	−1	1	7	7	−1	−1	−1	−1	−1	1	1	1	1	.	.	−1	−1	−1	−1	−1	−1	−1	1	.	.
X.10	8	−8	.	.	.	.	.	.	−4	4	.	.	.	.	.	.	−2	.	1	1	.	.	2	.	−2	.	.	.	−1	−1
X.11	8	8	8	8	.	.	.	−1	8	8	.	.	.	.	.	−1	−1	−1	1	1	.	.	2	.	−2	.	.	.	−1	1	1
X.12	14	14	−2	2	2	−1	−2	−2	2	2	2	2	2	.	.	−1	1	−1	.	.	−2	−2	.	−2	.	.	.	1	.	.
X.13	14	14	−2	−2	2	−2	−1	−10	6	2	2	2	2	2	−2	−1	1	1	.	.	−2	−2	.	2	.	.	.	−1	.	.
X.14	21	21	−3	3	1	1	.	9	1	1	1	3	5	−3	−1	.	.	.	.	.	3	−1	−1	−1	−1	1	1	.	.	.
X.15	21	21	−3	−3	−3	−3	.	9	1	5	5	1	1	1	1	.	.	.	.	.	3	−1	1	−1	1	−1	−1	.	.	.
X.16	21	21	−3	−3	−3	1	.	−15	9	1	1	−3	1	1	1	.	.	.	.	.	−1	3	−1	−1	−1	1	−1	.	.	.
X.17	21	21	−3	−3	1	−3	.	9	1	1	1	5	−3	1	−1	.	.	.	.	.	−1	3	−1	−1	1	−1	−1	.	.	.
X.18	21	21	−3	−3	1	1	.	−15	9	−3	−3	1	−3	1	−1	.	.	.	.	.	3	−1	1	−1	1	−1	1	.	.	.
X.19	21	21	21	−3	−3	1	.	−3	−3	1	1	1	−3	−3	1	.	.	.	.	.	1	1	−1	1	−1	−1	1	.	.	.
X.20	21	21	−3	−3	5	1	.	9	1	−3	−3	1	1	−3	1	.	.	.	.	.	−1	3	1	−1	1	−1	−1	.	.	.
X.21	21	21	−3	1	−3	.	.	−3	−3	−3	−3	3	−3	1	1	−1	.	.	.	.	1	1	1	1	1	1	−1	.	.	.
X.22	24	−24	.	.	.	.	.	.	.	4	−4	.	.	.	.	.	$\bar{A}$	A	.	.	.	.	2	.	−2	.	.	.	−$\bar{A}$	−A
X.23	24	−24	.	.	.	.	.	.	.	4	−4	.	.	.	.	.	A	$\bar{A}$	.	.	.	.	2	.	−2	.	.	.	−A	−$\bar{A}$
X.24	28	28	−4	4	4	.	1	−4	−4	4	4	−4	−4	.	.	1	1	−1	.	.	.	.	.	.	.	.	.	−1	.	.
X.25	28	28	−4	4	.	.	1	−4	−4	−4	−4	4	4	.	.	1	1	−1	.	.	.	.	.	.	.	.	.	−1	.	.
X.26	42	42	−6	−6	−2	2	.	18	2	−2	−2	−2	−2	2	.	.	.	.	.	.	−2	−2	.	2	.	.	.	.	.	.
X.27	48	−48	.	.	.	.	.	.	.	−8	8	.	.	.	.	.	−1	−1	.	.	.	.	.	.	.	.	.	.	1	1
X.28	56	56	−8	8	.	.	−1	−8	−8	.	.	.	.	.	.	−1	−1	1	.	.	.	.	.	.	.	.	.	1	.	.
X.29	56	−56	.	.	.	.	.	2	.	.	.	4	−4	.	.	.	−2	.	.	.	.	.	−2	.	2	.	.	.	.	.
X.30	64	−64	.	.	.	.	.	−2	.	.	.	.	.	.	.	.	2	.	.	.	1	1	.	.	.	.	.	.	−1	−1

$A = -\frac{1}{2}(1 + i\sqrt{7})$.

B.3. *Character table of* $H = \langle x, y, h \rangle$

	1a	2a	2b	2c	3a	3b	3c	4a	4b	4c	4d	4e	5a	6a	6b	6c	6d	6e	6f	6g	6h	7a	8a	8b	8c	8d	8e
2	10	10	9	7	5	4	3	9	9	8	8	7	2	5	4	3	4	4	4	2	2	1	6	6	5	5	5
3	4	4	2	1	3	4	3	2	1	1	1	.	1	3	4	3	2	2	2	2	1	.	1	1	.	.	.
5	1	1	.	.	1	.	.	1	.	.	.	.	1	.	.	.	.	.	.	.	.	.	.	.	.	.	.
7	1	1	.	.	.	.	.	.	.	.	.	.	.	1	.	.	.	.	.	.	.	1	.	.	.	.	.
2P	1a	1a	1a	1a	3a	3b	3c	2a	2a	2b	2b	2b	5a	3a	3b	3c	3b	3b	3a	3c	3c	7a	4b	4b	4b	4e	4e
3P	1a	2a	2b	2c	1a	1a	1a	4a	4b	4c	4d	4e	5a	2a	2a	2a	2b	2b	2b	2b	2c	7a	8a	8b	8c	8d	8e
5P	1a	2a	2b	2c	3a	3b	3c	4a	4b	4c	4d	4e	1a	6a	6b	6c	6d	6e	6f	6g	6h	7a	8a	8b	8c	8d	8e
7P	1a	2a	2b	2c	3a	3b	3c	4a	4b	4c	4d	4e	5a	6a	6b	6c	6d	6e	6f	6g	6h	1a	8a	8b	8c	8d	8e
X.1	1	1	1	1	1	1	1	1	1	1	1	1	1	1	1	1	1	1	1	1	1	1	1	1	1	1	1
X.2	7	7	−1	−1	4	−2	1	−5	3	3	3	−1	2	4	−2	1	2	2	2	−1	−1	.	−3	1	1	1	1
X.3	8	−8	.	.	−4	−1	2	.	−4	4	.	−2	4	1	−2	3	.	−3	.	.	1	.	.	−2	.	.	2
X.4	15	15	7	−1	.	−3	3	−5	3	−1	−1	3	.	−3	3	1	−2	1	1	−1	1	1	1	−3	−1	1	−1
X.5	21	21	5	−3	6	3	.	−11	5	1	1	1	1	6	3	.	−1	2	−1	2	.	−3	−3	−3	−1	1	−1
X.6	21	21	−3	−3	6	3	.	9	1	5	5	1	1	6	3	.	3	.	3	.	.	3	−1	1	−1	1	
X.7	27	27	3	3	9	.	.	15	7	3	3	−1	2	9	.	.	3	.	.	.	.	−1	5	1	−1	1	−1
X.8	35	35	11	3	5	−1	2	15	7	−1	−1	3	.	5	−1	2	−1	−1	−1	2	.	1	5	1	1	1	1
X.9	35	35	3	3	5	−1	2	−5	−5	7	7	−1	.	5	−1	2	3	−3	3	.	.	−1	−1	−1	1	−1	1
X.10	48	−48	.	.	−12	3	.	.	.	−8	8	.	−2	12	−3	.	3	.	−3	.	.	−1	.	.	.	.	.
X.11	56	56	−8	.	11	2	2	−24	8	.	.	.	1	11	2	2	−2	1	−2	−2	.	.	−4	4	.	.	.
X.12	64	−64	.	.	4	−8	−2	.	.	.	.	.	.	−1	−4	8	2	.	.	.	.	1	.	.	.	.	.
X.13	64	−64	.	.	4	−8	−2	.	.	.	.	.	.	−1	−4	8	2	.	.	.	.	1	.	.	.	.	.
X.14	70	70	−10	−2	−5	7	1	−10	6	2	2	2	.	−5	7	1	−1	−1	−1	−1	1	.	2	2	.	−2	.
X.15	84	84	20	4	−6	3	3	4	4	4	4	4	−1	−6	3	3	−1	2	−1	−1	1	.	.	.	.	.	.
X.16	105	105	−7	1	.	6	3	25	9	−3	−3	−3	.	.	6	3	2	−4	2	−1	1	.	−3	−3	−1	1	−1
X.17	105	105	17	−7	.	6	3	5	−3	−3	−3	1	.	.	6	3	2	2	2	−1	−1	.	−1	−3	−1	−1	−1
X.18	105	105	1	1	15	−3	−3	−35	5	5	5	1	.	15	−3	−3	1	1	1	1	1	.	−5	−1	−1	−1	−1
X.19	112	−112	.	.	4	4	4	.	.	−8	8	.	2	−4	−4	−4	.	.	.	.	.	−1	.	.	.	.	.
X.20	112	−112	.	.	−8	−5	−2	.	.	−8	8	.	2	8	5	2	3	.	−3	.	.	.	.	.	.	.	.
X.21	120	120	−8	.	15	−6	.	40	8	.	.	.	.	15	−6	.	−2	1	−2	−2	.	1	4	−4	.	.	.
X.22	120	−120	.	.	3	6	.	.	.	4	−4	.	.	.	−3	−6	3	.	−3	.	.	1	.	.	2	.	−2
X.23	168	168	8	8	6	−3	.	40	8	.	.	.	−2	6	−3	.	2	2	2	−1	−1	.	.	.	2	.	−2
X.24	168	−168	.	.	−24	−3	.	.	.	−4	4	.	−2	24	3	.	−3	.	3	.	.	.	.	.	2	.	−2
X.25	189	189	21	−3	9	.	.	−39	1	−3	−3	1	−1	9	.	.	.	.	3	.	.	.	−1	−5	1	−1	1
X.26	189	189	−3	−3	9	.	.	21	−11	9	9	1	−1	9	.	.	.	−3	.	.	.	.	1	1	−1	1	−1
X.27	189	189	−3	−3	9	.	.	−51	13	−3	−3	−3	−1	9	.	.	.	−3	.	.	.	.	1	1	1	1	1
X.28	210	210	2	−6	15	3	.	50	2	−2	−2	−2	.	15	3	.	−1	−1	−1	2	.	.	2	2	.	−2	.
X.29	210	210	−14	2	−15	−6	3	10	10	6	6	−2	.	−15	−6	3	−2	1	−2	1	−1	.	−2	−2	.	−2	.
X.30	216	216	24	.	−9	.	.	−24	8	.	.	.	1	−9	.	.	.	−3	.	.	.	−1	4	−4	.	.	
X.31	280	280	−8	8	10	10	1	−40	−8	.	.	.	.	10	10	1	−2	−2	−2	1	−1	.	−4	4	.	.	.
X.32	280	280	24	.	−5	−8	−2	40	8	.	.	.	.	−5	−8	−2	3	.	−3	.	.	.	−4	4	.	.	.
X.33	280	−280	.	.	−20	1	4	.	.	4	−4	.	.	20	−1	−4	−3	.	3	.	.	.	.	.	−2	.	2
X.34	315	315	−21	3	.	−9	.	−45	3	−5	−5	3	.	.	−9	.	3	3	.	.	.	.	3	3	−1	−1	−1
X.35	336	336	16	.	6	−6	.	.	−16	−16	.	.	1	6	−6	.	−2	−2	−2	−2	.	.	.	.	.	.	.
X.36	378	378	−6	−6	−9	.	.	−30	2	6	6	−2	−2	−9	.	.	.	3	.	.	.	.	2	2	.	2	.
X.37	405	405	−27	−3	.	.	.	45	−3	−3	−3	5	.	.	.	.	−3	3	.	.	.	.	−1	−3	−3	1	1	1
X.38	420	420	4	4	.	−3	3	20	−12	−4	−4	−4	.	−3	3	1	4	1	1	1	.	.	1	.	.	.	.
X.39	448	−448	.	.	16	16	−2	.	.	.	.	.	.	−2	−16	−16	2	.	.	.	.	.	1	.	.	.	.
X.40	512	−512	.	.	−16	8	−4	.	.	.	.	.	.	2	16	−8	4	.	.	.	.	1	.	.	.	.	.
X.41	512	512	.	.	−16	8	−4	.	.	.	.	.	.	2	−16	8	−4	.	.	.	.	1	.	.	.	.	.
X.42	560	−560	.	.	20	−7	2	.	.	−8	8	.	.	−20	7	−2	−3	.	3	.	.	.	.	.	.	.	.
X.43	720	−720	.	.	−9	.	.	.	.	8	−8	.	.	.	9	.	3	.	−3	.	.	−1	.	.	.	.	.

Character table of H (continued)

	8f	9a	10a	12a	12b	12c	12d	12e	14a	15a	18a	20a	20b	24a	24b	30a
2	4	1	2	4	4	2	3	3	1	1	1	2	2	3	3	1
3	.	2	1	2	1	2	1	1	.	1	2	.	.	1	1	1
5	.	.	1	.	.	.	.	.	.	1	.	1	1	.	.	1
7	.	.	.	.	.	.	.	.	1	.	.	.	.	.	.	.
2P	4c	9a	5a	6a	6a	6c	6d	6d	7a	15a	9a	10a	10a	12b	12b	15a
3P	8f	3b	10a	4a	4b	4a	4c	4d	14a	5a	6b	20a	20b	8a	8b	10a
5P	8f	9a	2a	12a	12b	12c	12d	12e	14a	3a	18a	4a	4a	24a	24b	6a
7P	8f	9a	10a	12a	12b	12c	12d	12e	2a	15a	18a	20a	20b	24a	24b	30a
X.1	1	1	1	1	1	1	1	1	1	1	1	1	1	1	1	1
X.2	-1	1	2	-2	.	1	.	.	.	-1	1	.	.	-2	.	-1
X.3	.	-1	2	.	.	.	1	-1	-1	1	1	.	.	.	.	-1
X.4	1	.	.	-2	.	1	-1	-1	1	.	.	.	.	.	-2	.
X.5	-1	.	1	-2	2	-2	1	1	.	1	.	-1	-1	.	.	1
X.6	-1	.	1	.	-2	.	-1	-1	.	1	.	-1	-1	2	.	1
X.7	1	.	2	3	1	.	.	.	-1	-1	.	.	.	1	-1	-1
X.8	-1	-1	.	3	1	.	-1	-1	.	.	-1	.	.	-1	1	.
X.9	1	-1	.	1	1	-2	1	1	.	.	-1	.	.	-1	-1	.
X.10	.	.	2	.	.	.	-1	1	1	-2	.	.	.	.	.	2
X.11	.	-1	1	-3	-1	.	.	.	1	-1	1	1	1	-1	.	1
X.12	.	1	1	.	.	.	.	.	-1	-1	-1	A	Ā	.	.	1
X.13	.	1	1	.	.	.	.	.	-1	-1	-1	Ā	A	.	.	1
X.14	.	1	.	-1	3	-1	-1	-1	.	.	1	.	.	-1	-1	.
X.15	.	.	-1	-2	-2	1	1	.	.	-1	.	-1	-1	.	.	-1
X.16	-1	.	.	4	.	1	.	.	.	.	.	.	.	.	2	.
X.17	1	.	2	.	-1	.	.	.	.	.	.	.	.	.	2	.
X.18	1	.	1	-1	1	-1	-1	.	.	.	.	.	.	-1	1	.
X.19	.	1	-2	.	.	.	2	-2	.	-1	-1	.	.	.	.	1
X.20	.	1	-2	.	.	.	-1	1	.	2	-1	.	.	.	.	-2
X.21	.	.	.	1	-1	-2	.	.	1	.	.	.	.	-1	1	.
X.22	.	.	.	.	.	.	-1	1	-1	.	.	.	.	.	.	.
X.23	.	.	-2	-2	2	1	.	.	1	.	.	.	.	.	.	1
X.24	.	.	2	.	.	.	1	-1	.	1	.	.	.	.	.	-1
X.25	-1	.	-1	3	1	.	.	.	.	-1	.	1	1	1	-1	-1
X.26	-1	.	-1	-3	1	.	.	.	.	-1	.	1	1	1	-1	-1
X.27	1	.	-1	-3	1	.	.	.	.	-1	.	-1	-1	1	1	-1
X.28	.	.	-1	-1	2	1	1	.	.	.	.	.	.	-1	-1	.
X.29	.	.	.	1	1	1	.	.	.	.	.	.	.	1	1	.
X.30	.	.	1	-3	-1	.	.	.	-1	1	1	1	-1	1	1	1
X.31	1	.	2	-2	-1	.	.	.	.	1	.	.	.	1	-1	.
X.32	1	.	1	-1	-2	.	.	.	.	1	.	.	.	1	-1	.
X.33	1	.	.	.	.	-1	1	.	.	-1	.	.	.	.	.	.
X.34	-1	.	.	.	.	1	1	.	.	.	.	.	.	.	.	.
X.35	.	.	1	2	2	2	.	.	.	1	.	.	-1	-1	.	1
X.36	.	.	-2	3	-1	.	.	.	.	1	.	.	.	-1	-1	1
X.37	1	.	.	.	.	.	.	.	-1	.	.	.	.	.	.	.
X.38	.	.	-4	.	-1	-1	-1	.	.	.	.	.	.	.	.	.
X.39	.	1	2	.	.	.	.	.	.	1	-1	.	.	.	.	-1
X.40	.	-1	-2	.	.	.	.	.	-1	-1	1	.	.	.	.	1
X.41	.	-1	2	.	.	.	.	.	1	-1	-1	.	.	.	.	-1
X.42	.	-1	.	.	.	.	-1	1	1	.	.	.	.	.	.	.
X.43	.	.	.	.	.	.	1	-1	1	.	.	.	.	.	.	.

$A = i\sqrt{5}.$

B.4. *Character table of simple group* $G = \langle x, y, h, e \rangle$

	1a	2a	2b	3a	3b	3c	4a	4b	5a	5b	6a	6b	6c	6d	6e	7a	8a	8b	8c	9a	9b	10a	10b
2	10	10	7	5	3	3	9	9	2	2	5	4	3	2	3	1	6	6	5	1	.	2	2
3	7	4	3	7	6	4	2	1	1	1	3	4	3	3	2	1	1	1	1	.	4	4	1
5	3	1	1	1	1	.	1	.	3	2	1	.	.	.	.	.	.	.	.	.	.	1	1
7	1	1	.	.	.	1	.	.	.	.	.	.	.	.	.	1	.	.	.	.	.	.	.
11	1	1	.	1	.	.	.	.	.	.	.	.	.	.	.	.	.	.	.	.	.	.	.
23	1	.	.	.	.	.	.	.	.	.	.	.	.	.	.	.	.	.	.	.	.	.	.
	1a	2a	2b	3a	3b	3c	4a	4b	5a	5b	6a	6b	6c	6d	6e	7a	8a	8b	8c	9a	9b	10a	10b
2P	1a	1a	1a	3a	3b	3c	2a	2a	5a	5b	3a	3a	3b	3b	3c	7a	4b	4b	4b	9a	9b	5a	5b
3P	1a	2a	2b	1a	1a	1a	4a	4b	5a	5b	2a	2a	2a	2b	2b	7a	8a	8b	8c	3a	3a	10a	10b
5P	1a	2a	2b	3a	3b	3c	4a	4b	1a	1a	6a	6b	6c	6d	6e	7a	8a	8b	8c	9a	9b	2a	2b
7P	1a	2a	2b	3a	3b	3c	4a	4b	5a	5b	6a	6b	6c	6d	6e	1a	8a	8b	8c	9a	9b	10a	10b
11P	1a	2a	2b	3a	3b	3c	4a	4b	5a	5b	6a	6b	6c	6d	6e	7a	8a	8b	8c	9a	9b	10a	10b
23P	1a	2a	2b	3a	3b	3c	4a	4b	5a	5b	6a	6b	6c	6d	6e	7a	8a	8b	8c	9a	9b	10a	10b
X.1	1	1	1	1	1	1	1	1	1	1	1	1	1	1	1	1	1	1	1	1	1	1	1
X.2	23	7	−1	−4	5	−1	−5	3	−2	3	4	−2	1	−1	−1	2	1	−3	1	−1	2	2	−1
X.3	253	13	−11	10	10	1	9	1	3	3	10	4	−2	−2	1	1	−1	3	−1	1	1	3	−1
X.4	253	29	−11	10	10	1	−11	5	3	3	2	2	2	−2	1	1	−3	−3	1	1	1	−1	−1
X.5	275	35	11	5	14	−1	15	7	.	5	5	−1	2	2	−1	2	1	5	1	−1	2	.	1
X.6	896	.	16	32	−4	−7	.	.	−4	1	.	.	.	−2	1	.	.	.	.	2	−1	.	1
X.7	896	.	16	32	−4	−7	.	.	−4	1	.	.	.	−2	1	.	.	.	.	2	−1	.	1
X.8	1771	−21	11	−11	16	7	−5	−5	−4	1	21	−3	.	2	−1	.	−1	−1	−1	−2	−2	4	1
X.9	2024	104	.	−1	26	8	−24	8	−1	4	−1	5	2	.	.	1	4	−4	.	−1	−1	−1	.
X.10	3520	−64	.	−44	10	−8	.	.	−5	.	−4	8	2	.	.	−1	.	.	.	1	1	1	.
X.11	3520	−64	.	−44	10	−8	.	.	−5	.	−4	8	2	.	.	−1	.	.	.	1	1	1	.
X.12	4025	105	1	−25	29	−7	−35	5	.	5	15	−3	−3	1	1	.	−1	−5	−1	2	2	.	1
X.13	5544	168	.	−45	36	.	40	8	−6	4	3	−3	.	.	.	.	−4	4	.	.	.	−2	.
X.14	7084	−84	44	10	19	−14	−4	−4	9	−1	18	6	3	−1	2	.	.	−4	4	4	−2	1	−1
X.15	8855	231	55	−1	35	−7	19	11	5	.	−9	−3	3	1	1	.	5	1	1	2	−4	1	.
X.16	9625	105	−55	40	−5	7	5	−3	.	.	6	3	−1	−1	.	3	−1	−1	−2	1	.	.	.
X.17	9625	105	−55	40	−5	7	5	−3	.	.	6	3	−1	−1	.	3	−1	−1	−2	1	.	.	.
X.18	20608	.	−16	−128	−20	7	.	.	8	3	.	.	.	2	−1	.	.	.	.	−2	−2	.	−1
X.19	20608	.	−16	−128	−20	7	.	.	8	3	.	.	.	2	−1	.	.	.	.	−2	−2	.	−1
X.20	23000	280	120	50	5	8	40	8	.	.	10	10	1	3	.	−2	.	.	.	−1	2	.	.
X.21	26082	−126	−54	81	.	.	−6	10	7	−3	9	9	.	.	.	2	2	−2	.	.	.	−1	1
X.22	31625	265	−55	35	35	−1	−55	9	.	.	−5	−5	−5	−1	−1	−1	1	1	1	1	−1	−1	.
X.23	31625	505	−55	35	35	−1	−35	5	.	.	−5	1	7	−1	−1	−1	−5	−1	−1	1	−1	−1	.
X.24	31625	−55	−55	35	35	−1	25	−7	.	.	35	−1	−1	−1	−1	−1	1	1	1	1	−1	−1	.
X.25	31878	294	−66	45	45	.	46	−2	3	3	−3	−3	−3	−3	.	.	2	2	−2	.	.	−1	−1
X.26	40250	−70	10	−115	−25	14	10	10	.	.	5	−7	−1	1	−2	.	−2	−2	−2	5	−1	.	.
X.27	57960	168	120	126	45	.	−40	−8	10	.	6	6	−3	3	.	.	.	.	.	.	.	−2	.
X.28	63250	210	−110	−65	−20	22	−30	2	.	.	15	3	.	−2	−2	−2	2	2	2	4	1	.	.
X.29	73600	.	144	160	16	13	.	.	−5	.	.	.	.	−3	2	.	.	.	.	4	1	.	−1
X.30	80960	−448	.	176	.	8	.	.	10	.	−16	−16	2	.	.	.	−2	.	.	.	−1	2	2
X.31	91125	405	45	.	.	27	45	−3	.	.	.	.	.	.	.	3	−1	−3	−3	1	.	.	.
X.32	93312	.	−144	.	.	27	.	.	.	.	12	−3	.	.	.	.	3	2	.	.	.	.	1
X.33	129536	512	.	−64	44	8	.	.	−14	−4	−16	8	−4	.	.	.	1	.	.	−1	−1	2	.
X.34	129536	−512	.	−64	44	8	.	.	−14	−4	16	−8	4	.	.	.	1	.	.	−1	−1	−2	.
X.35	177100	140	44	−20	−29	−14	−20	12	.	−5	20	−4	−1	−1	2	.	.	−5	1	.	.	.	−1
X.36	184437	405	−99	.	.	−27	45	−3	12	−3	.	.	.	.	.	−3	1	−3	−3	1	.	.	1
X.37	221375	735	55	−160	−25	−7	−25	−9	.	.	.	.	−12	3	1	1	3	3	−1	2	2	.	.
X.38	226688	.	−176	320	−40	−7	.	.	−12	3	.	.	.	4	1	.	3	3	−1	2	2	.	−1
X.39	246400	.	176	160	−56	7	.	.	.	5	.	.	.	−4	−1	.	.	.	.	−2	−2	.	1
X.40	249480	−504	.	−81	.	.	−24	8	5	.	−9	9	.	.	.	.	.	−4	4	.	.	1	.
X.41	253000	−440	.	−125	10	−8	40	8	.	.	−5	1	−2	.	.	.	−1	4	−4	.	1	1	.
X.42	255024	−336	.	−126	36	.	−16	−16	−1	4	−6	6	.	.	.	.	.	.	.	.	.	−1	.

Character table of the simple group $G = \langle x, y, h, e\rangle$ *(continued)*

	11a	11b	12a	12b	12c	14a	15a	15b	18a	20a	20b	21a	22a	22b	23a	23b	24a	24b	30a
2	1	1	4	4	2	1	1	.	1	2	2	.	1	1	.	.	3	3	1
3	.	.	2	1	2	.	1	1	2	.	.	1	.	.	.	.	1	1	1
5	.	.	.	.	.	.	1	1	.	1	1	.	1	1	.	.	.	.	1
7	.	.	.	.	.	1	.	.	.	.	.	.	.	.	.	.	.	.	.
11	1	1	.	.	.	.	.	.	.	.	.	.	.	.	1	1	.	.	.
23	.	.	.	.	.	.	.	.	.	.	.	.	.	.	1	1	.	.	.
2P	11b	11a	6a	6a	6c	7a	15a	15b	9a	10a	10a	21a	11a	11b	23a	23b	12b	12b	15a
3P	11a	11b	4a	4b	4a	14a	5a	5b	6b	20a	20b	7a	22a	22b	23a	23b	8a	8b	10a
5P	11a	11b	12a	12b	12c	14a	3a	3b	18a	4a	4a	21a	22a	22b	23b	23a	24a	24b	6a
7P	11b	11a	12a	12b	12c	2a	15a	15b	18a	20a	20b	3c	22b	22a	23a	23b	24a	24b	30a
11P	1a	1a	12a	12b	12c	14a	15a	15b	18a	20b	20a	21a	2b	2b	23b	23a	24a	24b	30a
23P	11a	11b	12a	12b	12c	14a	15a	15b	18a	20a	20b	21a	22a	22b	1a	1a	24a	24b	30a
X.1	1	1	1	1	1	1	1	1	1	1	1	1	1	1	1	1	1	1	1
X.2	1	1	-2	.	1	.	1	.	1	.	.	-1	-1	-1	.	.	.	-2	-1
X.3	.	.	.	-2	.	-1	.	.	1	-1	-1	1	.	.	.	.	.	2	.
X.4	.	.	-2	2	-2	1	.	.	-1	-1	-1	1	.	.	.	.	.	.	2
X.5	.	.	3	1	.	.	.	-1	-1	.	.	-1	.	.	.	-1	-1	-1	1
X.6	A	Ā	.	.	.	.	2	1	.	.	.	.	Ā	A	-1	-1	.	.	.
X.7	Ā	A	.	.	.	.	2	1	.	.	.	.	A	Ā	-1	-1	.	.	.
X.8	.	.	1	1	-2	.	-1	1	.	.	.	.	.	.	.	.	-1	-1	.
X.9	.	.	-3	-1	.	.	-1	-1	1	-1	1	1	1	.	.	.	-1	1	-1
X.10	.	.	.	.	.	.	-1	1	.	-1	B	B̄	-1	.	.	1	1	.	1
X.11	.	.	.	.	.	.	-1	1	.	-1	B̄	B	-1	.	.	1	1	.	1
X.12	-1	-1	1	-1	1	.	.	-1	.	.	.	.	1	1	.	.	1	-1	.
X.13	.	.	1	-1	-2	.	.	1	.	.	.	.	.	.	1	1	1	-1	-2
X.14	.	.	2	2	-1	.	.	-1	.	1	1	.	.	.	.	.	1	-1	-2
X.15	.	.	1	-1	1	.	-1	.	.	.	.	-1	-1	.	.	.	1	-1	1
X.16	.	.	2	.	-1	.	.	.	.	.	.	.	.	.	C	C̄	2	.	.
X.17	.	.	2	.	-1	.	.	.	.	.	.	.	.	.	C̄	C	2	.	.
X.18	A	Ā	.	.	.	.	2	.	.	.	.	.	-Ā	-A	.	.	.	.	.
X.19	Ā	A	.	.	.	.	2	.	.	.	.	.	-A	-Ā	.	.	.	.	.
X.20	-1	-1	-2	2	1	.	.	.	1	.	.	1	-1	-1	.	.	-1	-1	-1
X.21	1	1	-3	1	.	.	1	.	.	-1	-1	1	1	.	.	.	-1	-1	-1
X.22	.	.	-1	3	-1	-1	.	1	.	.	.	-1	.	.	.	.	1	1	.
X.23	.	.	1	-1	1	1	.	1	.	.	.	-1	.	.	.	.	1	1	.
X.24	.	.	-5	-1	1	1	.	-1	.	.	.	-1	.	.	.	.	1	1	2
X.25	.	.	1	1	1	.	.	.	1	1	.	.	.	.	.	.	-1	-1	2
X.26	1	1	1	1	1	.	.	-1	.	.	.	.	-1	-1	.	.	1	1	.
X.27	1	1	2	-2	-1	.	1	.	.	.	.	.	-1	-1	.	.	.	.	1
X.28	.	.	3	-1	.	.	.	.	.	.	.	1	.	.	.	.	-1	-1	.
X.29	-1	-1	.	.	.	.	1	.	.	.	.	-1	1	1	.	.	.	.	-1
X.30	.	.	.	.	.	.	1	.	-1	.	.	1	.	.	.	.	.	.	-1
X.31	1	1	.	.	.	-1	.	.	.	.	.	-1	1	1	1	-1	-1	.	.
X.32	-1	-1	.	.	.	.	.	.	.	.	.	-1	-1	-1	1	1	.	.	.
X.33	.	.	.	.	.	1	1	-1	-1	.	.	1	.	.	.	.	.	.	-1
X.34	.	.	.	.	.	-1	1	-1	1	.	.	1	.	.	.	.	.	.	1
X.35	.	.	4	.	1	.	1	-1	.	.	.	.	.	.	.	.	.	.	.
X.36	.	.	.	.	.	-1	.	.	.	.	.	1	.	.	.	.	.	.	.
X.37	.	.	-4	.	-1	.	.	.	.	.	.	.	.	.	.	.	.	.	.
X.38	.	.	.	.	.	.	-1	.	.	.	.	.	.	.	.	.	.	.	.
X.39	.	.	.	.	.	.	-1	.	.	.	.	.	.	.	1	1	.	.	.
X.40	.	.	-3	-1	.	.	-1	.	.	1	1	.	.	.	-1	-1	1	-1	1
X.41	.	.	1	-1	-2	1	.	1	.	.	.	-1	.	.	.	.	-1	1	.
X.42	.	.	2	2	2	.	-1	1	.	.	.	-1	-1	.	.	.	.	.	-1

$A = \tfrac{1}{2}(-1 + i\sqrt{11}), B = i\sqrt{5}, C = \tfrac{1}{2}(-1 + i\sqrt{23}).$

References

[1] J. Alperin, R. Brauer, D. Gorenstein. Finite groups with quasi-dihedral and wreathed Sylow 2-subgroups. *Transact. Amer. Math. Soc.*, **151**:1–262, 1970.

[2] H. Bender. Transitive Gruppen gerader Ordnung, in denen jede Involution genau einen Punkt festlaesst. *J. Algebra*, **17**:527–554, 1971.

[3] J. Cannon and C. Playoust. *An Introduction to* MAGMA. School of Mathematics and Statistics, University of Sydney, 1993.

[4] John J. Cannon, Derek F. Holt. Automorphism group computation and isomorphism testing in finite groups. *J. Symbolic Computat.*, **35**:241–267, 2003.

[5] R.W. Carter. *Simple groups of Lie type.* John Wiley and Sons, London, 1972.

[6] J.H. Conway. A group of order 8,315,553,613,086,720,000. *Bull. London Math. Soc.*,**1**:79–88, 1969.

[7] J.H. Conway. Three lectures on exceptional groups, in M.B. Powel, G. Higman (eds.)*Finite simple groups.* Academic Press, London, 1971, pp. 215–247.

[8] J.H. Conway, R.T. Curtis, S.P. Norton, R.A. Parker, and R.A. Wilson. *Atlas of finite groups.* Clarendon Press, Oxford, 1985.

[9] R.K. Dennis, M.R. Stein. Injective stability for K_2 of local rings. *Bulletin Amer. Math. Soc.*, **80**:1010–1013, 1974.

[10] W. Feit. On integral representations of finite groups. *Proc. London Math. Soc.*, **29**:633–683, 1974.

[11] D. Fendel. The characterization of Conway's group .3. *J. Algebra*, **24**:159–196, 1973

[12] B. Fischer. Finite groups generated by 3-transpositions. *Inventiones Math.*, **13**:232–246, 1971.

[13] D. Gorenstein, J. H. Walter. The characterization of the finite groups with dihedral Sylow 2-subgroups. *J. Algebra*, **2**:85–151, 218–270, 354–393, 1964.

[14] D. Gorenstein. The classification of finite simple groups I. Simple groups and local analysis. *Bull. (New Series) Amer. Math. Soc.*, **1**:43–199, 1979.

[15] D.F. Holt. The mechanical computation of first and second cohomology groups. *J. Symbolic Computation*, **1**:351–361, 1985.

[16] D.F. Holt. Cohomology and group extensions in Magma, in W. Bosma, J. Cannon, Discovering Mathematics with Magma. Springer, Berlin, 2006, pp. 221–241.

[17] Z. Janko. A new finite simple group of order 86,775,571,046,077,562,880 which possesses M_{24} and the full cover of M_{22} as subgroups. *J. Algebra*, **42**:564–596, 1976.

[18] H. Kim, G.O. Michler. Simultaneous construction of the sporadic groups Co_2 and Fi_{22}. This volume.

[19] J. Milnor. *Introduction to algebraic K-theory.* Princeton University Press, Princeton N.J., 1971.

[20] G. O. Michler. Theory of finite simple groups. Cambridge University Press, Cambridge, 2006.

[21] G. O. Michler. Theory of finite simple groups II. (In preparation).

[22] G. O. Michler, L. Wang. Another existence and uniqueness proof of the Tits group. *Algebra Colloquium* **15**:241–278, 2008.

[23] M. Suzuki. On a class of double transitive groups II.. *Annals of Math.*, **79**:514–589, 1964.

[24] J. G. Thompson. A simple subgroup of $E_8(3)$. In N. Iwahori (edt.) *Finite Groups* 113–116. Tokyo, 1976

[25] M. Weller, G. O. Michler, A. Previtali. Thompson's sporadic group uniquely determined by centralizer of a 2-central involution. *J. Algebra*, **298**:371–459, 2006.

DEPARTMENT OF MATHEMATICS, CORNELL UNIVERSITY, ITHACA, N.Y. 14853, USA

INSTITUTE FOR EXPERIMENTAL MATHEMATICS, UNIVERSITY OF DUISBURG-ESSEN, ELLERNSTR. 29, 45326 ESSEN, GERMANY

Contemporary Mathematics
Volume **470**, 2008

Dickson Polynomials and the Norm Map Between the Hecke Algebras of Gelfand-Graev Representations

Julianne G. Rainbolt

1. Background

Let $\tilde{G}$ denote a connected reductive algebraic group defined over a finite field $\mathbf{F}_q$ where q is a power of a prime. Recall a standard Frobenius endomorphism on $\tilde{G}$ is a map $F : \tilde{G} \rightarrow \tilde{G}$ such that $F(a_{ij}) = (a_{ij}^{q^m})$ for some positive integer m where here elements of $\tilde{G}$ are being represented as n by n matrices (a_{ij}) with a_{ij} in the algebraic closure of $\mathbf{F}_q$. A Frobenius endomorphism (not necessarily standard) is a homomorphism $F : \tilde{G} \rightarrow \tilde{G}$ such that some power of F is a standard Frobenius endomorphism. Fix a Frobenius endomorphism F on $\tilde{G}$. Let $G = \tilde{G}^F$, all the fixed points of F in $\tilde{G}$. That is, G is a finite group of Lie type. Let $\tilde{U}$ be the unipotent radical of a F-stable Borel subgroup in $\tilde{G}$. Let $U = \tilde{U}^F$, the fixed points in $\tilde{U}$. Let ψ be a nondegenerate linear character of U and let e denote the idempotent

$$e = |U|^{-1} \sum_{u \in U} \psi(u^{-1})u.$$

(A nondegenerate character is roughly a character of U that is nontrivial on the simple root subgroups. See [**2**] p. 519 for a precise definition of nondegenerate.) Let $\mathbf{C}$ denote the complex numbers. The Hecke algebra of the Gelfand-Graev representation, ψ^G, of G is the algebra $H = e\mathbf{C}Ge$. (See [**1**] Section 8.1, [**2**], [**3**] Section 11D, and [**6**] Chapter 14, for example, for more details on the construction of this Hecke algebra and the connection between this algebra and the Gelfand-Graev character of G.)

Given an F-stable maximal torus, $\tilde{T}$, of the algebraic group $\tilde{G}$ let $T = \tilde{T}^F$. The subgroup T of G is called a maximal torus of the finite group G. It is well known that the irreducible characters of the Hecke algebra H can by indexed by the pairs (T, θ) as $\tilde{T}$ runs over the F-stable maximal tori of $\tilde{G}$ and θ runs over the irreducible characters of T ([**2**], Theorem 3.1). In addition each irreducible character $f_{T,\theta}$ of H can be factored as $f_{T,\theta} = \theta \cdot f_T$ where here $f_T : H \rightarrow \mathbf{C}T$ is a homomorphism of algebras which is independent of θ and is uniquely determined by this factorization condition ([**2**], Theorem 4.2). (In this factorization the character θ of T has been extended to the group algebra $\mathbf{C}T$.) Since the irreducible characters θ of a torus are

---

2000 *Mathematics Subject Classification.* Primary 20C08, Secondary 20C40.

straightforward to determine, the problem of describing the irreducible characters $f_{T,\theta}$ of H reduces to describing the homomorphisms f_T. For example there are only three maximal tori (up to isomorphism) in the finite group of Lie type GL(3, q). Thus in this example in order to describe all the irreducible characters of the corresponding Hecke algebra, it suffices to determine the three homomorphisms f_T, as T runs over the three maximal tori of GL(3, q).

Note the composition of a Frobenius endomorphism with itself any finite number of times is also a Frobenius endomorphism. Denote by $f_{T,m}$ the homomorphism f_T described in the previous paragraph when $G = \tilde{G}^{F^m}$ for some fixed nonnegative integer m. Denote by $H_{F,m}$ the Hecke algebra corresponding (in the sense of the previous paragraph) to $\tilde{G}^{F^m}$. The norm map of finite field extensions extends to a map $N_{T,m} : \tilde{T}^{F^m} \to \tilde{T}^F$ given by $N_{T,m} = 1 + F + \cdots + F^{m-1}$. That is $N_{T,m}$ is a homomorphism from $\mathbf{C}\tilde{T}^{F^m}$ to $\mathbf{C}\tilde{T}^F$ which is the straightforward extension of the usual norm map of finite field extensions. In [5] Curtis and Shoji discovered a relationship between the homomorphisms $f_{T,m} : H_{F,m} \to \mathbf{C}\tilde{T}^{F^m}$ and $f_T : H_{F,1} \to \mathbf{C}\tilde{T}^F$. Namely, they proved there exists a homomorphism of algebras $\Delta_{F,m} : H_{F,m} \to H_{F,1}$ characterized as the unique linear map from $H_{F,m}$ to $H_{F,1}$ with the property that $f_T \cdot \Delta_{F,m} = N_{T,m} \cdot f_{T,m}$ for all F-stable maximal tori $\tilde{T}$ of $\tilde{G}$ ([5], Theorem 1). We will call $\Delta_{F,m}$ the norm map between the Hecke algebras $H_{F,m}$ and $H_{F,1}$.

In the following two sections we summarize some of the results where the norm maps between Hecke algebras have been determined. In particular we summarize situations where it has been shown that the image of these norm maps can be represented in terms of Dickson polynomials. Representing these images in terms of Dickson polynomials provides a very concrete representation. In particular if the image of all basis elements of the Hecke algebra could be represented in terms of Dickson polynomials it would then be very easy to represent these images on the computer and preform computations with these representations. Alternatively, computational means could be used to try to determine if the below results can be extended to other finite groups of Lie type or to other basis elements of a given Hecke algebra. We are presenting these results in the hopes of motivating a computational investigation of this theory. This area has not been investigated by computational means but at this point appears to be set up for such an investigation.

2. Evidence of Relationship of Norm Map to Dickson Polynomials

Let $P_m(x, y)$ be the polynomial:

$$P_m(x, y) = \sum_{j=0}^{[m/2]} (-1)^{m-j-1} \frac{m}{m-j} \binom{m-j}{j} x^{m-2j} y^j.$$

This is called a Dickson polynomial ([8], p. 228 and p. 355).

Let $\tilde{G} = GL(2, \bar{\mathbf{F}}_q)$, where $\bar{\mathbf{F}}_q$ denotes the algebraic closure of the finite field $\mathbf{F}_q$. Let $F : \tilde{G} \to \tilde{G}$ be the Frobenius endomorphism given by $F(a_{ij}) = (a_{ij}^q)$. Note that $\tilde{G}^{F^m} = GL(2, q^m)$. That is F^m is a standard Frobenius endomorphism for any positive m. In this section we summarize results on the images of the norm map between the Hecke algebra corresponding to the group $\tilde{G}^{F^m} = GL(2, q^m)$ and the Hecke algebra corresponding to the group $\tilde{G}^F = GL(2, q)$.

Recall $\tilde{U}$ denotes the unipotent radical of an F-stable Borel subgroup of $\tilde{G}$. Let

$$e^{(m)} = \frac{1}{q^m} \sum_{u \in \tilde{U}^{F^m}} \psi_m(u^{-1})u,$$

where ψ_m denotes a nondegenerate linear character of $\tilde{U}^{F^m}$. In [9] it was shown that a basis for the Hecke algebra $H_{F,m} = e^{(m)}CGL(2, q^m)e^{(m)}$ is

$$\left\{ q^m e^{(m)} \begin{pmatrix} 0 & -u \\ v & 0 \end{pmatrix} e^{(m)}, \quad \begin{pmatrix} u & 0 \\ 0 & u \end{pmatrix} e^{(m)} \;\middle|\; u, v \in \mathbf{F}_{q^m}^* \right\},$$

where $\mathbf{F}_{q^m}^*$ denotes the group of nonzero elements in $\mathbf{F}_{q^m}$. Denote the basis element

$$q^m e^{(m)} \begin{pmatrix} 0 & -u \\ v & 0 \end{pmatrix} e^{(m)}$$

by $c_{u,v}^{(m)}$. Denote the basis element

$$e^{(m)} \begin{pmatrix} u & 0 \\ 0 & u \end{pmatrix} e^{(m)}$$

by $c_u^{(m)}$.

For the remainder of this article $\Delta_{F,m}$ will denote the norm map between the Gelfand-Graev Hecke algebras for $GL(2, q^m)$ and $GL(2, q)$. That is for the remainder of this article $\Delta_{F,m} : H_{F,m} \to H_{F,1}$ is the norm map when the Frobenius endomorphism is taken to be the standard Frobenius map F^m. In [4] it was shown that the norm map applied to the basis element $c_{1,1}^{(m)}$ of $H_{F,m}$ can be written in terms of a Dickson polynomial for any m. In particular, it was shown in [4] that:

IDENTITY 1. $\Delta_{F,m}(c_{1,1}^{(m)}) = P_m(c_{1,1}^{(1)}, qc_{-1}^{(1)})$.

This relationship was proved using theoretical (noncomputational) techniques.

Identity 1 shows the image of the norm map of one particular basis element of $H_{F,m}$ can be expressed in terms of a Dickson polynomial. The question we want to investigate in this section is whether other basis elements of $H_{F,m}$ have images (under this norm map) expressible in terms of Dickson polynomials.

Recall the factorization

$$f_T \cdot \Delta_{F,m} = N_{T,m} \cdot f_{T,m}$$

described in the previous section. The maps $N_{T,m} : T^{F^m} \to T^F$ are the norm maps of the tori (which are just the extensions of the norm maps of finite field extensions) and are thus easy to describe and are a part of some computer algebra system, such as GAP [7]. The homomorphism $f_{T,m}$ and f_T are determined in [9] and are presented in such a way that they can be explicitly entered into GAP. Thus tests for generalizations of Identity 1 to other basis elements can be done using computational means.

Note that

$$
\begin{aligned}
P_1(x, y) &= x, \\
P_2(x, y) &= \sum_{j=0}^{1} (-1)^{1-j} \frac{2}{2-j} \binom{2-j}{j} x^{2-2j} y^j = -x^2 + 2y.
\end{aligned}
$$

Thus Identity 1 when $m = 2$ becomes $\Delta_{F,2}(c_{1,1}^{(2)}) = -(c_{1,1}^{(1)})^2 + 2qc_{-1}^{(1)}$.

Consider the basis element $c_u^{(2)}$ of the Hecke algebra $H_{F,2}$. In [9], Lemma 9.4 it was shown the image of $c_u^{(2)}$ under the map $\Delta_{F,2}$ is $c_{u^q+1}^{(1)}$. Thus we have:

IDENTITY 2. $\Delta_{F,2}(c_u^{(2)}) = P_1(c_{u^q+1}^{(1)}, 1)$, for all $u \in \mathbf{F}_{q^2}^*$.

The goal is to express the remaining basis elements of $H_{F,2}$ (those of type $c_{u,v}^{(2)}$) in terms of Dickson polynomials. Some evidence that we can extend Identity 1 to other basis elements of $H_{F,2}$ is provided by the following two results from [9]:

IDENTITY 3. $\Delta_{F,2}(c_{u,1}^{(2)}) = P_2(c_{u,1}^{(1)}, qc_{-u}^{(1)})$ for all $u \in \mathbf{F}_q^*$.

IDENTITY 4. $\Delta_{F,2}(c_{v,v^{-q}}^{(2)}) = c_{v^{-q-1}}^{(1)} P_2(c_{v^q+1,1}^{(1)}, qc_{-v^q+1}^{(1)})$ for all $v \in \mathbf{F}_{q^2}^*$.

Thus the image of $c_{v,v^{-q}}^{(2)}$ under $\Delta_{F,2}$ is expressed in Identity 4 as a central element $c_{v^{-q-1}}^{(1)}$ times a Dickson polynomial. Identities 3 and 4 were first discovered using GAP to reveal patterns and then proved independently of computational means.

Identities 2 and 4 do not give a complete description (in terms of Dickson polynomials) of the norm map $\Delta_{F,2} : H_{F,2} \to H_{F,1}$ since $\{c_u^{(2)}, c_{u,v}^{(2)} \mid u, v \in \mathbf{F}_{q^2}^*\}$ not $\{c_u^{(2)}, c_{v,v^{-q}}^{(2)} \mid u, v \in \mathbf{F}_{q^2}^*\}$ is a basis of $H_{F,2}$. Thus Identity 4 still needs to be generalized to an identity for all basis elements of type $c_{u,v}^{(2)}$ in order to provide a complete description of $\Delta_{F,2} : H_{F,2} \to H_{F,1}$.

Question 1: What is the image of $c_{u,v}^{(2)}$ under $\Delta_{F,2}$ in terms of Dickson polynomials? In other words, what is the generalization of Identity 4 to all basis elements of $H_{F,2}$ of type $c_{u,v}^{(2)}$?

Also note that Identities 2–4 are only for the case $m = 2$ (unlike Identity 1 which was proven for all $m \geq 2$). Thus another question is how to generalize Identities 2–4 for $m > 2$.

Question 2: What is the image of $c_u^{(m)}$ and $c_{u,v}^{(m)}$ under $\Delta_{F,m}$ in terms of Dickson polynomials? In other words, what is the generalizations of Identities 2 and 4 for $m > 2$.

Based on how the above identities were proven and the existence of the factorization formulas discussed in Section 1 above, these two questions could be investigated using computational means. More specifically, in [9], Section 8 the maps $f_{T,m}$ and f_T are provided explicitly and can thus be entered into a computer algebra system such as GAP. The technique would be to use this software to test whether a map $\delta : H_{F,m} \to H_{F,1}$ satisfies the factorization formula $f_T \delta = N_{T,m} f_{T,m}$ (discussed in Section 1 above). This would be enough to show $\delta = \Delta_{F,m}$ by Theorem 1 of [5].

In other words consider Figure 1 below. In [9], Section 8 the horizontal maps are given explicitly. The right vertical map is included in the software GAP and the left vertical maps are given for some basis elements by Identities 2–4 above. By [5], Theorem 1, $\Delta_{F,2}$ is the unique linear map of algebras that makes this diagram commute. Thus GAP could be used to test for the images of the remaining basis elements of $H_{F,2}$, answering Question 1.

Now consider Figure 2 below. As above in [9], Section 8 the horizontal maps are given explicitly and the right vertical map is included in the software GAP

$$H_{F,2} \xrightarrow{f_{\tilde{T}^F,2}} C\tilde{T}^{F^2}$$

$$\Delta_{F,2} \downarrow \qquad \qquad N_{\tilde{T}^F,2} \downarrow$$

$$H_{F,1} \xrightarrow{f_{\tilde{T}^F}} C\tilde{T}^F$$

FIGURE 1

$$H_{F,m} \xrightarrow{f_{\tilde{T}^F,m}} C\tilde{T}^{F^m}$$

$$\Delta_{F,m} \downarrow \qquad \qquad N_{\tilde{T}^F,m} \downarrow$$

$$H_{F,1} \xrightarrow{f_{\tilde{T}^F}} C\tilde{T}^F$$

FIGURE 2

(being the extension of the norm map of finite field extensions). Using again [5], Theorem 1, we have that $\Delta_{F,m}$ is the unique linear map of algebras that makes this diagram commute. Identity 1 provides the image $c_{1,1}^{(m)}$ under $\Delta_{F,m}$. GAP could be used to test for the images of the remaining basis elements of $H_{F,m}$ for $m > 2$, answering Question 2.

3. Further Evidence of Relationship of Norm Map to Dickson Polynomials

It is possible to extend the discussion following Identity 1 to the Gelfand-Graev Hecke algebra of the unitary group $U(2,q)$. In this section we will continue to use all the notation introduced in the previous section. Let $\alpha : \tilde{G} \to \tilde{G}$ be defined by $\alpha(a_{ij}) = (a_{ji}^q)^{-1}$. Note that $\alpha \cdot \alpha = F^2$, thus α is a (nonstandard) Frobenius endomorphism whose square is a standard Frobenius endomorphism. Note that $\tilde{G}^\alpha = U(2,q)$.

As above let $\tilde{U}$ denote the unipotent radical of an α-stable Borel subgroup of $\tilde{G}$ and let

$$e^* = \frac{1}{q} \sum_{u \in \tilde{U}^\alpha} \psi_*(u^{-1})u$$

where ψ_* denotes a nondegenerate linear character of $\tilde{U}^\alpha$. The Hecke algebra corresponding to $U(2,q)$ is $H_{\alpha,1} = e^* CU(2,q)e^*$.

It is straightforward to show that $\tilde{U}^F \cong \tilde{U}^\alpha$. Thus we may assume $e^* = e^{(1)}$. In [9], Section 5 it was shown that

$$\left\{ qe^* \begin{pmatrix} 0 & -v \\ v^{-q} & 0 \end{pmatrix} e^*, \ \begin{pmatrix} u & 0 \\ 0 & u \end{pmatrix} e^* \ \middle| \ u, v \in \mathbf{F}_{q^2}^*, u^{q+1} = 1 \right\}$$

is a basis for the Hecke algebra, $H_{\alpha,1}$, corresponding to $U(2,q)$. Denote the basis element

$$qe^* \begin{pmatrix} 0 & -v \\ v^{-q} & 0 \end{pmatrix} e^*$$

by $c^*_{v,v-q}$. Denote the basis element

$$\begin{pmatrix} u & 0 \\ 0 & u \end{pmatrix} e^*$$

by c^*_u.

Since $\alpha^2 = F^2$ the Hecke algebra constructed using α^2 and the Hecke algebra constructed using F^2 are the same. That is, $H_{F,2} = H_{\alpha,2}$. Thus the norm map $\Delta_{F,2} : H_{F,2} \to H_{F,1}$ (which we investigated in the previous section) between the Hecke algebras constructed using the Frobenius endomorphism F^2 and the Hecke algebra constructed using F has the same domain as the norm map between the Hecke algebras constructed using the Frobenius endomorphism α^2 and the Hecke algebra constructed using α. This norm map, $\Delta_{\alpha,2} : H_{\alpha,2} \to H_{\alpha,1}$, is the unique linear map (introduced in the first section) which satisfies the factorization formula $f_{\tilde{T}^\alpha} \Delta_{\alpha,2} = N_{\tilde{T}^\alpha,2} f_{\tilde{T}^F,2}$ where here $f_{\tilde{T}^\alpha} : H_{\alpha,1} \to C\tilde{T}^\alpha$ and $f_{\tilde{T}^F,2} : H_{F,2} \to C\tilde{T}^{F^2}$ are the homomorphisms of algebras described in the first section and $N_{\tilde{T}^\alpha,2} : C\tilde{T}^{F^2} \to C\tilde{T}^\alpha$ is the norm map of the finite field extension extended to the group algebras of these maximal tori.

As with the norm map $\Delta_{F,2}$, it appears that images of the norm map $\Delta_{\alpha,2} : H_{F,2} \to H_{\alpha,1}$ can be expressed in terms of Dickson polynomials. In [9], Theorem 10.4, it was shown that for $u \in \mathbf{F}^*_{q^2}$ when $u^{q+1} = 1$ that

$$\Delta_{\alpha,2}(c_u^{(2)}) = \begin{pmatrix} u^{-q+1} & 0 \\ 0 & u^{-q+1} \end{pmatrix} \Delta_{F,2}(c_u^{(2)}).$$

Thus, using Identity 2, we have

$$
\begin{aligned}
\Delta_{\alpha,2}(c_u^{(2)}) &= \begin{pmatrix} u^{-q+1} & 0 \\ 0 & u^{-q+1} \end{pmatrix} P_1(c_{u^{q+1}}^{(1)}, 1) \\
&= \begin{pmatrix} u^{-q+1} & 0 \\ 0 & u^{-q+1} \end{pmatrix} \begin{pmatrix} u^{q+1} & 0 \\ 0 & u^{q+1} \end{pmatrix} e^* \\
&= \begin{pmatrix} u^2 & 0 \\ 0 & u^2 \end{pmatrix} e^* \\
&= \begin{pmatrix} u^{-q+1} & 0 \\ 0 & u^{-q+1} \end{pmatrix} e^{(1)},
\end{aligned}
$$

since $u^{q+1} = 1$. Thus we have the following identity.

IDENTITY 5. For $u \in \mathbf{F}_{q^2}$ such that $u^{q+1} = 1$, $\Delta_{\alpha,2}(c_u^{(2)}) = P_1(c_{u^{-q+1}}^{(1)}, 1)$.

In addition the following identity from [9], Theorem 10.7 provides the image under $\Delta_{\alpha,2}$ of basis elements of type $c_{v,v-q}^{(2)}$ in terms of Dickson polynomials.

IDENTITY 6. For all $t \in \mathbf{F}^*_{q^2}$, $\Delta_{\alpha,2}(c_{t,t-q}^{(2)}) = P_2(c^*_{t,t-q}, qc^*_{t-q+1})$.

As with Identities 2 and 4, Identities 5 and 6 do not give a complete description of the norm map $\Delta_{\alpha,2} : H_{F,2} \to H_{\alpha,1}$. Identity 6 needs to be generalized to an identity for all basis elements of type $c_{u,v}^{(2)}$ in $H_{\alpha,2} = H_{F,2}$ in order to provide a complete description of $\Delta_{F,2} : H_{F,2} \to H_{\alpha,1}$.

Question 3: What is the image of $c_{u,v}^{(2)}$ under $\Delta_{\alpha,2}$ in terms of Dickson polynomials? In other words, what is the generalization of Identity 6 to all basis elements of $H_{F,2}$ of type $c_{u,v}^{(2)}$?

Consider Figure 3.

$$
\begin{array}{ccc}
H_{F,2} = H_{\alpha,2} & \xrightarrow{\;f_{\tilde{T}^\alpha,2}\;} & \mathbf{C}\tilde{T}^{F^2} = \mathbf{C}\tilde{T}^{\alpha^2} \\
\Delta_{\alpha,2} \downarrow & & N_{\tilde{T}^\alpha,2} \downarrow \\
H_{\alpha,1} & \xrightarrow{\;f_{\tilde{T}^\alpha}\;} & \mathbf{C}\tilde{T}^\alpha
\end{array}
$$

FIGURE 3

In [9], Sections 7 and 8 the horizontal maps $f_{\tilde{T}^\alpha,2}$ and $f_{\tilde{T}^\alpha}$ are provided explicitly and thus these as well as the map $N_{\tilde{T}^\alpha,2}$ can be entered into a computer algebra system such as GAP. Thus, using that $\Delta_{\alpha,2}$ is the unique linear map for which this diagram will commute, a similarly technique as described in the previous section could be used to test for the homomorphism $\Delta_{\alpha,2}$. This would provide an answer to Question 3.

More generally let β be a twisted Frobenius endomorphism. In other words suppose β is not a standard Frobenius endomorphism but there exists some positive power of β that is a standard Frobenius endomorphism. Let m be the smallest positive integer such that β^m is a standard Frobenius map. Then $\beta^m = F^k$, for some positive integer k, where F is the standard Frobenius map which raises every entry to the qth power. We have the commutative diagram in Figure 4.

$$
\begin{array}{ccc}
H_{F,k} = H_{\beta,m} & \xrightarrow{\;f_{\tilde{T}^\beta,m}\;} & \mathbf{C}\tilde{T}^{F^n} = \mathbf{C}\tilde{T}^{\beta^m} \\
\Delta_{\beta,m} \downarrow & & N_{\tilde{T}^\beta,m} \downarrow \\
H_{\beta,1} & \xrightarrow{\;f_{\tilde{T}^\beta}\;} & \mathbf{C}\tilde{T}^\beta
\end{array}
$$

FIGURE 4

Note (as Figure 4 illustrates) the norm map between the Hecke algebras $H_{\beta,m} = H_{F,k}$ and $H_{\beta,1}$ provides a relationship between the characters of the Hecke algebra of the Gelfand-Graev representation of a twisted finite group of Lie type in terms of the characters of the Hecke algebra of the Gelfand-Graev representation of $GL(n, q^k)$, for some positive integer n. For example in the above description where $\beta = \alpha$ and $m = n = k = 2$, Figure 3 illustrates how the characters of the Hecke algebra of the Gelfand-Graev representation of $U(2, q)$ can be expressed in terms of the characters of the Hecke algebra of the Gelfand-Graev representation of $GL(2, q)$. The main question is to describe this norm map $\Delta_{\beta,m} : H_{F,k} \to H_{\beta,1}$. Answering this question would provide this relationship between the twisted and untwisted Hecke algebras.

Main Question Use computational techniques to provide an explicit description of the norm map $\Delta_{\beta,m} : H_{\beta,m} \to H_{\beta,1}$ for a given fixed twisted Frobenius endomorphism $\beta : \tilde{G} \to \tilde{G}$ such that β^m is a standard Frobenius endomorphism.

4. Acknowledgement

Many of these results were determined while I was visiting the University of Oregon. I would like to thank this university for their hospitality. I would particularly like to thank Charles W. Curtis for the many interesting discussions we had on this and other material while I was in Oregon. I would also like to thank the referee for the suggestions and comments.

References

1. R. W. Carter, *Finite Groups of Lie Type: Conjugacy Classes and Complex Characters*, John Wiley and Sons, 1985.
2. C. Curtis, On the Gelfand-Graev Representations of a Reductive Group over a Finite Field, *Journal of Algebra* **157**, 517-533 (1993).
3. C. Curtis and R. Reiner, *Methods of Representation Theory with Applications to Finite Groups and Orders*, Volume I, John Wiley and Sons, 1981.
4. C. Curtis and K. Shinoda, Unitary Kloosterman Sums and Gelfand-Graev Representations of GL_2, *Journal of Algebra* **216**, 431-447 (1999).
5. C. Curtis and T. Shoji, A Norm Map for Endomorphism Algebras of Gelfand-Graev Representations, *Progr. Math.* **141**, 185-194 (1997).
6. F. Digne and J. Michel, *Representations of Finite Groups of Lie Type*, Cambridge University Press, 1991.
7. The GAP Group, GAP–Groups, Algorithms, and Programming, Version 4.4.9; 2006. (http://www.gap-system.org).
8. R. Lidl and H. Niederreiter, *Finite Fields*, Encyclopedia of Mathematics and its Applications 20, Cambridge, 1984.
9. J. Rainbolt, The Norm Map Between the Hecke Algebras of the Gelfand-Graev Representations of $GL(2, q^2)$ and $U(2, q)$, submitted.

DEPARTMENT OF MATHEMATICS AND COMPUTER SCIENCE, SAINT LOUIS UNIVERSITY, SAINT LOUIS, MO 63103, USA

E-mail address: rainbolt@slu.edu

Contemporary Mathematics
Volume **470**, 2008

On orbit equivalent, two-step imprimitive permutation groups

Ákos Seress and Keyan Yang

ABSTRACT. We prove that if a primitive group has a regular orbit on the power set of the permutation domain then with some exceptions the group must have at least four regular orbits on the power set. This result is used to investigate orbit equivalence of some transitive but imprimitive permutation groups.

1. Introduction

Two permutation groups $G, H \leq \mathrm{Sym}(\Omega)$ are called *orbit equivalent*, in notation $G \equiv H$, if they have the same orbits on the power set $\mathcal{P}(\Omega)$. It is easy to see that if $G \equiv H$ then G is transitive if and only if H is transitive and G is primitive if and only if H is primitive. All pairs of primitive orbit equivalent groups were determined in [**11**] (see Theorem 2.1). In this paper, we start the investigation of orbit equivalence of two-step imprimitive permutation groups.

For $G \leq \mathrm{Sym}(\Omega)$ and $\Gamma \subset \Omega$, we denote by G_Γ the setwise stabilizer of Γ in G. If G acts on a set Δ then G^Δ denotes this action. A transitive $G \leq \mathrm{Sym}(\Omega)$ is called *two-step imprimitive* if there exists a nontrivial block system $\Sigma = \{\Delta := \Delta_1, \Delta_2, \ldots, \Delta_n\}$ such that $A := G_\Delta^\Delta$ and $B := G^\Sigma$ are primitive.

If $G \equiv H$ then $G \equiv H \equiv \langle G, H \rangle$, so it is enough to determine orbit equivalent pairs of groups satisfying $H < G$. Another obvious but very useful observation is that if $G \leq \mathrm{Sym}(\Omega)$ has a regular a regular orbit O in $\mathcal{P}(\Omega)$ then O cannot be an orbit of a proper subgroup $H < G$ and so $H \not\equiv G$. The primitive groups with no regular orbit on $\mathcal{P}(\Omega)$ were also determined in [**11**] (see Theorem 2.2) and this list played a crucial role in the proof of Theorem 2.1. Here we shall prove the following extension of Theorem 2.2. The notation (G, n, k) means that G is a primitive group of degree n and the action of G on $\mathcal{P}(\Omega)$ has k regular orbits. Later, (G, n) means that G is a primitive group of degree n. Cyclic and dihedral groups of order n are denoted by C_n and D_n, respectively. We denote the set $\{1, 2, \ldots, n\}$ by $[n]$.

THEOREM 1.1. *Suppose that a primitive group $G \leq \mathrm{Sym}(\Omega)$ has a regular orbit, but no more than three regular orbits on $\mathcal{P}(\Omega)$. Then (G, n, k) must be one of*

2000 *Mathematics Subject Classification.* Primary 20B10, 20B15.
Key words and phrases. two-step primitive group, regular orbit on power set, orbit equivalent.
This research was supported in part by the NSF.

the following: $(C_2, 2, 1)$, $(C_3, 3, 2)$, $(D_{14}, 7, 2)$, $(\mathrm{AGL}(1, 8), 8, 3)$, $(\mathrm{PSL}(2, 11), 12, 2)$, $(\mathrm{PSL}(2, 13), 14, 2)$, $(A_7, 15, 2)$, or $(C_2^4.S_5, 16, 2)$.

We divide primitive permutation groups into the following three classes.

(α) Primitive groups with at least one regular orbit on the power set. By Theorem 1.1, with eight exceptions, any primitive group with a regular orbit has at least four regular orbits.

(β) Primitive groups with no regular orbits but not (A_n, n) or (S_n, n).

(γ) (A_n, n) and (S_n, n).

Our main result is the following. Let $A \leq \mathrm{Sym}(\Delta)$ and $B \leq \mathrm{Sym}([n])$ be primitive groups and let $G \leq \mathrm{Sym}(\Delta \times [n])$ be a two-step imprimitive group with block system $\Sigma = \{\Delta_1, \ldots, \Delta_n\}$, $\Delta_i := \{(\delta, i) \mid \delta \in \Delta\}$ for $i = 1, 2, \ldots, n$. Suppose further that $G^\Sigma \cong B$ and $G_{\Delta_1}^{\Delta_1} \cong A$.

THEOREM 1.2. *With the notation of the previous paragraph, let G be a two-step imprimitive group and suppose that $(A, |\Delta|) \in (\alpha)$ but $A \neq (C_2, 2)$ and $B \notin (\gamma)$. If $H < G$ and $H \equiv G$ then (G, H) must be one of the following:*

(i) $G = A \wr B$ with $(A, |\Delta|) = (D_{14}, 7)$ and $(B, n) = (D_{10}, 5)$, $(\mathrm{A\Gamma L}(1, 9), 9)$, $(\mathrm{AGL}(2, 3), 9)$, or $(\mathrm{P\Gamma L}(2, 9), 10)$. *Moreover, H is an index 2 subgroup of G.*

(ii) $G = A \wr B$ with $(A, |\Delta|) = (C_3, 3)$ and either $(B, n) = (\mathrm{A\Gamma L}(1, 8), 8)$ or $(\mathrm{P\Gamma L}(2, 8), 9)$. *Moreover, H is an index 3 subgroup of G.*

If G belongs to case (i) then H is unique while if G belongs to case (ii) then there are two possibilities for H.

We shall prove Theorem 1.1 in Section 3. Some general results about orbit equivalence are given in Section 4 and the proof of Theorem 1.2 is in Section 5. Computations in the computer algebra system GAP [6] play a crucial role in the proofs of both main theorems. When we refer to permutations and set stabilizers in primitive groups explicitly, we assume that the groups are given as in the primitive group library of GAP Version 4.4.9.

2. Previous results

THEOREM 2.1. [11] *For $n \geq 11$, the only orbit equivalent pairs of primitive permutation groups of degree n are A_n and S_n. For $n \leq 10$, the families of orbit equivalent groups are as follows. For $n = 3 : \{A_3, S_3\}$; for $n = 4 : \{A_4, S_4\}$; for $n = 5 : \{C_5, D_{10}\}$, $\{\mathrm{AGL}(1, 5), A_5, S_5\}$; for $n = 6 : \{\mathrm{PGL}(2, 5), A_6, S_6\}$; for $n = 7 : \{A_7, S_7\}$; for $n = 8 : \{\mathrm{AGL}(1, 8), \mathrm{A\Gamma L}(1, 8), \mathrm{ASL}(3, 2)\}$, $\{A_8, S_8\}$; for $n = 9 : \{\mathrm{AGL}(1, 9), \mathrm{A\Gamma L}(1, 9)\}$, $\{\mathrm{ASL}(2, 3), \mathrm{AGL}(2, 3)\}$, $\{\mathrm{PSL}(2, 8), \mathrm{P\Gamma L}(2, 8), A_9, S_9\}$; for $n = 10 : \{\mathrm{PGL}(2, 9), \mathrm{P\Gamma L}(2, 9)\}$, $\{A_{10}, S_{10}\}$.*

In the next theorem, we abbreviate C_k by k.

THEOREM 2.2. [11] *The primitive groups G of degree n with no regular orbit on the power set and not containing A_n, listed by their degree, are:*

(a) *With $n = 5$, (G, n) is equal to $(D_{10}, 5)$ or $(\mathrm{AGL}(1, 5), 5)$;*

(b) *With $n = 6$, (G, n) is equal to $(\mathrm{PSL}(2, 5), 6)$ or $(\mathrm{PGL}(2, 5), 6)$;*

(c) *With $n = 7$, (G, n) is equal to $(\mathrm{AGL}(1, 7), 7)$ or $(\mathrm{PSL}(3, 2), 7)$;*

(d) *With $n = 8$, (G, n) is equal to $(\mathrm{A\Gamma L}(1, 8), 8)$, $(\mathrm{ASL}(3, 2), 8)$, $(\mathrm{PSL}(3, 2), 8)$, or $(\mathrm{PSL}(3, 2).2, 8)$;*

(e) With $n = 9$, (G, n) is equal to $(3^2.D_8, 9)$, $(A\Gamma L(1, 9), 9)$, $(ASL(2, 3), 9)$, $(AGL(2, 3), 9)$, $(PSL(2, 8), 9)$, or $(P\Gamma L(2, 8), 9)$;

(f) With $n = 10$, (G, n) is equal to $(S_5, 10)$, $(A_6, 10)$, $(S_6, 10)$, $(PGL(2, 9), 10)$, $(M_{10}, 10)$, or $(P\Gamma L(2, 9), 10)$;

(g) With $n = 11$, (G, n) is equal to $(PSL(2, 11), 11)$ or $(M_{11}, 11)$;

(h) With $n = 12$, (G, n) is equal to $(PGL(2, 11), 12)$; $(M_{11}, 12)$, or $(M_{12}, 12)$;

(i) With $n = 13$, (G, n) is equal to $(PSL(3, 3), 13)$;

(j) With $n = 14$, (G, n) is equal to $(PGL(2, 13), 14)$;

(k) With $n = 15$, (G, n) is equal to $(A_8, 15)$;

(l) With $n = 16$, (G, n) is equal to $(2^4.(A_5 \times 3).2, 16)$, $(2^4.A_6, 16)$, $(2^4.S_6, 16)$, $(2^4.A_7, 16)$, or $(ASL(4, 2), 16)$;

(m) With $n = 17$, (G, n) is equal to $(PSL(2, 16).2, 17)$ or $(PSL(2, 16).4, 17)$;

(n) With $n = 21$, (G, n) is equal to $(P\Gamma L(3, 4), 21)$;

(o) With $n = 22$, (G, n) is equal to $(M_{22}, 22)$ or $(M_{22}.2, 22)$;

(p) With $n = 23$, (G, n) is equal to $(M_{23}, 23)$;

(q) With $n = 24$, (G, n) is equal to $(M_{24}, 24)$; and

(r) With $n = 32$, (G, n) is equal to $(ASL(5, 2), 32)$.

The *distinguishing number* $D(G, \Omega)$ of a permutation group $G \leq \text{Sym}(\Omega)$ is the minimal number r with the property that there exists an ordered partition P of Ω into r parts such that only the identity element of G fixes P. Equivalently, there exists a function $f : \Omega \to \{1, 2, \ldots, r\}$ such that if for some $g \in G$ we have $f(\omega^g) = f(\omega)$ for all $\omega \in \Omega$ then $g = 1$. The name "distinguishing number" was introduced in [1] but the notion itself, under different names, was investigated earlier [3, 4, 7, 10]. Clearly, $D(G, \Omega) = 2$ if and only if the action G on $\mathcal{P}(\Omega)$ has a regular orbit, $D(A_n, [n]) = n - 1$, and $D(S_n, [n]) = n$. For the primitive groups occurring in Theorem 2.2, the distinguishing number was determined by Dolfi [5].

LEMMA 2.3. [5] *The distinguishing number of the groups* $(M_{12}, 12)$, $(M_{11}, 11)$, $(ASL(3, 2), 8)$, $(PSL(3, 2), 7)$, *and* $(PGL(2, 5), 6)$ *is four. For all other groups listed in Theorem 2.2, the distinguishing number is three.*

LEMMA 2.4. [12] *Let* $H, G \leq \text{Sym}(\Omega)$ *be orbit equivalent. Then* H *is primitive if and only if* G *is primitive. Moreover,* H *and* G *have the same block systems of imprimitivity.*

3. Proof of Theorem 1.1

The proof mimics the proof of Theorem 2.2 in [11]. Primitive groups are classified by the O'Nan–Scott theorem. Following the notation of Liebeck, Praeger, and Saxl [8], we say that a primitive group is type I if $\text{Soc}(G)$ is abelian, type II if G is almost simple, type III(a) if G has simple diagonal action, type III(b) if G has wreath product action, and type III(c) if $\text{Soc}(G)$ is regular and nonabelian. For the purposes of this paper, no detailed knowledge of the actions of the different types is needed.

LEMMA 3.1. *Let* $\Omega = \Omega_1 \times \cdots \times \Omega_r$ *be the set of r-tuples* $(\alpha_1, \alpha_2, \ldots, \alpha_r)$ *where* $\Omega_i = [n]$, $\alpha_i \in \Omega_i$ *for some* $n \geq 5$ *and let* $r \geq 2$. *Then* $G = S_n \wr S_r$ *in its product action on* Ω *has at least four regular orbits on* $\mathcal{P}(\Omega)$.

PROOF. For $r = 2$, let

$$\Delta_1 = \{(1,1),(1,2),\ldots,(1,n),(2,2),\ldots,(n,n),(2,3),(3,4),\ldots,(n-1,n)\},$$
$$\Delta_2 = \{(1,2),\ldots,(1,n),(2,2),\ldots,(n,n),(2,3),(3,4),\ldots,(n-1,n)\},$$

and let $\Delta_3 = \Omega \setminus \Delta_1$ and $\Delta_4 = \Omega \setminus \Delta_2$. Then the four sets Δ_i are in different G-orbits in $\mathcal{P}(\Omega)$ because their cardinalities are different. Moreover, $G_{\Delta_i} = 1$ for $i = 1, 2, 3, 4$. We prove this for Δ_1; for Δ_2, the proof is similar, and obviously $G_{\Delta_i} = G_{\Delta_{i-2}}$ for $i = 3, 4$.

The elements of Δ_1 correspond to the edges of a subgraph Γ of the complete bipartite graph $K_{n,n}$ and we have to show that $\mathrm{Aut}(\Gamma) = 1$. Any automorphism of Γ must fix $1 \in \Omega_1$ as the unique vertex of valency n and so the set E of edges incident to this vertex is also fixed (setwise). Deleting E, the remaining edges of Γ form a path. This path has to be fixed pointwise, because its endpoints $2 \in \Omega_2$ and $n \in \Omega_1$ cannot be exchanged as only one of them is a neighbor of the fixed $1 \in \Omega_1$. Hence all vertices of Γ must be fixed.

We proceed inductively. Suppose that for some $r \geq 2$, we have four subsets $\Gamma_r^{(i)} \subset \Omega_1 \times \cdots \times \Omega_r$ of different cardinality such that $(S_n \wr S_r)_{\Gamma_r^{(i)}} = 1$ for $i = 1, 2, 3$, and 4. Let

$$\Gamma_{r+1}^{(i)} := \{(\alpha_1, \alpha_2, \ldots, \alpha_{r+1}) \mid \alpha_1 = 1, \ (\alpha_2, \alpha_3, \ldots, \alpha_{r+1}) \in \Gamma_r^{(i)}\}$$
$$\cup \{(\alpha, \alpha, \ldots, \alpha) \mid 2 \leq \alpha \leq n\}.$$

Let $g \in (S_n \wr S_{r+1})_{\Gamma_{r+1}^{(i)}}$ be arbitrary and consider the natural imprimitive action of $S_n \wr S_{r+1}$ on $\cup\Omega_i$, $1 \leq i \leq r+1$. In this action, g must fix $1 \in \Omega_1$ as the unique element covered by the most sequences in $\Gamma_{r+1}^{(i)}$. So g must fix setwise $\cup\Omega_i$, $2 \leq i \leq r+1$, as the set of elements occurring in sequences together with $1 \in \Omega_1$ in $\Gamma_{r+1}^{(i)}$ and so, by the inductive hypothesis, g fixes $\cup\Omega_i$, $2 \leq i \leq r+1$, pointwise. Finally, the $n-1$ sequences in $\Gamma_{r+1}^{(i)}$ not containing $1 \in \Omega_1$ uniquely determine the elements of Ω_1 and g must fix all of them. Hence $g = 1$ and so $(S_n \wr S_{r+1})_{\Gamma_{r+1}^{(i)}} = 1$. Thus $S_n \wr S_{r+1}$ has at least four regular orbits of different size on $\Omega_1 \times \cdots \times \Omega_{r+1}$. $\square$

COROLLARY 3.2. *Primitive groups of type* III(b) *or* III(c) *have at least four regular orbits on the power set.*

PROOF. Primitive groups of type III(b) are subgroups of some $S_m \wr S_r$, in the product action. Primitive groups of type III(c) are subgroups of groups of type III(b) [**8**, Remark 2(ii)]. $\square$

Let $\mu(G)$ be the minimal degree of $G \leq \mathrm{Sym}(\Omega)$ (that is, the minimal number of points in Ω moved by nonidentity elements of G).

LEMMA 3.3. *Let $n > 8$ and let $G \leq \mathrm{Sym}(\Omega)$, $|\Omega| = n$, be transitive. If G has at most three regular orbits on $\mathcal{P}(\Omega)$ then $|G| > 2^{\mu(G)/2}$.*

PROOF. We estimate the size of the set $A := \{(\Delta, g) \mid \Delta \subset \Omega, g \in G, \Delta^g = \Delta\}$ two different ways. For any nontrivial element $g \in G$, the number of cycles of g is at most $(n - \mu(G)) + \mu(G)/2 = n - \mu(G)/2$ and so the number of subsets fixed by g is at most $2^{n-\mu(G)/2}$. Thus $|A| \leq 2^{n-\mu(G)/2}(|G| - 1) + 2^n$.

On the other hand, G has at most three regular orbits on $\mathcal{P}(\Omega)$. For any set $\Gamma \subset \Omega$ not in the regular orbits, $|G_\Gamma| \geq 2$. Moreover, if $|\Gamma| = 1$ then $|G_\Gamma| = |G|/n$

and for $\Gamma \in \{\emptyset, \Omega\}$ we have $|G_\Gamma| = |G|$. Hence

$$(1) \qquad |A| \geq (3|G|) \cdot 1 + (2^n - 2 - 3|G| - n) \cdot 2 + n\frac{|G|}{n} + 2|G| = 2(2^n - 2 - n).$$

Comparing the estimates, we obtain

$$2(2^n - 2 - n) \leq 2^{n-\mu(G)/2}(|G| - 1) + 2^n.$$

Solving for $|G|$ yields $|G| \geq 2^{\mu(G)/2} + 1 - (\frac{2n+4}{2^n})2^{\mu(G)/2}$. Since $\mu(G) \leq n$ and $n \geq 9$, we have $1 > (\frac{2n+4}{2^n})2^{\mu(G)/2}$ and $|G| > 2^{\mu(G)/2}$. $\square$

LEMMA 3.4. *Primitive groups of type* III(a) *have at least four regular orbits on the power set.*

PROOF. It is shown in [**11**, Lemma 6] that if $G \leq \mathrm{Sym}(\Omega)$, $|\Omega| = n$, is of type III(a) then either $|G| < 2^{\sqrt{n/2}}$ or $|G| < 2^{\mu(G)/2}$. It is observed in [**3**] that $|G| < 2^{\sqrt{n/2}}$ implies $|G| < 2^{\mu(G)/2}$. Hence, in any case, Lemma 3.3 applies. $\square$

LEMMA 3.5. *If* $G \leq \mathrm{Sym}(\Omega)$ *is of type* I *and* $|\Omega| \geq 180$ *then* G *has at least four regular orbits on* $\mathcal{P}(\Omega)$.

PROOF. If $G \leq \mathrm{AGL}(d,p) \leq \mathrm{Sym}(\Omega)$, $|\Omega| = p^d$, then $\mu(G) \geq p^d - p^{d-1}$ because the fixed points of any $g \in \mathrm{GL}(d,p)$ constitute a subspace of the vector space $\mathrm{GF}(p)^d$. Moreover, $|G| = p^d \cdot |\mathrm{GL}(d,p)| < p^{d^2+d}$. Therefore, by Lemma 3.3, G has at least four regular orbits on $\mathcal{P}(\Omega)$ if

$$2^{(p^d - p^{d-1})/2} > p^{d^2+d}.$$

The only pairs (p,d) not satisfying this inequality are $(p,1)$ for $p \leq 17$, $(2,2)$, $(3,2)$, $(5,2)$, $(2,3)$, $(3,3)$, $(2,4)$, $(3,4)$, and $(2,d)$ for $5 \leq d \leq 8$. Among these pairs, only $(p,d) = (2,8)$ yields $|\Omega| \geq 180$.

In the case $p = 2$, $d = 8$, if $G = \mathrm{AGL}(8,2)$ has no more than three regular orbits on $\mathcal{P}(\Omega)$ then by equation (1) in the proof of Lemma 3.3, the set $A = \{(\Delta, g) \mid \Delta^g = \Delta, \Delta \subset \Omega, g \in G\}$ satisfies $|A| \geq 2(2^{256} - 2 - 2^8)$. However, we use a GAP program to compute the exact size of A (by first computing the conjugacy classes of G) and it turns out that $|A|$ is less than this bound. $\square$

LEMMA 3.6. *Let* $m \geq 9$, $2 \leq k \leq m/2$, *and let* $G = S_m$ *act on the* k-subsets of $\{1, 2, ..., m\}$. *Then* G *has at least four regular orbits on the power set.*

PROOF. For $n \geq 7$, we define a graph X_n as the union of three paths of length 1, 2, and $n - 4$, respectively (the length of a path is the number of edges). The three paths share one common endpoint and have no other common vertices. We also define a graph Y_n on $[n]$ with edge set $\{\{1,2\}, \{2,3\}, \ldots, \{n-2, n-1\}\} \cup \{\{2,n\}, \{3,n\}\}$.

For $k = 2$, take Δ_1 as the edge set of the graph X_m on $[m]$. Clearly, the automorphism group of X_m is trivial and so $G_{\Delta_1} = 1$.

For $k = 3$, take X_{m-1} on $[m-1]$ and denote the edges of X_{m-1} by $e_1, e_2, \ldots, e_p$, where $p = m - 2$. Let $\Delta_1 = \{e_1 \cup \{m\}, e_2 \cup \{m\}, \ldots, e_p \cup \{m\}\}$. Then $G_{\Delta_1} = 1$ because on its natural action on m points, any $g \in G_{\Delta_1}$ must fix m and the graph X_{m-1}.

For $k = 4$, take X_{m-2} on $[m-2]$ and denote the edges of X_{m-1} by $e_1, e_2, \ldots, e_p$, where $p = m - 3$. Let

$$\Delta_1 = \{e_1 \cup \{m-1, m\}, e_2 \cup \{m-1, m\}, \ldots, e_p \cup \{m-1, m\}, \{m-2, m, a, b\}\}$$

where $\{a, b\} \subset [m-2]$ but $\{a, b\}$ is not an edge of X_{m-2}. Then $G_{\Delta_1} = 1$ because on its natural action on m points, any $g \in G_{\Delta_1}$ must fix $\{m-1, m\}$ and the graph X_{m-2}; moreover, the last element in Δ_1 distinguishes $m-1$ and m.

For $m/2 \geq k \geq 5$, take X_{m-k+2} on the vertex set $\{k-1, k, ..., m\}$ and denote the edges of X_{m-k+2} by $e_1, e_2, ..., e_p$, where $p = m - k + 1$. Let

$$\begin{aligned}
\Gamma_1 &= \{e_1 \cup [k-2], e_2 \cup [k-2], ..., e_p \cup [k-2]\}, \\
\Gamma_2 &= \{\{k-1, k-2\} \cup \{m, m-1, ..., m-k+3\}, ..., \\
&\qquad \{2, 1\} \cup \{m, m-1, ..., m-k+3\}\}.
\end{aligned}$$

Let $\Delta_1 = \Gamma_1 \cup \Gamma_2$. Any $i \in [k-2]$ occurs $p+1$ or $p+2$ times in Δ_1. On the other hand, any $i > k-2$ occurs at most $k-2+3 = k+1$ times in Δ_1. Notice that $p+1 = m-k+2 > k+1$. So $[k-2]$ is fixed setwise by G_{Δ_1}. This implies that X_{m-k+1} is fixed, so in the natural action of G_{Δ_1} on m points, the set $\{k-1, ..., m\}$ is fixed pointwise. Then the elements of Γ_2 distinguish the points in $[k-2]$ and so $G_{\Delta_1} = 1$.

In the constructions above, we can replace X_n by Y_n to get another set Δ_2 whose stabilizer is trivial. Let $\Delta_3 = \Omega \setminus \Delta_1$ and $\Delta_4 = \Omega \setminus \Delta_2$. It is easy to see $\Delta_1, \Delta_2, \Delta_3$ and Δ_4 are of different cardinality and so they are from four different regular orbits on $\mathcal{P}(\Omega)$.   □

For primitive groups of type II, we use the following estimates of Liebeck and Saxl [9].

THEOREM 3.7. [9] *Let $G \leq \mathrm{Sym}(\Omega)$ be almost simple and primitive, and suppose that $G \neq A_m, S_m$ acting on the k-subsets of $[m]$. Then*

(a) $\mu(G) \geq n/3$.
(b) *If $\mathrm{Soc}(G)$ is alternating or sporadic then $\mu(G) \geq n/2$.*
(c) *If $\mathrm{Soc}(G)$ is a group of Lie type defined over $GF(q)$ and $\mathrm{Soc}(G)$ is not equal to $\mathrm{PSL}(4, 2), \mathrm{PSp}(4, 3), \mathrm{P\Omega}^-(4, 3), \mathrm{PSL}(2, q)$ then $\mu(G) \geq n(1 - \frac{4}{3q})$.*
(d) *If $\mathrm{Soc}(G) = \mathrm{PSL}(2, q)$ then, with the exception of an explicit list, $\mu(G) \geq n(1 - \frac{4}{3q})$.*

LEMMA 3.8. *If $G \leq \mathrm{Sym}(\Omega)$ is of type II, $|\Omega| \geq 180$, and G does not contain $\mathrm{Alt}(\Omega)$ then G has at least four regular orbits on $\mathcal{P}(\Omega)$.*

PROOF. Using Theorem 3.7, in [11, Lemma 12] it was established that with finitely many exceptions, almost simple primitive groups $G \leq \mathrm{Sym}(\Omega)$ not considered in Lemma 3.6 and not containing $\mathrm{Alt}(\Omega)$ satisfy $|G| \leq 2^{\mu(G)/2}$ (and so, by Lemma 3.3, they have at least four regular orbits on $\mathcal{P}(\Omega)$). In the exceptional cases, $\mathrm{Soc}(G)$ occurs on the following list:

(1) Higman-Sims group and Mathieu groups.
(2) A_m for some $m \leq 8$.
(3) $Sz(8)$.
(4) $\mathrm{PSL}(2, q)$ for some $q \leq 32$.
(5) One of the classical groups: $\mathrm{PSL}(5, 3), \mathrm{PSL}(4, 4), \mathrm{PSL}(3, 3), \mathrm{PSL}(3, 5),$
$\mathrm{PSL}(4, 3), \mathrm{PSL}(8, 2), \mathrm{PSL}(7, 2), \mathrm{PSL}(6, 2), \mathrm{PSL}(5, 2), \mathrm{PSL}(4, 2), \mathrm{PSL}(3, 2),$
$\mathrm{PSp}(6, 2), \mathrm{PSp}(4, 2), \mathrm{PSp}(4, 3), \mathrm{PSU}(3, 3), \mathrm{PSU}(3, 4)$.

For the above exceptions, we use Theorem 3.7(a) and Lemma 3.3 to establish a lower bound on $n := |\Omega|$ which guarantees at least four regular orbits on $\mathcal{P}(\Omega)$.

If $|G| \leq 2^{n/6}$, that is, if $n \geq 6 \log_2 |G|$, then $2^{\mu(G)/2} \geq 2^{n/6} \geq |G|$ and we are done. This estimate gives a lower bound less than 180 for n in all cases except $\mathrm{Soc}(G) = \mathrm{PSL}(6,2)$, $\mathrm{PSL}(7,2)$, or $\mathrm{PSL}(8,2)$. Moreover, for these three groups we only need to consider the natural action on the projective space due to the bounds for n. Hence in the first two cases $|\Omega| < 180$ and in the case $\mathrm{Soc}(G) = \mathrm{PSL}(8,2)$ we are done by Lemma 3.5. $\qquad\square$

PROOF OF THEOREM 1.1. We may assume that $G \neq \mathrm{Alt}(\Omega), \mathrm{Sym}(\Omega)$ with $|\Omega| \geq 4$ because these groups have no regular orbit on $\mathcal{P}(\Omega)$. By Corollary 3.2 and Lemmas 3.4, 3.5, 3.6, and 3.8, it is enough to consider primitive groups of degree less than 180.

Primitive groups G of degree $n < 180$ are listed in the GAP library. For $32 < n < 180$, we compute the conjugacy classes of G and then the exact size of the set $A = \{(\Delta, g) \mid \Delta^g = \Delta, \Delta \subset \Omega, g \in G\}$. It turns out that in each case $|A| < 2(2^n - 2 - n)$ and so equation (1) in the proof of Lemma 3.3 implies that G has at least four regular orbits on $\mathcal{P}(\Omega)$.

For $17 < n < 33$, we also compute $|A|$ and eliminate those groups where $|A| < 2(2^n - 2 - n)$. Only the following groups survive this test: $(G, n) = (\mathrm{PGL}(2,17), 18)$, $(S_7, 21)$, $(\mathrm{PSL}(3,4), 21)$, $(\mathrm{P}\Sigma\mathrm{L}(3,4), 21)$, $(\mathrm{PGL}(3,4), 21)$, $(\mathrm{P}\Gamma\mathrm{L}(3,4), 21)$, $(M_{22}, 22)$, $(M_{22}.2, 22)$, $(M_{23}, 23)$, $(M_{24}, 24)$, $((S_5 \times S_5).2, 25)$, $(\mathrm{AGL}(3,3), 27)$, $(\mathrm{PSp}(4,3).2, 27)$, $(\mathrm{PSp}(6,2), 28)$, $(\mathrm{PSL}(5,2), 31)$, and $(\mathrm{ASL}(5,2), 32)$.

Some of these groups are listed in Theorem 2.2 and hence have no regular orbits on $\mathcal{P}(\Omega)$. For all of the other groups, we compute the stabilizers of randomly chosen subsets of Ω. In each case, we can construct four regular orbits on $\mathcal{P}(\Omega)$ explicitly.

For $n \leq 17$, we compute the full orbit structure of G on $\mathcal{P}(\Omega)$ by brute force, yielding the examples listed in Theorem 1.1. $\qquad\square$

4. Orbit equivalent permutation groups

In this section we prove some general results about orbit equivalent permutation groups. *Throughout this section, we suppose that* $G, H \leq \mathrm{Sym}(\Omega)$, $H < G$, *and* $H \equiv G$. Moreover, $\Sigma = \{\Delta_1, \Delta_2, \ldots, \Delta_n\}$ is a block system of imprimitivity for G (and so, by Lemma 2.4, a block system for H).

LEMMA 4.1. $H^\Sigma \equiv G^\Sigma$.

PROOF. Let O be an orbit of G^Σ in $\mathcal{P}(\Sigma)$ and let $\Gamma \in O$ and $\Gamma' \in O$ be two subsets of Σ. Then there exists $g \in G$ such that $\Gamma^g = \Gamma'$. Let $\overline{\Gamma}$ be the union of all blocks in Γ and $\overline{\Gamma'}$ be the union of all blocks in Γ'. Hence we have $\overline{\Gamma}^g = \overline{\Gamma'}$. Since H and G are orbit equivalent, there exists $h \in H$ such that $\overline{\Gamma}^h = \overline{\Gamma'}$, which implies $\Gamma^h = \Gamma'$. So O is also an orbit of H^Σ in $\mathcal{P}(\Sigma)$. $\qquad\square$

LEMMA 4.2. Let $\Delta \in \Sigma$ (that is, Δ is a block for G and H). Then $G_\Delta^\Delta \equiv H_\Delta^\Delta$.

PROOF. Let Γ be a subset of Δ. We want to show that $\Gamma^{G_\Delta^\Delta} = \Gamma^{H_\Delta^\Delta}$.

Since Δ is a block, for any $g \in G$ we have $\Gamma^g \subset \Delta$ or $\Gamma^g \cap \Delta = \emptyset$, with the first of these possibilities occurring if and only if $g \in G_\Delta$. Hence $\Gamma^{G_\Delta^\Delta} = \Gamma^{G_\Delta} = \{\Gamma' \in \Gamma^G \mid \Gamma' \subset \Delta\}$. Similarly, $\Gamma^{H_\Delta^\Delta} = \{\Gamma' \in \Gamma^H \mid \Gamma' \subset \Delta\}$. However, $H \equiv G$, so $\Gamma^G = \Gamma^H$ and so $\Gamma^{G_\Delta^\Delta} = \Gamma^{H_\Delta^\Delta}$. $\qquad\square$

LEMMA 4.3. [12] *For any $\Theta \subset \Omega$, $|G_\Theta : H_\Theta| = |G : H|$ and $|G : H|$ is a divisor of $|G_\Theta|$.*

PROOF. Since $H \equiv G$, we have $|G : G_\Theta| = |\Theta^G| = |\Theta^H| = |H : H_\Theta|$. Hence $|G : H| = |G_\Theta : H_\Theta|$ and so $|G : H|$ divides $|G_\Theta|$.   □

COROLLARY 4.4. $|G : H|$ *divides* $\gcd\{|G_\Theta| \mid \Theta \subset \Omega\}$.   □

LEMMA 4.5. [2, Ex. 1.1] *Let $K, G \leq \mathrm{Sym}(\Omega)$ with $K < G$. Then K and G are orbit equivalent on Ω if and only if for all $\Delta \subset \Omega$, $KG_\Delta = G$.*

PROOF. Clearly, $KG_\Delta \leq G$. Suppose $K \equiv G$ and let $\Delta \subset \Omega$. Then, since K is transitive on Δ^G, for any $g \in G$ there exists $k \in K$ such that $\Delta^g = \Delta^k$. So $k^{-1}g \in G_\Delta$ and $g \in KG_\Delta$, implying $KG_\Delta = G$. Conversely, if $KG_\Delta = G$ for $\Delta \subset \Omega$, then K is transitive on Δ^G, by reversing the steps of the previous argument.   □

We say that the block system Σ is *maximal* for G if and only if G^Σ is primitive (and then, by Lemma 2.4, H^Σ is primitive).

THEOREM 4.6. *If Σ is a maximal block system then either $(H^\Sigma, G^\Sigma) = (A_n, S_n)$ or $H^\Sigma = G^\Sigma$.*

PROOF. Since Σ is a maximal block system, H^Σ and G^Σ are primitive on Σ. Moreover, by Lemma 4.1, $H^\Sigma \equiv G^\Sigma$. Hence, if the conclusion of the theorem does not hold then (H^Σ, G^Σ) must be one of the orbit equivalent pairs of primitive groups listed in Theorem 2.1. We process the pairs of groups listed in Theorem 2.1 one-by-one, and in each case we reach a contradiction. We suppose that the primitive groups are given as in the GAP library.

Case 1: $n = 5$ and $(H^\Sigma, G^\Sigma) = (C_5, D_{10})$. Choose $S \subset \Omega$ such that $S = S_1 \cup S_2$ with $S_1 \subset \Delta_1$, $S_2 \subset \Delta_2$, $|S_1| = 1$, $|S_2| = 2$. Then G_S must fix Δ_1 and Δ_2 setwise. Notice that $G^\Sigma = D_{10}$, and so $(G_S)^\Sigma = 1$. Therefore, $(HG_S)^\Sigma = H^\Sigma(G_S)^\Sigma < G^\Sigma$, implying $HG_S \neq G$. However, this contradicts Lemma 4.5, and so $(H^\Sigma, G^\Sigma) \neq (C_5, D_{10})$.

Case 2: $n = 5$ and $(H^\Sigma, G^\Sigma) = (\mathrm{AGL}(1,5), A_5)$ or $(\mathrm{AGL}(1,5), S_5)$. Choose $S \subset \Omega$ such that $S = S_1 \cup S_2 \cup S_3 \cup S_4$ with $S_i \subset \Delta_i$ $(i = 1, 2, 3, 4)$, $|S_1| = |S_2| = 2$, $|S_3| = |S_4| = 1$. Then G_S^Σ fixes $\{\Delta_1, \Delta_2\}$ and $\{\Delta_3, \Delta_4\}$, and so $(G_S)^\Sigma \leq C_2^2$.

By assumption $G = HG_S$ and so

$$|G^\Sigma| = |H^\Sigma(G_S)^\Sigma| = \frac{|H^\Sigma||(G_S)^\Sigma|}{|(G_S)^\Sigma \cap H^\Sigma|}.$$

However, the right side of this equation is a number not divisible by 3, contradicting that $|G^\Sigma|$ is divisible by 3.

Case 3: $n = 6$ and $(H^\Sigma, G^\Sigma) = (\mathrm{PGL}(2,5), A_6)$ or $(\mathrm{PGL}(2,5), S_6)$. Note that $\mathrm{PGL}(2,5)$ is not a subgroup of A_6, so the case $(H^\Sigma, G^\Sigma) = (\mathrm{PGL}(2,5), A_6)$ does not occur at all. For the case $(H^\Sigma, G^\Sigma) = (\mathrm{PGL}(2,5), S_6)$, choose $S \subset \Omega$ such that $S = S_1 \cup S_2 \cup S_3 \cup S_4$ with $S_i \subset \Delta_i$ $(i = 1, 2, 3, 4)$, $|S_1| = |S_2| = 2$, $|S_3| = |S_4| = 1$. Then G_S^Σ fixes $\{\Delta_1, \Delta_2\}$, $\{\Delta_3, \Delta_4\}$, and $\{\Delta_5 \Delta_6\}$. So $(G_S)^\Sigma \leq C_2^3$ and 3 does not divide $|(G_S)^\Sigma|$. Thus 9 does not divide $|(HG_s)^\Sigma| = |H^\Sigma(G_S)^\Sigma| = |H^\Sigma||(G_S)^\Sigma|/|(G_S)^\Sigma \cap H^\Sigma|$, which contradicts the fact that $|G^\Sigma|$ is divisible by 9.

Case 4: $n = 8$ and $(H^\Sigma, G^\Sigma) = (\mathrm{AGL}(1,8), \mathrm{A\Gamma L}(1,8))$. Choose $S \subset \Omega$ such that $S = S_1 \cup S_2 \cup S_3 \cup S_4$ with $S_i \subset \Delta_i$ $(i = 1, 2, 3, 4)$, $|S_1| = |S_2| = 2$, $|S_3| = |S_4| = 1$.

Then G_S^Σ fixes $A = \{\Delta_1, \Delta_2\}$ and $B = \{\Delta_3, \Delta_4\}$. Let $(G^\Sigma)_{AB}$ be the subgroup of G^Σ which fixes A and B setwise. Clearly $(G_S)^\Sigma \le (G^\Sigma)_{AB}$.

By a GAP computation, we have $|(G^\Sigma)_{AB}| = 4$. So $|(HG_S)^\Sigma| = |H^\Sigma (G_S)^\Sigma| = |H^\Sigma||(G_S)^\Sigma|/|(G_S)^\Sigma \cap H^\Sigma|$ is not divisible by 3 which implies that $|G^\Sigma| \ne |(HG_S)^\Sigma|$.

Case 5: $n = 8$ and $(H^\Sigma, G^\Sigma) = (\mathrm{AGL}(1,8), \mathrm{ASL}(3,2))$ or $(\mathrm{A\Gamma L}(1,8), \mathrm{ASL}(3,2))$. Choose $S \subset \Omega$ such that $S = S_1 \cup S_2 \cup S_3 \cup S_4 \cup S_5 \cup S_6$ with $S_i \subset \Delta_i$, $i = 1, 2, \ldots, 6$, $|S_1| = |S_2| = |S_3| = 2$, $|S_4| = |S_5| = |S_6| = 1$. Then G_S^Σ fixes $A = \{\Delta_1, \Delta_2, \Delta_3\}$ and $B = \{\Delta_4, \Delta_5 \Delta_6\}$.

By a GAP computation, we have $|(G^\Sigma)_{AB}| = 4$. Hence

$$|(HG_S)^\Sigma| = |H^\Sigma (G_S)^\Sigma| \le |H^\Sigma||(G_S)^\Sigma| < |G^\Sigma|,$$

a contradiction.

Case 6: $n = 9$ and $(H^\Sigma, G^\Sigma) = (\mathrm{AGL}(1,9), \mathrm{A\Gamma L}(1,9))$. Choose $S \subset \Omega$ such that $S = S_1 \cup S_2 \cup S_4 \cup S_5$ with $S_i \subset \Delta_i$ $(i = 1, 2, 4, 5,)$, $|S_1| = |S_2| = 2$, $|S_4| = |S_5| = 1$. Then G_S^Σ fixes $A = \{\Delta_1, \Delta_2\}$ and $B = \{\Delta_4, \Delta_5\}$.

By a GAP computation, we have $|(G^\Sigma)_{AB}| = 1$. Hence $|(HG_S)^\Sigma| = |H^\Sigma| < |G^\Sigma|$, a contradiction.

Case 7: $n = 9$ and $(H^\Sigma, G^\Sigma) = (\mathrm{ASL}(2,3), \mathrm{AGL}(2,3))$ or $(\mathrm{PSL}(2,8), \mathrm{P\Gamma L}(2,8))$. Choose $S \subset \Omega$ such that $S = S_1 \cup S_2 \cup S_4 \cup S_6 \cup S_7 \cup S_9$ with $S_i \subset \Delta_i$ $(i = 1, 2, 4, 6, 7, 9)$, $|S_1| = |S_2| = |S_7| = 2$, $|S_4| = |S_6| = |S_9| = 1$. Then G_S^Σ fixes $A = \{\Delta_1, \Delta_2, \Delta_7\}$ and $B = \{\Delta_4, \Delta_6, \Delta_9\}$.

By a GAP computation, we have $|(G^\Sigma)_{AB}| = 1$. Hence $|(HG_S)^\Sigma| = |H^\Sigma| < |G^\Sigma|$, a contradiction.

Case 8: $n = 9$ and $(H^\Sigma, G^\Sigma) = (\mathrm{PSL}(2,8), A_9)$, $(\mathrm{PSL}(2,8), S_9)$, $(\mathrm{P\Gamma L}(2,8), A_9)$, or $(\mathrm{P\Gamma L}(2,8), S_9)$. Choose $S \subset \Omega$ such that $S = S_1 \cup S_2 \cup S_3 \cup S_4 \cup S_5 \cup S_6$ with $S_i \subset \Delta_i$ $(i = 1, 2, 3, 4, 5, 6)$, $|S_1| = |S_2| = |S_3| = 2$, $|S_4| = |S_5| = |S_6| = 1$. Then G_S^Σ fixes $A = \{\Delta_1, \Delta_2, \Delta_3\}$ and $B = \{\Delta_4, \Delta_5, \Delta_6\}$.

We have $|(G^\Sigma)_{AB}| \in \{108, 216\}$. Hence

$$|(HG_S)^\Sigma| = |H^\Sigma (G_S)^\Sigma| = \frac{|H^\Sigma||(G_S)^\Sigma|}{|(G_S)^\Sigma \cap H^\Sigma|}$$

is not divisible by 5, contradicting the fact that 5 divides $|G^\Sigma|$.

Case 9: $n = 10$ and $(H^\Sigma, G^\Sigma) = (\mathrm{PGL}(2,9), \mathrm{P\Gamma L}(2,9))$. Choose $S \subset \Omega$ such that $S = S_1 \cup S_2 \cup S_3 \cup S_4 \cup S_5 \cup S_7$ with $S_i \subset \Delta_i$ $(i = 1, 2, 3, 4, 5, 7)$, $|S_1| = |S_2| = |S_3| = 2$, $|S_4| = |S_5| = |S_7| = 1$. Then G_S^Σ fixes $A = \{\Delta_1, \Delta_2, \Delta_3\}$ and $B = \{\Delta_4, \Delta_5, \Delta_7\}$.

By a GAP computation, we have $|(G^\Sigma)_{AB}| = 1$. Hence $|(HG_S)^\Sigma| = |H^\Sigma| < |G^\Sigma|$, a contradiction.

Therefore, the only possibility is that $H^\Sigma = G^\Sigma$ or $(H^\Sigma, G^\Sigma) = (A_n, S_n)$ for some n. $\square$

5. Proof of Theorem 1.2

In this section we use the following notation. Let $\Omega = \Delta \times [n]$, and $G \le \mathrm{Sym}(\Omega)$ with block system $\Sigma = \{\Delta_1, \ldots, \Delta_n\}$, $\Delta_i := \{(\delta, i) \mid \delta \in \Delta\}$ for $i = 1, 2, \ldots, n$. For $\Gamma \subset \Delta$, we shall abbreviate $\{(\gamma, i) \mid \gamma \in \Gamma\}$ by (Γ, i). Let $A \le \mathrm{Sym}(\Delta)$ and $B \le \mathrm{Sym}([n])$ such that $G_{\Delta_1}^{\Delta_1} \cong A$ and $G^\Sigma \cong B$. Moreover, let K be the kernel of the action of G on Σ.

We write the elements of $W := A \wr B$ as $(n+1)$-tuples $(a_1, \ldots, a_n; b)$ where $a_i \in A$ for $1 \le i \le n$ and $b \in B$. For $a := (a_1, \ldots, a_n; 1) \in A^n$, we define the *support* supp(a) of a as the set of indices $I \subset [n]$ with $a_i \neq 1$ for $i \in I$ and $a_i = 1$ for $i \notin I$. The following lemma is basic.

LEMMA 5.1. [**2**, Theorem 1.8] *The group G can be identified with a subgroup of W acting on $\Delta \times [n]$.*

LEMMA 5.2. *If A has at least k regular orbits on $\mathcal{P}(\Delta)$ and the distinguishing number of B is k then G has a regular orbit on $\mathcal{P}(\Omega)$.*

PROOF. Let $\mathcal{S} = (S_1, S_2, \ldots, S_k)$ be an ordered partition of $[n]$ such that only the identity element in B fixes $\mathcal{S}$. Let $R_1, R_2, \ldots, R_k$ be k subsets of Δ, in k different regular orbits of A on $\mathcal{P}(\Delta)$. Let $\Gamma = \bigcup_{j=1}^{k}(\bigcup_{i \in S_j}(R_i, i))$. Then only the identity of $A \wr B$ fixes Γ setwise. So, by Lemma 5.1, G has a regular orbit on $\mathcal{P}(\Omega)$.   □

We need the following variant of Lemma 2.3.

LEMMA 5.3. *We divide the groups (H, k) listed in Theorem 2.2 into the following three categories.*

Type I: $(D_{10}, 5)$, $(AGL(1, 5), 5)$, $(AGL(1, 7), 7)$, $(A\Gamma L(1, 8), 8)$, $(PSL(3, 2), 8)$, $(AGL(1, 9), 9)$, $(A\Gamma L(1, 9), 9)$, $(ASL(2, 3), 9)$, $(PSL(2, 8), 9)$; $(S_5, 10)$, $(PGL(2, 9), 10)$, $(M_{10}, 10)$, $(A_6, 10)$, $(PSL(2, 11), 11)$, $(PGL(2, 11), 12)$, $(PGL(2, 13), 14)$, $(2^4.(A_5 \times 3).2, 16)$, $(2^4.A_6, 16)$, $(2^4.S_6, 16)$, $(2^4.A_7, 16)$, $(PSL(2, 16).2, 17)$, $(PSL(2, 16).4, 17)$, $(P\Gamma L(3, 4), 21)$, $(M_{22}, 22)$, $(M_{22}.2, 22)$, $(ASL(5, 2), 32)$.

Type II: $(PSL(2, 5), 6)$, $(PGL(2, 5), 6)$, $(PSL(3, 2), 7)$, $(PSL(3, 2).2, 8)$, $(AGL(2, 3), 9)$, $(P\Gamma L(2, 8), 9)$, $(P\Gamma L(2, 9), 10)$, $(S_6, 10)$; $(M_{11}, 11)$, $(M_{11}, 12)$, $(PSL(3, 3), 13)$, $(A_8, 15)$, $(M_{23}, 23)$.

Type III: $(ASL(3, 2), 8)$, $(M_{12}, 12)$, $(ASL(4, 2), 16)$, $(M_{24}, 24)$.

For each group H of type I, *there is an ordered partition $\mathcal{Q} = (Q_1, Q_2, Q_3)$ of the permutation domain with $|Q_1| = 1$ such that only the identity of H fixes $\mathcal{Q}$. For each group H of type* II, *there is an ordered partition $\mathcal{Q} = (Q_1, Q_2, Q_3, Q_4)$ of the permutation domain with $|Q_1| = |Q_2| = 1$ such that only the identity of H fixes $\mathcal{Q}$. For each group H of type* III, *there is an ordered partition $\mathcal{Q} = (Q_1, Q_2, Q_3, Q_4, Q_5)$ of the permutation domain with $|Q_1| = |Q_2| = |Q_3| = 1$ such that only the identity of H fixes $\mathcal{Q}$.*

PROOF. It is done by a GAP computation. In the cases $k \geq 18$, we list the partitions below; for $k \leq 17$, a brute force search yields appropriate partitions. As usual, the set stabilizers are taken in the permutation representation given in the GAP library.

If $(H, k) = (P\Gamma L(3, 4), 21)$, take $Q_1 = \{1\}$ and $Q_2 = \{2, 4, 5, 8, 9, 10, 11, 15\}$. If $(H, k) = (M_{22}.2, 22)$ or $(M_{22}, 22)$, $Q_1 = \{1\}$, $Q_2 = \{2, 4, 5, 8, 9, 10, 11, 15, 16, 19\}$. If $(H, k) = (ASL(5, 2), 32)$, the one-point stabilizer is primitive but it is not in the list of Theorem 2.2, so the required partition (Q_1, Q_2, Q_3) exists. If $(H, k) = (M_{23}, 23)$, take $Q_1 = \{23\}$, $Q_2 = \{1\}$, and $Q_3 = \{2, 4, 5, 8, 9, 10, 11, 15, 16, 19\}$. Finally, if $(H, k) = (M_{24}, 24)$, we take $Q_1 = \{24\}$, $Q_2 = \{23\}$, $Q_3 = \{1\}$, and $Q_4 = \{2, 4, 5, 8, 9, 10, 11, 15, 16, 19\}$.   □

LEMMA 5.4. *For the groups (H, k) listed in Theorem 2.2, let $M_H(q)$ denote the k-dimensional permutation module of H over $\mathrm{GF}(q)$, for $q \in \{2, 3\}$. Then $M_H(q)$ has a unique $(k-1)$-dimensional H-invariant submodule. Moreover, if*

$(H, k) \neq (\mathrm{PSL}(3, 2), 7)$ then $M_H(2)$ has no $(k - 3)$ or $(k - 2)$-dimensional H-invariant submodule.

PROOF. For each group of type (β), we construct the permutation matrices of generators and compute the submodule structure by a GAP program. Checking the dimensions of submodules, we obtain the statement of the lemma. $\square$

LEMMA 5.5. Let $(A, |\Delta|) = (C_2^4.S_5, 16)$ or $(D_{14}, 7)$ and $B \in (\beta)$. Then, if G has no regular orbit on Ω then $G = W$ or $|W : G| = 2$.

PROOF. We identify Δ with $\{1, 2, \ldots, |\Delta|\}$. Let R_1 and R_2 be two subsets of Δ, from two different regular orbits of A on $\mathcal{P}(\Delta)$. We also define three subsets T_1, T_2, and T_3 of Δ the following way. In the case $(A, |\Delta|) = (C_2^4.S_5, 16)$, let $T_1 = \{1, 2, 3, 4, 5, 6, 9\}$, $T_2 = \Delta \setminus T_1$, and $T_3 = \{1, 2, 3, 5, 7, 8, 9\}$. In the case $(A, |\Delta|) = (D_{14}, 7)$, let $T_1 = \{1\}$, $T_2 = \Delta \setminus T_1$, and $T_3 = \{2, 5\}$.

A small GAP computation verifies the following facts. First of all, the T_i are in three different orbits of A on $\mathcal{P}(\Delta)$ and $|A_{T_i}| = 2$ for each i. Moreover, in the case $A = D_{14}$, for each i we have $A_{T_i} = \langle a_1 \rangle$ for the permutation $a_1 = (2, 5)(3, 6)(4, 7)$, and $a_1 \in A \setminus A'$. The permutation $a_2 = (1, 3)(2, 4)(5, 6)$ is an A-conjugate of a_1, the a_j are noncommuting, and there is no nontrivial element of A centralizing both a_1 and a_2. Finally, the A-conjugates of a_1 generate A, and any two A-conjugates of a_1 either commute, or the normal closure of their commutator is A'.

In the case $A = C_2^4.S_5$, for each i we have $A_{T_i} = \langle a_1 \rangle$ for the permutation $a_1 = (3, 5)(4, 6)(11, 13)(12, 14)$, and $a_1 \in A \setminus A'$. The permutations

$$
\begin{aligned}
a_2 &= (5, 9)(6, 10)(7, 11)(8, 12), \\
a_3 &= (5, 16)(6, 15)(7, 14)(8, 13), \\
\text{and } a_4 &= (2, 5)(4, 7)(10, 13)(12, 15)
\end{aligned}
$$

are A-conjugates of a_1, the a_j are pairwise noncommuting for $1 \leq j \leq 4$, and there is no nontrivial element of A centralizing all a_j. Finally, the A-conjugates of a_1 generate A and any two A-conjugates of a_1 either commute, or the normal closure of their commutator is A'.

Suppose first that B is of type I in Lemma 5.3. Then there exists an ordered partition $\mathcal{Q} = (Q_1, Q_2, Q_3)$ of $[n]$, with $Q_1 = \{j\}$ for some $j \leq n$, such that only the identity of B fixes $\mathcal{Q}$. Let $\Gamma = \bigcup_{i \in Q_1}(T_1, i) \cup \bigcup_{i \in Q_2}(R_1, i) \cup \bigcup_{i \in Q_3}(R_2, i)$. Then $|W_\Gamma| = 2$ and $W_\Gamma = \langle x_1 := (1, 1, \ldots, 1, a_1, 1, \ldots, 1; 1) \rangle$ for the element $x_1 \in A^n$ with support $\{j\}$. Since G has no regular orbit, $G_\Gamma = W_\Gamma$ and $x_1 \in G$. The elements of $x_1^{G_{\Delta_j}} \subset G$ also have support $\{j\}$. Since $G_{\Delta_j}^{\Delta_j} \cong A$, these conjugates generate a subgroup $A_j \leq G$ isomorphic to A by the last observations in the previous paragraphs and the nontrivial elements of A_j all have support $\{j\}$. Also, since G acts transitively on Σ, the G-conjugates of A_j generate A^n. Hence $K = A^n$ and $G = W$.

Suppose next that B is of type III. In this case, B is 3−transitive on $[n]$ and there exists an ordered partition $\mathcal{Q} = (Q_1, Q_2, Q_3, Q_4, Q_5)$ of $[n]$ with $|Q_1| = |Q_2| = |Q_3| = 1$ such that only the identity of B fixes $\mathcal{Q}$. First, we show that $K \geq C_7^n$ if $A = D_{14}$ and $K \geq (C_2^4.A_5)^n$ if $A = C_2^4.S_5$.

Let $\Gamma = \bigcup_{i \in Q_1}(T_1, i) \cup \bigcup_{i \in Q_2}(T_2, i) \cup \bigcup_{i \in Q_3}(T_3, i) \cup \bigcup_{i \in Q_4}(R_1, i) \cup \bigcup_{i \in Q_5}(R_2, i)$. Then $W_\Gamma \cong C_2 \times C_2 \times C_2$, and the supports of the elements of W_Γ are subsets of $Q_1 \cup Q_2 \cup Q_3$. In each $w \in W_\Gamma$, the nontrivial coordinates are equal to a_1. Let

$x_1 \in G_\Gamma$ be a nontrivial element of G_Γ; since A_{T_i} is the same for all i, without loss of generality we can suppose that the jth coordinate of x_1 is a_1, for $Q_1 = \{j\}$.

Conjugating x_1 by appropriate elements of G_{Δ_j}, we obtain $x_2, x_3, x_4 \in G$ with support size at most three such that the jth coordinate of x_i is a_i, for $2 \le i \le 4$. If there exists $i \in \{2, 3, 4\}$ such that $\text{supp}(x_1) \cap \text{supp}(x_i) = \{j\}$ then the commutator $[x_1, x_i] \ne 1$ and has support $\{j\}$. Its conjugates by G_{Δ_j} generate a subgroup $A'_j \le G$, $A'_j \cong A'$, and the nontrivial elements of A'_j all have support $\{j\}$. Also, since G acts transitively on Σ, the G-conjugates of A'_j generate $(A')^n$. Hence $K \ge (A')^n$. On the other hand, if $|\text{supp}(x_1) \cap \text{supp}(x_i)| \ge 2$ for all $i \le 4$ then the union of the four supports have size at most six. Since we are in case III, $n \ge 8$ and B is 3-transitive. Hence there exists a G_{Δ_j}-conjugate y of x_1 with $\text{supp}(y) \cap \text{supp}(x_i) = \{j\}$ for all $i \le 4$. Since the jth coordinate of y cannot commute with all four a_i, there exists $i \le 4$ with $[x_i, y] \ne 1$ and $\text{supp}([x_i, y]) = \{j\}$. So $A'_j \le G$ and $K \ge (A')^n$, as above.

The factor group $K/(A')^n$ can be identified with an n-dimensional vector space over $\text{GF}(2)$. The element $x_1 \in K$ corresponds to a vector with at most three nontrivial coordinates; if it has three nontrivial coordinates then the product of x_1 with an appropriate G-conjugate has only two (we use the 3-transitivity of B in this step). Hence, using 2-transitivity of B, $K/(A')^n$ contains all vectors with two nontrivial coordinates and these vectors generate an $(n-1)$-dimensional subspace. Therefore $K \ge (A')^n . C_2^{n-1}$ and $|W : G| \le 2$.

The proof in the case when B is of type II is similar. We start with an ordered partition $Q = (Q_1, Q_2, Q_3, Q_4)$ of $[n]$ with $|Q_1| = |Q_2| = 1$ such that only the identity of B fixes Q and with $\Gamma = \bigcup_{i \in Q_1}(T_1, i) \cup \bigcup_{i \in Q_2}(T_2, i) \cup \bigcup_{i \in Q_3}(R_1, i) \cup \bigcup_{i \in Q_4}(R_2, i)$. Then $W_\Gamma \cong C_2 \times C_2$, and the supports of the elements of W_Γ are subsets of $Q_1 \cup Q_2$. We define x_1, x_2, x_3, x_4 as in the type III case, and if all four x_i have the same 2-element support then, noting that all type II groups B are 2-transitive, we can construct $y \in G$ with $\text{supp}(x_i) \cap \text{supp}(y) = Q_1$ for all i. The rest of the proof is identical to the type III case.   □

LEMMA 5.6. *(i) Let $(A, |\Delta|) = (D_{14}, 7)$ and $B \in (\beta)$. Then W has an index 2 subgroup H orbit equivalent with W if and only if B has an index 2 subgroup C orbit equivalent with B. Moreover, if such orbit equivalent H exists then it is unique.*

(ii) Let $(A, |\Delta|) = (C_3, 3)$. Then W has an index 3 subgroup H orbit equivalent with W if and only if B has an index 3 subgroup C orbit equivalent with B. Moreover, if such orbit equivalent H exists then there are exactly two possibilities for H.

PROOF. We prove parts (i) and (ii) simultaneously. Let $m = 2$ if $A = D_{14}$ and $m = 3$ if $A = C_3$. Let R_1 and R_2 be two subsets of Δ in different regular orbits of A on $\mathcal{P}(\Delta)$. For $D \le B$, we denote by D^* the subgroup $\{(1, 1, \ldots, 1; b) \mid b \in D\}$ of W.

Suppose that there exists an index m subgroup H of W that is orbit equivalent to W. Let $\Gamma = \bigcup_{i=1}^n (R_1, i)$. Then $W_\Gamma = B^*$ and $H_\Gamma = B^* \cap H = C^*$ for some subgroup $C \le B$. By Lemma 4.3, $|W_\Gamma : H_\Gamma| = |W : H| = m$ and so $|B : C| = m$. We shall prove that $B \cong C$.

For any subset $S \subset [n]$, take $\Gamma(S) = \bigcup_{i \in S}(R_1, i) \cup \bigcup_{i \notin S}(R_2, i)$. Then $W_{\Gamma(S)} = (B_S)^*$ and $H_{\Gamma(S)} = H \cap (B_S)^* = (C_S)^*$. Using Lemma 4.3 again, $|W_{\Gamma(S)} : H_{\Gamma(S)}| =$

m and so $|B_S : C_S| = m$. This implies $|S^B| = |S^C|$. Since S is an arbitrary subset of $[n]$, we obtain $B \equiv C$.

Conversely, suppose that an index m subgroup C of B is orbit equivalent to B. We construct an index m subgroup of H of W the following way. The factor group $A^n/(A')^n$ can be identified with an n-dimensional vector space over $\mathrm{GF}(m)$. Let V be the $(n-1)$-dimensional subspace consisting of vectors whose sum of coordinates is 0 (in $\mathrm{GF}(m)$) and let L be the index m subgroup of A^n containing $(A')^n$ and $L/(A')^n$ corresponding to V. Let $a \in A \setminus A'$ and $b \in B \setminus C$. We define H as $H = \langle L, C^*, (a, 1, 1, \ldots, 1; b) \rangle$.

We claim that $H \equiv W$. Let $\Gamma \subset \Omega$ be arbitrary. By Lemma 4.5, it is enough to prove that $HW_\Gamma = W$; since $|W : H|$ is a prime number, this is equivalent to the existence of $w \in W_\Gamma \setminus H$, with $w^m \in H$. We distinguish two cases.

Case 1: There exists some $i \leq n$ such that $\Gamma_i := \Gamma \cap \Delta_i$ has nontrivial stabilizer in A. In this case, there exists $w := (1, \ldots, 1, t, 1, \ldots, 1; 1) \in W_\Gamma$ with support $\{i\}$. The crucial observation is (and this is the feature that distinguishes these two groups A from the other groups listed in Theorem 1.1) that t can be chosen in $A \setminus A'$. This fact is obvious for $A = C_3$. In the case $A = D_{14}$, if $\Gamma_i \in \{\emptyset, \Delta_i\}$ then its stabilizer is D_{14} so t can be chosen as desired. If Γ_i is a proper subset of Δ_i with nontrivial stabilizer then $|(D_{14})_{\Gamma_i}| = 2$ and the nontrivial element of the stabilizer is a good choice for t. In any case, since $w \notin L$ and $H \cap A^n = L$, we have $w \in W_\Gamma \setminus H$ and $w^m = 1 \in H$.

Case 2: For all $i \leq n$, $\Gamma \cap \Delta_i$ has trivial stabilizer in A. Since A has only two regular orbits on $\mathcal{P}(\Delta)$, this means that for any i, $\Gamma \cap \Delta_i$ is in the same orbit as R_1 or R_2. Define the set $S \subset [n]$ so that for $i \in S$, $\Gamma \cap \Delta_i = (R_1^{a_i}, i)$ and for $i \in [n] \setminus S$, $\Gamma \cap \Delta_i = (R_2^{a_i}, i)$, for some $a_i \in A$. Then $W_\Gamma = (B_S)^*$ and $H_\Gamma = (C_S)^*$. Since $B \equiv C$, there exists $b \in B_S \setminus C_S$. This implies that $w = (1, 1, \ldots, 1; b) \in W_\Gamma \setminus H$ and $w^m \in C^* \leq H$.

Finally, we prove the claims about the possibilites for H. In the case $A = C_3$, $\langle L, C^*, (a, 1, 1, \ldots, 1; b) \rangle$ and $\langle L, C^*, (a^2, 1, 1, \ldots, 1; b) \rangle$ are two different $H < W$ that are orbit equivalent with W. Conversely, suppose $H \equiv W$ and $|W : H| = m$. Examining the list of Theorem 2.1, we see that for all groups $B \in (\beta)$ which have an orbit equivalent subgroup C of order m, $C = B'$ so B determines C uniquely. Moreover, we have shown above that H must contain C^* (by examining H_Γ for $\Gamma = \bigcup_{i=1}^n (R_1, i)$). Also, since H acts as B on the vector space $(A/A')^n$ by Theorem 4.6 and $(A/A')^n$ has a unique $(n-1)$-dimensional B-invariant subspace by Lemma 5.4, H must contain L. Since $W/\langle C^*, L \rangle \cong C_m \times C_m$, we have to determine which subgroups of order m of this factor group can be included in H. The group $C_m \times C_m$ has $m + 1$ subgroups of order m. In our setting, we cannot add $(a, 1, 1, \ldots, 1; 1)$ to $\langle C^*, L \rangle$ because H does not contain A^n. We also cannot add $(1, 1, \ldots, 1; b)$ to $\langle C^*, L \rangle$ because H does not contain B^*. This leaves $m - 1$ subgroups, and we have seen that these $m - 1$ possibilities lead to orbit equivalent $H < W$. $\square$

PROOF OF THEOREM 1.2. Suppose that $A \in (\alpha)$ but $(A, |\Delta|) \neq (C_2, 2)$ and $B \notin (\gamma)$. If A has at least four regular orbits on $\mathcal{P}(\Delta)$ then, by Lemmas 2.3 and 5.2, W has a regular orbit on $\mathcal{P}(\Omega)$. Similarly, if $B \in (\alpha)$ then the distinguishing number of B is 2 and so by Lemma 5.2 W has a regular orbit on $\mathcal{P}(\Omega)$. Hence in these cases there is no $H < G \leq W$ with $H \equiv G$ and it is enough to consider the case when A is listed in Theorem 1.1 and $B \in (\beta)$.

From now on, suppose that $(A, |\Delta|)$ is one of $(C_3, 3)$, $(D_{14}, 7)$, $(\mathrm{AGL}(1,8), 8)$, $(\mathrm{PSL}(2,11), 12)$, $(\mathrm{PSL}(2,13), 14)$, $(A_7, 15)$, $(C_2^4.S_5, 16)$ and $B \in (\beta)$. Moreover, let $H < G$ and $H \equiv G$ (note that this implies that G has no regular orbit on $\mathcal{P}(\Omega)$). Let M be the kernel of action of H on Σ. By Theorem 4.6, $H^\Sigma = G^\Sigma$, so $|K : M| = |G : H|$. Let R_1 and R_2 be two subsets of Δ in two different regular orbits of A on $\mathcal{P}(\Delta)$ and let $\mathcal{S} = (S_1, S_2, S_3, S_4)$ be an ordered partition of $[n]$ such that only the identity of B fixes $\mathcal{S}$.

Case 1: $(A, |\Delta|)$ *is one of* $(\mathrm{PSL}(2,11), 12)$, $(\mathrm{PSL}(2,13), 14)$, $(A_7, 15)$. A small GAP computation shows that there exist $T_1 \subset \Delta$ and $T_2 \subset \Delta$ such that $|T_i| < |\Delta|/2$ for $i = 1, 2$, $|A_{T_1}| = 2$, and $|A_{T_2}| = 3$. Let

$$\Gamma_1 = \bigcup_{i \in S_1} (T_1, i) \cup \bigcup_{i \in S_2} (\Delta \setminus T_1, i) \cup \bigcup_{i \in S_3} (R_1, i) \cup \bigcup_{i \in S_4} (R_2, i).$$

Then $|G_{\Gamma_1}|$ divides $2^{|S_1 \cup S_2|}$. If we replace T_1 by T_2, we obtain a subset $\Gamma_2 \subset \Omega$ such that $|G_{\Gamma_2}|$ divides $3^{|S_1 \cup S_2|}$. So Corollary 4.4 implies $|G : H| = 1$, a contradiction.

Case 2: $(A, |\Delta|) = (\mathrm{AGL}(1,8), 8)$. If $T_1, T_2 \subset \Delta$, $|T_j| = j$ for $j = 1, 2$ then $|A_{T_1}| = 7$ and $|A_{T_2}| = 2$. Hence for the sets

$$\Gamma_j = \bigcup_{i \in S_1} (T_j, i) \cup \bigcup_{i \in S_2} (\Delta \setminus T_j, i) \cup \bigcup_{i \in S_3} (R_1, i) \cup \bigcup_{i \in S_4} (R_2, i),$$

we have that $|G_{\Gamma_1}|$ divides $7^{|S_1 \cup S_2|}$ and $|G_{\Gamma_2}|$ divides $2^{|S_1 \cup S_2|}$. So Corollary 4.4 leads to a contradiction.

Case 3: $(A, |\Delta|)$ *is* $(D_{14}, 7)$ *or* $(C_2^4.S_5, 16)$. By Lemma 5.5, either $G = W$ or $|W : G| = 2$. Let T_1, T_2, and T_3 be the three subsets of Δ defined in the first paragraph of the proof of Lemma 5.5. By Lemma 5.3, there exists an ordered partition $\mathcal{Q} = (Q_1, Q_2, Q_3, Q_4, Q_5)$ of $[n]$ with $|Q_1| = |Q_2| = |Q_3| = 1$ such that only the identity of B fixes $\mathcal{Q}$. Let

$$\Gamma = \bigcup_{i \in Q_1} (T_1, i) \cup \bigcup_{i \in Q_2} (T_2, i) \cup \bigcup_{i \in Q_3} (T_3, i) \cup \bigcup_{i \in Q_4} (R_1, i) \cup \bigcup_{i \in Q_5} (R_2, i).$$

Then $|W_\Gamma| = 8$. Therefore, by Lemma 4.3, if $G = W$ then $|W : H|$ divides 8. If $|W : G| = 2$ then $|G_\Gamma| = 4$ (because $K \cong (A')^n.C_2^{n-1}$ is uniquely determined by Lemma 5.4 and this unique K does not contain elements from $A^n \setminus (A')^n$ with one nontrivial coordinate) and so $|G : H|$ divides 4 by Lemma 4.3. Thus $|W : H|$ is a divisor of 8 in this case as well.

Since $|W : H| = |A^n : M|$, we obtain that $M \cong (A')^n.C_2^k$, for some $k \geq n - 3$. However, if $(B, n) \neq (\mathrm{PSL}(3,2), 7))$ then, by Lemma 5.4, there is no $(n - 2)$ or $(n - 3)$-dimensional B-invariant subspace of $\mathrm{GF}(2)^n$. Hence the only possibility is that $k = n - 1$, which means $G = W$ and $|G : H| = 2$. If $(B, n) = (\mathrm{PSL}(3,2), 7))$ then let

$$\Gamma(\mathcal{S}) = \bigcup_{i \in S} (R_1, i) \cup \bigcup_{i \in [n] \setminus S} (R_2, i) \quad \text{and} \quad d(B) = \gcd\{|W_\Gamma|, |W_{\Gamma(\mathcal{S})}| \mid S \subset [n]\}.$$

We obtain $d(B) = 2$ by a GAP computation. Now $|G : H|$ divides $d(B)$, so $|G : H|$ divides 2 and $|W : H|$ is a divisor of 4. There is no 5-dimensional B-invariant subspace of $GF(2)^7$, which means the only possibility is $W = G$ and $|G : H| = 2$.

In the case $A = D_{14}$, we are in the situation described in Lemma 5.6. Examining the list of Theorem 2.1, we obtain the four $B \in (\beta)$ with an orbit equivalent index 2 subgroup as listed in Theorem 1.2(i).

In the case $A = C_2^4.S_5$ there are no examples. Identify Δ with $\{1, 2, \ldots, 16\}$ and let $U_1 = \{1, 2, 3, 6, 11\}$, $U_2 = \Delta \setminus U_1$, $U_3 = \{1, 2, 3, 6, 11, 13\}$, and $U_4 = \Delta \setminus U_3$. The U_i are in different orbits of A on $\mathcal{P}(\Delta)$, and a small GAP computation shows that $A_{U_i} \leq A' = C_2^4.A_5$ for $1 \leq i \leq 4$. Hence for the set

$$\Gamma = \bigcup_{i \in S_1} (U_1, i) \cup \bigcup_{i \in S_2} (U_2, i) \cup \bigcup_{i \in S_3} (U_3, i) \cup \bigcup_{i \in S_4} (U_4, i),$$

we have $G_\Gamma \leq (A')^n \leq M \leq H$. This implies $G_\Gamma = H_\Gamma$, a contradiction.

Case 4: $(A, |\Delta|) = (C_3, 3)$. Let $\Gamma = \bigcup_{i \in S_1} (\Delta, i) \cup \bigcup_{i \in S_3} (R_1, i) \cup \bigcup_{i \in S_4} (R_2, i)$. Then $|W_\Gamma|$ divides $3^{|S_1 \cup S_2|}$. Moreover, for $S \subset [n]$, let

$$\Gamma(S) = \bigcup_{i \in S} (R_1, i) \cup \bigcup_{i \in [n] \setminus S} (R_2, i).$$

We define $d(B)$ as $d(B) = \gcd\{|W_\Gamma|, |W_{\Gamma(S)}| \mid S \subset [n]\}$.

We claim that $d(B) = 3$ if $(B, n) = (\mathrm{ASL}(3, 2), 8)$, $(\mathrm{A\Gamma L}(1, 8), 8)$, $(\mathrm{P\Gamma L}(2, 8), 9)$, $(M_{12}, 12)$, $d(B)$ divides 3 if $(B, n) = (M_{24}, 24)$, and $d(B) = 1$ for all other groups $B \in (\beta)$. We prove this claim by a GAP computation. For $n \leq 17$, we compute $|W_{\Gamma(S)}|$ for a representative S from each B-orbit on the power set of $[n]$. For $n \geq 18$, it is enough to observe that for $S = \{1, 2, 3, 4, 5, 8\}$, $|W_{\Gamma(S)}| = 8, 8, 16$ for $(B, n) = (\mathrm{P\Gamma L}(3, 4), 21)$, $(M_{22}, 22)$, and $(M_{22}.2, 22)$, respectively and for $S = \{1, 4, 5, 6, 8, 10, 11, 13, 15, 17, 18, 20\}$, $|W_{\Gamma(S)}| = 20, 660, 8$ for $(B, n) = (M_{23}, 23)$, $(M_{24}, 24)$, and $(\mathrm{ASL}(5, 2), 32)$, respectively.

Since $\gcd\{|G_\Theta| \mid \Theta \subset \Omega\}$ divides $d(B)$, we must have $d(B) = 3$ by Corollary 4.4. Hence it is enough to consider the cases $(B, n) = (\mathrm{ASL}(3, 2), 8)$, $(\mathrm{A\Gamma L}(1, 8), 8)$, $(\mathrm{P\Gamma L}(2, 8), 9)$, $(M_{12}, 12)$, and $(M_{24}, 24)$. Moreover, in these five cases, $|G : H| = 3$ and so $|K : M| = 3$. The subgroups K and M correspond to B-invariant subspaces of $\mathrm{GF}(3)^n$, and the dimensions of these subspaces differ by 1. We check by a GAP computation that in all five cases there are exactly four B-invariant subspaces, of dimensions $0, 1, n - 1$, and n, respectively. So either $|K| = 3$ or $K = A^n$, the latter implying $G = W$.

In the case $|K| = 3$, the nontrivial elements of K have support Ω. Hence $K_\Gamma = G_\Gamma = 1$ for the set Γ defined in the first line of Case 4, contradicting the fact that G has no regular orbit. So the only possibility is $G = W$ and $|G : H| = 3$. This is the the situation described in Lemma 5.6. Examining the list of Theorem 2.1, we obtain the two $B \in (\beta)$ with orbit equivalent index 3 subgroups as listed in Theorem 1.2(i). $\qquad\square$

References

[1] Michael O. Albertson and Karen L.Collins, *Symmetry breaking in graphs*. Electron. J. of Comb. **3** (1996), Research Paper #18.

[2] Peter J. Cameron, *Permutation Groups*. London Math. Soc. Student Texts #45, Cambridge Univ. Press, 1999.

[3] Peter J.Cameron, Peter M. Neumann and Jan Saxl, *On groups with no regular orbits on the set of subsets*. Arch. Math. **43** (1984), 295–296.

[4] G. H. Chan, *A characterization of minimal (k)-groups of degree $n \leq 3k$*. Lin. and Multilin. Alg. **4** (1976/77), 285–305.

[5] Silvio Dolfi, *Orbits of permutation groups on the power set*. Arch. Math. **75** (2000), 321–327.

[6] The GAP Group, GAP – *Groups, Algorithms, and Programming*, Version 4.4.9. Aachen–St Andrews, 2006. http://www.gap-system.org.

[7] David Gluck, *Trivial set-stabilizers in finite permutation groups*. Can. J. Math. **XXXV** (1983), 59–67.

[8] Martin W. Liebeck, Cheryl E. Praeger and Jan Saxl, *On the O'Nan-Scott theorem for finite primitive permutation groups.* J. Austral. Math. Soc. **44** (1988), 389–396.

[9] Martin W. Liebeck and Jan Saxl, *Minimal degree of primitive permutation groups with an application to monodromy groups of covers of Riemann surfaces.* Proc. London Math. Soc. **63** (1991), 266–314.

[10] Ákos Seress, *The minimal base size of primitive solvable permutation groups.* J. London Math. Soc. **53** (1996), 243–255.

[11] Ákos Seress, *Primitive groups with no regular orbits on the set of subsets.* Bull. London Math. Soc. **29** (1997), 697–704.

[12] Johannes Siemons, *On partitions and permutation groups on unordered sets.* Arch. Math. **38** (1982), 391–403.

DEPARTMENT OF MATHEMATICS, THE OHIO STATE UNIVERSITY, COLUMBUS, OHIO 43210
E-mail address: akos@math.ohio-state.edu

DEPARTMENT OF MATHEMATICS, THE OHIO STATE UNIVERSITY, COLUMBUS, OHIO 43210
E-mail address: kyyang@math.ohio-state.edu

Titles in This Series

TITLES IN THIS SERIES

For a complete list of titles in this series, visit the
AMS Bookstore at **www.ams.org/bookstore/**.